Algebraic and Discrete Mathematical Methods for Modern Biology

Algebraic and Discrete Mathematical Methods for Modern Biology

Edited by

Raina S. Robeva
Department of Mathematical Sciences, Sweet Briar College, Sweet Briar, VA, USA

AMSTERDAM • BOSTON • HEIDELBERG • LONDON
NEW YORK • OXFORD • PARIS • SAN DIEGO
SAN FRANCISCO • SINGAPORE • SYDNEY • TOKYO
Academic Press is an imprint of Elsevier

Academic Press is an imprint of Elsevier
32 Jamestown Road, London NW1 7BY, UK
525 B Street, Suite 1800, San Diego, CA 92101-4495, USA
225 Wyman Street, Waltham, MA 02451, USA
The Boulevard, Langford Lane, Kidlington, Oxford OX5 1GB, UK

First edition 2015

Copyright © 2015 Elsevier Inc. All rights reserved.

No part of this publication may be reproduced or transmitted in any form or by any means, electronic or mechanical, including photocopying, recording, or any information storage and retrieval system, without permission in writing from the publisher. Details on how to seek permission, further information about the Publisher's permissions policies and our arrangements with organizations such as the Copyright Clearance Center and the Copyright Licensing Agency, can be found at our website: www.elsevier.com/permissions.

This book and the individual contributions contained in it are protected under copyright by the Publisher (other than as may be noted herein).

Notices
Knowledge and best practice in this field are constantly changing. As new research and experience broaden our understanding, changes in research methods, professional practices, or medical treatment may become necessary.

Practitioners and researchers must always rely on their own experience and knowledge in evaluating and using any information, methods, compounds, or experiments described herein. In using such information or methods they should be mindful of their own safety and the safety of others, including parties for whom they have a professional responsibility.

To the fullest extent of the law, neither the Publisher nor the authors, contributors, or editors, assume any liability for any injury and/or damage to persons or property as a matter of products liability, negligence or otherwise, or from any use or operation of any methods, products, instructions, or ideas contained in the material herein.

Library of Congress Cataloging-in-Publication Data
A catalog record for this book is available from the Library of Congress

British Library Cataloguing in Publication Data
A catalogue record for this book is available from the British Library

For information on all Academic Press publications
visit our website at http://store.elsevier.com/

Printed and bound in the USA

ISBN: 978-0-12-801213-0

Contents

Companion website: http://booksite.elsevier.com/9780128012130/.

Contributors

Numbers in parentheses indicate the pages on which the authors' contributions begin.

Réka Albert (65), Pennsylvania State University, University Park, PA, USA

Todd J. Barkman (261), Department of Biological Sciences, Western Michigan University, Kalamazoo, MI, USA

Mary Ann Blätke (141), Otto-von-Guericke University, Magdeburg, Germany

Hannah Callender (193,217), University of Portland, Portland, OR, USA

Margaret (Midge) Cozzens (29), Rutgers University, Piscataway, NJ, USA

Kristina Crona (51), Department of Mathematics and Statistics, American University, 4400 Massachusetts Ave NW, Washington, DC 20016

Robin Davies (93,321), Department of Biology, Sweet Briar College, Sweet Briar, VA, USA

Monika Heiner (141), Brandenburg University of Technology, Cottbus, Germany

Terrell L. Hodge (261), Department of Mathematics, Western Michigan University, Kalamazoo, MI, USA

Qijun He (93,321), Department of Mathematical Sciences, Clemson University, Clemson, SC, USA

John R. Jungck (1), Center for Bioinformatics and Computational Biology, University of Delaware, Newark, DE, USA

Winfried Just (193,217), Department of Mathematics, Ohio University, Athens, OH, USA

Bessie Kirkwood (237), Sweet Briar College, Sweet Briar, VA, USA

M. Drew LaMar (193,217), The College of William and Mary, Williamsburg, VA, USA

Matthew Macauley (93,321), Department of Mathematical Sciences, Clemson University, Clemson, SC, USA

Wolfgang Marwan (141), Otto-von-Guericke University, Magdeburg, Germany

David Murrugarra (121), Department of Mathematics, University of Kentucky, Lexington, KY, USA

Christian M. Reidys (347), Department of Mathematics and Computer Science, University of Southern Denmark, Odense M, Denmark

Raina Robeva (65), Sweet Briar College, Sweet Briar, VA, USA

Janet Steven (237), Christopher Newport University, Newport News, VA, USA

Blair R. Szymczyna (261), Department of Chemistry, Western Michigan University, Kalamazoo, MI, USA

Natalia Toporikova (193), Washington and Lee University, Lexington, VA, USA

Alan Veliz-Cuba (121), Department of Mathematics, University of Houston, and Department of BioSciences, Rice University, Houston, TX, USA

Rama Viswanathan (1), Beloit College, Beloit, WI, USA

Grady Weyenberg (293), Department of Statistics, University of Kentucky, Lexington, KY

Emilie Wiesner (51), Department of Mathematics, Ithaca College, 953 Danby Rd, Ithaca, NY 14850

Ruriko Yoshida (293), Department of Statistics, University of Kentucky, Lexington, KY

Preface

In the last 15 years, the field of modern biology has been transformed by the use of new mathematical methods, complementing and driving biological discoveries. Problems from gene regulatory networks and genomics, RNA folding, infectious disease and drug resistance modeling, phylogenetics, and ecological networks and food webs have increasingly benefited from the application of discrete mathematics and computational algebra. Modern algebra approaches have proved to be a natural fit for many problems where the use of traditional dynamical models built with differential equations is not appropriate or optimal.

While the use of modern algebra methods is now in the mainstream of mathematical biology research, this trend has been slow to influence the undergraduate mathematics and biology curricula, where difference and differential equation models still dominate. Several high-profile reports have been released in the past 5 years, including Refs. [1–3], calling urgently for broadening the undergraduate exposure at the interface of mathematics and biology, and including methods from modern discrete mathematics and their biological applications. However, those reports have been slow to elicit the transformative change in the undergraduate curriculum that many of us had hoped for. The anemic response may be attributed to a relative lack of educational undergraduate resources that highlight the critical impact of algebraic and discrete mathematical methods on contemporary biology. It is this niche that our book seeks to fill.

The format of this volume follows that of our earlier book, *Mathematical Concepts and Methods in Modern Biology: Using Modern Discrete Methods*, Robeva and Hodge (Editors), published in 2013 by Academic Press. At the time of its planning, we considered the modular format of that text (with chapters largely independent from one another) experimental, but we felt reassured when the book was selected as 1 of 12 contenders for the 2013 Society of Biology Awards in its category. We have adopted the same format here, as we believe that it provides readers and instructors with the independence to choose biological topics and mathematical methods that are of greatest interest to them.

Due to the modular format, the order of the chapters in the volume does not necessarily imply an increased level of difficulty or the need for more prerequisites for the later chapters. When chapters are connected by a common biological thread, they are grouped together, but they can still be used independently. Each chapter begins with a question or a number of related questions from modern biology, followed by the description of certain mathematical methods and theory appropriate in the search of answers. As in our earlier book, chapters can be viewed as fast-track pathways through the problem, which start by presenting the biological foundation, proceed by covering the relevant mathematical theory and presenting numerous examples, and end by highlighting connections with ongoing research and current publications. The level of presentation varies among chapters—some may be appropriate for introductory courses, while others may require more mathematical or biological background. Exercises are embedded within the text of each chapter, and their execution requires only material discussed up to that point. In addition, many chapters feature challenging open-ended questions (designated as projects) that provide starting points for explorations appropriate for undergraduate research, and supply references to relevant publications from the recent literature. In their most general form, some of the projects feature truly open questions in mathematical biology.

The book's companion website (http://booksite.elsevier.com/9780128012130) contains solutions to the exercises, as well all figures and relevant data files for the examples and exercises in the chapters. In addition, the site hosts software code, project guidelines, online supplements, appendices, and tutorials for selected chapters. The specialized software utilized throughout the book highlights the critical importance of computing applications for visualization, simulation, and analysis in modern biology. We have been careful to feature software that is in the mainstream of current mathematical biology research, while also being mindful of giving preference to freely available software.

We hope that the book will be a valuable resource to mathematics and biology programs, as it describes methods from discrete mathematics and modern algebra that can be presented, for the most part, at a level completely accessible to undergraduates. Yet the book provides extensions and connections with research that would also be helpful to graduate students and researchers in the field. Some of the material would be appropriate for mathematics courses such as finite mathematics, discrete structures, linear algebra, abstract/modern algebra, graph theory, probability, bioinformatics, statistics, biostatistics, and modeling, as well as for biology courses such as genetics, cell and molecular biology, biochemistry, ecology, and evolution.

The selection of topics for the volume and the choice of contributors grew out of the workshop "Teaching Discrete and Algebraic Mathematical Biology to Undergraduates" organized by Raina Robeva, Matthew Macauley, and Terrell Hodge and funded and hosted by the Mathematical Biosciences Institute (MBI) on July 29-August 2, 2013 at The Ohio State University. The editor and contributors of this volume greatly appreciate the encouragement and assistance received from the MBI's leadership and staff. Without their support, this volume would not have been possible. We also acknowledge with gratitude the support of the National Institute for Mathematical and Biological Synthesis (NIMBioS) in providing an opportunity to further test selected materials as part of the tutorial "Algebraic and Discrete Biological Models for Undergraduate Courses" offered on June 18-20, 2014 at NIMBioS.

I would like to express my personal thanks to all contributors who embraced the project early on and committed time and energy into producing the chapter modules for this unconventional textbook. Your enthusiasm for the project was remarkable, and you have my deep gratitude for the dedication and focus with which you carried it out. My special thanks also go to Daniel Hrozencik and Timothy Comar for providing feedback on a few of the chapter drafts. I am indebted to the editorial and production teams at Elsevier and particularly to the book's editors, Paula Callaghan and Katey Birtcher, our editorial project managers, Sarah Watson and Amy Clark, and our production manager, Vijayaraj Purushothaman. It has been a pleasure and a privilege to work with all of you. Finally, I would like to thank my husband, Boris Kovatchev, for his patience and support throughout.

Raina S. Robeva
October 20, 2014

REFERENCES

[1] Committee on a New Biology for the 21st Century: Ensuring the United States Leads the Coming Biology Revolution, Board on Life Sciences, Division on Earth and Life Studies, National Research Council. A new biology for the 21st century. Washington, DC: The National Academies Press; 2009.

[2] Brewer CA, Anderson CW (eds). Vision and change in undergraduate biology education: a call to action. Final report of a National Conference organized by the American Association for the Advancement of Science with support from the National Science Foundation, July 15-17, 2009, Washington, DC. The American Association for the Advancement of Science; 2011. http://visionandchange.org/files/2013/11/aaas-VISchange-web1113.pdf (accessed March 1, 2015).

[3] Committee on the Mathematical Sciences in 2025, Board on Mathematical Sciences and Their Applications, Division on Engineering and Physical Sciences, National Research Council. The mathematical sciences in 2025. Washington, DC: The National Academies Press; 2013.

Companion website: http://booksite.elsevier.com/9780128012130/

Supplementary Resources for Instructors

The website features the following additional resources available for download:

- All figures from the book
- Solutions to all exercises
- Computer code, data files, and links to software and materials carefully chosen to supplement the content of the textbook
- Appendices, tutorials, and additional projects for selected chapters

Chapter 1

Graph Theory for Systems Biology: Interval Graphs, Motifs, and Pattern Recognition

John R. Jungck[1] and Rama Viswanathan[2]
[1]*Center for Bioinformatics and Computational Biology, University of Delaware, Newark, DE, USA,* [2]*Beloit College, Beloit, WI, USA*

1.1 INTRODUCTION

Systems thinking is perceived as an important contemporary challenge of education [1]. However, *systems biology* is an old and inclusive term that connotes many different subareas of biology. Historically two important threads were synchronic: (a) the systems ecology of the Odum school [2–4], which was developed in the context of engineering principles applied to ecosystems [5, 6], and (b) systems physiology that used mechanical principles [7] to understand organs as mechanical devices integrated into the circulatory system, digestive system, anatomical system, immune system, nervous system, etc. For example, the heart could be thought of as a pump, the kidney as a filter, the lung as a bellows, the brain as a wiring circuit (or later as a computer), elbow joints as hinges, and so on. It should be noted that both areas extensively employed *ordinary* and *partial differential equations* (ODEs and PDEs). Indeed, some systems physiologists argued that all mathematical biology should be based on the application of PDEs. On the other hand, evolutionary biologists argued that these diachronic systems approaches too often answered only "how" questions that investigated optimal design principles and did not address "why" questions focusing on the constraints of historical contingency.

Not surprisingly, one of the leading journals in the field—*Frontiers in Systems Biology*—announces in its mission statement, [8] "Contrary to the reductionist paradigm commonly used in Molecular Biology, in Systems Biology the understanding of the behavior and evolution of complex biological systems need not necessarily be based on a detailed molecular description of the interactions between the system's constituent parts." Therefore, in this chapter we emphasize two major macroscopic and global aspects of contemporary systems biology: (i) the graph-theoretic relationships between components in networks and (ii) the relationship of these patterns to the historical contingencies of evolutionary constraints. Numerous articles and several books [9, 10] exist on graph theory and its application to systems biology, so the reader may ask what are we doing in this chapter that is different. Our main purpose is to help biologists, mathematicians, students, and researchers recognize which graph-theoretic tools are appropriate for different kinds of questions, including quantitative analyses of interactions for mining large data sets, visualizing complex relationships, modeling the consequences of perturbation of networks, and testing hypotheses.

Every network construct in systems biology is a hypothesis. For example, Rios and Vendruscolo [11] describe the network hypothesis as the assumption "according to which it is possible to describe a cell through the set of interconnections between its component molecules." They then conclude, "it becomes convenient to focus on

Algebraic and Discrete Mathematical Methods for Modern Biology. http://dx.doi.org/10.1016/B978-0-12-801213-0.00001-0
Copyright © 2015 Elsevier Inc. All rights reserved.

these interactions rather than on the molecules themselves to describe the functioning of the cell." In this chapter, we go a step further. We believe that a mathematical biology perspective also studies such questions as: Which molecules are involved? What do they do functionally? What is their three-dimensional structure? Where are they located in a cell? We stress that every network and pathway that we discuss is a useful construct from a biological perspective. They do not exist *per se* inside of cells. Imagine a series of biological macromolecules (proteins, nucleic acids, polysaccharides) that are crowded and colliding with one another in a suspension. The networks and pathways for the interactions between these molecules constructed by biologists may represent preferred associations defined by tighter bindings of specific macromolecules or the product of a reaction catalyzed by one macromolecule (an enzyme) as the starting material (substrate) of another enzyme. Thus, biologists have already drawn mathematical diagrams and graphs in the sense that they have abstracted, generalized, and symbolized a set or relationships.

Too often biologists produce networks as visualizations without further analysis. In this chapter, using Excel and Java-based software that we have developed, we show readers how to make mathematical measurements (average degree, diameter, clustering coefficient, etc.) and discern holistic properties (small world versus scale-free, see Hayes [12] for a complete overview) of the networks being studied and visualized, and obtain insights that are relevant and meaningful in the context of systems biology. We show how the network hypothesis can be investigated by complementary and supplementary mathematical and biological perspectives to yield key insights and help direct and inform additional research.

Palsson [10] suggests that twenty-first century biology will focus less on the reductionist study of components and more on the integration of systems analysis. He identifies four principles in his "systems biology paradigm": "First, the list of biological components that participated in the process of interest is enumerated. Second, the interactions between these components are studied and the 'wiring diagrams' of genetic circuits are constructed .... Third, reconstructed network[s] are described mathematically and their properties analyzed.... Fourth, the models are used to analyze, interpret, and predict biological experimental outcomes." Here, we assume that the first two steps exist in databases or published articles; this allows us to focus on the mathematics of the third step as a way that allows biologists to better direct their work on the fourth step. Thus, the goals for this chapter are as follows.

- Learn how graph theory can be used to help obtain meaningful insights into complex biological data sets.
- Analyze complex biological networks of diverse types (restriction maps, food webs, gene expression, disease etiology) to detect patterns of relationships.
- Visualize ordering of modules/motifs within complex biological networks by first testing the applicability of simple linear approaches (interval graphs).
- Demonstrate that even when strict mathematical assumptions do not apply fully to a given biological data set, there is still benefit in applying an analytical approach because of the power of the human mind to discern prominent patterns in data rearranged through the application of mathematical transformations.
- Show that the visualizations help biologists obtain insights into their data, examine the significance of outliers, mine databases for additional information about observed associations, and plan further experiments.

To accomplish this, we first emphasize *how* graph theory is a natural fit for biological investigations of relationships, patterns, and complexity. Second, graph theory lends itself easily to questions about *what* biologists should be looking for among representations of relationships. We introduce concepts of hubs, maximal cliques, motifs, clusters, interval graphs, complementary graphs, ordering, transitivity, Hamiltonian circuits, and consecutive ones in adjacency matrices. Finally, graph theory helps us interrogate *why* these relationships are occurring. Basically, we examine the triptych of form, function, and phylogeny to differentiate between evolutionary and engineering constraints.

The chapter is structured as follows. We begin by introducing some background concepts from graph theory that will be utilized later in the chapter. We then introduce interval graphs through two biological examples related to chromosome sequencing and food webs. The rest of the chapter is devoted to two extended examples of biological questions related to recently published studies on gene expression and disease etiology. The analyses for those examples demonstrate how graph theory can help illuminate concepts of biological importance. Each of the examples is followed by suggestions for open-ended projects in pursuit of similar analyses of related biological questions and data.

1.2 REVISUALIZING, RECOGNIZING, AND REASONING ABOUT RELATIONSHIPS

Graph theory has enormous applicability to biology. It is particularly powerful in this era of terabytes of data because it allows a tremendous topological reduction in complexity and investigation of patterns. The applications of graph theory in mathematical biology, several of which are illustrated in this chapter, include subcellular localization of coordinated metabolic processes, identification of hubs central to such processes and the links between them, analysis of flux in a system, temporal organization of gene expression, the identification of drug targets, determination of the role of proteins or genes of unknown function, and coordination of sequences of signals. Medical applications include diagnosis, prognosis, and treatment. We will see below that by reducing a biological exploration to a relevant graph representation, we are able to examine, study, and measure various quantitative and meta-properties of the resulting graph and to obtain insights into why particular biological processes such as gene-gene, protein-protein, signal-detector-effector, and predator-prey occur.

1.2.1 Basic Concepts from Graph Theory

A *graph* in mathematics is a collection of *vertices* connected by *edges*. Graphs are often used in biology to represent networks and, more generally, to represent relationships between objects. The objects of interest are the vertices of the network, usually depicted as geometrical shapes such as dots, circles, or squares, while the connections between them are represented by the edges. In an applied context, the vertices are generally labeled. Vertices u and v that are directly connected by an edge are called *adjacent vertices* or *neighbors*. A subgraph that consists of all vertices adjacent to a vertex u and all edges connecting any two such vertices forms the *neighborhood of the vertex u*. For each graph, one can construct its *complementary graph:* a graph that has the same vertices as the original graph, but such that vertices u and v are adjacent in the complementary graph if and only if u and v are not adjacent in the original graph.

If a vertex u is related to itself, the edge connecting u with itself is called a *loop*. A *path* is a sequence of edges connecting neighboring vertices and the *length of a path* is the number of edges it uses. Loops could be considered paths of length 1 that start and end in the same vertex. We say that a vertex u in a graph is *connected* to vertex v if there is a path from u to v. An undirected graph is a *connected graph* if a pathway exists from every vertex to every other vertex. Otherwise, the graph is *disconnected*. A *Hamiltonian path* is a path that goes through all vertices in the graph and visits each vertex exactly once. A graph in which any two vertices are connected by a unique path is called a *tree*.

If there is directional dependence (e.g., "u activates v"; "u is the parent of v" as opposed to "u and v are friends"), then the direction is represented by an arrow. Graphs with directional dependencies are called *directed graphs*. Paths in directed graphs must follow the direction of the edges. The number of edges connected to a vertex u represents the *degree of the vertex* (loops are usually counted twice). In a directed graph, a vertex is called a *source* when all of its edges are outgoing edges; it is called a *sink* when all of its edges are incoming edges. The *in-degree* of a vertex is the number of incoming edges to the vertex and the *out-degree* is defined by the number of outgoing edges. Thus, the degree of each vertex in a directed graph is the sum of the in-degree and out-degree. Vertices with degrees among the top 5% in a network are often characterized as *hubs*. As hubs have a large number of neighbors, they often perform important roles in many biological networks.

Additional graph-theoretical definitions and properties that we will use in a substantive way in the chapter are:

- *Clique*—a subgraph is a graph in which every vertex is connected by an edge to any other vertex in the subgraph; a *maximal clique* is a clique that cannot be extended by including an additional adjacent vertex; in other words a maximal clique is a clique that is not a subset of a larger clique.

- *Diameter of a graph*—the maximum number of edges that have to be traversed to go from one vertex to another in a graph using shortest paths, i.e., the *longest* shortest path in the graph.
- *Degree distribution of a graph*—the probability distribution of the vertex degrees over the whole graph. It is represented as a histogram, in which the probability p_k that a vertex has degree k is represented by the proportion of the nodes in the graph with degree k. When $p_k = Ck^{-a}$, where a is a constant and C is a normalizing factor, the degree distribution follows a *power law*.
- *Connectivity of a graph*—the minimum number of edges that need to be removed from a connected graph to obtain a graph that is no longer connected. The connectivity of a graph is an important measure of its robustness as a network.
- *Clustering coefficient of a network*—we will not provide a mathematically rigorous definition here but, heuristically speaking, it represents the degree to which nodes in a graph tend to cluster together. In its local version, the clustering coefficient of a vertex quantifies how close its neighbors are to being a clique. The mathematical definition and further details can be found in Chapter 5 [13]
- *Transitively oriented graphs*—directed graphs in which if three vertices are connected in a triangle, and two successive edges are in the same direction, then a third edge must be present and go from the first to the third vertex.
- *Small world network*—A large graph with a relatively small number of neighbors in which any two vertices are connected by a path of relatively short length.

It has been hypothesized [12] that many real world networks, including biological networks, are *small world networks* that are in between lattice (highly ordered) and completely random networks, with properties that promote efficient information transfer and retrieval. In particular, such networks exhibit three unique properties: (a) they are usually sparse, i.e., they have relatively few edges compared to vertices; (b) they have large average clustering coefficients; and (c) a relatively small diameter on the order of $\log N$, where N is the number of vertices in the network [12]. The usual popularization of small world networks draws attention to two features: (a) every vertex is connected to every other vertex through relatively few edges ("six degrees of separation," "the Kevin Bacon problem," "what is your Erdös number?") and (b) it only takes a few "weak" links (i.e., edges that connect distant clusters) to create this effect. Much attention in mathematical biology has been paid to the question of why small world networks are manifested and have evolved at so many different levels of biological systems.

- *Interval graphs*—a special class of graphs that can be depicted as a family of intervals positioned along the real line.

Interval graphs are an interesting case because a biologist first developed them, and the formal mathematics to explore them was developed later. Interval graphs have a variety of biological applications across broad samplings of phylogenetic diversity, spatial and temporal scales, and diverse biological mechanisms.

In order to understand how interval graphs are constructed, we begin from the experimental biological determination of which intervals of finite lengths (fragments, sequences, deletions, etc.) overlap one another. Consider a hypothetical dataset with eight overlapping fragments (I_1 through I_8) as the intervals. All pairwise overlap relations are determined and an "adjacency" matrix is constructed (Figure 1.1a). The entry in the ith row and jth column is 1, if the vertices i and j are adjacent (fragments overlap) and 0 otherwise. The *adjacency matrix* is a square symmetric matrix. Next, we generate an undirected graph called the *intersection graph* (Figure 1.1b) in which the rows and the columns are labeled by the graph's vertices in the following way: each interval corresponds to a vertex and two vertices u and v are connected with an edge if and only if the intervals u and v overlap. Note that this property of interval graphs also has another interesting matrix formulation. As we will see later, it is equivalent to the *consecutive ones* property of matrices.

Finally, we determine the maximal cliques from the intersection graphs—in this case we determine that there are five such cliques (A, B, C, D, and E) by visual inspection—and set up a different binary matrix M where the rows represent the maximal cliques in the graph and the entry at the kth row and the rth column of M is 1, if vertex r belongs to the kth maximal clique, and is 0 otherwise. The line representation of the resulting interval graph is shown in Figure 1.1c.

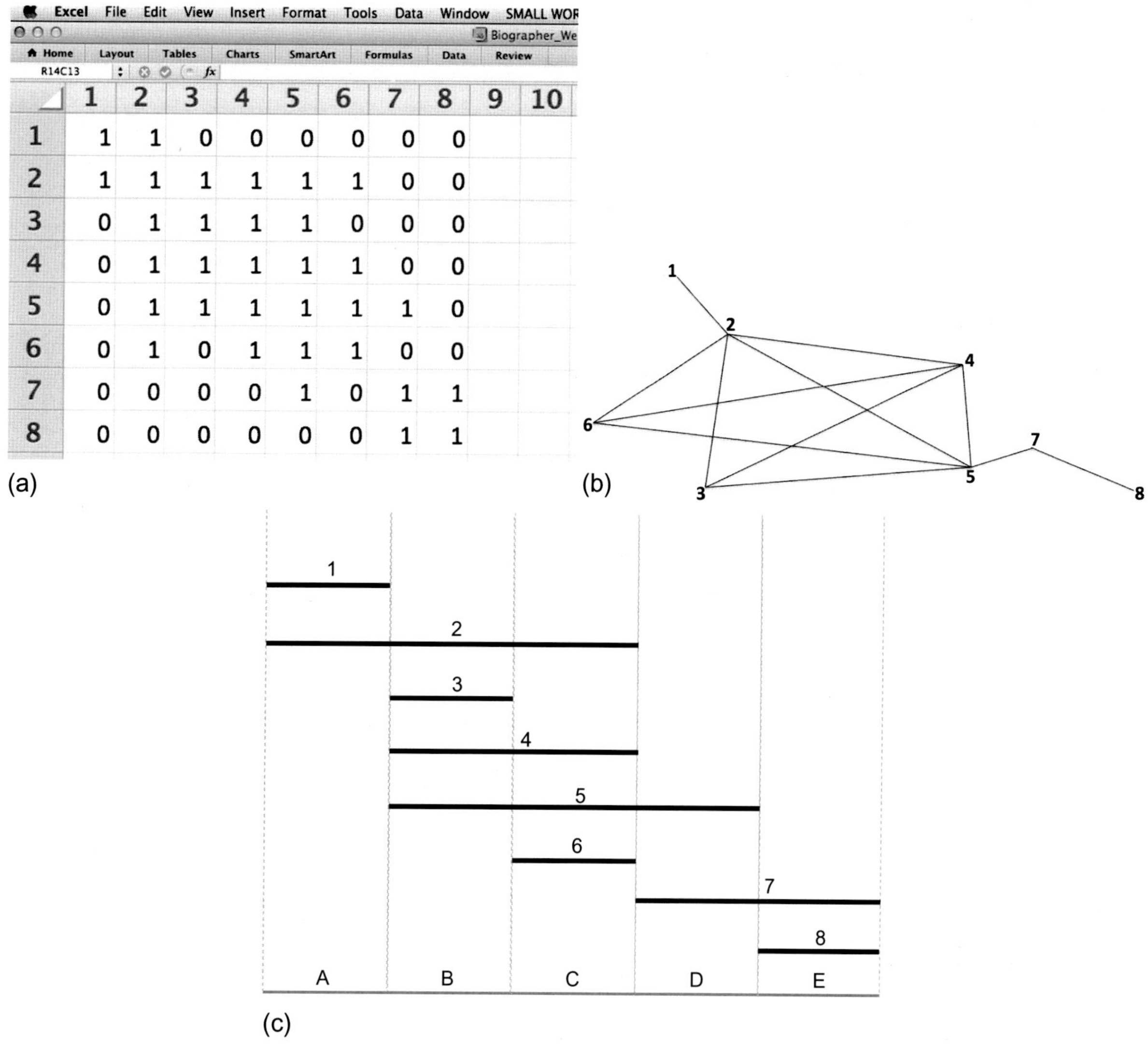

FIGURE 1.1 The connection between adjacency matrices, intersection graphs, and interval graphs. (a) The adjacency matrix from the experimental biological determination of which intervals of finite lengths (fragments, sequences, deletions, etc.) overlap one another. (b) The corresponding intersection graph of the adjacency matrix connects two vertices u and v with an edge if and only if the intervals u and v overlap. (c) An interval graph is a set of intervals of finite lengths arranged along a line where the rows represent the intervals of finite lengths (fragments, sequences, deletions, etc.) and the columns are labeled at the bottom according to the maximal cliques in the intersections graph (Maximal Clique A: 1 and 2; Maximal Clique B: 2, 3, 4, and 5; Maximal Clique C: 2, 4, 5, and 6; Maximal Clique D: 5 and 7; Maximal Clique E: 7 and 8). Note that there should be no horizontal gap between adjacent maximal cliques as that would mean that we have information that cliques which are not maximal exist in such regions (e.g., in Panel c, if interval 5 were shortened on its right end and interval 8 were shortened on its left end, there would be a clique between intervals 5 and 8 that only contained interval 7 which is obviously not maximal because it is contained in both D and E, each of which contains more members).

An important property of interval graphs is that their maximal cliques can be ordered in sequence in such a way that for any vertex (interval) v, the maximal cliques containing v occur consecutively in the sequence. Consider Figure 1.1c, where the maximal cliques for the interval graph are represented by regions between vertical line segments. The five maximal cliques A, B, C, D, E are ordered in a way where, for example, the three cliques containing interval I_5 (B, C, D) appear in a sequence; the three cliques containing I_2 (A, B, C) appear in a sequence; the two cliques containing I_7 (E, D) appear in a sequence, and so on. The matrix M is called the *clique matrix* for the intersection graph, and is shown below with row labels corresponding to the cliques and column labels corresponding to vertices (intervals) added for clarity:

$$\begin{array}{c} \\ A \\ B \\ C \\ D \\ E \end{array} \begin{array}{c} \begin{matrix} 1 & 2 & 3 & 4 & 5 & 6 & 7 & 8 \end{matrix} \\ \begin{pmatrix} 1 & 1 & 0 & 0 & 0 & 0 & 0 & 0 \\ 0 & 1 & 1 & 1 & 1 & 0 & 0 & 0 \\ 0 & 1 & 0 & 1 & 1 & 1 & 0 & 0 \\ 0 & 0 & 0 & 0 & 1 & 0 & 1 & 0 \\ 0 & 0 & 0 & 0 & 0 & 0 & 1 & 1 \end{pmatrix} \end{array}$$

A binary matrix is said to have the *consecutive ones property* for columns if its rows can be permuted in a way that ensures that the 1's in each column occur consecutively. The following clique matrix has the consecutive one's property:

$$\begin{pmatrix} 1 & 0 & 0 & 1 \\ 1 & 1 & 1 & 0 \\ 0 & 1 & 0 & 0 \\ 1 & 0 & 1 & 1 \\ 1 & 1 & 0 & 0 \end{pmatrix} \rightarrow \begin{pmatrix} 1 & 0 & 0 & 1 \\ 1 & 0 & 1 & 1 \\ 1 & 1 & 1 & 0 \\ 1 & 1 & 0 & 0 \\ 0 & 1 & 0 & 0 \end{pmatrix}$$

On the other hand, the following clique matrix does not have the property (try it!):

$$\begin{pmatrix} 1 & 1 & 1 & 1 \\ 1 & 0 & 0 & 0 \\ 0 & 1 & 0 & 0 \\ 0 & 0 & 1 & 0 \\ 0 & 0 & 0 & 1 \end{pmatrix}$$

It can be shown that an undirected graph is an interval graph if and only if its clique matrix M has the consecutive ones property for columns [14]. Note that the clique matrix corresponding to the interval graph in Figure 1.1c shown above (matrix with *A-E* rows and 1-8 columns) has this property.

As we will see in the examples below, these important and interesting properties of interval graphs make the problem of finding the maximal cliques of an interval graph relatively easy because all one needs to do is go from source to sink by decreasing degrees of vertices at each successive step [15]. In general, finding the maximal cliques for a graph is a computationally challenging problem, one that is referred to as NP-complete [16] since the computational time taken in a brute force enumeration of all the cliques increases very rapidly (more than a polynomial increase) as the size of the graph increases. However, a variety of good algorithms that reduce the time exist for finding the maximal cliques even in a very large network. For general graphs, a class of computational methods based on data structures called PQ trees and PC trees has been developed, and Hsu [17] extended a PQ approach [18, 19] to develop PC Trees, which can generate interval graphs from adjacency matrices algorithmically (instead of heuristically) in linear time. Our colleague, Noppadon Khirpet, in Thailand has implemented this algorithm for us in a program appropriately named PC-Tree [20], which we hope to extend to handle larger data sets in the near future.

We have developed four software packages for handling biological data using interval graph-theoretic techniques: (a) javaBenzer [21, 22] and (b) BioGrapher [23, 24], with only passing reference to (c) PC-tree [20] (currently lacking a data interface), and (d) javaBenzer FoodWeb, which is more relevant to material presented by Cozzens [25] in the second chapter of this book. In the examples that follow, we demonstrate that many biological problems can indeed be addressed from a discrete perspective to yield meaningful insights.

1.2.2 Interval Graphs in Biology

In the late 1950s, Seymour Benzer [26] studied the fine structure of genes. Prior to his work genes were defined as units of structure, function, and recombination. His work occurred shortly after it was discovered that the structure of DNA consisted of two antiparallel and complementary linear strands of nucleotides arranged in a double helix that encodes genetic information as a sequences of four nucleobases: adenine, cytosine, guanine, and thymine, usually abbreviated as *A*, *C*, *G*, and *T*, respectively. Mutation could be as small as a substitution, deletion, or addition (point mutations) of one of these four nucleotides (letters) in a DNA sequence. Recombination, the process in which two DNA molecules may exchange genetic information by breakage and reunion of the respective fragments, may occur between two adjacent letters. Function may be encoded in a sequence of letters that is comparatively long – several hundreds of these letters. Benzer unpacked these terms by experimentally mapping the DNA of a virus that infected a bacterium. In order to speed up the process of fine mapping genes, he used deletion mutations (identified as nonreversible mutations because of the improbability of reinserting an appropriate length of DNA with letters of the appropriate sequence) because he could map a series of point mutations to a short region covered by one small interval defined by the extent of a deletion. Although Benzer's deletion mapping method is regularly taught in undergraduate genetics courses [21], students sometimes struggle to visualize it. Therefore, we will introduce interval graphs with a biological example that is easier to visualize, namely, restriction mapping of DNA, or the contig assembly problem.

Since 1978, we have been able to manually sequence relatively short DNA strands (hundreds to thousands of letters long) but the techniques are insufficient to sequence a whole chromosome, which may be millions or even billions of letters long. One approach to solve this problem is to cut chromosomes with molecular scissors (enzymes, called restriction endonucleases) into many small pieces that can be sequenced. The problem then remains: how do we order these many small fragments to reconstruct the sequence of letters of the entire chromosome? We will begin by simply determining which fragments overlap with one another. We can do this in two different ways: (a) if the fragments are sequenced, we can examine whether they have long subsequences in common—since sequencing has become so easy to do and the price has dropped so considerably, sequencing centers normally do sixfold reads of the sequences of fragments to obtain a very high accuracy in reading the composition and the letter orders and then comparing sequences for overlap—and this is the approach currently used and (b) if we cut a fragment with a second restriction endonuclease, we can determine whether the daughter fragments are fragments that have already been identified; this is the older approach used in the literature and is referred to as the "double digest" problem. Because the binary logic of qualitatively deciding whether two fragments overlap or not is more transparent in this second approach, we describe it below for the pedagogical purposes of focusing on the logic inherent in the problem.

In the simple example that follows, we begin with twelve fragments that we have labeled with Greek letters and four restriction endonucleases: EcoR1, Pst1, Sp1, and XmaIII. Twenty-five experiments (Table 1.1) were conducted to determine which fragments overlap or do not overlap one another.

The results of these experiments can now be used to construct a symmetrical adjacency matrix specifying which fragments overlap with one another. The rows and columns of the adjacency matrix are labeled with the Greek letters corresponding to the fragments. First, it is important to note that each fragment overlaps with itself, which leads to all cells along the main diagonal being filled. Second, fragments generated in the right hand column in Table 1.1 cannot overlap one another, as they are subsections from a larger linear segment from which they were generated. Third, the parent fragments in the second column overlap all the progeny fragments in the fourth column. With these insights, it is possible to construct a complete symmetrical adjacency matrix of overlapping fragments. The software package javaBenzer can be downloaded from the Biological ESTEEM Project website [27], and inferences about which fragments overlap can be input (Figures 1.2a-c) using the experimental data in Table 1.1.

We first solve this interval graph problem of determining how many maximal cliques exist in our data, visualizing them, and then finding the order of these maximal cliques by using the BioGrapher Excel-based graphical visualization software [24] that we developed (before moving to a formal solution using javaBenzer), because BioGrapher illustrates well the formal graph-theoretic concepts involved. We begin by arranging the vertices of the network in an evenly distributed radial fashion, then construct an intersection graph based on the matrix of data in Figure 1.1c by entering that matrix as a series of zeroes and ones (the "adjacency" matrix) in BioGrapher, which then automatically generates the intersection graph of connected interrelated vertices (Figure 1.3a) from the adjacency matrix. If we flip the zeroes and ones, we can also use BioGrapher to generate the complementary graph (Figure 1.3b).

TABLE 1.1 Simulated Experimental Data for a contig Experiment that Generated 25 Fragments Using Four Restriction Enzymes

Experiment #	Initial Fragment	Enzyme	Fragment(s) Generated
1	Lambda	None	Lambda
2	Lambda	EcoR1	Alpha, theta, gamma
3	Lambda	Pst1	Delta, mu
4	Lambda	Pst1 and Sp1	Episilon, kappa, mu
5	Lambda	EcoR1 and Sp1	Alpha, epsilon, eta, theta
6	Lambda	XmaIII and Sp1	Alpha, beta, epsilon
7	Lambda	EcoR1, XmaIII, and Pst1	Alpha, iota, theta
8	Lambda	All four enzymes	Alpha, epsilon, kappa, theta, zeta
9	Delta	Sp1	Epsilon, kappa
10	Mu	EcoR1	Alpha, theta, zeta
11	Gamma	Pst1	Epsilon, kappa, zeta
12	Iota	Pst1	Epsilon, kappa, zeta
13	Gamma	Sp1	Epsilon, eta
14	Iota	Sp1	Epsilon, eta
15	Gamma	Pst1	Zeta, delta
16	Iota	Pst1	Zeta, delta
17	Beta	EcoR1	Eta, theta
18	Beta	EcoR1 and Pst1	Kappa, theta, zeta
19	Eta	Pst1	Kappa, zeta
20	Alpha	All four enzymes	Alpha
21	Epsilon	All four enzymes	Epsilon
22	Kappa	All four enzymes	Kappa
23	Theta	All four enzymes	Theta
24	Zeta	All four enzymes	Zeta
25	Lambda	All four enzymes	Alpha, epsilon, kappa, theta, zeta

Fragments are labeled with Greek letters. The four restriction endonucleases used are Sp1, Pst1, EcoR1, and XMaIII.

To make sure that the matrix corresponds to an interval graph, we need to check two requirements:

(a) There can be no Z4s in the intersection graph. A Z4 is a polygon (face) of four or more connected vertices with no interior chords (adjacency matrix in Figure 1.4A) because such polygons would be topologically congruent with a circle instead of a line (Figure 1.4B).

(b) The transitively oriented complementary graph (Figure 1.5) is called a *comparability graph* [28] because there is a flow from sources at one end of a line to sinks at the other end along with the transitive property,

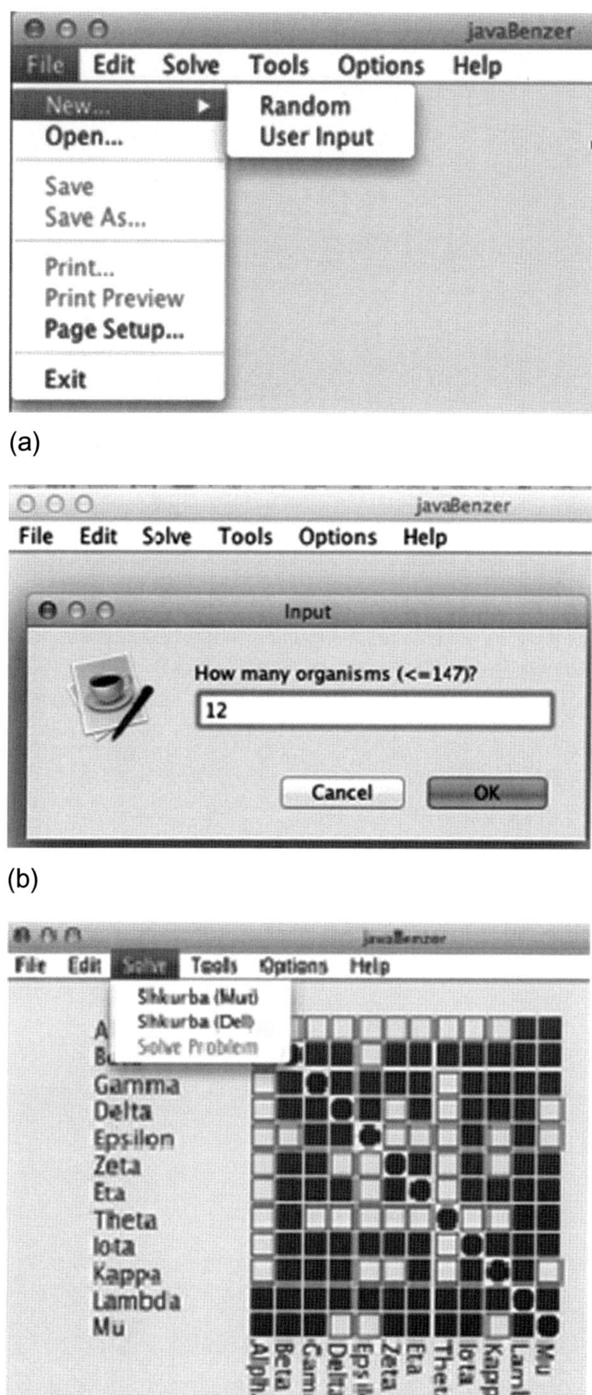

FIGURE 1.2 The construction of the symmetrical matrix that shows which fragments overlap with one another in javaBenzer [21]. (a) Choice of user input. (b) Choice of the number of "organisms" (these could be the fragments as above or they could be genes, proteins, metabolites, etc.; mathematically, in graph theory, these are vertices of a graph). (c) The symmetrical matrix is the adjacency matrix for the graph and was generated by filling any cell in row i, column j if the fragments overlap with one another according to logic described above. The empty cells correspond to non-overlapping fragments.

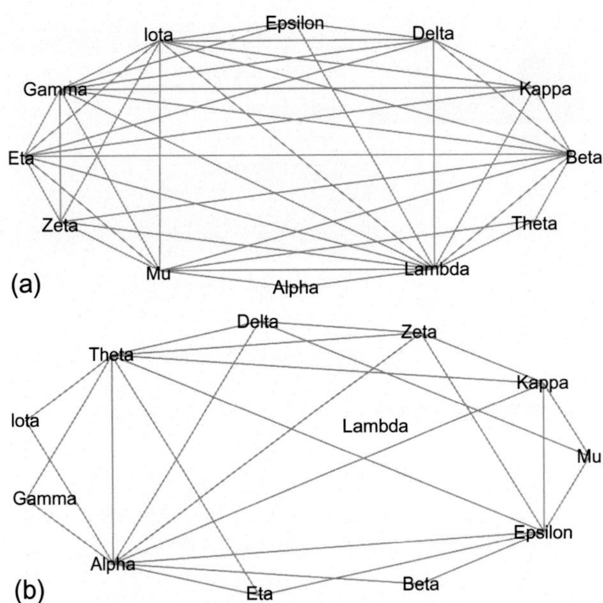

FIGURE 1.3 Intersection and complementary graphs. (a) Intersection graph based on hypothetical sample graph data generated by javaBenzer. (b) Complementary graph of the intersection graph in Figure 1.3a.

	a	b	c	d
a	1	1	0	1
b	1	1	1	0
c	0	1	1	1
d	1	0	1	1

(A)

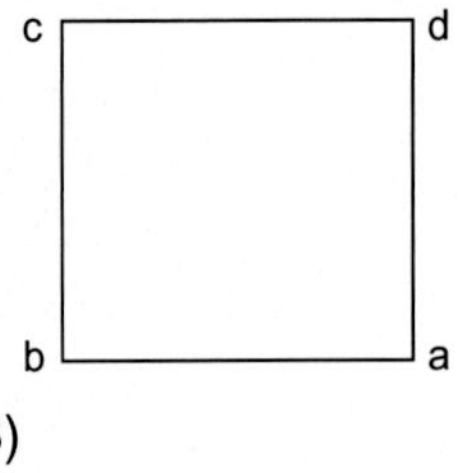

c d

b a

(B)

FIGURE 1.4 Z4 implies a circular structure. (A) The successive overlapping of four fragments implied by the adjacency matrix on the left (a) is represented by the intersection graph, a Z4, on the right. (B) The intersection graph of the overlaps in A is a Z4 (on the right). Such representations imply a circular structure.

such that if interval A is before interval B and interval B is before interval C, then interval A must be before interval C. This is mathematically equivalent to positing that if there is a triangular face in the complementary graph, one can label all edges as directed in such a fashion that if vertex A is directed to vertex B and vertex B is directed to vertex C, then the third edge of the triangular face must exist and be directed A to C.

Next, we need to find the maximal cliques in the original intersection graph. There are five highlighted in Figure 1.6: (1) Alabama = beta, delta, eta, gamma, iota, kappa, and lambda; (2) Alaska = beta, eta, gamma, iota, lambda, mu, and zeta; (3) Arkansas = delta, epsilon, gamma, iota, and lambda; (4) Arizona = beta, lambda, mu, and theta; and (5) California = alpha, mu, and lambda.

The maximal cliques are then ordered by consulting the comparability graph (Figure 1.5). We are able to do this iteratively by looking at all relationships between two maximal cliques as shown below.

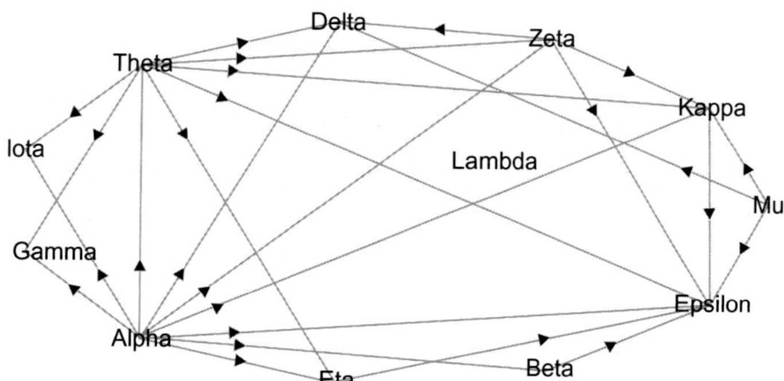

FIGURE 1.5 The directed graph with transitively oriented labeling (arrows) superimposed upon the undirected complementary graph of Figure 1.3b. Such graphs are called *comparability graphs*; see Ref. [28].

(a) Find the elements (restriction fragments) in one maximal clique not in common with another.
(b) Look at the ordering in the comparability graph of such pairs of maximal cliques.
(c) Construct a complete directed graph of the maximal cliques as vertices and the directed edges as the direction between the two maximal cliques determined in the previous step.
(d) Find the Hamiltonian path in the directed graph constructed in step (c) by tracing from the source (California, the vertex with all outgoing edges) and sequentially moving to each subsequent maximal clique of one degree of outgoing edges less until one reaches the sink (Arkansas) with zero outgoing edges (Figure 1.7a).
(e) Convert the Hamiltonian path of maximal cliques into an interval graph that shows where each restriction fragment resides (Figure 1.7b).

After the construction of the restriction map of the DNA fragments as an interval graph, it is important to review all of the data in Table I to double-check whether the resulting map corresponds with the inferences made from the data.

Summary of the steps used to produce an interval graph from adjacency matrix data [27]:

1. Convert each restriction fragment into a vertex.
2. Construct the intersection graph by placing an edge between each pair of vertices that represent overlapping restriction fragments.
3. Construct the complement of the intersection graph.
4. Check for the absence of Z4s in the intersection graph.
5. Determine whether the complement of the intersection graph can be made transitive (i.e., a comparability graph).
6. Find all the maximal cliques in the intersection graph.
7. Order these maximal cliques in the same way as in the comparability graph.
8. Find the Hamiltonian path of all the ordered maximal cliques.
9. Construct the interval graph by assigning restriction fragments to each interval of the line, which sequentially orders the maximal cliques, for all the cliques to which the restriction fragment vertex belongs.

The steps above are performed manually. However, the resulting adjacency matrices in Steps 2, 3, 5, and 6 are entered into the BioGrapher Excel spreadsheet and visualized, as shown in Figures 1.3, 1.5, and 1.6. These visualizations make it easy to follow the steps through the entire process. A number of different "layouts"—spring, radial, circular, all of which attempt to minimize edge crossings and reduce clutter and vertex overlaps in the visualization—can be specified for the graphical display. For example, the "spring" layout treats the edges as mechanical springs and rearranges the vertices in the display in a fashion that minimizes the total "energy." The "circular" layout depicted in this specific exercise is a visualization that attempts to arrange the vertices on the circumference of a circle or ellipsoid. A "radial" layout that positions a high-degree vertex at the center, with connections radiating outward to other vertices arranged roughly in concentric circles or ellipsoids is also available. These three different formats may help a user more easily see patterns such as motifs, maximal

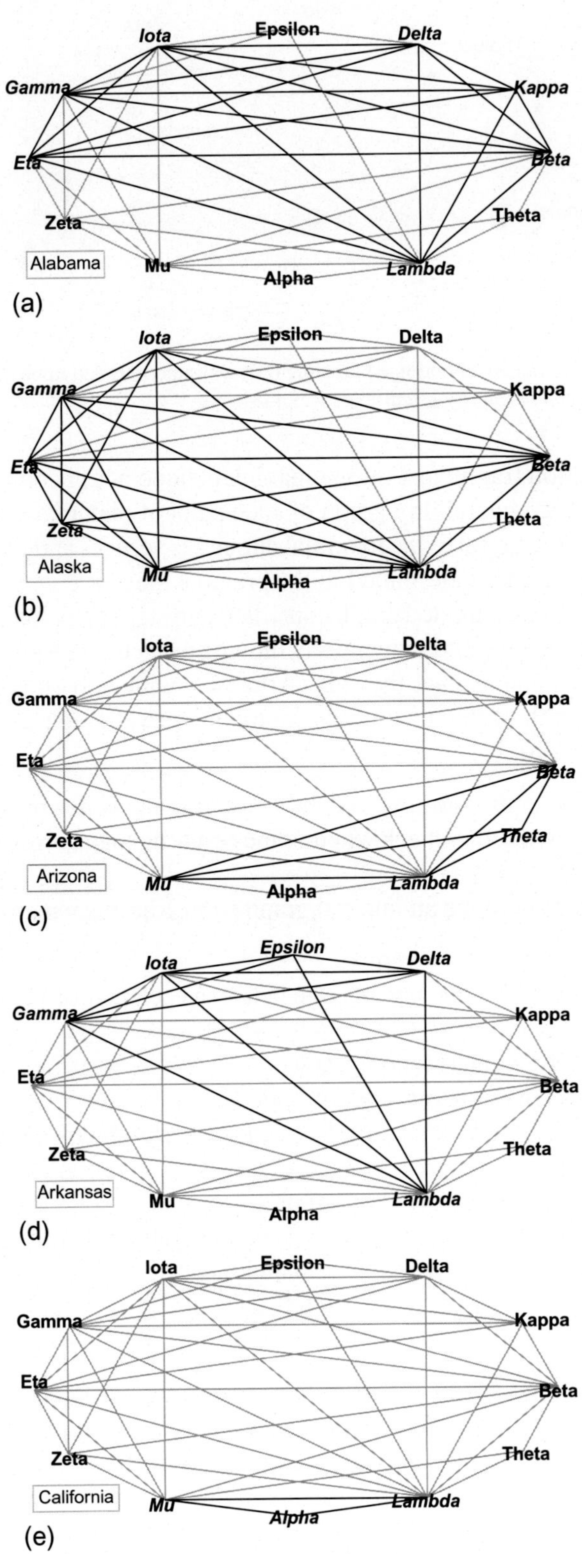

FIGURE 1.6 The five maximal cliques in the original intersection graph of connected vertices (Figure 1.3a). (a) Alabama (maximal clique) is a complete subgraph of the intersection graph containing the vertices beta, delta, eta, gamma, iota, kappa, and lambda. (b) Alaska = beta, eta, gamma, iota, lambda, mu, and zeta. (c) Arkansas = delta, epsilon, gamma, iota, and lambda. (d) Arizona = beta, lambda, mu, and theta. (e) California = alpha, mu, and lambda.

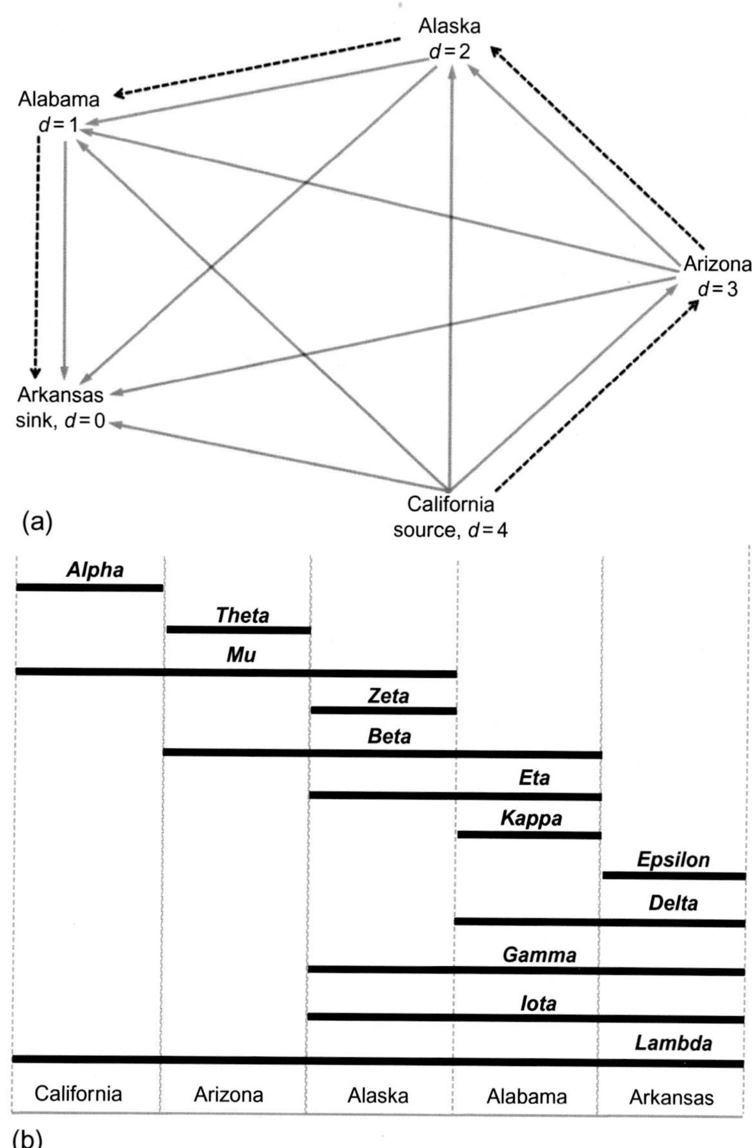

FIGURE 1.7 Graphical representation of the Hamiltonian pathway of maximum cliques and interval graph constructed from the Hamiltonian pathway. (a) The interior complete directed graph of five vertices shows the binary relations between each pair of maximal cliques determined by the ordering of nonshared vertices in the maximal cliques from the arrows in the comparability graph (the transitively ordered complementary graph). The dotted line is a Hamiltonian path of maximal cliques that is constructed by moving from source (degree of outgoing edges equals 4) successively to each maximal clique vertex with one smaller outgoing degree until the sink (outgoing degree equals zero) is reached. (b) The interval graph is constructed from the Hamiltonian path of maximal cliques. Each maximal clique is represented at the bottom. Elements of each maximal clique are represented by horizontal line segments that cover each maximal clique region to which they belong as a member. Please note that all such segments should be contiguous and thus no holes should occur within any horizontal line. Note that the divisions between maximal cliques are demarcated at the top of the diagram by the specificity of where particular restriction endonucleases cut the DNA.

cliques, hubs, small worldness, etc. For a detailed description of layout schemes and other features of BioGrapher, along with a helpful web resource bibliography, please see the *BioGrapher Description and Tutorial* document available on the supporting website.

In summary, BioGrapher allows a user to simply input an adjacency matrix and then use a drop-down menu to generate the intersection graph and complementary graph in any of three different configurations: spring, radial, or circular. The identification of maximal cliques and the transitive ordering of the complementary graph to produce a comparability graph are done by hand. If a matrix is developed for a complete graph of maximal cliques and/or for a comparability graph, BioGrapher can draw the respective directed graphs. The determination of the Hamiltonian path corresponding to the order of maximal cliques in the final interval graph is easily done by counting outgoing degrees from a source to a sink.

Now that we have insight into the graph-theoretic manipulations necessary for using the BioGrapher Excel-based graphical visualization tool, let us return to the javaBenzer program to streamline and automate the process. Shkurba [29] simply rearranged the symmetrical adjacency matrix, such as the one originally shown in Figure 1.1c, so that all recombinants were tightly connected to the main diagonal (Figure 1.8a; this is achieved by automatically performing the rearrangements using the *Solve* option in javaBenzer). This rearrangement is often referred to as the *Shkurba form* of the adjacency matrix and is explained in more mathematical detail in a later section (Example I). The recombinants belonging to the squares on the main diagonal supplemented by those recombinants belonging to arrowheads that span those squares are isomorphic to the maximal cliques identified in Figure 1.6. The Hamiltonian path runs along the main diagonal from the source of all arrows to their sink. Once the Shkurba form of the matrix is achieved, the interval graph (Figure 1.8b, which is isomorphic to Figure 1.7b) can be generated by going to the *Solve* option in javaBenzer and choosing *deletions* instead of *recombinants*.

Because of its simplicity, most students prefer to use javaBenzer (rather than going through all of the logical steps of the underlying graph theory approach) to solve problems related to deletion mapping, complementation mapping, restriction mapping, assembling contigs of DNA, ordering protein sequence fragments, or determining the niche space of food webs in community ecology. However, we find that if we require students to work with both tools they develop a better understanding of the underlying principles and why these techniques work.

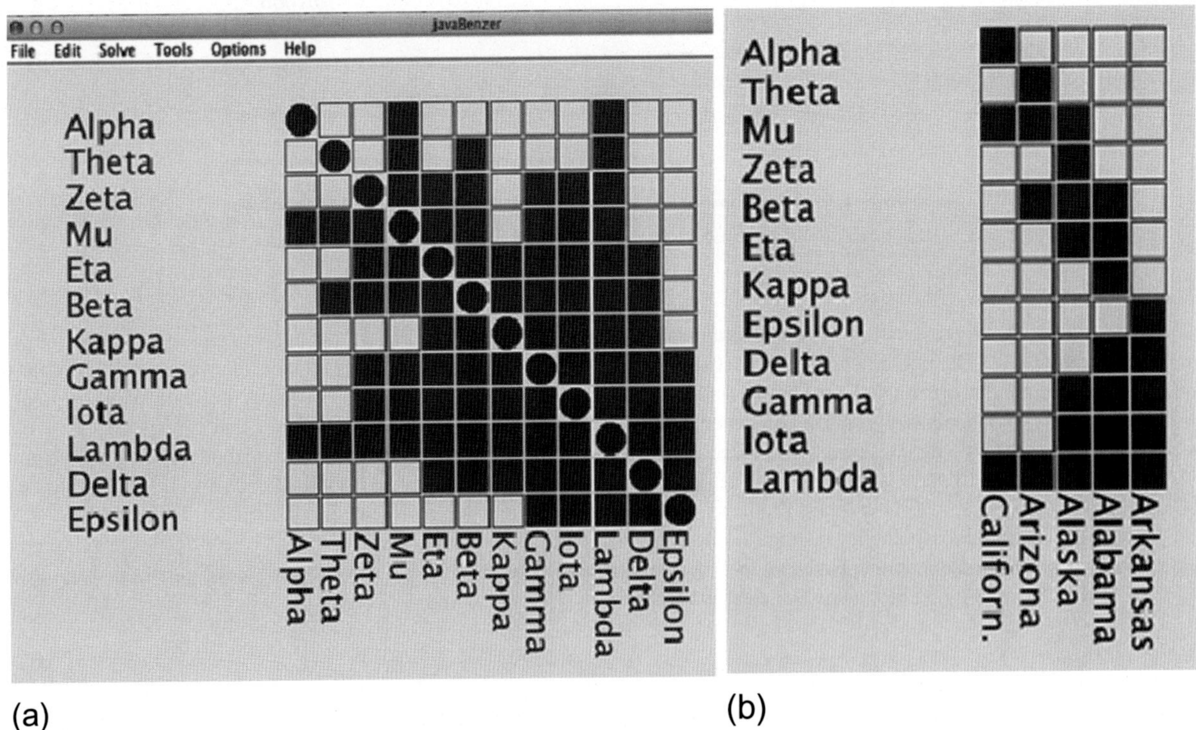

FIGURE 1.8 Shkurba form [29] of adjacency matrix and the interval graph of maximum cliques. (a) The adjacency matrix in Figure 1.2c has been rearranged to the Shkurba form. (b) The interval graph of the maximal cliques. Note that order is from the source maximal clique California in the upper left of Figure 1.8a to the sink maximal clique Arkansas in the lower right corner of Figure 1.8a. Also, note that this is a clique matrix.

Interval graphs apply to all sorts of problems that involve the linear ordering of segments of a line. We have illustrated this above in terms of the restriction map/contig assembly problem and discussed it as an easier illustration of Benzer's original deletion mapping problem. Similar problems involve the assembly of protein sequences from overlapping fragments of short peptides [15] and the complementation mapping of cistrons (segments of DNA that correspond to the coding sequences of individual polypeptides). However, a very counterintuitive example in the community ecology primary literature is the niche space of food webs. Cozzens, a pioneer mathematician in interval graphs and their extension to circular arc graphs, lays out a variety of mathematical aspects of interval graphs in Chapter 2 [25]. Note that in graph theory, the term *circular* describes the topology overall of maximum clique intervals (not to be confused with the *circular layout* visualization option available in BioGrapher described above). In our next example, we elaborate on the biological ramifications of extending this approach to asking questions in community ecology.

Disney-like cartoons of food webs abound in biology textbooks. Students are often confused because they consider that top predators are the most evolutionarily fit rather than that there is variation in fitness within each species, they think of the organismal icons as representing individuals rather than populations, and they do not understand the direction of arrows in the directed graph as the flow of energy and/or biomass. The diagrams also lack any sense of the frequency with which relative prey appear in the diet of the predator or the sampling within the population, and avoid such issues as cannibalism and infanticide. However, we want to demonstrate an even subtler pattern missed in these cartoon representations. We start with a subtropical aquatic food web—a popularized version of a food web studied by Paine [30]—shown in Figure 1.10a. We used the data to generate a symmetrical predator-predator adjacency matrix (where a filled cell in the matrix represents two predators competing since they eat a common prey; such matrices are sometimes referred to as competition matrices) and determine whether such a matrix has an interval graph structure. The Shkurba form of the symmetrical predator-predator matrix was constructed using javaBenzer and is depicted in Figure 1.9b. Finally, the predators are arranged vertically according to their trophic level. A predator is considered at a higher trophic level than its prey. If in turn, that predator is eaten by another predator, the second predator is considered to have an even higher trophic level than the first predator. The maximal cliques then represent communities of predators that compete for common prey (Figure 1.9c). Such communities of predators are referred to as ecological niches.

Some of the questions in community ecology that have arisen since the initial application of interval graphs to examine food webs are: Why one dimension? What gradient accounts for that dimension? Is the partition of niche space driven by biotic or physical variables or both? Is the interval graph property of many food webs due to oversimplification or taxonomic lumping? What is different about those food webs that cannot be represented by interval graphs (i.e., they break the Z4 assumption of Figure 1.4 or the transitive orientability criterion of Figure 1.5)?

1.3 EXAMPLE I—DIFFERENTIATION: GENE EXPRESSION

Single linkage cluster analysis is a widely used tool for identifying associations in biological data such as those based on microarrays. The basic purpose of such microarray experiments is to identify coordinated patterns of sequential clusters of genes being synchronously turned on or off (activated or inhibited). A beautiful study by Rives and Galitski [31] studied genes in yeast over their life cycle. They focused on a "filamentation-network protein set ...derived from a search of the Yeast Protein Database [32] and other published sources [33, 34] for proteins with mutant phenotypes or expression patterns associated with the filamentous form." They produced the cluster analysis result shown in Figure 1.10.

Unfortunately, the mathematical details of how they achieved the final pattern in Figure 1.10 above are not given; the authors [31] only state: "Clusters were delimited manually by using the cluster tree ...as a guide." We now demonstrate two mathematically explicit ways (previously described) for generating such patterns which are easily learned and more easily lend themselves to visually emphasizing the subpatterns that those authors identified.

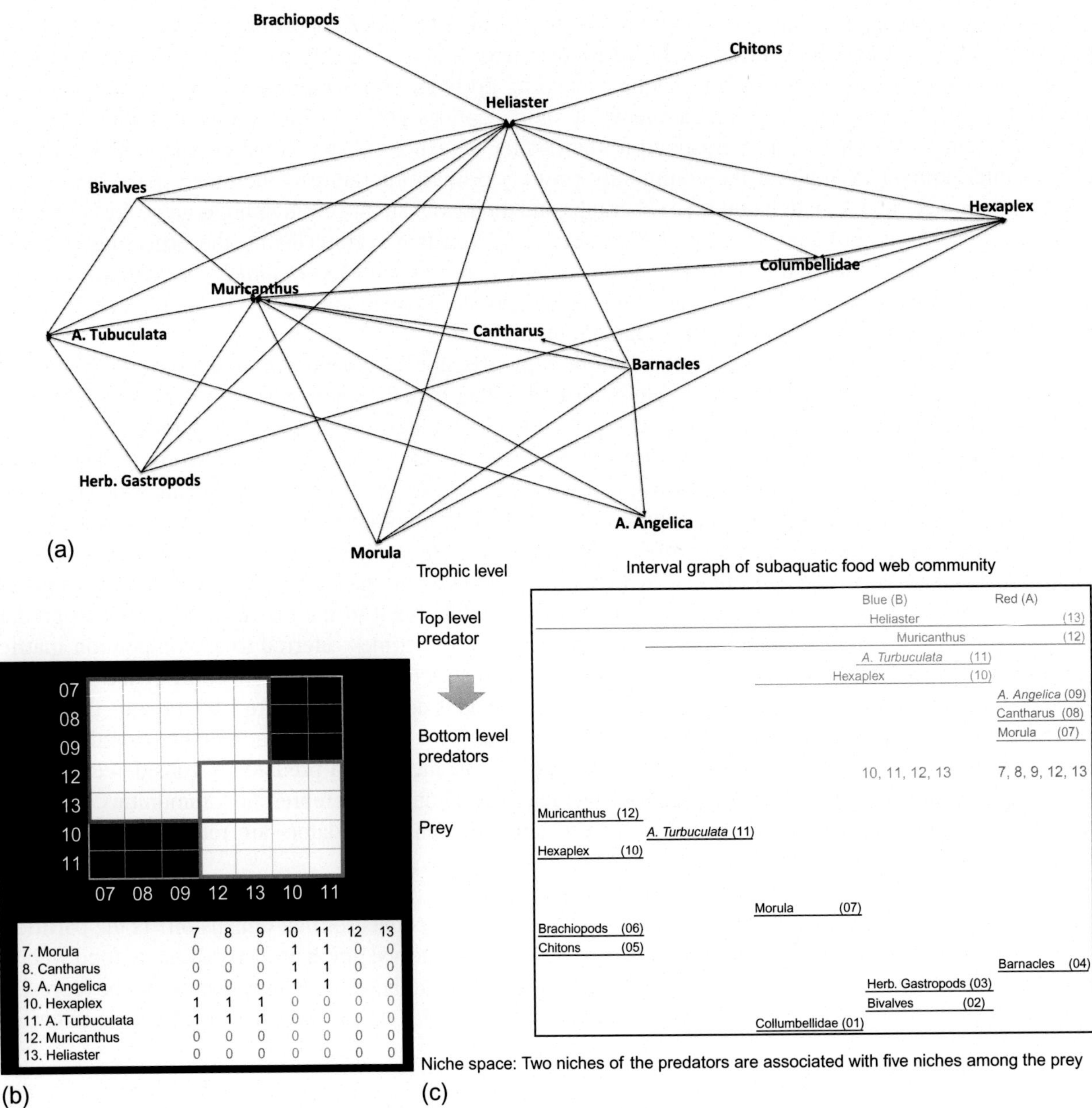

	7	8	9	10	11	12	13
7. Morula	0	0	0	1	1	0	0
8. Cantharus	0	0	0	1	1	0	0
9. A. Angelica	0	0	0	1	1	0	0
10. Hexaplex	1	1	1	0	0	0	0
11. A. Turbuculata	1	1	1	0	0	0	0
12. Muricanthus	0	0	0	0	0	0	0
13. Heliaster	0	0	0	0	0	0	0

FIGURE 1.9 Foodwebs and their interval graphs. (a) A subtropical aquatic food web from Paine [30]. (b) The symmetrical predator-predator matrix is constructed (bottom) and the rearrangement in the Shkurba form of the symmetrical predator-predator matrix (top). (c) Interval graph: the predators are arranged vertically according to their trophic level and the maximal cliques represent communities of predators (upper half) that compete for common prey (lower half). While there are seven predators, they only occupy two niches in this representation. Furthermore, this interval graph representation draws attention to the difference between generalist predators (*Heliaster* and *Meracanthus* span both niches) and specialists (the other five only belong to one of the two maximal cliques. However, when we add the structure among the prey, we see that *Hexaplex* consumes a greater variety of prey than *A. turbuculata*.

In particular, we ask the reader to note two things: First, the symmetrical matrix has been arranged such that well-connected clusters are associatied with nearly solid white blocks along the main diagonal of the matrix along with arrowheads that overlap some adjacent solid blocks. Second, there are a variety of vertical and horizontal white stripes that are not well connected to these blocks. The observed pattern resembles the Shkurba form [29] of the adjacency matrix of graphs that possess interval properties. This form, proposed by the Russian mathematician Viktor Shkurba, is a rearrangement of the rows and columns of the $n \times n$ adjacency matrix $M = (\delta_{ij})$ (where $\delta_{ij} = 1$, if vertices i and j are adjacent and $\delta_{ij} = 0$, if they are not) into a form

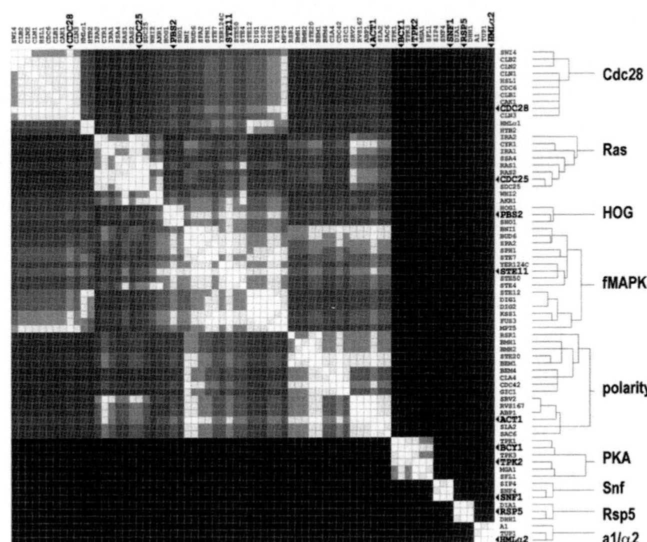

FIGURE 1.10 Reproduced from the paper on modular organization of cellular networks by Rives and Galitski [31], with caption. Clustering of the yeast filamentation network. Proteins of the yeast filamentation network were clustered. A tree-depth threshold was set. Tree branches with three or more leaves (clusters with three or more proteins) below the tree threshold are shown. Bullets and large bold labels indicate proteins of highest intracluster connectivity. Copyright (2003) National Academy of Sciences, U.S.

$$\overline{M} = \left(\overline{\delta_{ij}}\right) = \begin{pmatrix} \delta_{i_1 i_1} & \delta_{i_1 i_2} & \cdots & \delta_{i_1 i_n} \\ \delta_{i_2 i_1} & \delta_{i_2 i_2} & \cdots & \delta_{i_2 i_n} \\ \cdots & \cdots & \cdots & \cdots \\ \delta_{i_n i_1} & \delta_{i_n i_2} & \cdots & \delta_{i_n i_n} \end{pmatrix},$$

with the following "basic property": for any integer value l such that $k \leq l \leq n$, the entries of the rearranged matrix satisfy the inequalities $\delta_{i_k i_l} \geq \delta_{i_k i_{l+1}}$. This simply means that the rows and the columns are rearranged in a way that ensures that the 1's are grouped together to the right of the main diagonal along the rows, and that the block of 1's is followed by a block of 0's. Since the matrix is symmetric, the same applies to the entries of the matrix columns below the main diagonal. Clearly, performing such a rearrangement, especially for large-size matrices, by simply using observation and/or heuristic methods would be difficult and unreliable. Shkurba's paper [28] contains a detailed algorithm that accomplishes this task. For the purposes of this chapter, the exact details of the algorithm are not essential; it has been implemented in our software package javaBenzer [20, 21]. The user can obtain the Shkurba form of the adjacency matrix (one of the options available) but only after a designated number of trials. Note that in javaBenzer, the adjacent vertices are represented by white cells while the non-adjacent vertices are black. In order to see an original unordered javaBenzer adjacency matrix and a javaBenzer adjacency matrix in Shkurba form, refer to the earlier Figures 1.2c and 1.8a.

While there is no *a priori* reason that gene expression microarray data should conform to the strict requirements of interval graphs, when we examined Figure 1.10 above from Rives and Galitski [31], we noticed the presence of complete blocks along the main diagonal and a variety of Shkurba arrowheads that pointed from the lower righthand corner to the upper lefthand corner. However, if we just consider the least grey (most white) elements in Figure 1.10, both the blocks and arrowheads are incomplete. Thus, it appears that we could both improve on the clustering and ordering and provide an explicit rationale for our procedures. Here, we have applied both the javaBenzer and BioGrapher software tools to the Rives and Galitski data with a simplified threshold requirement, specifying only binary [white(connected) or black(disconnected)] elements, and ignoring shades of grey (representing the strengths of connections) shown in the original visualization. The result is shown in Figure 1.11.

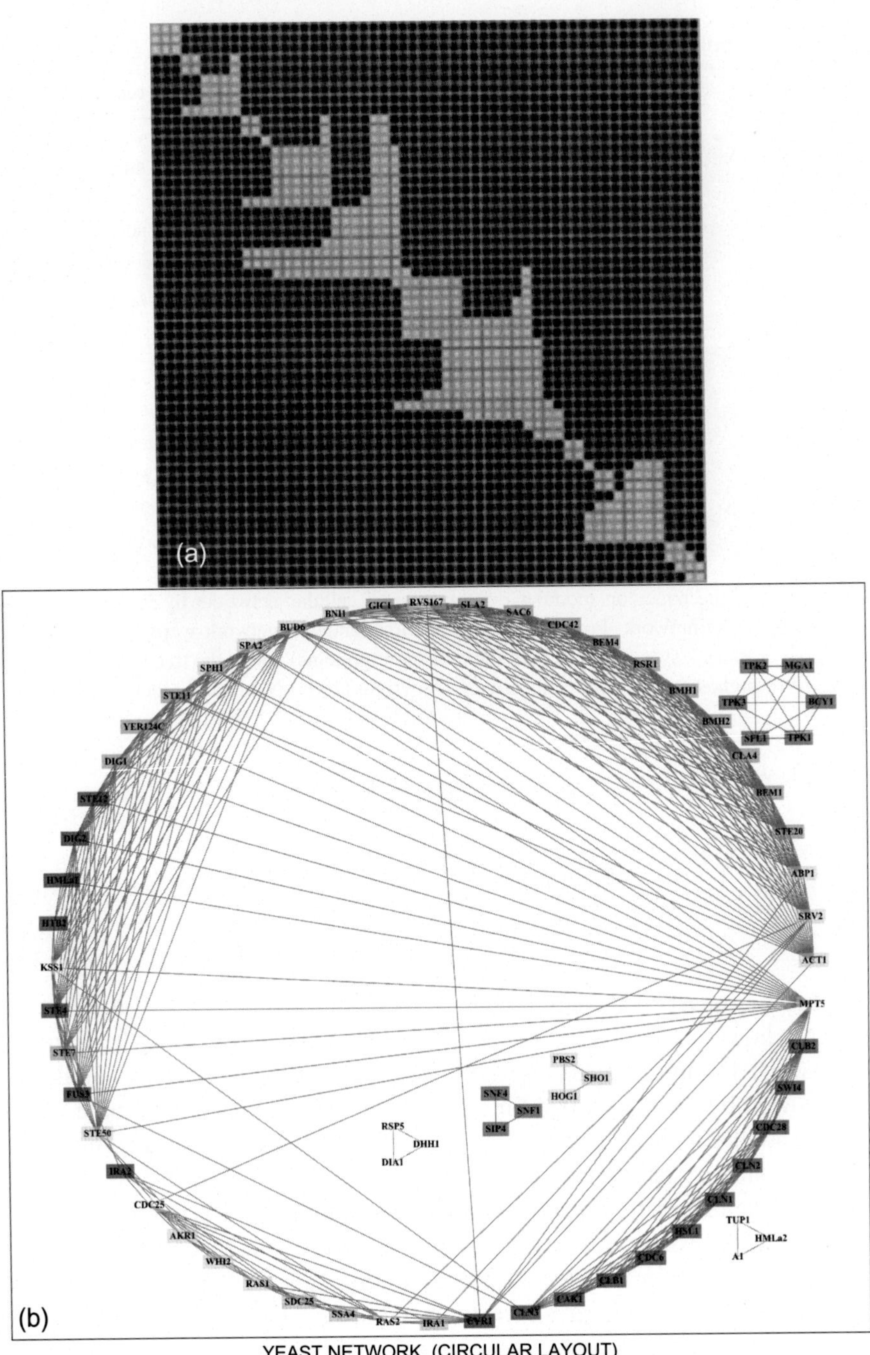

FIGURE 1.11 javaBenzer [21] and BioGrapher [24] representations of the data in Figure 1.10. (a) javaBenzer rearrangement of Rives and Galitski's adjacency matrix [31] illustrated in Figure 1.10. (b) Spring graph produced by BioGrapher with the entry of this adjacency matrix and coloring of maximal cliques. Note that in a, arrowheads do not overlap all successive blocks. Thus, we could rearrange the six independent blocks in 6!/2 distinct orders and still maintain Shkurba form. This is more easily seen in the spring graph, as those small maximal cliques which are not connected to one another are clearly separate.

The BioGrapher software includes tools to analyze networks for small world properties. Using those tools yields a diameter of 10 for the network containing 70 nodes, with a mean free (shortest) path of 2 and an average vertex degree (number of direct connections per node) of 10. While the average clustering coefficient is 0.85 (close to 1), it is the diameter of the graph that is a key indicator of whether the network exhibits small world properties, in which case it should be close to log N for large values of N, where N is the number of nodes.

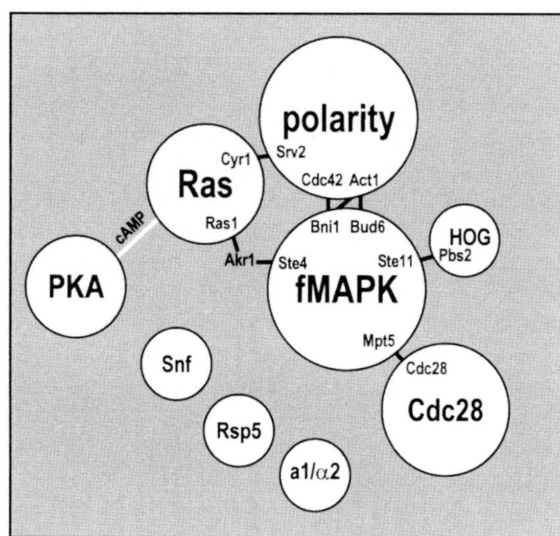

FIGURE 1.12 Rives and Galitski's modular model of the yeast filamentous network. The only differences between their topology and ours are due to our higher threshold for interaction, causing us to miss the connection of two of our free-standing maximal cliques to the larger aggregate. We do not know whether the biological association is weaker for these links than others displayed. Copyright (2003) National Academy of Sciences, U.S.

For this particular example, $N = 70$ and $\log_{10}(70) \approx 2$, while the actual diameter, which is equal to 10, is much larger. Additionally, a log-log plot of the degree count versus the number of nodes with a given degree is not linear, indicating that the power law distribution characteristic of small world networks is not manifested, at least not for the yeast data that were analyzed.

The nine maximal cliques illustrated by BioGrapher in the circular layout graph in Figure 1.11b also correspond to the nine circles of biologically meaningful networks of mutually expressed genes that Rives and Galitski [31] identified (Figure 1.12).

In conclusion, this example illustrates that the exploration of complex biological data with interval graph theory extends beyond the traditional applications to linear molecules (deletion mapping, complementation mapping, restriction mapping, assembling contigs of DNA, and ordering protein sequence fragments) to gene expression data such as those from Rives and Galitski [31]. In Chapter 2, Cozzens [25] shows the value of interval graphs in determining the niche space of food webs in community ecology. □

PROJECTS

Circular graphs are represented by circular arcs along a circle. They can easily be represented as ordinary undirected graphs by considering the following rules: Each arc is represented by a vertex and two vertices are connected by an edge if and only if the circular arcs overlap. An example of a circular graph is shown in Figure 1.13.

Project 1.1. Two beautiful examples that would lend themselves to extending the interval graphs to handle circular arc graphs [36] are shown in Figure 1.11. □

Project 1.2. Another appropriate example is based on data in Breyne et al. [37].

While the analysis of each network in both these projects could become a full-fledged research project, we suggest that you begin by analyzing these networks with the concepts described above.

- Develop an adjacency matrix for each network.
- Construct intersection and complementary graphs and visualize them in the three layouts—spring, radial, and circular—available in the Excel BioGrapher workbook.
- Identify maximal cliques.
- Check for Z4s.

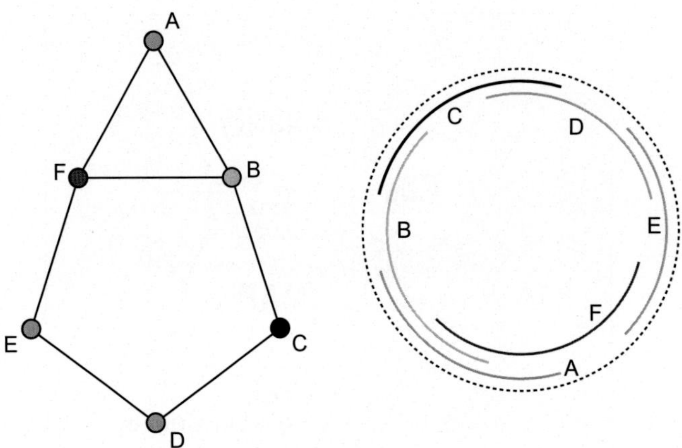

FIGURE 1.13 Circular arc graph with associated matrix [35].

- Determine whether the complementary graphs satisfy the transitive orientability property to construct a comparibility graph.
- Export the adjacency matrix for each network into javaBenzer and rearrange into Shkurba form as much as possible.
- Check whether you can construct something close to an interval graph or a circular arc graph from the above elements.
- Look for a causal biological relationship for the maximal cliques and their connections and orders such as time or location of expression, collision or binding between molecules, metabolic flux, or signal transmission. Each of the three networks has a substantially different biological context. In Figure 1.14a, Smith et al. [38] examined responses to iron limitation in a globally distributed marine bacterium by measuring transcriptional (reading of DNA into messenger RNA) and translational (reading of messenger RNA to construct protein sequences) regulatory responses. In Figure 1.14b [39], "two normally incompatible activities fixing nitrogen in the dark and photosynthesizing in the light" were examined in an "ocean-dwelling cyanobacteria that is capable of it mak[ing] this switch inside the same cell every light/dark cycle (normally about 12 h). This makes it interesting from the standpoint of bioenergy ...production [as well as its] regulation [because] the process of how it is able to drastically rearrange its machinery every 12 h is not well understood." The authors [39] made "measurements of levels of gene expression" and were surprised that the circular clock of coexpressions of clusters of genes corresponded so well to phases in the light/dark cycle. Breyne et al. [37] did similar analyses of gene expression during cell division in plants. It is important to check the biological context of each of these networks in order to relate the topological ordering to underlying causal relationships among physical materials.
- Consider whether the network is resilient to perturbation (particularly useful if investigators have conducted "knockout" experiments, which are equivalent to removing vertices from a network, or to examine the stable evolutionary strategy for robustness in the face of perturbation). □

1.4 EXAMPLE II—DISEASE ETIOLOGY

We now look at another systems biology example with a complex data set that does not have many overlapping maximal cliques in order to demonstrate that the use of javaBenzer and BioGrapher is still very profitable in terms of visualizing other aspects of complex networks. In the examples discussed above, the maximal cliques are clearly associated with biological processes and subcellular organelles.What do gene associations mean if these biological connections are not clearly associated with network features? As our next example illustrates, many network visualizations obfuscate important connections that should be pursued, or do not highlight them. We now discuss research that examined a specific chemotherapeutic anticancer drug, Herceptin, significant because

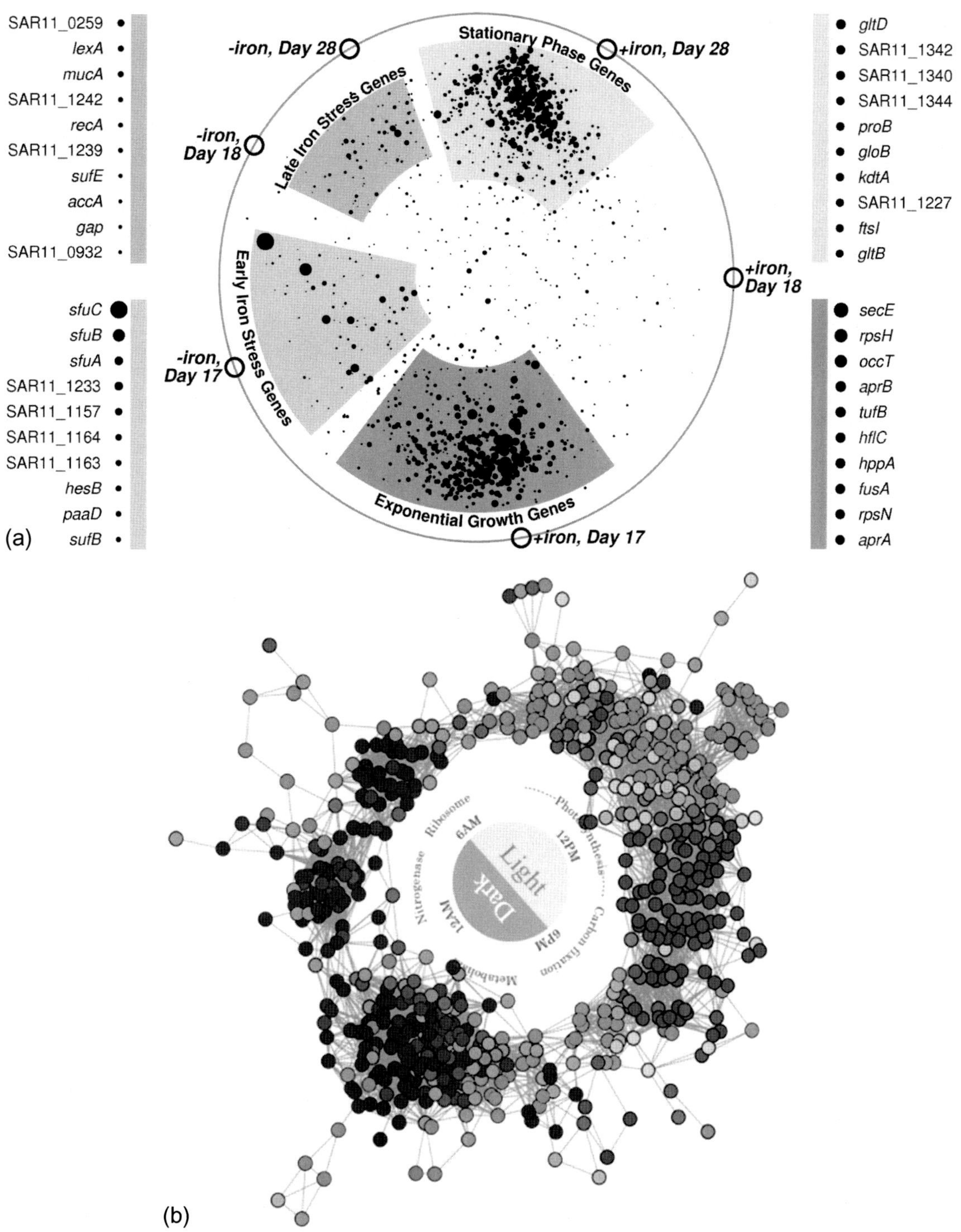

FIGURE 1.14 Examples of circular arc graphs. (a) Transcriptional and translational regulatory responses to iron limitation in the globally distributed marine bacterium *Candidatus Pelagibacter ubique* [38]. (b) A model of cyclic transcriptomic behavior in *Cyanothece* species ATCC 51142 [39].

only about 50% of breast cancer patients respond to Herceptin and the remainder only experience the toxic effects of the medication. This is potentially a major public health issue since many physicians and other health providers as well as members of the public may not realize that resistance to cancer chemotherapy is as big or a bigger problem than the evolution of resistance to antibiotics, pesticides, and herbicides. Neapolitan et al. [40] recently published an analysis of genes involved in breast cancer. They were able to process 22 signal transduction

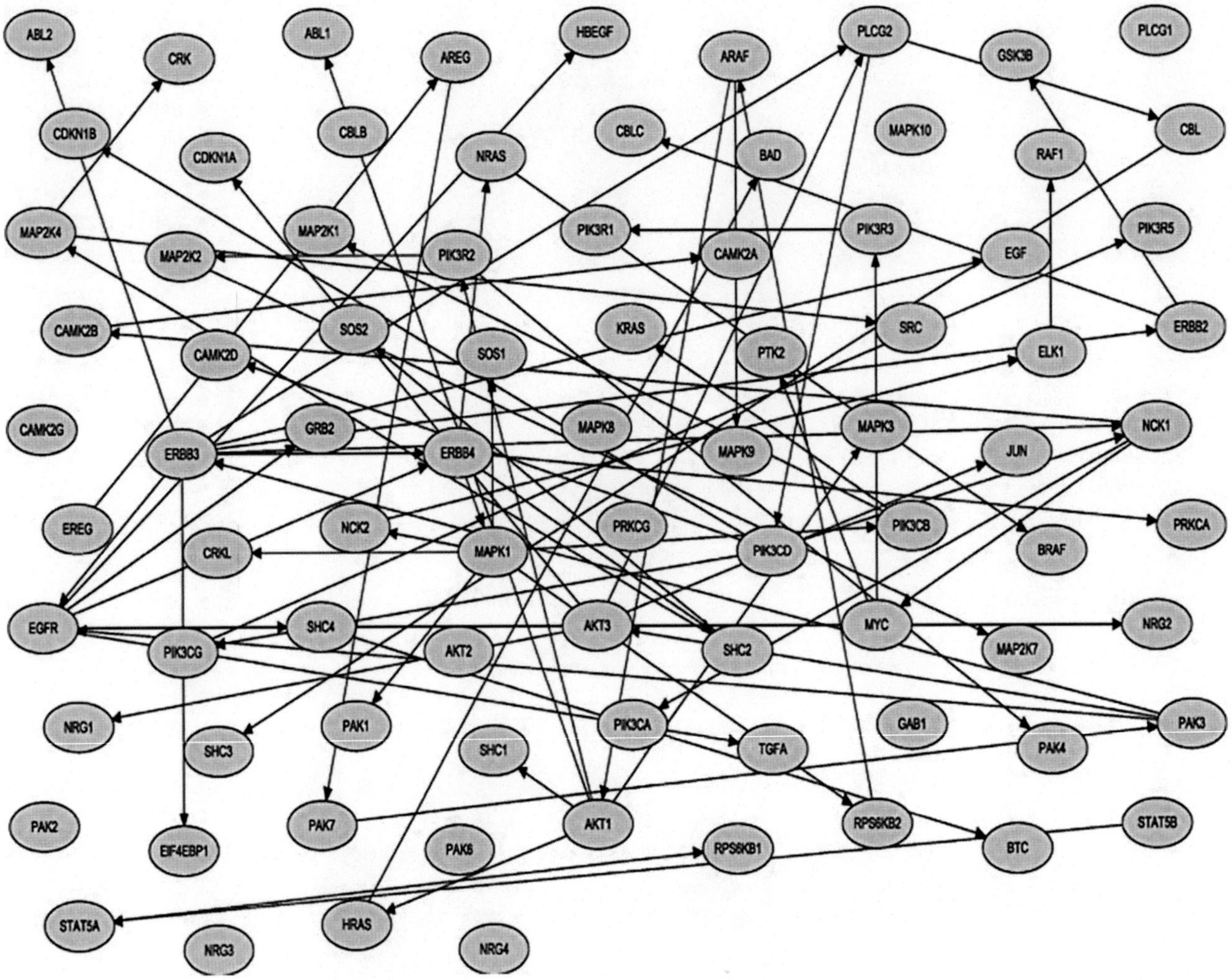

FIGURE 1.15 The causal Bayesian network published by Neapolitan et al. [40]. The vertices are genes in the ErbB signal tranduction pathway associated with modified gene expression in breast cancer patients.

pathways from the comprehensive online databases in the Kyoto Encyclopedia of Genes and Genomes (KEGG) [41] and relate these to gene expression profiles collected from 529 breast cancer patients and 61 controls. They scored the gene expression levels as low, medium, or high. They only published the results (Figure 1.15) for the ErbB signal tranduction pathway, one of the 22 networks that they constructed. We are not told if these results are the best, the worst, or randomly chosen from the 22 networks constructed. They do state that "using such a network, we can learn about possible driver genes and the effect of genetic variants on these driver genes and therefore on the network."

We argue that while such network visualizations and results may be optimized to compactly fit on a published page, they are inadequate because: (a) it is difficult to see who is connected to whom (notice that some pairs are connected by long lines that traverse the whole image rather than being proximal); (b) hubs are not easily identified as having much higher degree than vertices spoked from them; (c) maximal cliques of varying size are hard to find in such a crowded nest; and (d) no interconnections between motifs are visible. We therefore attempted an alternative visualization of the data using javaBenzer and BioGrapher, to see if we could obtain further insights. We removed all vertices that were freestanding and created an adjacency matrix of the remaining 75 genes. From this we generated a javaBenzer matrix (Figure 1.16a and b) and used BioGrapher to

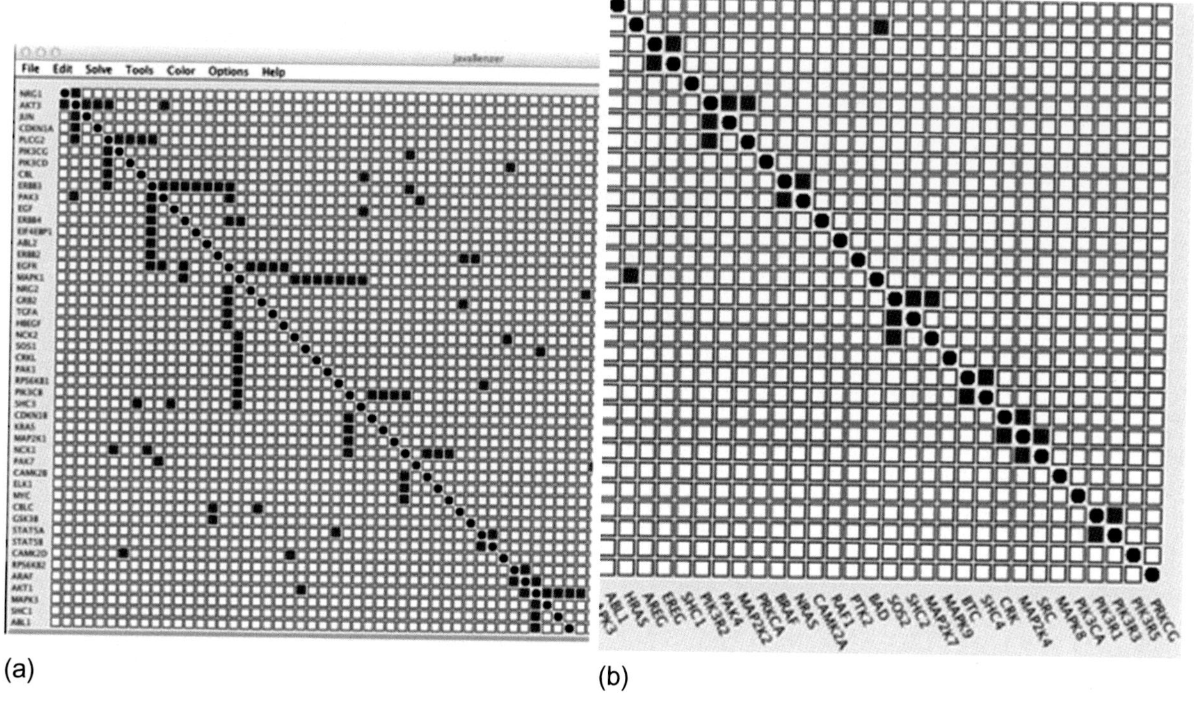

FIGURE 1.16 javaBenzer representation of the Bayesian network of Figure 1.15 reorganized to diagonalize the data as much as possible (the Shkurba form). The whole matrix was bigger than the screen, so it was split into two parts. (a) The upper left portion of the matrix. (b) The lower right portion with the overlap at gene ABL1.

generate visualizations of the resulting graph using different visual layout schemes, spring, radial, and circular (Figure 1.17a-c). We also used BioGrapher to compute the clustering coefficient and the diameter of the network.

Compare the reorganized matrix in Figure 1.16a with the matrices in Figures 1.8a and Figure 1.11a. Note how sparse it is. The maximal cliques are insignificant because maximal cliques of size two are just connected pairs. The arrowheads do let us see that there are hubs with three to seven spokes. These are much more visible in the BioGrapher images in Figure 1.17.

The overall graph is broken into many small pieces and those pieces that are connected are more like a chain than a crowded neighborhood. In the radial version, we can also see that this network is probably not a small world network because there are not many links across the center. Note the absence of any maximal cliques of four or more vertices in each of these representations. The diameter is 9, comparable to the yeast protein network [31], with approximately the same number of vertices ($N = 75$) and a similar average clustering coefficient of 0.72, assuming the graph is undirected. However the average degree is 3, significantly less than the yeast degree of 10. Additionally, the average mean free path is 4, twice that of the yeast network.

The different graphical visual layouts are interesting because the best one, in our opinion, is the *radial*, not the circular or spring, in that the vertices are spread out the best. However, if we were to remove terminal edges (i.e., edges with a vertex connected to only one other vertex), the circular representation would show a topological reduction to a circle of numerous edges; this may be a biologically important feature of the Bayesian analysis. Thus, being able to draw different graphical visualizations with the same data points may elicit visual connections and insights that would otherwise be lost or not apparent. In terms of examining the assumptions of the authors [40], the inferences about hubs are very different than those about motifs. In hubs, a single gene's expression is connected to a variety of other genes, but, unlike a motif, there is no large set of genes that is coordinately controlled spatially or temporally.

Since resistance to cancer chemotherapy is such a major public health issue, the identification of "possible driver genes and the effect of genetic variants on these driver genes and therefore on the network" is a crucial

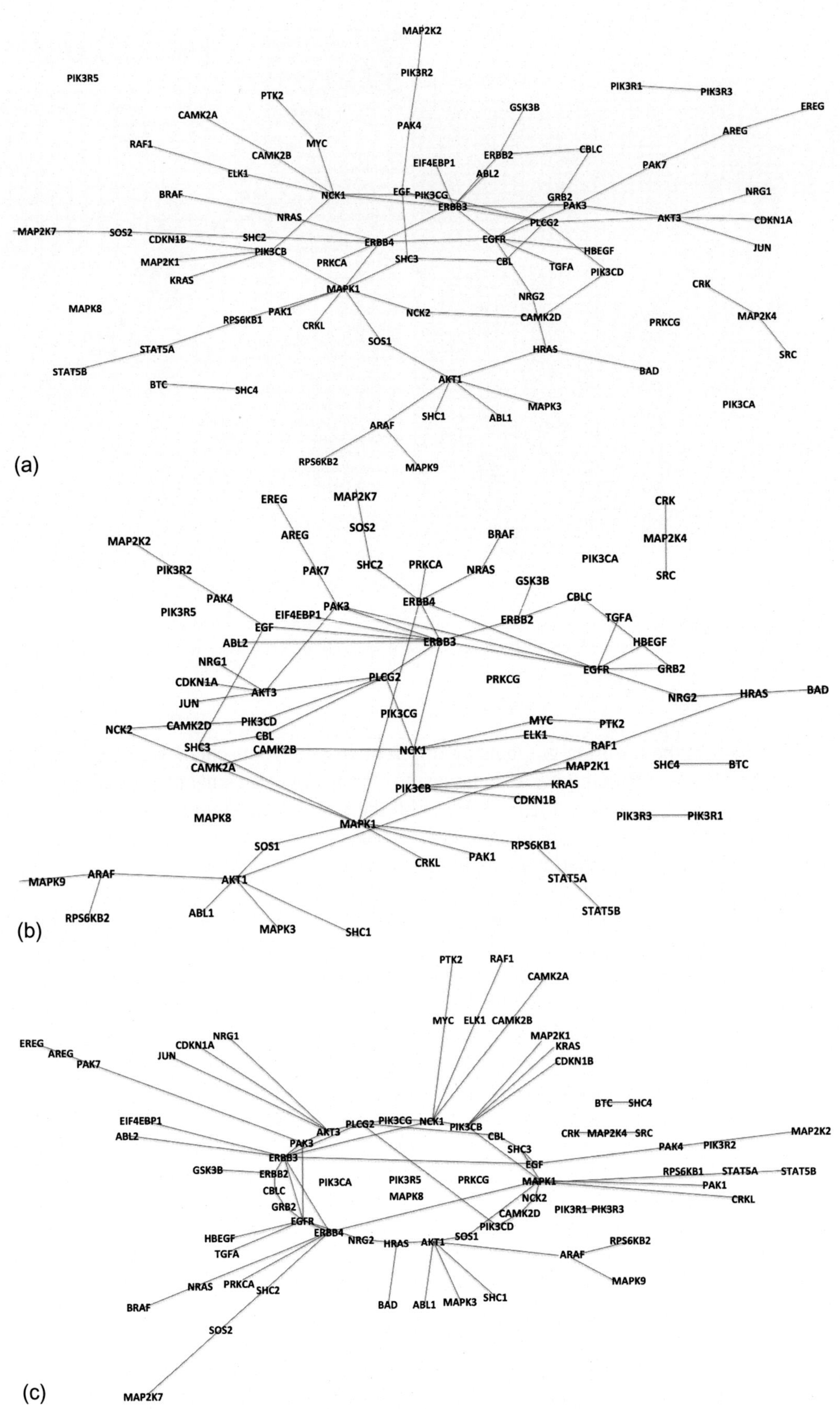

FIGURE 1.17 BioGrapher-based visualizations of the adjacency matrix for the Bayesian network in Figure 1.16, using various layouts. (a) Spring layout. (b) Radial layout. (c) Circular layout.

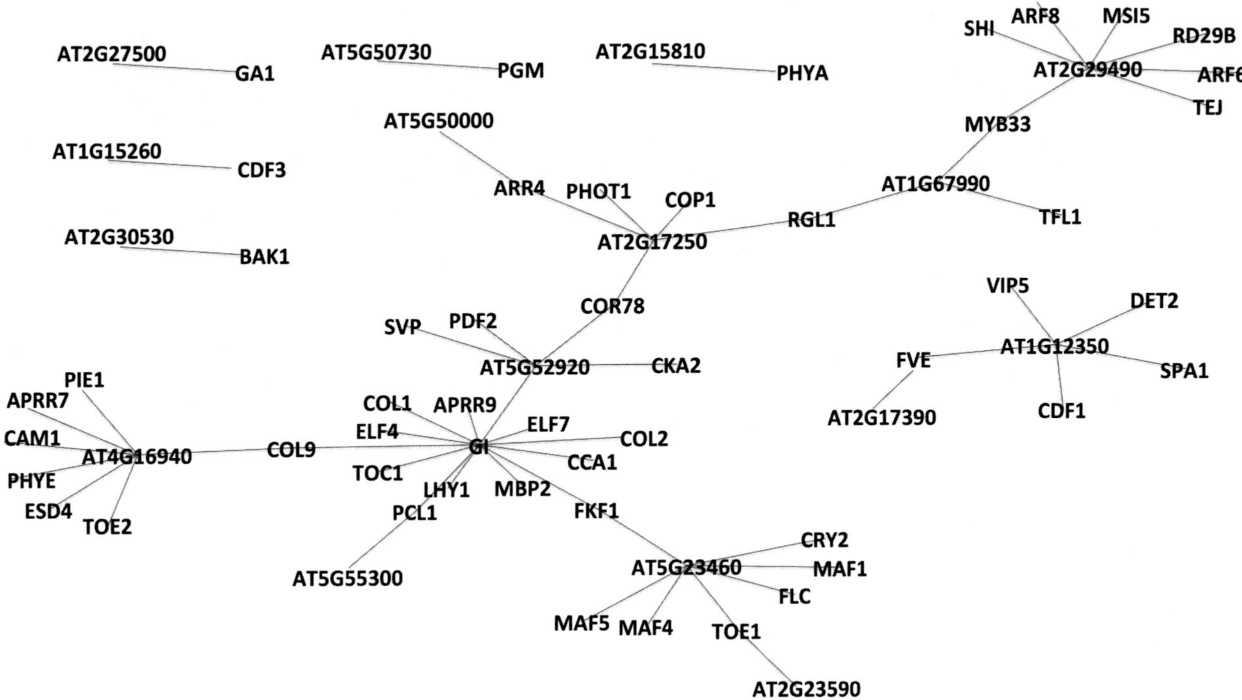

FIGURE 1.18 BioGrapher spring layout visualization of the *Arabidopsis* gene data set published recently by Keurentjes et al. [42]. This graph is a tree and has some high-degree hubs.

problem. We think that our reanalysis of the recently published Neapolitan, Xue, and Jiang [40] analysis of genes involved in breast cancer draws into question some of their conclusions. Namely, they argue that the network helps identify "driver genes." However, even the hubs of Figure 1.17 do not have a high degree. An inference that might be drawn from this observation is that without large maximal cliques or large hubs, the Bayesian analysis has primarily only identified pairs of relationships rather than more complex associations. □

Project 1.3. There are many hub-like networks published in the primary literature. These do not all share the obfuscation critiqued above. For example, the network in Figure 1.18 clearly demonstrates a chain of hubs.

Keurentjes et al. [42] came to an important conclusion based upon their analysis of this network. "Distant gene expression regulation occurs more frequently but local regulation is stronger." Their inference is that *cis*-regulation of closely linked genes on a chromosome is more important than *trans*-regulation of distantly separated genes or genes on different chromosomes. We invite you to explore further their data set of 176 of the 192 genes known to be involved in flowering *Arabidopsis*. In particular we want to draw your attention to a graph-theoretic property of the network visualization based on their data (Figure 1.18), namely, that it is a tree and hence a planar graph. Some vertices have a high degree (the hubs) but most vertices only have one neighbor (called spokes). Now re-examine Figure 1.17c: while there are a few overlapping edges that cannot be rearranged into a planar form, the overall appearance of the graph shares much of the hub and spoke aspect of Figure 1.18. Therefore, we might ask, How tree-like is a network without many maximal cliques within it? That is, if we were to eliminate a few edges or collapse small polygons (traingles, squares, or maximal cliques) to form vertices, would the transformed network now be a tree? The topological reduction of graphs is frequently useful for finding linear or circular associations, as we previously highlighted with regard to interval and circular arc graphs. □

1.5 CONCLUSION

Applications of graph theory in mathematical biology span a diverse class of applications. They include subcellular localization of coordinated metabolic processes, identification of hubs central to such processes and the links between them, analysis of flux in a system, temporal organization of gene expression, the identification of drug targets, determining the role of proteins or genes of unknown function, and the coordination of sequences of signals. Medical applications include diagnosis, prognosis, and treatment. Public health ramifications deal with epidemics and the evolutionary consequences of resistance to antibiotics, pesticides, herbices, and chemotherapy.

Graph theory provides an approach to systematically testing the structure of and exploring connections in various types of biological networks. In particular, interval graph properties such as the ordering of maximal cliques via a transitive ordering along a Hamiltonian path are useful in detecting patterns in complex networks. In this chapter, we have illustrated several different types of biological networks based upon the restriction mapping problem, community ecology relationships in food webs, gene expression patterns in development or in response to external environmental changes, and Bayesian analysis of disease associations. In gene expression, we focused on how the identification of maximal cliques could be associated with subcellular location of associated activities and how the directed Hamiltonian path of maximal cliques could be used to infer a vectorial flow between such subcellular activities. In the second example, we demonstrated that some published networks lack sufficient substructure to inform further investigations, and implicitly emphasized the importance of easy access to raw data. Unlike the textbook mathematics used in most of undergraduate biology that is usually associated with small data sets, these graph-theoretic approaches and the software that we have developed empower students to explore real-world problems and complex data sets from contemporary primary research.

The data used for the examples and some of the projects in the chapter are available for download from the volume's companion website.

ACKNOWLEDGMENTS

Two investigators have sustained a major and fruitful collaboration with the authors for a long time in the development of software: Vince Strief in the Department of Biostatistics at the University of Wisconsin, Madison, and Noppadon Khiripet, Director of the Knowledge Elicitation Laboratory at NECTEC, Bangkok, Thailand. We deeply thank our student research assistants for their help in analyzing several data sets, adding features to the BioGrapher software package, and producing visualizations. Yang Yang and Sijia Liang, Statistics, University of Minnesota; Tia Johnson, Chemistry, Johnny Franco, Program in Biochemistry, and Tony Abell, Min Thu Aung, Hlaing Lin, and Khalid Qumsieh, Department of Mathematics and Computer Science, Beloit College; Han Lai, Bioinformatics, Carnegie-Mellon; and, Laura Cauhill, Quantitative Biology, Damir Creecy, Ecology, Andre Freleigh, Biological Sciences, and Ross Sousa, Mathematical Sciences, University of Delaware.

REFERENCES

[1] Boersma KT, Waarlo AJ, Klaassen CWJM. The feasibility of systems thinking in biology education. J Biol Educ 2011;45(4):190–7.
[2] Odum EP. The new ecology. Bioscience 1964;14:14–6.
[3] Odum HT, Odum EC. Modeling for all scales: an introduction to system simulation. San Diego, CA: Academic Press; 2000.
[4] Craige BJ. Eugene odum. Ecosystem ecologist and environmentalist. Athens, GA: The University of Georgia Press; 2001.
[5] Hagen J. An entangled bank: the origins of ecosystem ecology. New Brunswick, NJ: Rutgers University Press; 1992.
[6] Taylor P. Technocratic optimism, H.T. Odum and the partial transformation of ecological metaphor after World War 2. J Hist Biol 1988;21:213–44.
[7] Loeb J. The mechanistic conception of life. Chicago, IL: The University of Chicago Press; 1912, reprinted 1964.
[8] http://www.frontiersin.org/Systems_Biology/about. Accessed September 13, 2014.
[9] Buchanan M, Caldarelli C, De Los Rios P, Rao F, Vendruscolo M, editors. Networks in cell biology. New York, NY: Cambridge University Press; 2010.
[10] Palsson BO. Systems biology: properties of reconstructed networks. New York, NY: Cambridge University Press; 2006.
[11] De Los Rios P, Vendruscolo M. Network views of the cell. In: Buchanan M, Caldarelli C, De Los Rios P, Rao F, Vendruscolo M, editors. Networks in cell biology. New York, NY: Cambridge University Press; 2010. p. 4–13.
[12] Hayes B. Graph theory in practice: part I. Am Scient 2000;88:9–13. Hayes B. Graph theory in practice: part II. Am Scient 2000;88:104–9, and references therein.
[13] He Q, Macauley M, Davies R. Dynamics of complex Boolean networks: canalization, stability, and criticality. In: Robeva R, editor. Algebraic and discrete mathematical methods for modern biology. New York: Academic Press; 2015, Chapter 5.
[14] Alan Tucker. A structure theorem for the consecutive 1's property. J Combinatorial Theory 1972;B12(2):153–62.

[15] Jungck JR, Dick G, Dick AG. Computer-assisted sequencing, interval graphs, and molecular evolution. BioSystems 1982;15: 259–73.

[16] Abello J, Parados PM, Resende MGC. On maximum clique problems in very large graphs. In: Abello J, Vitter J, editors. External memory algorithms. DIMACS series. Providence, RI: American Mathematical Society; 1999. p. 119–30.

[17] Hsu WL, McConnell RM. PC Trees and circular ones arrangements. Theor Comput Sci 2003;296:99–116.

[18] Landau GM, Parida L, Weimann O. Gene proximity across whole genomes via PQ trees. J Comput Biol 2005;12(10):1289–306.

[19] Kratsch D, McConnell RM, Mehlhorn K, Spinrad JP. Certifying algorithms for recognizing interval graphs and permutation graphs. Siam J Comput 2006;36(2):326–53.

[20] Jungck JR, Khiripet N, Viruchpinta R, Maneewattanapluk J. Evolutionary bioinformatics: making meaning of microbes, molecules, maps. Microbe 2006;1(8):365–71.

[21] Jungck JR, Streif V. Deletion mapping of genetic "fine structure": supplementing ad hoc problem solving approaches with algorithms and heuristics. Bioscene 1986;12(2):13–27.

[22] Jungck JR, Streif V, Jungck P, Everse S. javaBenzer. A module of the Biological ESTEEM Collection, published by the BioQUEST Curriculum Consortium. Available at: http://bioquest.org/esteem/esteem_details.php?product_id=202;2004. Accessed September 13, 2014.

[23] Johnson T, Viswanathan R, Farbotko A, Jungck JR. Manipulatives, models, and mathematics: Biomolecular visualization via analog and digital modalities for measuring and making meaning of 3D geometry and topology. Abstracts Molecular Biology of the Cell 2004;15:234a.

[24] Viswanathan R. BioGrapher and WDisplay: excel modules for data visualization. BioQuest Notes 2006;15:6–7.

[25] Cozzens M. Food webs and graphs. In: Robeva R, editor. Algebraic and discrete mathematical methods for modern biology. New York: Academic Press; 2015, Chapter 2.

[26] Benzer S. On the topology of genetic fine structure. Proc Natl Acad Sci U S A 1959;45:1607–20.

[27] The Biological ESTEEM Collection: A project of the BioQUEST Curriculum Consortium. Available at: http://bioquest.org/esteem. Accessed September 13, 2014.

[28] Gilmore PC, Hoffman AJ. A characterization of comparability graphs and of interval graphs. Can J Math 1964;16:539–48.

[29] Shkurba VV. Mathematical processing of a class of biochemical experiments. Kibernetika 1965;1:62–7.

[30] Paine RT. Food Web complexity and species diversity. Am Nat 1966;100(910):65–75.

[31] Rives A, Galitski T. Modular organization of cellular networks. PNAS 2003;100:1128–33.

[32] Costanzo MC, Crawford ME, Hirschman JE, Kranz JE, Olsen P, Robertson LS, et al. YP™, PombeP™ and WormPD™: model organism volumes of the BioKnowledge™ Library, an integrated resource for protein information. Nucleic Acids Res 2001;9:75–9.

[33] Lengeler KB, Davidson RC, D'Souza C, Harashima T, Shen WC, Wang P, et al. Signal transduction cascades regulating fungal development and virulence. Microbiol Mol Biol Rev 2000;64:746–85.

[34] Madhani HD, Fink GR. The control of filamentous differentiation and virulence in fungi. Trends Cell Biol 1998;8:348–53.

[35] http://en.wikipedia.org/wiki/Circular-arc_graph. Accessed October 6, 2014.

[36] Cozzens MB, Roberts FS. On dimensional properties of graphs. Graphs and Combinatorics 1989;5:29–46.

[37] Breyne P, Dreesen R, Vandepoele K, Veylder L, Breusegem F, Callewaert L, et al. Transcriptome analysis during cell division in plants. PNAS 2002;99(23):14825–30.

[38] Smith DP, Kitner JB, Norbeck AD, Clauss TR, Lipton MS, Schwalbach MS, et al. Transcriptional and translational regulatory responses to iron limitation in the globally distributed marine bacterium Candidatus Pelagibacter ubique. PLoS One 2010;5(5):e10487.

[39] McDermott JE, Oehmen C, McCue LA, Hill H, Choi DM, Stöckel J, et al. A model of cyclic transcriptomic behavior in Cyanothece species ATCC 51142. Mol Biosystems 2011;7(8):2407–18, Accessed September 13, 2014. Image made openly available at: http://jasonya.com/wp/five-minute-explanation-cyanothece-transcriptional-model, and data available from: http://jasonya.com/wp/wp-content/uploads/2013/03/SupplementalInformation_04_18_2011.pdf.

[40] Neapolitan R, Xue D, Jiang X. Modeling the altered expression levels of genes on signaling pathways in tumors as causal Bayesian networks. Cancer Informat 2014;13:77–84.

[41] Kyoto Encyclopedia of Genes and Genomes. Available at: http://www.genome.ad.jp/kegg/. Accessed September 13, 2014. (Also see Kanehisa M. Post-Genome Informatics. Oxford, UK: Oxford University Press; 2001.).

[42] Keurentjes JJB, Fu J, Terpstra IR, Garcia JM, van den Ackerveken G, Snoek LB, et al. Regulatory network construction in Arabidopsis by using genome-wide gene expression quantitative trait loci. PNAS 2007;104(5):1708–13.

Chapter 2

Food Webs and Graphs

Margaret (Midge) Cozzens

Rutgers University, Piscataway, NJ, USA

2.1 INTRODUCTION

The study of food webs has occurred over the last 50 years, generally by ecologists working in natural habitats with specific relatively narrow interests in mind. At the outset, a few mathematicians became interested in the graph-theoretic properties of food webs and their corresponding competition graphs; however, linkages between ecologists', mathematicians', and conservationists' interests and results were few and far between. This chapter introduces food webs and various corresponding graphs and parameters to those interested in important research areas that link mathematics and ecology. A basic background on food webs and graphs is provided, with exercises to further illustrate the concepts. Each section has exercises which reinforce the newly introduced concepts. These exercises, sometimes open-ended, together with additional references to preliminary work may be used as springboards to numerous research questions presented in the chapter. Research questions accessible to biology and mathematics students of all levels, with references to previous work, are provided. We should note that all specific research questions are, at least in part, open questions, thus providing ample opportunities for student involvement with actual research. The citations accompanying each research question give more background and starting points for exploration.

The main goals for this chapter are to provide the necessary background that will allow the reader to:

- Recognize various relationships between organisms, and look for patterns in food webs.
- Use graphs and directed graphs (digraphs) to model complex trophic relationships.
- Determine trophic levels and status within a food web, and the significance of these levels in calculating the relative importance of each species (vertices) and each relationship (arcs) in a food web.
- Use a food web to create the corresponding competition (predator or niche) overlap graph, and projection graphs to determine the dimensions of a community's habitat.
- Determine the competition number of a graph and its significance for a community's ecological health.
- Inform conservation policy decisions by determining what happens to the whole food web and habitat if a species becomes extinct (nodes are removed) or prey relationships change (arcs).

2.2 MODELING PREDATOR-PREY RELATIONSHIPS WITH FOOD WEBS

Have you ever played the game Jenga? It's a game where towers are built from interwoven wooden blocks, and each player tries to remove a single block without the tower falling. The player who crashes the tower of blocks loses the game. *Food webs* are towers of organisms. Each organism depends for food on one or many other organisms in an ecosystem. The exceptions are the *primary producers*—the organisms at the foundation of the ecosystems that produce their energy from sunlight through photosynthesis or from chemicals through chemosynthesis. Factors that limit the success of primary producers are generally sunlight, water, or nutrient

Algebraic and Discrete Mathematical Methods for Modern Biology. http://dx.doi.org/10.1016/B978-0-12-801213-0.00002-2
Copyright © 2015 Elsevier Inc. All rights reserved.

availability. These are physical factors that control a food web from the "bottom up." On the other hand, certain biological factors can also control a food web from the "top down." For example, certain predators, such as sharks, lions, wolves, or humans, can suppress or enhance the abundance of other organisms. They can suppress them *directly* by eating their prey or *indirectly* by eating something that would eat something else. Understanding the difference between direct and indirect interactions within ecosystems is critical to building food webs. For example, suppose your favorite food is a hamburger. The meat came from a cow, but a cow is not a primary producer—it can't photosynthesize! But a cow eats grass, and grass is a primary producer. So, you eat cows, which eat grass. This is a simple food web with three players. If you were to remove the grass, you wouldn't have a cow to eat. So, the availability and growth of grass *indirectly* influences whether or not you can eat a hamburger. On the other hand, if cows were removed from the food web, then the *direct* link to your hamburger would be gone, even if grass persisted.

Primary producers, also called *basal species*, are always at the bottom of the food web. Above the primary producers are various types of organisms that exclusively eat plants. These are considered to be *herbivores*, or grazers. Animals that eat herbivores, or each other, are *carnivores*, or predators. Animals that eat both plants and other animals are *omnivores*. An animal at the very top of the food web is called a *top predator*.

Through the various interactions in a food web, energy gets transferred from one organism to another. Food webs, through both direct and indirect interactions, describe the flow of energy through an ecosystem. By tracking the energy flow, you can derive where the energy from your last meal came from, and how many species contributed to your meal. Understanding food webs can also help to predict how important any given species is, and how ecosystems change with the addition of a new species or removal of a current species.

Food webs are complex! In this chapter, we explore the complexity of food webs in mathematical terms using a physical model, called a *directed graph (digraph)*, to map the interactions between organisms. A digraph represents the species in an ecosystem as points or vertices (singular = vertex) and puts arrows for arcs from some vertices to others, depending on the energy transfer, that is, from a prey species to a predator of that prey.

The species that occupy an area and interact either directly or indirectly form a *community*. The mixture and characteristics of these species define the biological structure of the community. These include parameters such as feeding patterns, abundance, population density, dominance, and diversity. Acquisition of food is a fundamental process of nature, providing both energy and nutrients. The interactions of species as they attempt to acquire food determine much of the structure of a community. We use food webs to represent these feeding relationships within a community.

Example 2.1. In the partial food web depicted in Figure 2.1, *sharks eat sea otters*, *sea otters eat sea urchins and large crabs*, *large crabs eat small fishes*, and *sea urchins and small fishes eat kelp*. Said in another way, sea urchins and large crabs are eaten by sea otters (both are prey for sea otters) and sea otters are prey for sharks. These relationships are modeled by the food web shown in Figure 2.1: there is an arrow from species A to species B if species B preys on species A. (In earlier depictions of food webs, some mathematicians reversed the arcs; this may appear in the literature.) □

Exercise 2.1. Create a food web from the predator-prey table (Table 2.1 is related to the food web shown in Figure 2.1). □

There are various tools online that are used to construct food webs from data of this nature, but this time try it by hand. Examples of online resources can be found at http://bioquest.org/esteem.

Question. Are there any species that are only predators and not prey, or any that are prey that are not predators? What are they? How can they be identified by looking at the food web? Justify your answers. □

2.3 TROPHIC LEVELS AND TROPHIC STATUS

Now let's consider where a species is located in the food chain or specifically in the food web. For example, producers, such as kelp in Figure 2.1, are at the bottom of the food chain; sharks appear at the top. If it is a *food chain*, where one and only one species is above the other, it is easy, but what if the food web is more complex, as in our first example, and represents a whole community?

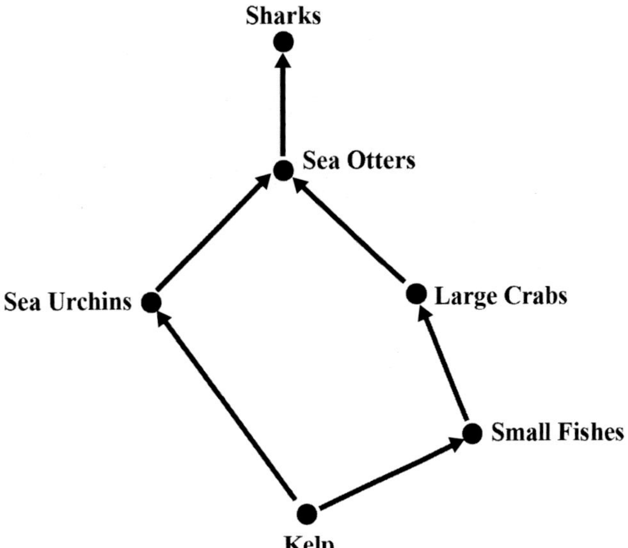

FIGURE 2.1 Simple food web.

2.3.1 Background and Definitions

Trophic levels in food webs provide a way of organizing species in a community food web into feeding groups. Scientists have used various methods in classifying species in a food web into these various feeding groups. The most elementary method is to divide them into the following categories: primary producers, secondary producers and consumers, the last of which are then divided into herbivores and carnivores based on their consumption of plant or animal products. This makes the assumption that there are at most three (trophic) levels in a food web. It also opens the question of how to classify consumers such as black bears or grizzly bears, who are omnivores. Most books group omnivores with carnivores, but is there a scientific basis for doing so? Others consider the positioning in the food web to be most important: a consumer is at a higher level than what it consumes. They then define trophic levels as follows: a species that is never a predator is at trophic level 0, one who preys on that species is at level 1, etc. Determining trophic level in a *food chain* is easy under this latter definition. For example, for the food chain in Figure 2.2, kelp is at trophic level 0, sea urchins are at trophic level 1, sea otters are at trophic level 2, and sharks are at trophic level 3.

Food webs, however, are not necessarily food chains like in Figure 2.2, but a mixture of multiple food chains meshed together. Determining trophic level for more complex food webs is more difficult. In fact, *the number of trophic levels* in a food web is sometimes used as a measure of complexity, as we will see in the next section. But since food webs can be represented as directed graphs (digraphs), we can use some properties of these digraphs to determine trophic level in any kind of food web. Fortunately or unfortunately, there are a number of possible definitions of trophic level. We illustrate two of them here.

The *length of a path* in a digraph is the number of arcs included in the path. The *shortest path* between two vertices x and y in a digraph is the path between x and y of shortest length among all such paths. The *longest path* between two vertices x and y in a digraph is the path between x and y of longest length. We use these definitions of path length to give various definitions of trophic level in a food web. The first is a commonly used definition of trophic level, derived from the trophic level of species in a chain:

Definition 2.1 (Option 1:)**.** The *trophic level* of species X is:

i. 0, if X is a primary producers in the food web (a species that does not consume any species in the food web).
ii. k, if the shortest path from a level 0 species to X is of length k. □

An example from the food web shown in Figure 2.1 is given in Table 2.2.

TABLE 2.1 Predator-Prey Relationships

Species	Species They Feed on
Shark	Sea otter
Sea otter	Sea stars, sea urchins, large crabs, large fishes and octopus, abalone
Sea stars	Abalone, small herbivorous fishes, sea urchins
Sea urchins	Kelp, sessile invertebrates, organic debris
Abalone	Organic debris
Large crabs	Sea stars, smaller predatory fishes and invertebrates, organic debris, small herbivorous fishes and invertebrates, kelp
Smaller predatory fishes	Sessile invertebrates, planktonic invertebrates
Small (herbivorous) fishes and invertebrates	Kelp
Kelp	–
Large fishes and octopus	Smaller predatory fishes and invertebrates
Sessile invertebrates	Microscopic planktonic algae, planktonic invertebrates
Organic debris	–
Planktonic invertebrates	Microscopic planktonic algae
Microscopic planktonic algae	

FIGURE 2.2 A food chain.

Exercise 2.2. Expand Table 2.2 to include the trophic level of all species given in Table 2.1. Answer the following questions before moving on:

1. Do you see any challenges to using the shortest path for computing trophic level?
2. Large crabs are direct prey of sea otters. Would they have different trophic levels?
3. Can you think of an alternative definition of trophic level? □

TABLE 2.2 Trophic Levels Using Option 1 Definition for the Food Web in Figure 2.1

Species	Trophic Level	Shortest Path
Kelp	0	
Sea urchins	1	Kelp-sea urchins
Small fishes	1	Kelp-small fishes
Large crabs	2	Kelp-sea urchins-large crabs
Sea otters	2	Kelp-sea urchins-sea otters
Sharks	3	Kelp-sea urchins-sharks

Definition 2.1 (Option 2:)**.** The *trophic level* of species X is:

i. 0 if X is a primary producer in the food web (a species that does not consume any species in the food web).
ii. k if the longest path from a level 0 species to X is of length k. □

An example using Option 2 to determine trophic level of the food web in Figure 2.1 is shown in Table 2.3.

Notice that the highest trophic level is now four, that sharks have a higher level using this definition than they did with the shortest path definition, and that large crabs and sea otters now have different trophic levels.

Exercise 2.3.

1. Complete your table of trophic levels from the complete food web shown in Table 2.1 to include the trophic level using the longest path definition.
2. An ecological rule of thumb is that about 10 percent of the energy passes from one trophic level to another. Compare these two definitions of trophic level with regard to energy loss. For example, if kelp starts with 1 million units of energy, how much energy is left for sharks using each definition? □

Neither of these trophic level definitions is entirely satisfactory, since each has its limitations in determining the hierarchical structure of the food web. For instance, neither definition reflects the number of species that are direct or indirect prey of a species. In addition, our intuitive understanding of trophic levels makes the following assumption reasonable:

Assumption. *If species X is a predator of species Y, then the trophic level of species X is greater than the trophic level of species Y.*

TABLE 2.3 Trophic Level Using the Option 2 Definition for the Food Web in Figure 2.1

Species	Trophic Level	Longest Path
Kelp	0	
Sea urchins	1	Kelp-sea urchins
Small fishes	1	Kelp-small fishes
Large crabs	2	Kelp-sea urchins-large crabs
Sea otters	3	Kelp-sea urchins-large crabs-sea otters
Sharks	4	Kelp-sea urchins-large crabs-sea otters-shark

However, when using Option 1 of the definition (considering the shortest path), this intuitive assumption may not apply in general. It does hold when Option 2 of the definition is used.

To solve this problem of inconsistency, we combine the length of the longest path and the number of species that are direct or indirect prey and call it the *trophic status of a species*.

Definition 2.2. The trophic status of a species u is defined as:

$$T(u) = \sum kn_k,$$

where n_k is the number of species whose longest path to u has length k and the sum is taken over all k. □

Example 2.2. Let's compute the trophic status of sea otters from the information included in Table 2.3:

- the longest path to sea otters from kelp is 3, thus when $k = 3$, $n_k = 1$
- the longest path from small fishes to sea otters is 2, thus when $k = 2$, $n_k = 1$
- the longest path from sea urchins and from large crabs to sea otters is 1, thus when $k = 1$, $n_k = 2$ □

Therefore, the trophic status of sea otters = T(sea otters) = 3(1) + 2(1) + 1(2) = 7

The trophic statuses for kelp, sea urchins, small fishes, large crabs, and sharks are shown in Table 2.4.

Exercise 2.4. Find the trophic status of each of the species in Table 2.1.

There are many ways of describing *dominant species* in a food web. One way is to always regard *keystone species* (a predator species whose removal causes additional species to disappear—species which effectively control the nature of the community) as dominant species.

We can use our digraph model to recognize dominant species in a food web by considering what happens when an arc is removed from the digraph representing the food web. □

Definition 2.3. We say that *species A is dominant* if the removal of any arc from a species B to A in the food web allows B to have uncontrolled growth, and thus become a new "dominant" species.

For example, when we remove the arc from sea urchin to sea otter in Figure 2.1, sea urchins will have uncontrolled growth and they then become a new "*dominant*" species in the food web. □

Exercise 2.5. Determine the dominant species in the food web corresponding to Table 2.1 using the definition of arc removal. □

Various other definitions of dominant species exist, some of which are:

a. the most numerous species in a food web;
b. the species which occupies the most space;
c. the species with the highest total body mass;
d. the species that contributes the most energy flow.

TABLE 2.4 Trophic Status for Species in the Food Web in Figure 2.1

Species	Trophic Level Definition 2.2	Trophic Status
Kelp	0	0
Sea urchins	1	1
Small fishes	1	1
Large crabs	2	3
Sea otters	3	7
Sharks	4	12

One example of a dominance definition, which incorporates both the number of species that are direct or indirect prey and the extent of energy transfer, is based on the trophic status of the species. This definition resembles the definition of status for people in a community or social network.

Definition 2.4. Trophic Status Dominant Species

A species is dominant in a food web if its trophic status is greater than the number of species in the food web *above* level 0. □

Exercise 2.6.

1. Determine which species are dominant in the food web corresponding to Table 2.1 using the trophic status definition.
2. Compare your results using trophic status with the results you obtain using the definition involving arc removal. □

2.3.2 Adding Complexity: Weighted Food Webs and Flow-Based Trophic Levels

Not all relationships are the same. For example, you interact differently with your family than with strangers. Or perhaps you love to eat ice cream, but eat as few brussels sprouts as possible. While both ice cream and brussels sprouts are foods, given an unlimited source of both, would you eat an equal amount of both? Probably not! Similarly, species may eat much more of one species of prey than another. We model this by putting numbers, or weights, on the arcs to indicate food preferences. Consider the food web in Figure 2.3.

The *weight of an arc* between vertex i and vertex j, denoted w_{ij}, is the proportional food contribution of vertex i to vertex j in the food web. For example, the weight 0.6 on the arc from rodents to snakes reflects that snakes eat "rodents" more than "other lizards" in a ratio of 6 to 4. Specifically, 60% of a snake's diet comes from "rodents," while 40% of its diet comes from "other lizards." The sum of the weights of the ingoing arcs to a species is 1 since the set of ingoing arcs represents the full diet of the species.

Weighting is more important than you might realize. Looking at hawks, it's clear that they eat both snakes and lizards. However, since snakes are weighted so much more heavily, the removal of snakes from this digraph would shift hawk diets dramatically. Instead of eating lizards 30% of the time, they would eat lizards 100% of the time. Poor lizards! Lucky insects! When lizards go extinct or are removed, foxes suffer a secondary extinction since, if the food web indicates all possible prey, their sole source of food disappears. When snakes are removed, hawks only eat lizards, which would dramatically decrease the lizard population. This in turn would vastly decrease lizard predation on insects, allowing insects to grow uncontrolled. Now consider what would happen if lizards were initially 90% of hawk diets (instead of 30%). Would the insect population increase so dramatically with the removal of snakes? Probably not! This is important if you are an insect, or anything that eats an insect, or anything that an insect eats. Therefore, it is important to consider the arc weight when predicting indirect changes that trickle down through a food web. In other words, arc weight matters!

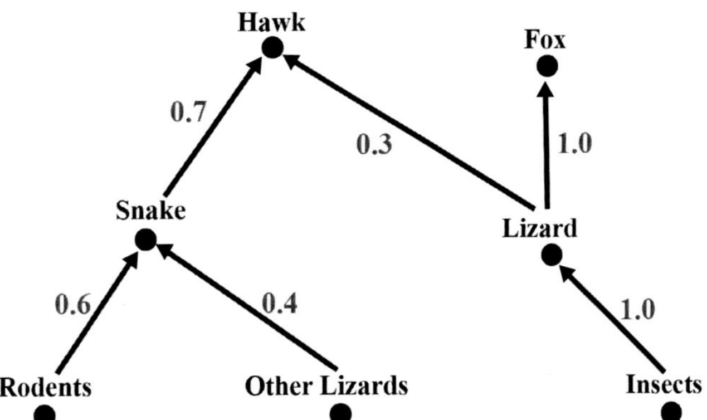

FIGURE 2.3 A weighted food web.

2.3.3 Flow-Based Trophic Level

Weighting the arcs of a food web or corresponding digraph gives new alternative definitions of trophic level that take these weights into account. One such definition is called the flow-based trophic level, or TL.

Definition 2.5. *Flow-based trophic level* (TL) is defined as:

$$\text{TL (species } i) = 1 + \sum \text{(weight of each food source for } i) \times \text{TL (food source } j)$$
$$= \text{TL}(i) = 1 + \sum (w_{ij}) \times \text{TL (food source } j). \qquad \square$$

Note that TL(primary producer) = 1.

Example 2.3. For example, based on the information in Figure 2.3,

$$\text{TL (snake)} = 1 + 0.6\,(1) + 0.3\,(1) = 1.9,$$

whereas

$$\text{TL (lizard)} = 1 + 1\,(1) = 2.$$

Notice here that even though under either the shortest path or longest path definitions, the snake and the lizard have the same trophic level, under the *flow-based trophic level* definition the lizard has a slightly higher flow-based trophic level than the snake. □

Exercise 2.7.

1. Finish the calculations for flow-based trophic level for the food web given in Figure 2.3.
2. What happens to the flow-based trophic level for various species if an arc is removed? Try it by removing the arc from snake to hawk. □

Exercise 2.8. Answer the following questions for the weighted food web shown in Figure 2.4:

1. If you ignore the weights on the arcs, describe the effect on the food web from the removal of prairie dogs.
2. Assume that a species can survive on 50% of its normal diet. Use the weights to describe the effect on the food web from the removal of jackrabbits and small rodents.

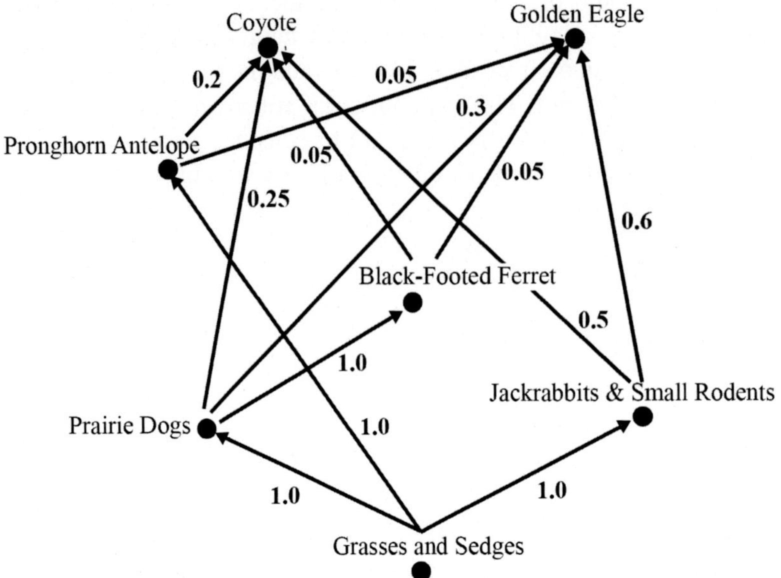

FIGURE 2.4 Weighted food web for the coyote.

3. Assume that a species can survive on 75% of its normal diet. Use the weights to describe the effect on the food web from the removal of black-footed ferrets.
4. Compute the flow-based trophic level of the species in the food web. □

RESEARCH QUESTIONS

2.1 What is the best all-round definition of trophic status and why? Provide many examples to back up your claim.

2.2 A community food web consists of all predation relationships among species. How can we be sure we have all of these relationships? How do we measure the percentage of a species diet that comes from another species—is it a constant?

2.3 Grizzly bears and wolves are species, among others, that biologists want to conserve, yet they can cause problems for people and their livestock. How could you use the material in the beginning of this chapter to design ways to maintain a species, for example, wolves, and yet preserve human and animal life? Choose a location and a species and design a habitat for this species in that location.

2.4 COMPETITION GRAPHS AND HABITAT DIMENSION

We will now use food webs to create additional models to help determine the dimension of community habitats from these relationships.

2.4.1 Competition Graphs (also Called Niche Overlap Graphs and Predator Graphs)

Given a food web, its *competition graph* is a new (undirected) graph that is created as follows. The vertices are the species in the community and there is an edge between species a and species b if and only if a and b have a common prey, that is, if there is some x so that there are arcs from x to a and x to b in the food web.

If we take our canonical example from Figure 2.1, we get a very simple competition graph with one edge and four independent vertices as shown in Figure 2.5.

Exercise 2.9. Draw the competition graph for the food web in Figure 2.6 and *describe* the isolated vertices in the competition graph. □

2.4.2 Interval Graphs and Boxicity

Definition 2.6. A graph is an *interval graph* if we can find a set of intervals on the real line so that each vertex is assigned an interval and two vertices are joined by an edge if and only if their corresponding intervals overlap. Interval graphs have been very important in genetics. They played a crucial role in the physical mapping of DNA and more generally in the mapping of the human genome. Given a competition graph, we want to determine if it is an interval graph. We need to find intervals on the real line for each vertex so that the intervals corresponding to two vertices overlap if and only if there is an edge between the two vertices (Figure 2.7). □

If G is the competition graph corresponding to a real community food web, and G is an interval graph, then the species in the food web have one-dimensional habitats or niches because each species can be mapped to the real line with overlapping intervals if they have common prey, and this single dimension applies to each species in the web. This one dimension might be determined by temperature, moisture, pH, or a number of other things.

We have not described the endpoints of the intervals in Figure 2.7. It would be easy to do so; for example, consider the possible correspondence:

c to [-2,0]
d to [-1,2]
b to [1,4]

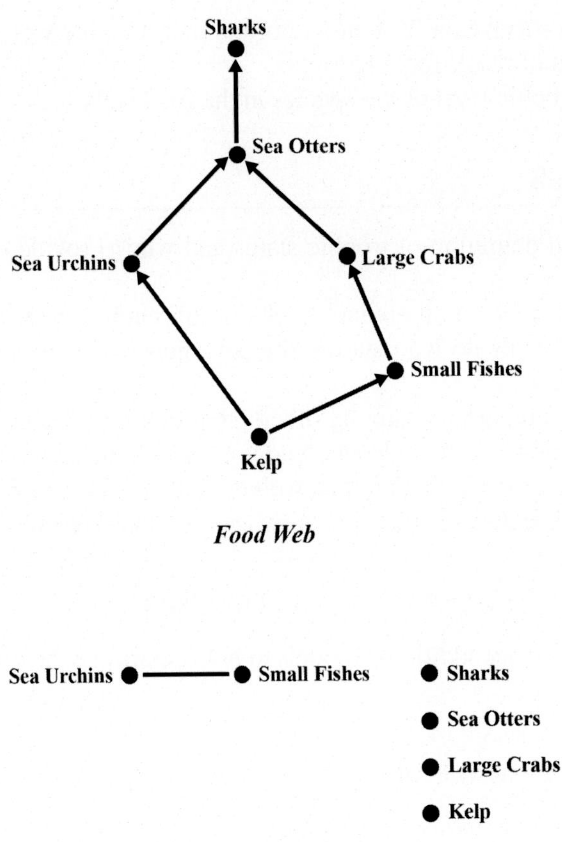

FIGURE 2.5 A food web and its corresponding competition graph.

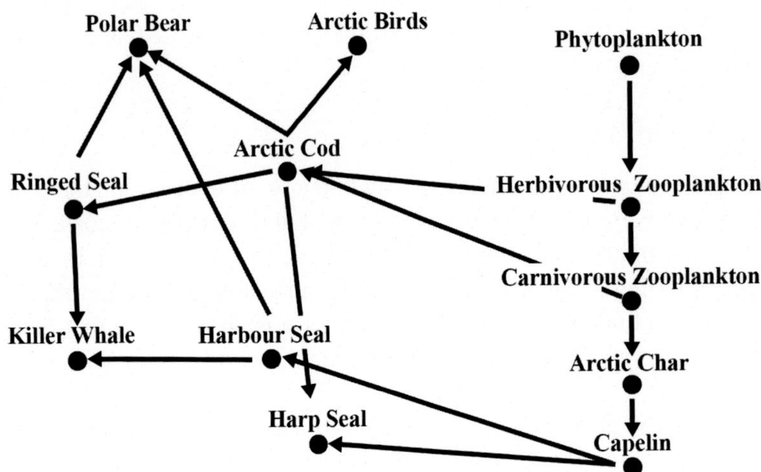

FIGURE 2.6 Food web for the polar bear.

e to [1.5, 6]
f to [5, 7]
g to [9,11]

We should note here that the size of the interval makes no difference, nor does whether the intervals are open (do not include the endpoints) or are closed (include the endpoints).

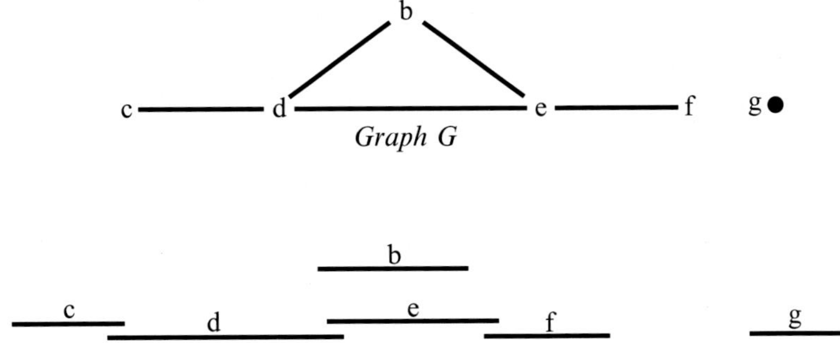

FIGURE 2.7 Graph G and a demonstration that G is an interval graph.

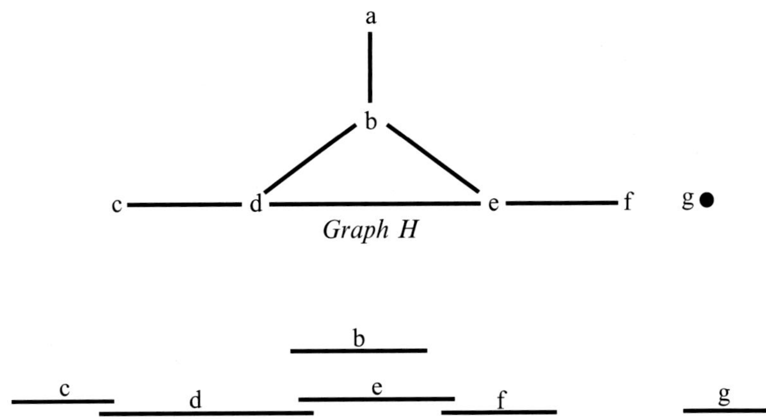

FIGURE 2.8 An attempt to find intervals corresponding to graph H. H is not an interval graph.

If we change the graph in Figure 2.7 slightly to H, by adding vertex a and edge $\{a, b\}$ (Figure 2.8), we don't get an interval graph. There is no place to put an interval corresponding to a which overlaps the interval for b, but which overlaps the intervals for neither d nor e.

Exercise 2.10. Is the competition graph from Exercise 2.9 an interval graph? If so, show the interval representation; if not, why not? □

Exercise 2.11. Give some examples of graphs that are not interval graphs. □

Exercise 2.12. Can you find a real community food web that has a competition graph that is not an interval graph? □

Exercise 2.13. Some graphs, like H shown in Figure 2.8, cannot be represented by intervals on the real line. Can H be represented by intersecting rectangles in the plane (two-dimensional space)? That is, can you find rectangles in the plane around each of the vertices of H such that for any two vertices i and j of H, the rectangles covering i and j overlap if and only if there is an edge between the vertices i and j? □

More generally, we can consider ways to represent graphs where the edges correspond to intersections of boxes in Euclidean space. An interval on the line is a one-dimensional box. If we find a representation of a graph where the vertices correspond to rectangles in two-dimensional space so that two rectangles intersect if and only if the corresponding vertices are connected by an edge, then we have a two-dimensional representation of the graph as shown in Figure 2.9. In general, we give the following definition.

Definition 2.7. The *boxicity* of G is the smallest positive integer p that we can assign to each vertex of G, a box in Euclidean p-space, so that two vertices are connected by an edge if and only if their corresponding boxes overlap. □

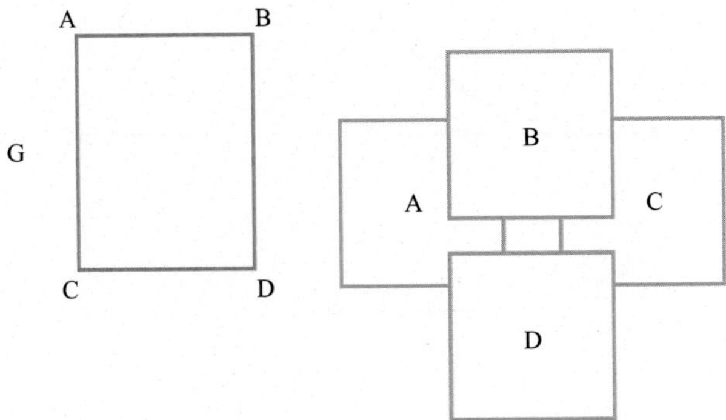

FIGURE 2.9 A two-dimensional box representation of graph G that is a square.

Notice that since a "box" in the one-dimensional Euclidean space is an interval, any interval graph has boxicity $p = 1$ and any graph with boxicity $p = 1$ is an interval graph. For graphs that are not interval graphs, the term boxicity is well-defined [1] and is hard to compute [2]. There are fast algorithms to test if a graph is an interval graph, but if it is not an interval graph then there are no fast ways of telling the boxicity of the graph [1],[2]. The example in Figure 2.9 shows graph G of boxicity 2 and the overlapping two-dimensional rectangles.

Exercise 2.14. Find the boxicity of graph H in Figure 2.8. □

2.4.3 Habitat Dimension

Different factors determine a species' normal healthy environment, such as moisture, temperature, and pH. We can use each such factor as a dimension. Then the range of acceptable values on each dimension is an interval. In other words, each species can be represented as a box in Euclidean space; the box represents its ecological niche.

For example, in Figure 2.10, the niche of the species might be determined by the three dimensions, temperature between 10 and 15 °C, moisture level between 1 and 2, and pH between 7 and 8.

Exercise 2.15. Can you find a competition graph with boxicity 3? In other words, find a graph where there is no rectangular representation of the vertices, but there is a three-dimensional box representation of the vertices so that there is an edge between two vertices if and only if the boxes overlap in three-dimensional space. □

In the 1960s, Joel Cohen found that food webs arising from "single habitat ecosystems" (homogeneous ecosystems) generally have competition graphs that are interval graphs. This remarkable empirical observation of Cohen [3], that real-world competition graphs are usually interval graphs, has led to a great deal of research on the structure of competition graphs and on the relation between the structure of digraphs and their corresponding competition graphs. It has also led to a great deal of research in ecology to determine just why this might be

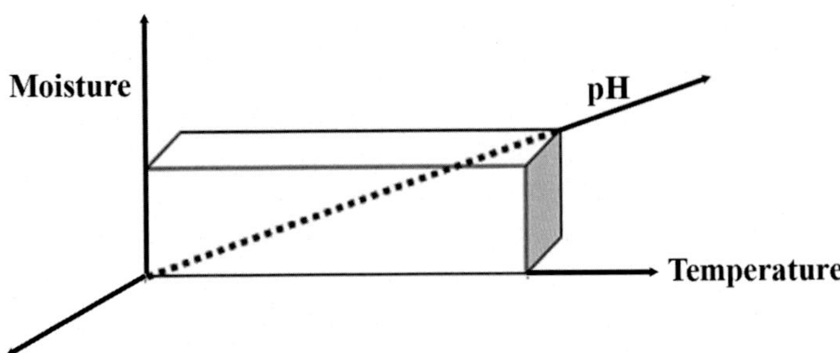

FIGURE 2.10 A three-dimensional species habitat niche represented by temperature, moisture, and pH.

the case. Is it that there is really only one overriding dimension that controls habitat (niche) formation for a community of species?

Statistically, maybe this should not be surprising. Models for randomly generated food webs have been created and the probability that the corresponding competition graph is an interval graph has been calculated. Much of Cohen's *Food Webs and Niche Space* [3] takes this approach. But Cohen et al. [4] showed that, with this model, the probability that a competition graph is an interval graph goes to 0 as the number of species increases. In other words, it should be highly unlikely that competition graphs corresponding to food webs are interval graphs.

Recall that a community food web contains all predatory relationships among species. In practice, however, one can rarely obtain data on all these relationships. Most often researchers do the best they can to get as much data as possible and draw a food web based on the data they collect. Given a species, a vertex W in the food web, the *sink food web* corresponding to W is the subgraph defined by all vertices (species) that are prey of W with their arcs. The *source food web for W* is defined by all vertices (species) that are predators of W, with their arcs. Cohen showed that a food web has a competition graph that is an interval graph if and only if each sink food web contained in it is an interval graph [3]. However, the same is not true for source food webs.

Exercise 2.16. Find an example of a food web that is an interval graph, but where there is a source food web contained in it that is not an interval graph. □

We close this section with a comment that competition graphs as models of food webs have various analogues. For example, an analogue to competition graphs is a *common enemy graph* where there is an edge between two species if they have a common predator. These graphs are sometimes called *prey graphs*.

RESEARCH QUESTIONS

2.4 Analyze the properties of competition graphs to better understand the underlying food web (directed graph). Consider the following questions: Can we characterize the directed graphs whose corresponding competition graphs are interval graphs? This is a fundamental open question in applied graph theory. Indeed there is no forbidden list of digraphs (finite or infinite) such that when these digraphs are excluded, one automatically has a competition graph which is an interval graph. [5],[6]

2.5 What are the ecological characteristics of food webs that seem to lead to their competition graphs being interval graphs? Most directed graphs do not have interval graph competition graphs, yet statistically most actual food webs have interval competition graphs. (This is a BIG unsolved problem described by Cohen, with no answers to date [4].)

2.6 It appears that most food webs have competition graphs that are interval graphs. A second fundamental question, an ecology question, is whether or not it is possible that the habitat or niche of the species in a community food web whose competition graph is an interval graph is truly based on one overriding component, such as temperature or moisture or pH. If so, can one determine what that overriding component is for a specific community food web?

2.7 What is the relationship between the boxicity of the competition graph of a community food web and the boxicity of the competition graph of its source food web? Note: we know they can be different [3].

2.8 It was shown a few years ago that it is possible to determine if a graph has boxicity 2, and that it is difficult to determine if a graph has boxicity k for $k > 2$ (NP-complete) [2]. Yet there are no nice characterizations of graphs with boxicity 2 as there are for interval graphs. See if you can find a forbidden subgraph characterization of graphs of boxicity 2.

2.5 CONNECTANCE, COMPETITION NUMBER, AND PROJECTION GRAPHS

There has been considerable attention paid lately to creating models for better understanding of predator-prey relationships and habitat formation, especially to inform conservation policy makers. This section includes a number of different parameters for food webs and competition graphs, describes a weighted version of competition graphs and the analogue common enemy graphs, and offers suggestions for further study linked to conservation.

2.5.1 Connectance

Early researchers believed community stability was proportional to the logarithm of the number of arcs, L, in the food web [7]. Later, the notion of *directed connectance*, which considers the number of arcs relative to the number of vertices (species) in a food web, has been used as a measure of robustness (implying stability) [8–10].

Definition 2.8. If there are S species and L edges in the directed graph, the *directed connectance* of the digraph is defined to be $C = L/S^2$. It is also considered to be the *density of arcs* in the food web. □

Note that if all possible arcs exist in the digraph, there would be a maximum of $S(S-1)$ edges, so the maximum connectance is $1-(1/S)<1$, and the minimum connectance is 0 if there are no arcs. However, since producers have no outgoing arcs, and the food web has no cycles, the maximum is much less than $1-(1/S)$. Trophic food webs also follow the pattern that if any two species have exactly the same predators and exactly the same prey, then the two species are merged and become one vertex. So in actual fact there are more species in a community than the number of vertices in the food web. It was long believed that the higher the connectance the more stable the food web. In fact, using real or simulated models and population dynamics on the food web and competition graph, the steady state (another term for stability) is achieved only for small S and C, particularly when the product $SC<2$, and these results are robust under the change of initial conditions and ecological parameters [11]. So, although connectance is an easy parameter to compute, it may not be very good for analyzing a large food web's potential stability.

Exercise 2.17. Find the connectance of the food web shown in Figure 2.11 [13],[14]. Using the connectance parameter does the shallow water Hudson River food web appear to be stable? Why or why not? □

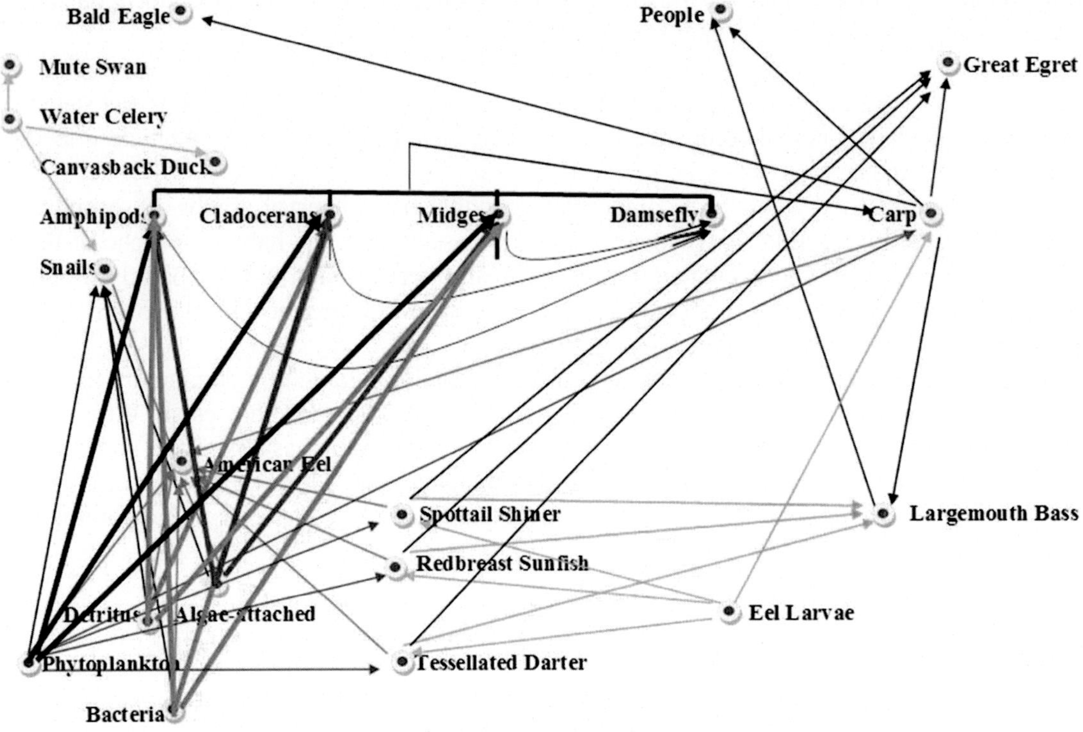

FIGURE 2.11 Partial food web of the shallow water Hudson River food web. Reference [12] with permission.

2.5.2 Competition Number

If D is a directed graph with no cycles (acyclic) there must be an isolated vertex in its corresponding competition graph, equal to one vertex with no incoming arcs. Note that there are four isolated vertices in the competition graph shown in Figure 5 sharks, sea otters, large crabs, and kelp. Kelp has no incoming arcs, only outgoing arcs. If kelp had an incoming arc there would be a cycle.

Exercise 2.18. Using the partial food web of the Hudson River in Figure 2.11, what vertices will be isolated in the competition graph? What type of species corresponds to an isolated vertex in the competition graph? □

Exercise 2.19. Draw the competition graph for the food web in Figure 2.11. How many isolated vertices are in the competition graph? □

Definition 2.9. *The competition number for a graph G*, denoted $k(G)$, is the least number of isolated vertices that need to be added to G so that G is the competition graph for an acyclic directed graph. □

Indeed, any graph can be the competition graph for an acyclic directed graph by adding adding a sufficient number of isolated vertices to it. As an example, consider the graph shown in Figure 2.12.

Build a directed graph (food web) as shown in Figure 2.13.

K and H and J are isolated vertices added to the graph shown in Figure 2.12. After they are added, it becomes the competition graph of the food web in Figure 2.13.

Exercise 2.20. In this example, three isolated vertices were added to make the graph a competition graph. Is this the least number of isolated vertices that are needed, or could we add only one or only two isolated vertices? Experiment! □

Exercise 2.21. Provide a proof by construction of the statement "Any graph can be the competition graph for an acyclic directed graph by adding to it a sufficient number of isolated vertices." □

Determining the competition number of a graph is a hard problem [5], but many theorems have been proven that help one find the competition numbers for smaller graphs.

Definition 2.10. Recalling that a cycle in a graph is a sequence of vertices $a_1 - a_2 \cdot a_k - a_1$ with no edges $a_i - a_j \; j > i + 1$, a *hole* in a graph is an induced n-cycle for $n > 3$. For example $ABCD$ is a 4-cycle hole, but DEF is a 3-cycle which is not a hole in Figure 2.12 since holes are 4 or bigger. □

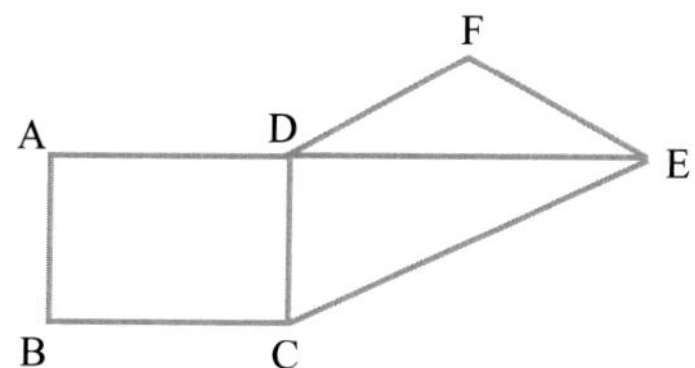

FIGURE 2.12 A graph.

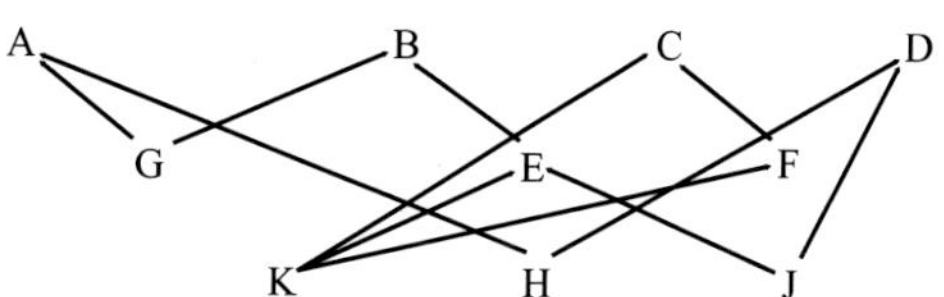

FIGURE 2.13 Food web related to Figure 2.12.

Theorem 2.1. *Cho and Kim [12]*

If G has exactly one hole, $k(G) \leq 2$.

Using this theorem, we know that we did not need to add three isolated vertices to the graph in Figure 2.12, we could have added only two. Did you construct a competition graph/food web by adding only two isolated vertices in Exercise 2.20? How about only one vertex?

Indeed, the previous theorem can be extended to more than one hole:

Theorem 2.2. *MacKay et al. [6]*

$k(G) \leq$ number of holes $+ 1$.

Mathematically, determining the competition number is an interesting question, and mathematicians have tried to find theorems that would precisely determine the competition number of a graph, or at least classes of graphs, with the hopes of characterizing directed graphs that give rise to actual food webs. Biologically, the competition number is not as interesting, however. If there are fewer vertices (species) that don't compete with any other species, then there will be a low number of isolated vertices in the competition graph. But there will always be producers at the bottom that don't compete and there is usually at least one species at the top that doesn't compete, so a minimal number of isolated vertices is unrealistic for graphs corresponding to actual food webs. Survival of a species often depends on more competition and dispersal of resources, so one species does not grow without bound.

RESEARCH QUESTIONS

(Virtually no research has been done on these questions.)

2.9 The *outdegree* of a vertex in a digraph is the number of arcs leaving the vertex, and the *indegree* of a vertex in a digraph is the number of arcs coming into the vertex. If you limit the indegree or outdegree, or both, of all vertices in a digraph, is the competition number of the associated graph lower, or higher, or neither? If you limit the indegree or outdegree, or both, of all vertices in a digraph, is it more likely to have a competition graph that is an interval graph?

2.10 What is the biological significance of limiting the indegrees or outdegrees in a food web; in other words, what does limiting competition mean? Is it reasonable to do so?

2.11 Is there any relationship between the connectance of the food web and the corresponding competition graph? If so, what? Is it possible to identify secondary extinctions from the competition graph of a food web?

2.5.3 Projection Graphs

Projection graphs are weighted graphs, which are versions of competition graphs and common enemy graphs. If we use T for top species (no outward arcs), B for basal species (no inward arcs), I for everything else (or intermediate species), and let S be the total number of species in the food web, we can define such graphs as follows.

Definition 2.11. Predator projection graph (PP graph). There is an edge between two species, A_i and A_j, in the PP graph if they have a common prey and the weight of edge a_{ij} (denoted A_{ij}) between them reflects the number of common prey.

$$A_{ij} = \frac{1}{S(B+I)} \sum_{k \in B+I} a_{ki} a_{kj}$$

□

Note that a PP graph is a weighted competition graph minus the basal species, since they have no prey.

Definition 2.12 (*Prey projection graph (EP graph)*). There is an edge between two species, B_i and B_j, in the EP graph if they have a common predator and the weight of the edge b_{ij} (denoted B_{ij}) reflects the number of common predators.

$$B_{ij} = \frac{1}{S(T+I)} \sum_{k \in T+I} a_{ki} a_{kj}$$

□

Note that the EP graph is a weighted common enemy graph minus the top species, since they have no predators.

Let's look at an example of the PP and EP graphs determined from the food web in Figure 2.6.

We can also use the competition graph shown in Figure 2.14 as a starting point for the PP graph, remove any basal species, and then label the edges with the appropriate weights, corresponding to the formula for weights. See Exercise 2.23.

The adjacency matrix for the food web is given by Table 2.5.

The predator projection graph is an undirected graph with weights corresponding to the number of prey each predator has divided by a normalization factor of $S(B + I) = 12(9) = 108$.

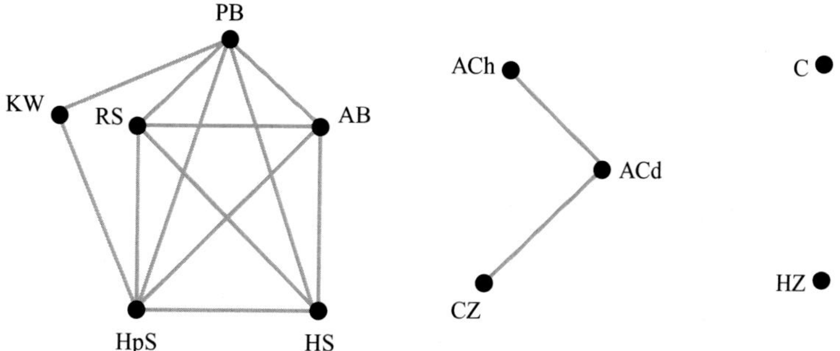

FIGURE 2.14 PP projection graph for food web in Figure 2.6 with weights given in the matrix shown in Table 2.7.

TABLE 2.5 Adjacency Matrix for PP Graph

	PB	AB	RS	Acd	KW	HS	HpS	P	H	CZ	Ach	C
PB	0	0	0	0	0	0	0	0	0	0	0	0
AB	0	0	0	0	0	0	0	0	0	0	0	0
RS	1	0	0	0	1	0	0	0	0	0	0	0
Acd	1	1	1	0	0	1	1	0	0	0	0	0
KW	0	0	0	0	0	0	0	0	0	0	0	0
HS	1	0	0	0	1	0	1	0	0	0	0	0
HpS	0	0	0	0	1	0	0	0	0	0	0	0
P	0	0	0	0	0	0	0	0	1	0	0	0
HZ	0	0	0	1	0	0	0	0	0	1	0	0
CZ	0	0	0	1	0	0	0	0	0	0	1	0
Ach	0	0	0	0	0	0	0	0	0	0	0	1
C	0	0	0	0	0	1	1	0	0	0	0	0

TABLE 2.6 Matrix Used to Calculate PP Graph

	PB	AB	RS	Acd	KW	Hs	Hps	HZ	CZ	Ach	C
PB	0	1	1	0	2	2	1	0	0	0	0
AB	1	0	1	0	0	1	1	0	0	0	0
RS	1	1	0	0	0	0	1	0	0	0	0
Acd	0	0	0	0	0	0	0	0	1	1	0
KW	2	0	0	0	0	0	2	0	0	0	0
Hs	2	1	1	0	0	0	2	0	0	0	0
ps	1	1	1	0	2	2	0	0	0	0	0
HZ	0	0	0	0	0	0	0	0	0	0	0
CZ	0	0	0	1	0	0	0	0	0	0	0
Ach	0	0	0	1	0	0	0	0	0	0	0
C	0	0	0	0	0	0	0	0	0	0	0

TABLE 2.7 Weights for PP Graph

	PB	AB	RS	Acd	KW	Hs	Hps	HZ	CZ	Ach	C
PB	0	0.00926	0.00926	0	0.01852	0.01852	0.00926	0	0	0	0
AB	0.00926	0	0.00926	0	0	0.00926	0.00926	0	0	0	0
RS	0.00926	0.00926	0	0	0	0.00926	0.00926	0	0	0	0
Acd	0	0	0	0	0	0	0	0	0.00926	0.00926	0
KW	0.01852	0	0	0	0	0	0.01852	0	0	0	0
Hs	0.01852	0.00926	0.00926	0	0	0	0.01852	0	0	0	0
Hps	0.00926	0.00926	0.00926	0	0.01852	0.01852	0	0	0	0	0
HZ	0	0	0	0	0	0	0	0	0	0	0
CZ	0	0	0	0.00926	0	0	0	0	0	0	0
Ach	0	0	0	0.00926	0	0	0	0	0	0	0
C	0	0	0	0	0	0	0	0	0	0	0

The matrix used to calculate the weights of the predator projection graph for our example is shown in Table 2.6 without normalization.

Normalizing gives the actual weight for edge; (Ach,ACd) = 1/108 = .00926 for our example.

With normalized actual weights we get the matrix shown in Table 2.7.

The prey projection graph, EP, for this example is given by the matrix in Table 2.8 and the graph shown in Figure 2.15.

Note that the weights have a divisor of [11], [12] or 132 this time.

Question. What similarities do you see common to both PP and EP graphs? What differences do you see? Can you interpret these ecologically? □

TABLE 2.8 Weights for EP Graph

	RS	Acd	HS	HpS	P	HZ	CZ	Ach	C
RS	0	0.00758	0.01515	0.00758	0	0	0	0	0
Acd	0.00758	0	0.01515	0	0	0	0	0	0.01515
HS	0.01515	0.01515	0	0.00758	0	0	0	0	0
HpS	0.01515	0	0.00758	0	0	0	0	0	0
P	0	0	0	0	0	0	0	0	0
HZ	0	0	0	0	0	0	0.00758	0	0
CZ	0	0	0	0	0	0.00758	0	0	0
Ach	0	0	0	0	0	0	0	0	0
C	0	0.01515	0	0	0	0	0	0	0

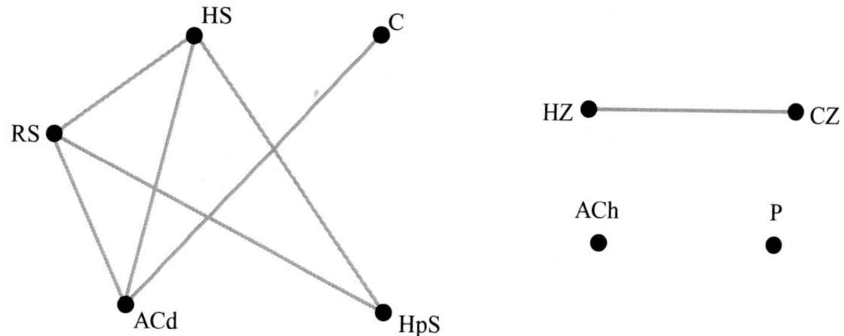

FIGURE 2.15 EP Projection graph for food web in Figure 2.6 with weights given in the matrix shown in Table 2.8.

Exercise 2.22. Compute PP and EP graphs for the food web in Figure 2.1. □

Exercise 2.23. Compute the PP and EP graphs for the food web in Figure 2.6. Can you suggest some hypotheses as a result of the network structure of these two graphs? □

Exercise 2.24. If the competition graph is an interval graph does it mean the PP and EP graphs are interval graphs, or vice versa? □

RESEARCH QUESTIONS

2.12 Does using multidimensional representations (boxes), rather than intervals, make sense ecologically even though interval representations exist? Does such a representation provide more guidance for managing communities and restoration of communities under stress? Similarly, if we allow more than one interval to represent a species (vertex) in an interval representation of a competition graph does it provide a better representation of the species interactions within a habitat? Is there a "magic" number? It has been suggested that using multiple intervals in connection with PP and EP graphs might be preferable to using single intervals to represent a species [11].

2.13 How are food webs and their corresponding competition and projection graphs of urban habitats different from marine and terrestrial habitats previously studied? [8].

2.6 CONCLUSIONS

In this chapter we presented the basics of using digraphs to model food webs. However, the construction of real food webs is not easy, and not uniform. Construction of real food webs usually includes coalescence of some species and not others. Often species are lumped together if they have common predators and common prey, but in calculations where the richness is important, i.e., the number of species S in the computation of connectance, this lumping together distorts numerical parameters. For instance, parasites as a species are almost never included in a food web, but have enormous impact on the survivability of other species. Kuris et al. [15] demonstrate that the biomass of parasites typically exceeds the biomass of all conventional vertebrate predators, with serious implications for the structure of any network of interactions.

Real food webs are not randomly assembled; they occur in nature and as such are subject to perturbations from a variety of sources, including various evolutionary processes. How food webs respond to the removal of particular species under these perturbations is and should be the primary focus of research, using food webs and their corresponding graphs as tools. Most of the interest in food webs belongs to conservation biology, especially since over the past century the number of documented extinctions of bird and mammal species has occurred at a rate of 100-1000 times the average extinction rates seen over the half billion years of existent fossil records or other evidence [16],[17]. What is learned from food webs and their corresponding graphs concerning systemic risk from perturbations and removal of vertices provides knowledge usable by those managing other network systems, such as financial systems, communication networks, and energy systems [18]. As Robert May, long-time researcher of food webs, recently indicated, "better understanding of ecosystem assembly and collapse is arguably of unparalleled importance for ourselves and for other living things as humanity's unsustainable impacts on the planet continue to increase" [18].

FURTHER RESEARCH QUESTIONS

2.14 How do different types of agriculture affect biodiversity and how does biodiversity provide ecosystem services (products of network interactions between species) such as pollination and pest control? Is the role of parasites different from free-living species, especially when considering biodiversity? [19],[20].

2.15 How will changes under stress, such as global change, be manifested in the food webs, and how will natural communities respond to this stress at the landscape level? What happens with habitat fragmentation? [8],[21],[22].

2.16 What is the effect of cross-habitat food webs in community level restoration, for example of marine and shore birds? Parallels exist between cross-habitat food webs and communications network structures where much is done within a section of the network then linked to other sections. For example, are there differences between terrestrial and aquatic ecological networks that make analysis of cross-habitat restoration more difficult? What types of habitats can be linked and which cannot? [10],[23].

2.17 How do we map ecosystem services onto trophic levels so that food web theory can explain how species extinction or decline leads to reductions in the delivery of these ecosystem services? Can and what would happen if ecosystem service declines before species loss? Why? Are there appropriate models? [24],[25].

2.18 Repeated failures of single-species fisheries to sustain a profitable harvest have led to an increased popularity of marine-protected areas. Fished species increase in abundance, leading to many indirect effects, which could be predicted by using food web networks to describe these marine areas. [25],[26].

2.19 Invasive species pose significant threats to habitats and biodiversity. Studies to date consider the effects of alien species at a single trophic level, yet linked extinctions and indirect effects are not recognized from this approach. Food webs can often identify these indirect effects and the impact of the addition of an alien species into the network, such as competition due to shared enemies. The removal of an invasive species is a priority for conservationists and the public, so these species provide opportunities for manipulative field experiments testing the removal of a node from a food web. Consider the effects of an invasive species on the whole food web [8].

2.20 Urban areas are one of the fastest growing habitats in the world, with numerous species of plants and invertebrates. Projects developing urban food webs that consider the biodiversity of urban habitats are needed [8].

REFERENCES

[1] Roberts FS. On the boxicity and cubicity of a graph. In: Tutte WT, editor. Recent progress in combinatorics. New York, NY: Academic Press; 1968. p. 301–10.

[2] Cozzens MB, Roberts FS. Computing the boxicity of a graph by covering its complement by cointerval graphs. Dis Appl Math 1983; 6: 217–28.

[3] Cohen JE. Food webs and niche space. Princeton, New Jersey: Princeton University Press; 1978.

[4] Cohen JE, Komlos J, Mueller M. The probability of an interval graph and why it matters. In: Proceedings of Symposium on Pure Mathematics, vol. 34; 1979. p. 97–115.

[5] Opsut R. On the competition number of a graph. SIAM J Algeb Dis Methods 1982;3:470–8.

[6] MacKay B, Schweitzer P, Schweitzer P. Competition numbers, quasi-line graphs, and holes. SIAM J Dis Math 2014;28(1):77–91.

[7] Mac Arthur RH. Fluctuations of animal populations and a measure of community stability. Ecology 1955;36:533–6.

[8] Memmott J. Food webs: a ladder for picking strawberries or a practical tool for practical problems? Trans R Soc B 2009;364(1574):1693–9.

[9] Dunne JA, Williams RJ, Martinez ND. Network structure and biodiversity loss in food webs: robustness increases with connectance. Ecol Lett 2002;5:558–67.

[10] Dunne J, Williams RJ, Martinez ND. Network structure and robustness of marine food webs, www.santafe.edu/media/workingpapers/03-04-024.pdf; 2003.

[11] Palamara GM, Zlatic' V, Scala A, Caldarelli G. Population dynamics on complex food webs. Adv Complex Syst 2011;16(4):635–47.

[12] Cho HH, Kim SR. A class of acyclic digraphs with interval competition graphs. Discret Appl Math 2005;148:171–80.

[13] Ghosh-Dastidar U, Fiorini G, Lora SM. Study and analysis of the robustness of Hudson river species. In: Conference Proceedings of the Business and Applied Sciences: Academy of North America; 2014.

[14] Department of Environmental Conservation, Hudson river estuary, wildlife and habitat conservation framework, http://www.dec.ny.gov/lands/5096.html (accessed on 4/14/14); 2014.

[15] Kuris AM, et al. Ecosystem energetic implications of parasites and free living biomass in three estuaries. Nature 2008;454:515–8.

[16] Kambhu J, Weldman S, Krishnan N. A report on a conference cosponsored by the Federal Reserve Bank of New York and the National Academy of Sciences. Washington D.C.: National Academy of Sciences Press; 2007.

[17] Millennium Ecosystem Assessment. Ecosystems and human well-being synthesis. Washington, D.C: Island Press; 2005.

[18] May R. Food web assembly and collapse: mathematical models and implications for conservation. Trans R Soc B 2009;364(1574):1643–6.

[19] Owen J. *The ecology of garden: the first fifteen years*. Cambridge UK: Cambridge University Press; 1991.

[20] Price JM. Stratford-upon-avon: a flora and fauna. Wallingford, UK: Gem Publishing; 2002.

[21] Fortuna MA, Bascompte J. Habitat loss and the structure of plant-animal mutualistic networks. Ecol Lett 2006;9:278–83.

[22] Forup ML, Henson KE, Craze PG, Memmott J. The restoration of ecological interactions: plant-pollinator networks on ancient and restored heathlands. J of Applied Ecology 2008;45:742–52.

[23] Knight TM, McCoy MW, Chase JM, McCoy KA, Holt RD. Trophic cascades across ecosystems. Nature 2005;437:880–3.

[24] Kremen C. Managing ecosystem services: what do we need to know about their ecology? Ecol Lett 2005;8:468–79.

[25] Dobson A, Allesina S, Lafferty K, Pascual M. The assembly, collapse and restoration of food webs. Transactions of the Royal Society B: Biological Sciences 2009;364(1574):16.

[26] Hilborn R, Stokes K, Maguire JS, Smith T, Botoford LW, Mangel M, et al. When can maritime reserves improve fisheries management? Ocean Coast Manag 2004;47:197–205.

[27] Cozzens MB. Food webs, competition graphs, and habitat formation. In: Jungck JR, Schaefer E, editors. Mathematical modeling of natural phenomena. Cambridge UK: Cambridge University Press; 2011.

[28] Gilbert AJ. Connectance indicates the robustness of food webs when subjected to species loss. Ecol Indic 2008;9:72–80.

Chapter 3

Adaptation and Fitness Graphs

Kristina Crona[1] **and Emilie Wiesner**[2]
[1]*Department of Mathematics and Statistics, American University, 4400 Massachusetts Ave NW, Washington, DC 20016,* [2]*Department of Mathematics, Ithaca College, 953 Danby Rd, Ithaca, NY 14850*

3.1 INTRODUCTION

Understanding evolutionary processes is important in the study of medicine and agriculture. In many instances, the underlying cause for novel pathogens, pests and antimicrobial drug resistance are complex evolutionary processes. Such processes may initially go unnoticed, if the first mutations have moderate effects. However, the end result may be a radically altered organism. This chapter presents discrete methods, or methods from combinatorics, for analyzing complex evolutionary processes involving multiple mutations.

Evolutionary processes of practical interest typically depend on several mutations. For example, the development of antibiotic and HIV drug resistance is a process in which mutations accumulate in the genome, sometimes resulting in a completely resistant genotype. For such a resistant genotype, the drug is of no use, and the infection needs a different treatment. The TEM family, which will be discussed in later sections, is a medically important family of enzymes associated with antibiotic resistance; it includes 200 clinically found mutants.

Another example of practical importance is the process by which a virus crosses the species barrier. Host changes from birds, bats, or pigs to humans may cause fatal diseases. Host changes for plant viruses can cause problems in agriculture. A substantial number of mutations can be involved when a virus adapts to a new host. For instance, the turnip mosaic virus acquired over 100 mutations during the transition from its original host to the radish [1].

An evolutionary process that affects many of us concerns seasonal flu. The flu virus mutates rapidly, and although we may have had the flu previously or been vaccinated, our immune systems do not recognize the mutated virus. As a result, a new flu vaccine is required every year.

Whenever several mutations are involved, one needs to consider gene interactions because the effect of a mutation may depend on background. For instance, two mutations may individually increase fitness (i.e., provide a survival benefit), whereas the combination of these mutations may have lower fitness or may no longer be viable. In addition to evolutionary processes, gene interactions are important for human diseases. In particular, some genetic disorders that cause blindness or deafness depend on multiple mutations.

Most approaches in classical biology concern one or two mutations at a time. One example is the study of recombination, or exchange of genetic material resulting in new genotypes. The literature on recombination is dominated by the study of systems involving two mutation sites and pairwise gene interactions, and there are few theoretical results for larger systems [2].

However, recent empirical studies of antimicrobial drug resistance demonstrate complex interactions between mutations (see Section 3.2). In general, there is no hope that one can reduce important questions in this area to an analysis of systems involving only two mutation sites.

New quantitative tools are necessary for analyzing systems involving complex gene interactions. Fitness graphs constitute an elementary and yet effective approach to coarse aspects of such systems. Analysis of fitness

Algebraic and Discrete Mathematical Methods for Modern Biology. http://dx.doi.org/10.1016/B978-0-12-801213-0.00003-4
Copyright © 2015 Elsevier Inc. All rights reserved.

graphs has been used for relating local and global properties of gene interactions, as well as for analyzing the impact of recombination. An important clinical application is to predict, prevent, and manage antimicrobial drug resistance. Recent applications concern HIV drug treatments and strategies for antibiotic cycling in hospitals.

This chapter focuses on the development of fitness graphs as a tool for understanding systems involving multiple mutations. Section 3.2 provides a more formal presentation of the basic theory of fitness graphs, including definitions and references. The remaining sections develop methods for addressing important questions in the field. Fitness graphs are applied to recombination in Section 3.3 and to drug cycling in Section 3.4. In addition, we propose projects that rely on the geometric theory of gene interactions, another recent approach to such systems.

Exercises designed to facilitate readers' understanding of the material are included at the end of each section. In addition, a sampling of further directions for exploration and research questions are highlighted the form of projects; they require significant independent work and are designed to serve as topics for student research endeavors.

3.2 FITNESS LANDSCAPES AND FITNESS GRAPHS

We begin this section by establishing terminology to describe mutations. We then define the concept of fitness and introduce fitness landscapes and fitness graphs. Finally we discuss epistasis, an important local property of fitness landscapes that can be analyzed via fitness graphs.

3.2.1 Basic Terminology and Notation

A *genotype* is the genetic makeup of an organism or cell and can be thought of as a word in the 20-letter alphabet of amino acids. A single mutation (i.e., a single amino acid substitution) then corresponds to a change of one letter in the word representing a genotype. A *population* is defined by its set of genotypes and the proportions of these genotypes. (A population is *monomorphic* if only one genotype occurs.) In many applications, we have a well-defined *wild type*: a genotype that dominates the population at the starting point of an evolutionary process. A mutant can be described in terms of how it differs from the wild type.

An important empirical example is the TEM family of β-lactamases, a group of enzymes that confers resistance to β-lactam antibiotics, including penicillin. TEM stands for Temoneira, the name of the patient from whom the enzyme was first identified. TEM beta-lactamases have been found in *Escherichia coli*, *Klebsiella pneumoniae*, and other gram-negative bacteria. TEM-1 is considered the wild type. The length of TEM-1 is 287, that is, TEM-1 can be represented as a sequence of 287 letters in the 20-letter alphabet. TEM-2 is a single mutant with the mutation Q39K, which means that the amino acid denoted Q (glutamine) at position 39 of TEM-1 is substituted by the amino acid denoted K (lysine). Approximately 200 mutants have been found clinically; 20% are single mutants, and 90% have at most four amino acid substitutions. The Lahey Clinic has a record of all clinically found TEM mutants, which can be found at their website (http://www.lahey.org/Studies/temtable.asp).

A *locus* refers to a particular position in the genetic code of a genotype, or the word representing that genotype. We describe the possible mutations that occur in that position as *alleles*. In many real populations at most two alleles occur at each locus, so that a biallelic assumption is reasonable. For simplicity, we restrict to biallelic systems throughout the chapter.

Under the biallelic assumption, genotypes can be represented as sequences of 0s and 1s rather than the 20-letter alphabet. Whenever there is a well-defined wild type, we denote it by a sequence of zeros. For instance, in a two-loci system, 00 denotes the wild type, 10 and 01 the single mutants, and 11 the double mutant combining the two single mutations. In a system with no well-defined wild type, the labels of genotypes are arbitrary.

In general, consider a biallelic L-locus system. Let $\Sigma = \{0, 1\}$, representing the two possible alleles at a given locus and let Σ^L be the set of bit strings of length L, representing the genotype space. For example,

$$\Sigma^2 = \{00, 10, 01, 11\} \text{ and}$$
$$\Sigma^3 = \{000, 100, 010, 001, 110, 101, 011, 111\}.$$

3.2.2 Fitness, Fitness Landscapes, and Fitness Graphs

A key concept in this chapter is fitness. *Absolute fitness*, W, is a measure of the expected reproductive success of a genotype and depends on both survival and reproductive success. For instance, in vertebrates one can typically think of absolute fitness as the expected contribution to the next generation. In a simplified scenario (assuming asexual reproduction, a constant generation length, and no overlap between generations), we may define absolute fitness precisely as the mean number of offspring reaching reproductive age for an individual of the given genotype. In this case, for an initial genotype size N_0, the population after n generations will be $N = N_0 W^n$.

Microbes require a different analysis. Because bacteria reproduce by cell division (where one cell always divides into two), it does not make sense to consider mean number of offspring per individual. However, bacteria will divide at different frequencies under different conditions. As a result, a more meaningful measure is growth rate. A basic model for population growth gives the number of individuals N as a function of time t by

$$N = N_0 e^{\alpha t},$$

where N_0 denotes the initial number of individuals and the parameter α is the *growth rate*. (The parameter α is also called the Malthusian parameter.) Growth rate and absolute fitness are closely related (see Exercise 3.15). In the study of antimicrobial drug resistance, the growth rate of a genotype is essentially determined by the degree of drug resistance. There are laboratory techniques for identifying resistant bacteria and estimating the degree of resistance.

To understand adaptation, it is often useful to compare the fitness of different genotypes. If W_0 is the absolute fitness of the wild type, then the *relative fitness* w_g of a genotype g is defined as

$$w_g = \frac{W_g}{W_0}.$$

Notice that the relative fitness of the wild type is 1. For a genotype g where $w_g > 1$, the genotype g is expected to grow more quickly than the wild type and thus will tend to dominate the wild type over time.

A famous metaphor for adaptation is the fitness landscape [3]. A *fitness landscape* is a function $w : \Sigma^L \mapsto \mathbb{R}$, which assigns a fitness value w_g to each genotype g. We measure "distance" (i.e., a metric) in a fitness landscape by the *Hamming distance*: the distance between two genotypes equals the number of positions where the genotypes differ. For example, the Hamming distance between the strings 001 and 111 is 2. Two genotypes are *mutational neighbors* if they differ at exactly one position.

Fitness landscapes provide a good illustration of the Darwinian process for populations that can be modeled using the strong-selection weak-mutation regimen (SSWM). The *SSWM* assumes that the time interval from when a beneficial mutation first appears to when it has taken over the population is extremely short and that the time between the appearance of beneficial mutations is extremely long. Under SSWM, we can think of beneficial mutations as occurring one at a time so that beneficial mutations do not compete with each other. (See [4, 5] for a more detailed description.)

Whenever mutations are sufficiently rare in a population, it is realistic to assume that a beneficial mutation will go to fixation before another beneficial mutation occurs. For a sufficiently large population, it is also reasonable to assume that deleterious mutations disappear within a short time span after appearing. These characteristics make SSWM reasonable (although not valid for all populations), and we assume it holds throughout the remainder of the section.

Under the assumption of SSWM, the population is monomorphic almost all the time. In this case, an adaptation process can be viewed as a sequence of beneficial mutations starting from the wild type. We visualize this in the corresponding fitness landscape as the population "walking" from genotype to genotype: each step of the walk corresponds to a single beneficial mutation, creating a path of increasing fitness from the wild type.

More formally, a *general walk* in a fitness landscape is a sequence of genotypes in which pairs of consecutive genotypes are mutational neighbors. A *shortest walk* between nodes is a general walk of minimal length (in the sense that the number of steps coincides with the Hamming distance between the nodes). An *adaptive step* corresponds to a change at exactly one locus so that the fitness strictly increases. An *adaptive walk* or an *uphill walk* is a sequence of adaptive steps.

An important property of a fitness landscape is its number of peaks. A genotype is at a *peak* if all its mutational neighbors have lower fitness. As an illustration, consider the following example. Let 00 denote the wild type. Suppose that the single mutants 10 and 01 have higher fitness than the wild type, and that the double mutant 11 has the highest fitness of the four genotypes. The two possible scenarios for an adaptation process are the adaptive walks

$$00 \mapsto 10 \mapsto 11, \quad 00 \mapsto 01 \mapsto 11.$$

In this example, there is a single peak: 11.

If there are several peaks, an adaptive walk may end up at a suboptimal peak. Again consider a two-loci system. Suppose that 01 and 10 have greater fitness than both 00 and 11, and 10 has higher fitness than 01. In this case, both 01 and 10 are peaks but 01 is suboptimal, and an adaptive process may terminate at either peak. This model of adaptation has been widely used and relies on approaches developed in [4–6].

A fitness graph provides a simplified representation of the information encapsulated in a fitness landscape. To formally define fitness graphs, we introduce some graph-theoretical concepts. A *directed graph* is a collection of nodes and arrows connecting pairs of nodes. A *cycle* in a directed graph is a sequence of arrows forming a closed loop, and a directed graph is *acyclic* if it has no loops. A *fitness graph* is a directed acyclic graph where each node corresponds to a string of Σ^L; arrows connect each pair of mutational neighbors, directed toward the node representing the more fit genotype. Figure 3.1 shows two possible fitness graphs for a two-locus system.

Fitness graphs are typically arranged into $L + 1$ levels, where a node appears on level i if the corresponding string contains exactly i 1s. For example, in Figure 3.1, 00 appears in the 0th row (at the bottom), and 10 appears in the 1st row. Within a row, the nodes are ordered from left to right according to the *lexicographic order*: alphabetical order, where 1 is treated as "a" and 0 is treated as "b."

Although fitness graphs are fairly coarse representations of fitness landscapes, they still maintain important properties of the landscapes. For example, in Figure 3.1, we can see that $00 \to 10 \to 11$ is a possible uphill walk (and hence evolutionary path) in the left graph but not in the right. A fitness graph also displays information about peaks in the landscape. In Figure 3.1, the left fitness graph has a single peak (11), whereas the right has two peaks (10 and 01). Similarly, one can identify uphill walks and peaks for the three-locus graphs displayed in Figure 3.2.

Some of the ideas used to describe fitness landscapes can be translated to graph-theoretical terms. A *path* between two nodes in a graph is a sequence of edges (and nodes) that connect the two nodes. (In a directed graph, we view the arrows as edges by ignoring direction.) Moreover, the graph-theoretical distance between two nodes is the minimal number of edges in a path connecting the nodes. We conclude that the Hamming distance between two nodes equals the graph-theoretical distance. Thus, the shortest walk in a fitness landscape is a path of minimal graph theoretical distance in the fitness graph.

Readers familiar with Hasse diagrams should notice the following fact: If we view each nonzero string as an event (i.e., a mutation has occurred), then the fitness graph coincides with the Hasse diagram of the power set of events, except that each edge in the Hasse diagram is replaced with an arrow toward the string with greater fitness.

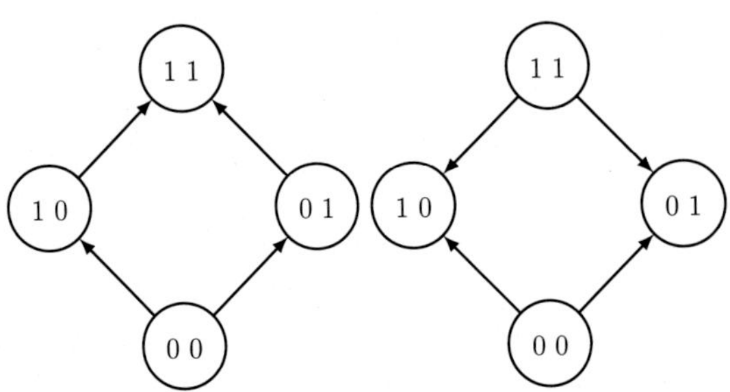

FIGURE 3.1 Example fitness graphs for a two-locus system.

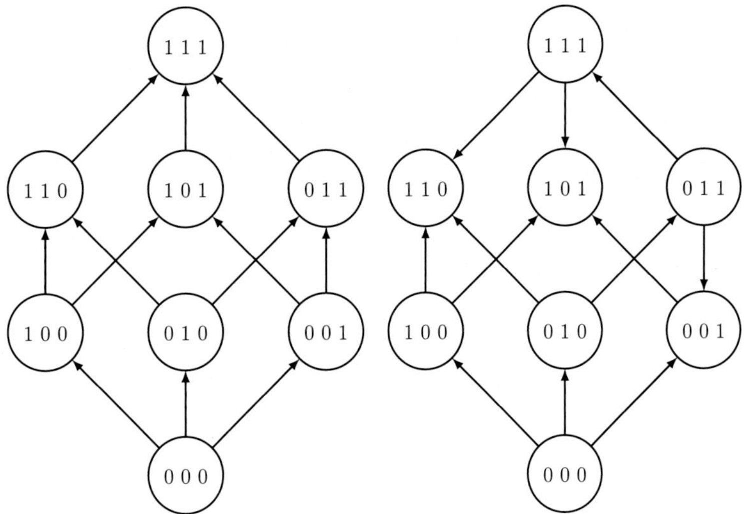

FIGURE 3.2 Example fitness graphs for a three-locus system.

3.2.3 Epistasis

Fitness graphs are useful for relating local properties of fitness landscapes—which can be observed in a laboratory—with global properties that may be of interest. A useful local property can be described in terms of gene interactions. In particular, *epistasis* describes any type of gene interaction.

To make this idea precise, we introduce the following notation to describe pairs of mutations. For $L \geq 2$, given a string and two positions, exactly four strings can be obtained that coincide with the original string except (at most) at the two positions. Denote such a set of four strings

$$ab, Ab, aB, AB,$$

according to the two positions of interest, and assume that the relative fitness, w_{ab}, is minimal. Then, a common condition to check for independence of the a/A and b/B mutations is that relative fitness is *multiplicative*:

$$w_{AB}w_{ab} - w_{Ab}w_{aB} = 0.$$

This condition implies that if Ab and aB have higher fitness than ab, then AB has even higher fitness.

It is convenient to recast this condition in terms of the logarithm $\log W$ of absolute fitness. We define *fitness*, w, to denote $\log W$. (Note that relative fitness and fitness use the same notation; throughout the remainder of this section, we use w to represent fitness.) Then mutations are independent if their fitness satisfies the additive condition

$$w_{AB} - w_{Ab} - w_{aB} + w_{ab} = 0.$$

(Additive fitness is defined analogously for any number of loci.) If the left side of the expression differs from zero, there is epistasis. *Sign epistasis* means that

$$w_{AB} < w_{Ab} \text{ or } w_{AB} < w_{aB}.$$

Reciprocal sign epistasis interactions means that

$$w_{AB} < w_{Ab} \text{ and } w_{AB} < w_{aB}.$$

See Figure 3.3 for the four possibilities under our assumption that w_{ab} is minimal.

Informally, epistasis means that the fitness effect of a mutation depends on background. Sign epistasis means that whether the mutation is beneficial or deleterious depends on background (i.e., on what other

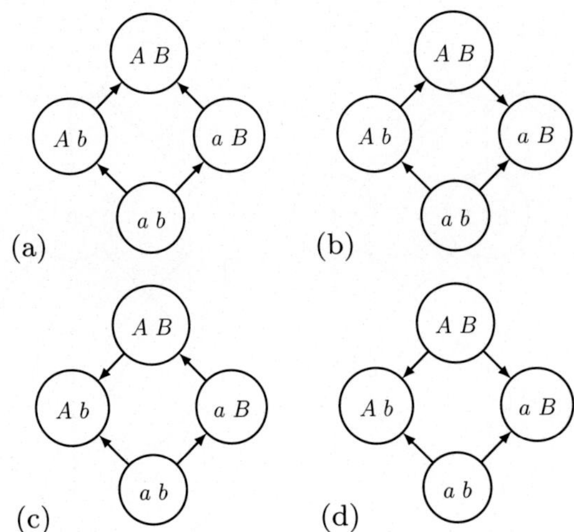

FIGURE 3.3 Two-locus systems.

mutations have occurred). Epistasis may occur for many reasons. For instance, suppose that two mutations have similar effects. Having one of them may be an advantage, whereas having both of them may be a disadvantage.

One way that epistasis can be used to understand global properties is in terms of all arrows up landscapes. An *all arrows up landscape* is a fitness landscape that can be represented by a fitness graph in which all arrows point up (after relabeling the nodes, if necessary). It is straightforward to verify that for a fitness landscape such that all arrows point up there is no sign epistasis (see Exercise 3.1).

This case is of some interest because one can hardly think of a more favorable case for adaptation. All arrows up landscapes are single peaked. Informally, adaptation is easy in an all arrows up landscape because any walk toward the unique peak is an uphill walk (see Exercise 3.9). In other words, an adapting population is never forced to move "backward" to reach the peak.

On the other hand, a single-peaked landscape is not necessarily an all arrows up landscape. For some single-peaked landscapes, adaption may proceed by *reversed* mutations, meaning that adaptive walks from the wild type to the peak contain steps toward the wild type.

For fitness graphs of some biallelic TEM-systems, see [7, 8]. Those graphs differ considerably from all arrows up landscapes. Sign epistasis and several peaks are common. For other empirical examples of fitness graphs, see, for example, [9–11] and references.

Theoretical work on fitness graphs concerns peaks and adaptive walks, as well as the relation between global and local properties of fitness landscapes [12, 13]. (See also [14, 15] for results on the relation between global and local properties of fitness landscapes.)

The graphs reflect qualitative information, such as if "good + good = better" or "good + good = not good" for two single mutations. As long as we restrict ourselves to single mutations (as defined here), the graphs provide complete information about the adaptive potential.

However, the graphs do not reveal fine-scaled properties of fitness landscapes. In addition to information about how fitness increases or decreases, it is desirable to have some notion of "curvature" for a fitness landscape [16]. Consequently, as described in Section 3.3, fitness graphs are insufficient for a complete analysis of recombination.

We will end the discussion by working out the following problem on epistasis, as a preparation for the exercises during which the reader will apply and interpret fitness graphs:

Example 3.1. Choose the correct statement and explain your answer: It is not rare that two beneficial mutations combine poorly, so that the corresponding double mutant is less fit than the wild type. The phenomenon requires

(A) epistasis, but not necessarily sign epistasis
(B) sign epistasis, but not necessarily reciprocal sign epistasis
(C) reciprocal sign epistasis.

Solution. We use fitness graphs, where 00 denotes the wild type. By assumption, the arrows are directed from 00 to 10 and from 00 to 01, because the single mutations are beneficial. The double mutant being less fit than the wild type, it is also less fit than the single mutants. The remaining arrows are directed from 11 to 10 and 11 to 01. The resulting graph, with bottom arrows up and top arrows down, is an example of reciprocal sign epistasis. The answer is C. □

Exercise 3.1. Explain why for a fitness landscape such that all arrows point up, there is no sign epistasis. □

Exercise 3.2. Explain why fitness graphs are acyclic. □

Exercise 3.3. *Choose the correct statement and explain your answer*: Antimicrobial drug resistance usually depends on several mutations. Sometimes there are constraints for the order in which mutations occur. For instance, a mutation A may only be beneficial if a mutation B has already occurred. One can take advantage of such constraints for managing resistance problems. Constraints on the order in which mutations occur require
(A) epistasis but not sign epistasis
(B) sign epistasis, but not reciprocal sign epistasis
(C) reciprocal sign epistasis. □

Exercise 3.4. *Choose the correct statement and explain your answer*: Some fitness landscapes have multiple peaks. Restrict to three-loci systems. Multiple peaks require
(A) epistasis but not sign epistasis
(B) sign epistasis, but not reciprocal sign epistasis
(C) reciprocal sign epistasis.

The result you showed for three-loci systems holds true in general. See Project 3.1 for a study of the theory. □

Exercise 3.5. How many fitness graphs can one construct for two loci? □

Exercise 3.6. How many mathematically different types of fitness graphs are there for two loci? (Graphs belong to the same type whenever they differ only in the labeling of the vertices.) □

Exercise 3.7. How many biologically distinct fitness graphs are there for two loci? That is, how many types of graphs are there, if the special status of 00 as the wild type is taken into consideration? Motivate your answer. □

Exercise 3.8. Explain why additive fitness landscapes are all arrows up landscapes. □

Exercise 3.9. Explain why any shortest walk to the peak is an adaptive walk in an all arrows up landscape. □

Exercise 3.10. Consider single-peaked landscapes. Find an example where there is an adaptive walk to the peak that contains a reversion. □

Exercise 3.11. Discuss the role of reversed mutations in adaptation in view of the previous two problems. □

Exercise 3.12. Consider an all arrows up graph for $L = 2$. We say that the corresponding fitness landscape has *negative epistasis* if $w_{11} < w_{10} + w_{01} - w_{00}$. Positive epistasis is defined similarly. Give examples showing that the fitness landscape may have positive, negative, or no epistasis. You may use relative fitness values in your examples. □

Exercise 3.13. For $L = 2$, and assume that both single mutants have higher fitness as compared to the wild type. Construct all possible fitness graphs under the assumption that epistasis is negative (in the sense of the previous problem) and give possible relative fitness values for each graph. □

Exercise 3.14. Express the relation between absolute fitness, growth rate, and generation time (i.e., the time between generations), based on the formula

$$N = N_0 e^{\alpha t}.$$

Growth rates of bacteria can be measured in the laboratory, so the relation is important for interpretations. □

Project 3.1 (Global and local properties of fitness landscapes)**.** Study global and local properties of fitness landscapes, with focus on peaks and sign epistasis. Your study should include the following results. For original proofs, see [12, 14, 15]. For more background, see also [13].

(A) Multiple peaks implies reciprocal sign epistasis.
(B) If reciprocal sign epistasis occurs, but no type-1 systems (a system with sign epistasis but no reciprocal sign epistasis), then the landscape has multiple peaks.
(C) The following conditions are equivalent:
  (i) Each general step toward the fitness peak is an adaptive step.
  (ii) Each shortest general walk to the fitness peak is an adaptive walk.
  (iii) The fitness landscape is an all arrows up landscape.
(D) If the equivalent conditions (i), (ii), and (iii) in (C) are satisfied, then each adaptive walk to the fitness peak is a shortest general walk. □

Project 3.2 (Complexity and speed of adaptation)**.** It seems plausible that population will tend to adapt slower if the fitness graph is complex, even if the landscape is single peaked. Quantify the relation between graphs and the speed of adaptation. For simplicity, restrict to single-peaked landscapes. Keep in mind that fitness graphs provide coarse information only, so that it is reasonable to use a simple model. In particular, assume the SSWM regimen, and that all available beneficial mutations are equally likely to occur. Model mutations (by a "mutation" we mean that a mutation occurs and goes to fixation) as a Poisson process. One can start with a systematic study of small examples, that is, few loci.

Compare your results with empirical fitness graphs. For survey articles with many references to empirical studies see [10, 11]. □

Project 3.3 (Classification of fitness landscapes)**.** Sign epistasis tends to be prevalent for landscapes associated with antibiotic resistance, whereas some other landscapes have little sign epistasis. Is there a good way to classify fitness graphs? Develop measures for comparing fitness landscapes.

One approach would be to focus on prevalence of sign epistasis and reciprocal sign epistasis in the fitness graphs (see also [10]). Alternatively, one could focus on deviations from all arrows up landscapes (how many arrows point in the "wrong direction"?).

Apply your classification for comparisons of empirical studies. For survey articles with many references to empirical studies, see [10, 11]. □

3.3 FITNESS GRAPHS AND RECOMBINATION

Genetic recombination, or the exchange of genetic material resulting in new genotypes, is widespread in nature. Recombination for vertebrates typically occurs when an individual produces *gametes*, that is, eggs or sperm, by combining genetic material from both parents. As a result, offspring inherit genetic material from all four grandparents. Recombination occurs for bacteria as well, when genetic material is transferred from one individual to another: the recipient individual thus combines genetic material from two individuals. The theory on recombination is extensive, and we will focus on some interesting aspects from a discrete perspective. For more background, see, for example, [17–19].

A possible advantage of recombination is that one may obtain entirely new genotypes. An early proposal was that recombination could be beneficial in the following situation: two single mutations are deleterious, but the double mutant combining the single mutations has very high fitness. The argument was immediately criticized

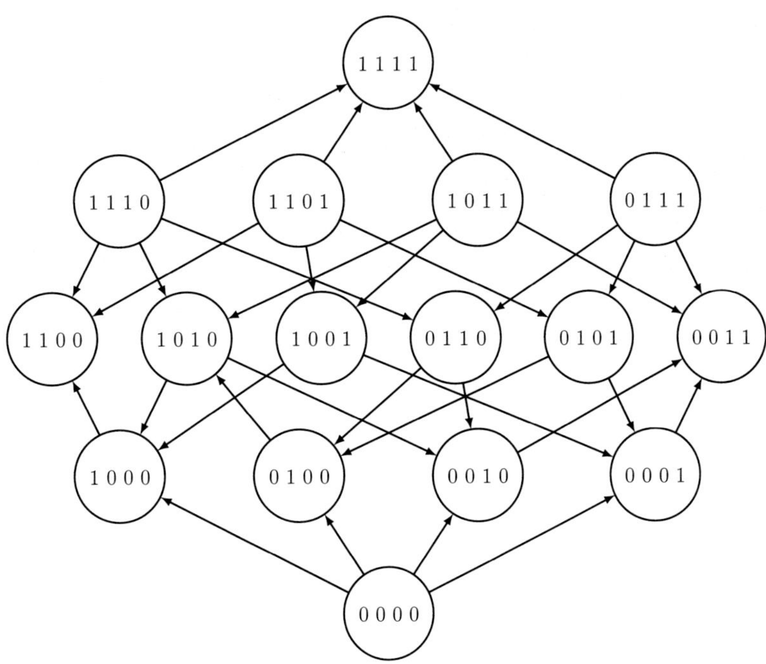

FIGURE 3.4 A four-locus system.

and described as a widespread fallacy [20]. (It has been argued that the mechanism could make recombination advantageous under some circumstances [21], but the mechanism is not considered important.)

The following example is interesting to analyze from the perspective of recombination. Consider a four-loci system, where 0000 is the wild type, and where 1111 has highest fitness. All single mutants have higher fitness than the wild type. The double mutants 1100 and 0011 have higher fitness than the single mutants, and the remaining double mutants have lower fitness. All triple mutants have low fitness compared to their mutational neighbors. The fitness graph for this system is shown in Figure 3.4.

In this system, recombination has the potential to speed up adaptation. Without recombination, an 1100 population could reach the peak 1111 only by a double mutation. However, recombination between 1100 and 0011 populations could produce the 1111 genotype. Although one might reasonably ask how plausible such a system is, the TEM family (discussed in Section 3.2) provides an intriguing example. The genotypes TEM-50, TEM-1, and all intermediates constitute a four-loci system. Out of the 16 genotypes in the system, exactly 8 have been found clinically. The 8 genotypes can be represented as

$$0000, 1000, 0100, 0010, 0001, 1100, 0011, 1111.$$

In particular, none of the triple mutants has ever been found in hospitals. It follows that the clinically found mutants are compatible with the fitness graph discussed.

Although fitness graphs provide some intuition for the effect of recombination, they are insufficient for a complete analysis. For instance, consider an all arrows up landscape for two loci, where 00 denotes the wild type (illustrated in the left fitness graph in Figure 3.1). Independent of recombination, we see from the fitness graph that an evolutionary process beginning with a homogeneous 00 population will eventually fixate at a 11 population. However, the fitness graph does not provide information about whether recombination will speed up or slow down this process.

For a well-mixed population, it is known that recombination speeds up the evolutionary process if

$$w_{11} - w_{10} - w_{01} + w_{00} < 0,$$

and slows down the evolutionary process if

$$w_{11} - w_{10} - w_{01} + w_{00} > 0,$$

whereas recombination has no effect if the left-hand expression equals zero [17]. In other words, recombination has no effect if fitness is additive, which holds true for any number of loci.

An intuitive explanation to the different effects of recombination relies on observing the frequencies of the four genotypes 00, 10, 01, and 11. In the first case (usually called negative epistasis), selection will have the effect that 10 and 01 are overrepresented as compared to a case with additive fitness. For that reason, recombination will tend to produce 00 and 11 genotypes from 10 and 01 genotypes (more frequently than vice versa), and the result is faster adaptation. A similar argument explains why the second case (positive epistasis) slows down adaptation. In particular, it follows that the effect of recombination may be different for fitness landscapes with the same fitness graph.

More generally, a central problem is to determine under what circumstances recombination provides an advantage. The following questions represent two important aspects of this issue:

A. Does recombination help the organism to adapt?
B. Is recombination itself selected for, in the sense that the mechanism of recombination provides a selective advantage?

Interestingly, even if recombination speeds up adaptation, it may not be selected for. It follows that Problems A and B are fundamentally different. Problem B is usually modeled by *recombination modifiers*, or genes that impact the recombination frequency. For instance, a particular gene may shut off recombination. One can then ask if such a recombination modifier will be selected for. Although Problem B has been studied extensively for two loci, there are few theoretical results on recombination for multiple loci [2]. Problem A for multiple loci may present a more tractable problem for study (see Project 3.5).

Exercise 3.15. The introduction to this section described the following situation: two single mutations are deleterious, but the double mutant combining the single mutations has very high fitness. Draw a fitness graph representing this two-loci system. □

Exercise 3.16. Analyze the argument described earlier as a "widespread fallacy" by Joseph Muller [20]. Explain why the argument seems fairly weak in most situations. □

Exercise 3.17. Earlier, we discussed the impact of recombination for a special four-loci system. An analogous situation for three loci would be a system such that 100, 010, and 001 have higher fitness than the wild type, the double mutants have lower fitness than the single mutants, and 111 has maximal fitness in the system. Analyze the fitness graph in this situation, and explain why recombination is not an obvious advantage. □

Exercise 3.18. Verify the claim about the four-loci system consisting of TEM-1, TEM-50, and intermediates. List the single and double mutants using standard notation. □

Project 3.4 (Recombination and the geometric theory of gene interaction)**.** Consider a biallelic population, where several genotypes are represented. Can we increase population fitness by recombination? A necessary and sufficient condition is given in [16]. The condition is phrased in terms of triangulations of polytopes. For mathematical background see [22, 23]. □

Project 3.5 (Analyze the effect of recombination for multiloci systems)**.** Study conditions for when recombination will speed up or slow down adaptation for multilocus systems. One may use fitness graphs or the geometric theory of gene interactions [16]. Very little is known in this field. Consequently, even small case studies would be of interest. For general background, see [17–19]. For evidence that some simple conditions are poor predictors of the effect of recombination, see [2]. □

3.4 FITNESS GRAPHS AND DRUG CYCLING

Antibiotic resistance is a serious health problem. Drug cycling, or drug rotation, can be used as a strategy for managing resistance problems in hospitals. The general method is to use one drug at a time in a given setting, periodically switching the drug in use. (Complications, such as an individual patients who are unable to use a particular drug, may prevent full implementation of this strategy.) Historically, drug cycling trials have had mixed results, perhaps because treatment plans were not designed in a systematic way [24, 25].

A recent approach to design treatment plans includes a careful analysis of evolutionary scenarios [7, 8]. A desirable outcome of a treatment (i.e., a sequential application of a set of drugs) is that the population returns to the wild type. Such an outcome is medically beneficial because there tend to be many treatments for the wild type. For instance, physicians have decades of experience of treating TEM-1 (the wild type in the TEM family). In contrast, Darwinian processes where several mutations accumulate may produce very problematic novel genotypes.

In concrete terms, suppose that the starting point is a 11-population. Then the scenarios

$$11 \mapsto 10 \mapsto 00$$

and

$$11 \mapsto 01 \mapsto 00$$

return the population to the wild type. Our goal is to find an optimal two-drug treatment plan by maximizing, over all sequences of two drugs, the sum of the probabilities for the two adaptive walks shown.

In general, given a set of genotypes Σ^L and available drugs, consider all treatment plans where we use a sequence of exactly k available drugs (potentially using the same drug several times). For a given population, our optimization problem is to find the treatment plan that maximizes the probability to return to the wild type. Following [8], we assume that one drug in the treatment plan induces exactly one mutation.

The empirical part of this type of study is to estimate fitness (largely determined by the degree of drug resistance) for each genotype via laboratory experiments. The resulting set of fitness landscapes is the starting point for a quantitative analysis.

We continue to assume SSWM, so that the probability that a deleterious mutation goes to fixation is zero. The probabilities for beneficial mutations can be estimated using the correlated fixation model (CPM) [26] and the equal fixation model (EPM) [27]. According to EPM all beneficial mutations are equally likely to occur. However, according to CPM, mutants of higher fitness are more likely to go to fixation; in this case formulas giving the exact probabilities are known [26]. (Whether CPM or EPM gives better estimate depends on the context: CPM is valid for moderate fitness differences, whereas EPM may be more accurate in a case where fitness differences are substantial.) For both models, a given genotype undergoes some mutation with probability 1, unless the genotype is at a peak. (A genotype that is at a peak will not undergo mutations.)

Using either EPM or CPM, the probabilities for each available mutation is determined given the genotype and drug. The effect of a drug a can be expressed as a transition matrix $M(a)$, where rows and columns are labeled by the genotypes Σ^L, ordered by level, and using the lexicographic ordering within levels. The matrix entry $M_{u,v}(a)$ represents the probability that genotype u is replaced by genotype v under the presence of the drug a.

For instance, consider a biallelic three-locus system and suppose that the mutation probabilities corresponding to a given drug are represented by the following matrix. (This drug is also discussed in Exercise 3.23.)

$$M = \begin{bmatrix} 1 & 0 & 0 & 0 & 0 & 0 & 0 & 0 \\ 0.9 & 0 & 0 & 0 & 0.1 & 0 & 0 & 0 \\ 1/3 & 0 & 0 & 0 & 1/3 & 0 & 1/3 & 0 \\ 1/3 & 0 & 0 & 0 & 0 & 1/3 & 1/3 & 0 \\ 0 & 0 & 0 & 0 & 0 & 0 & 0 & 1 \\ 0 & 1/2 & 0 & 0 & 0 & 0 & 0 & 1/2 \\ 0 & 0 & 0 & 0 & 0 & 0 & 0 & 1 \\ 0 & 0 & 0 & 0 & 0 & 0 & 0 & 1 \end{bmatrix}$$

In this case, the rows and columns are ordered by the genotypes

$$000, 100, 010, 001, 110, 101, 011, 111.$$

From the second row we conclude that the probability for the mutations

$$100 \mapsto 000 \text{ and } 100 \mapsto 110,$$

is 0.9 and 0.1, respectively; no other mutations are predicted to go to fixation for the 100 genotype. Notice that the first and last rows are special in that mutations will not occur and go to fixation. Thus, in the environment determined by the drug, 000 and 111 have higher fitness than their mutational neighbors.

To understand a treatment plan, we must take into account the effects of all drugs involved. We assume that the effect of a drug on a population depends on the genotype of the population and the drug only, regardless of evolutionary history. This implies that the probability for an adaptive walk is the product of the probabilities for each mutation. Because the probability that a genotype u mutates to a genotype v via k mutations is the sum of probabilities over all possible adaptive walks between u and v of length k, it follows that the effect of a the application of drugs $a_1, \ldots, a_k$ can be described as the matrix product $M_{a_1}, \ldots, M_{a_k}$. In particular, for a genotype u, our optimization problem is equivalent to finding a sequence of drugs $a_1, \ldots, a_k$ which maximizes the matrix entry

$$(M_{a_1}, \ldots, M_{a_k})_{u,\mathbf{0}}$$

where $\mathbf{0} = (0, \ldots, 0)$ represents the wild type.

One could approach this optimization problem with by comparing the relevant matrix entries for each sequence of k drugs. However, as the number of possible drugs grows, this approach quickly becomes computationally difficult. Better algorithms would be valuable, but this is currently an open problem.

Exercise 3.19. The probabilities for mutations can be expressed in matrix form, as described in the text. Explain why all entries on the main diagonal are 0 or 1. □

Exercise 3.20. Recall that the trace of a matrix is the sum of the entries on the main diagonal. Interpret the trace in terms of the fitness landscape associated with the drug. □

Exercise 3.21. Consider four different drugs and a two-loci system. Construct fitness graphs so that a sequence of four drugs induce the mutational trajectory

$$00, 10, 11, 01, 00.$$

This means that successive mutations lead back to the wild type. □

Exercise 3.22. Discuss why a treatment plan as in the previous problem may be valuable for managing resistance problems in hospitals. □

Exercise 3.23. For any biallelic system and set of drugs, the maximum probabilities for returning to the wild type depend on how many steps one allows in the treatment plan. The following example demonstrates that the maximum probabilities may increase by the number of steps indefinitely.

Consider a three-loci system where the genotypes are ordered as

$$000, 100, 010, 001, 110, 101, 011, 111.$$

Suppose there are two available drugs, A and B, with transition matrices as given following. Starting at the genotype 100, we see that the probability for returning to 000 by applying A is 0.9. One can calculate that the probability for ending at 000 for A-B-A 0.99, for A-B-A-B-A 0.999, and so forth.

Compressive Behavior of Ti-6Al-4V/TiC Layered Composites: Experiments and Modeling

A.J. Wagoner Johnson, C.L. Briant, C.W. Bull, and K.S. Kumar

Brown University
Division of Engineering, Box D
Providence, RI 02912

Abstract

The Ti-6Al-4V/TiC composite system was studied in compression at strain rates of $0.1s^{-1}$ and $1000s^{-1}$ at room temperature. Symmetric three-layered structures were successfully fabricated with equal layer thickness by diffusion bonding individual layers. Each layer consisted of either monolithic Ti-6Al-4V or a particulate reinforced composite of Ti-6Al-4V and 10 volume %TiC. Layer-interfaces were macroscopically flat and free of voids; the matrix was continuous through the structure. The strength of the layered structures was significantly improved over the homogeneous Ti64. In addition, the structures were more damage resistant than the homogeneous particulate composite. Cracks were blunted and deflected within the composite layer, approximately parallel to the hard/soft material interface, thereby arresting the crack before catastrophic failure. A simple finite element model accurately represented the engineering stress-strain behavior as well as other macroscopic features observed in the deformed specimens. The model also served as a tool to understand the influence of friction on the deformation behavior of the layered structures.

Lightweight Alloys for Aerospace Applications
Edited by Kumar Jata, Eui Whee Lee,
William Frazier and Nack J. Kim
TMS (The Minerals, Metals & Materials Society), 2001

LIGHTWEIGHT ALLOYS FOR AEROSPACE APPLICATION

Edited by:
Dr. Kumar Jata, Dr. Eui Whee Lee,
Dr. William Frazier and Dr. Nack J. Kim

COMPOSITES

Compressive Behavior of Ti-6Al-4V/TiC Layered Composites: Experiments and Modeling

A.J. Wagoner Johnson, C.L. Briant, C.W. Bull and K.S. Kumar

Pgs. 261-271

184 Thorn Hill Road
Warrendale, PA 15086-7514
(724) 776-9000

$$M(A) = \begin{bmatrix} 1 & 0 & 0 & 0 & 0 & 0 & 0 & 0 \\ 0.9 & 0 & 0 & 0 & 0.1 & 0 & 0 & 0 \\ 1/3 & 0 & 0 & 0 & 1/3 & 0 & 1/3 & 0 \\ 1/3 & 0 & 0 & 0 & 0 & 1/3 & 1/3 & 0 \\ 0 & 0 & 0 & 0 & 0 & 0 & 0 & 1 \\ 0 & 1/2 & 0 & 0 & 0 & 0 & 0 & 1/2 \\ 0 & 0 & 0 & 0 & 0 & 0 & 0 & 1 \\ 0 & 0 & 0 & 0 & 0 & 0 & 0 & 1 \end{bmatrix}$$

$$M(B) = \begin{bmatrix} 1 & 0 & 0 & 0 & 0 & 0 & 0 & 0 \\ 1/2 & 0 & 0 & 0 & 0 & 1/2 & 0 & 0 \\ 1/3 & 0 & 0 & 0 & 1/3 & 0 & 1/3 & 0 \\ 1/3 & 0 & 0 & 0 & 0 & 1/3 & 1/3 & 0 \\ 0 & 1 & 0 & 0 & 0 & 0 & 0 & 0 \\ 0 & 0 & 0 & 0 & 0 & 1 & 0 & 0 \\ 0 & 0 & 0 & 0 & 0 & 0 & 0 & 1 \\ 0 & 0 & 0 & 0 & 1/2 & 1/2 & 0 & 0 \end{bmatrix}$$

(i) Construct two fitness graphs that are compatible with the matrices A and B,
(ii) Describe the possible evolutionary scenarios in terms of the fitness graphs you found. Use the 100 starting point. □

REFERENCES

[1] Ohshima K, Akaishi S, Kajiyama H, Koga R, Gibbs AJ. Evolutionary trajectory of turnip mosaic virus populations adapting to a new host. J Gen Virol 2010;91:788-801. doi:10.1099/vir.0.016055-0.
[2] Misevic D, Kouyos RD, Bonhoeffer S. Predicting the evolution of sex on complex fitness landscapes. PLoS Comput Biol 2009;5(9):e1000510. doi:10.1371/journal.pcbi.1000510. Epub 2009 Sep 18.
[3] Wright S. Evolution in Mendelian populations. Genetics 1931;16:97-159.
[4] Gillespie JH. A simple stochastic gene substitution model. Theor Pop Biol 1983;23:202-15.
[5] Gillespie JH. The molecular clock may be an episodic clock. Proc Natl Acad Sci USA 1984;81:8009-13.
[6] Maynard Smith J. Natural selection and the concept of protein space. Nature 1970;225:563-4.
[7] Goulart CP, Mahmudi M, Crona K, Jacobs Stephen D, Kallmann M, Hall GH, et al. Designing antibiotic cycling strategies by determining and understanding local adaptive landscapes. PLoS ONE 2013;8(2):e56040. doi:10.1371/journal.pone.0056040.
[8] Mira P, Crona K, Greene D, Meza J, Sturmfels B, Barlow M. Rational design of antibiotic treatment plans. 2014; arXiv:1406.1564.
[9] De Visser JA, Park SC, Krug J. Exploring the effect of sex on an empirical fitness landscapes. Am Nat 2009;174:S15-30.
[10] Szendro IG, Schenk MF, Franke J, Krug J, de Visser JAGM. Quantitative analyses of empirical fitness landscapes. J Stat Mech 2013;P01005.
[11] De Visser JAGM, Krug J. Empirical fitness landscapes and the predictability of evolution. Nat Rev Genetic 2014;15(7):480-90. doi:10.1038/nrg3744.
[12] Crona K, Greene D, Barlow M. The peaks and geometry of fitness landscapes. J Theor Biol 2013;317:1-13.
[13] Crona K. Graphs, polytopes and fitness landscapes. In: Engelbrecht A, Richter H, editors. Recent advances in the theory and application of fitness landscapes. Springer series in emergence, complexity, and computation. Berlin: Springer; 2013. Chapter 7, p. 177-206.
[14] Poelwijk FJ, Sorin T-N, Kiviet DJ, Tans SJ. Reciprocal sign epistasis is a necessary condition for multi-peaked fitness landscapes. J Theor Biol 2011;272(1):141-4.
[15] Weinreich DM, Watson RA, Chao L. Sign epistasis and genetic constraint on evolutionary trajectories. Evolution 2005;59:1165-74.
[16] Beerenwinkel N, Pachter L, Sturmfels B. Epistasis and shapes of fitness landscapes. Stat Sin 2007;17:1317-42.
[17] Otto SP, Lenormand T. Resolving the paradox of sex and recombination. Nat Rev Genet 2002;3:252-61.

[18] Agrawal AF. Evolution of sex: why do organisms shuffle their genotypes? Curr Biol 2006;16(17):R696-704. doi:10.1016/j.cub.2006.07.063.
[19] Otto SP. The evolutionary enigma of sex. Am Nat 174(Suppl 1):S1-14.
[20] Muller HJ. The relation of recombination to mutational advance. Mutat Res 1964;1:2-9.
[21] Lenski RE, Ofria C, Pennock RT, Adami C. The evolutionary origin of complex features. Nature 2003;423:139-44.
[22] De Loera JA, Rambau J, Santos F. Triangulations: applications, structures and algorithms. Number 25 in algorithms and computation in mathematics. Heidelberg: Springer-Verlag; 2010.
[23] Ziegler G. Lectures on polytopes (graduate texts in mathematics, 152). New York: Springer-Verlag; 1995.
[24] Brown EM, Nathwani D. Antibiotic cycling or rotation: a systematic review of the evidence of efficacy. J Antimicrob Chemother 2005;55:6-9. doi:10.1093/jac/dkh482.
[25] Bergstrom CT, Lo M, Lipsitch M. Ecological theory suggests that antimicrobial cycling will not reduce antimicrobial resistance in hospitals. Proc Natl Acad Sci USA 2004;101:13285-90. doi:10.1073/pnas.0402298101.
[26] Gillespie JH. Molecular evolution over the mutational landscape. Evolution 1984;38:1116-29.
[27] Kimura M. On the probability of fixation of mutant genes in a population. Genetics 1962;47:715.

Chapter 4

Signaling Networks: Asynchronous Boolean Models

Réka Albert[1] and Raina Robeva[2]

[1]Pennsylvania State University, University Park, PA, USA, [2]Sweet Briar College, Sweet Briar, VA, USA

4.1 INTRODUCTION TO SIGNALING NETWORKS

Living cells receive various external stimuli and convert them into intracellular responses. This process is collectively known as *signal transduction*, and involves a collection of interacting (macro)molecules such as enzymes, proteins, and second messengers [1]. Signal transduction is an important part of a cell's communication with its surroundings. Signal transduction is crucial to the maintenance of cellular homeostasis and for cell behavior (growth, survival, apoptosis, movement). Many disease processes such as developmental disorders, diabetes, vascular diseases, autoimmunity, and cancer [2, 3] arise from mutations or alterations in the expression of signal transduction pathway components.

Figure 4.1 illustrates the characteristic steps of signal transduction. Signal transduction processes are activated by extracellular signaling molecules that bind to receptor proteins located in the cell membrane. The signals are transferred inside the cell through changes in the shape of the receptor proteins and trigger a sequence of biochemical reactions leading to the production of small molecules called second messengers. The signals are amplified through additional biochemical reactions or protein-protein interactions in the cytoplasm, for example phosphorylation of a protein by another protein called a kinase. The information can be passed to the nucleus and can lead to changes in the expression of certain genes. Other signal transduction processes lead to a cellular response at the protein level, such as opening of ion channels. At every step of the signal transduction process feedbacks are possible and are often important.

Many signal transduction processes involve numerous and diverse components and interactions. For this reason it is beneficial to represent them with a network, or graph. The components (e.g., biomolecules) are represented by *nodes* (also called *vertices*), whereas the interactions and processes among the nodes are denoted by *edges* (also called *links*). Edges in the network can be directed, indicating the orientation of mass transfer or of information propagation, and can also have a positive or negative sign to represent activation or inhibition. The totality of the nodes and edges of a network form the *network topology*. This network representation, called a *signal transduction network* or *signaling network*, provides a basis for structural analysis and dynamic modeling of the underlying signal transduction process. These mathematical analyses enable us to trace the propagation of information in the network, to determine the key mediators, and to determine the system's responses under normal circumstances and in the case of perturbations.

Figure 4.2 depicts an example of a real signal transduction network, involved in activation-induced cell death of white blood cells called cytotoxic T cells [5],[6]. This network is of interest because the process of

Algebraic and Discrete Mathematical Methods for Modern Biology. http://dx.doi.org/10.1016/B978-0-12-801213-0.00004-6
Copyright © 2015 Elsevier Inc. All rights reserved.

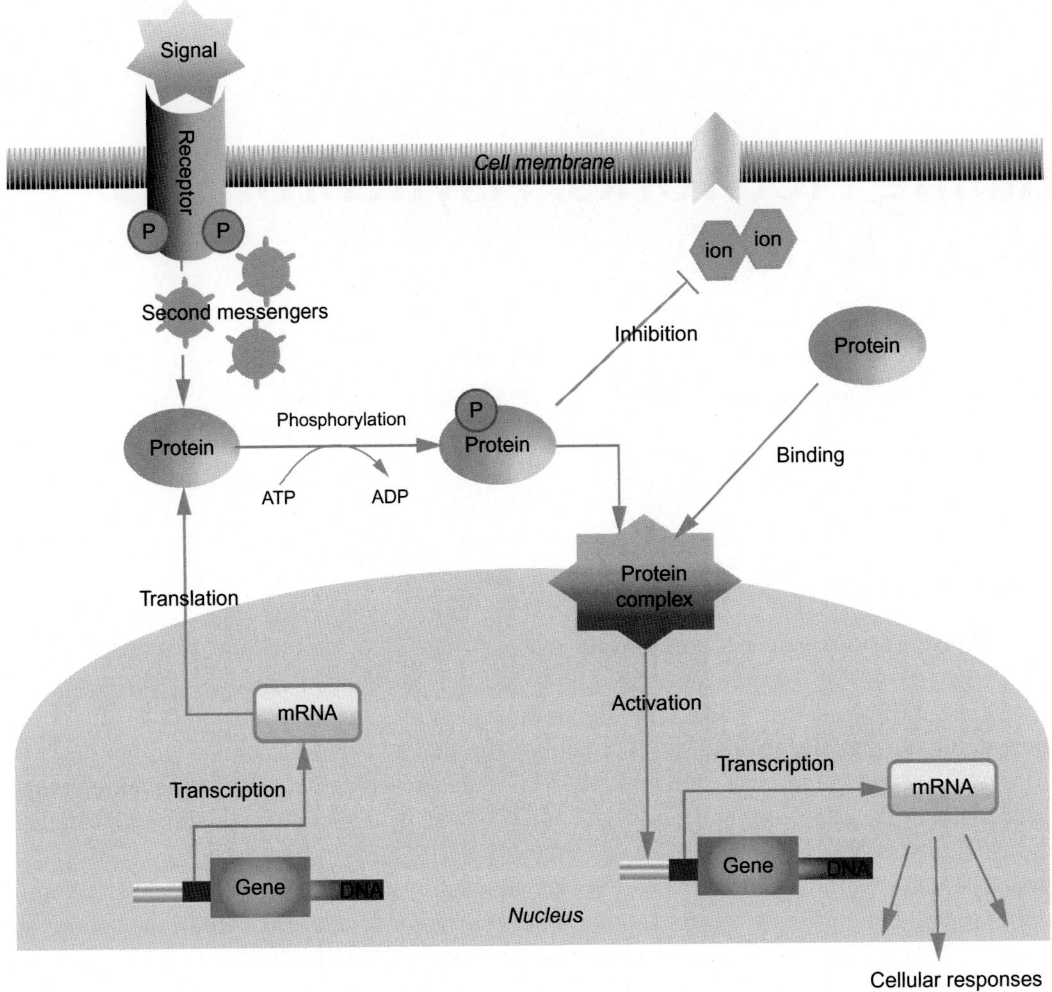

FIGURE 4.1 Scheme of a hypothetical signal transduction process involving diverse interactions of cellular components. Figure reproduced from Ref. [4].

activation-induced cell death is disrupted in the disease T-LGL leukemia, causing the survival of a fraction of activated T cells, which later start attacking healthy cells. In Figure 4.2, the shape of the nodes indicates their cellular location: rectangles indicate intracellular components, ellipses indicate extracellular components, and diamonds indicate receptors. In addition, as this network represents only a small part of the cellular signaling processes, hexagonal nodes are used to summarize its connections with other signal transduction mechanisms or cell behaviors. Such nodes, called conceptual nodes, encapsulate behaviors that are relevant to the network functions. For more details about this specific example see the legend of Figure 4.2.

4.2 A BRIEF SUMMARY OF GRAPH-THEORETIC ANALYSIS OF SIGNALING NETWORKS

A network representation of a signaling mechanism contains essential information, which can be incorporated into its initial analysis. This analysis includes the use of graph-theoretic measures, such as *centrality measures*, *network motifs*, and *shortest paths*, to describe the organization of the network [7].

In signal transduction networks, like in all directed networks, we can categorize the nodes by their incoming and outgoing edges. The nodes with only outgoing edges are called *sources*, and nodes with only incoming edges are *sinks* of the network. Source nodes generally correspond to the signals, while sink nodes denote the outcomes

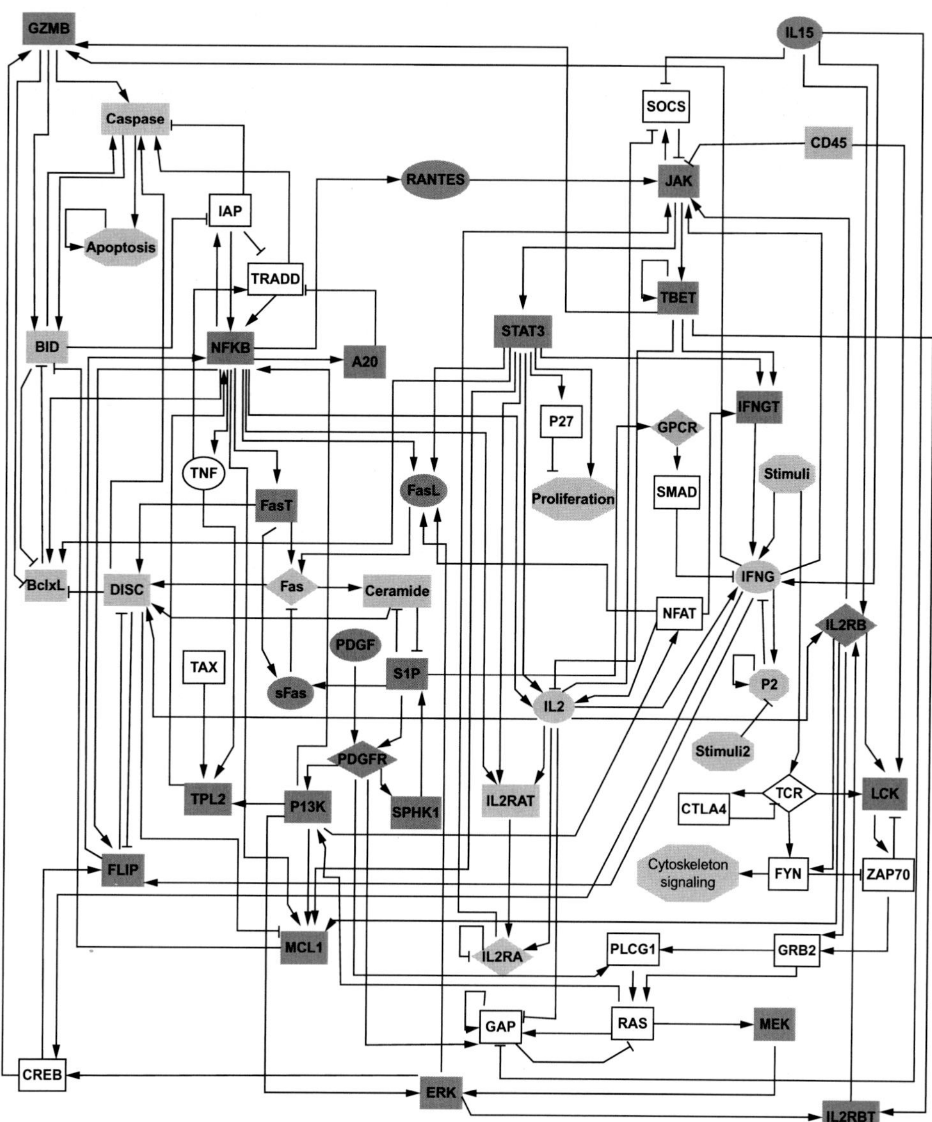

FIGURE 4.2 A signal transduction network involved in activation-induced cell death of white blood cells called cytotoxic T cells. The key signals are Stimuli (representing stimulus of the cell by the presence of pathogens) together with the external molecules interleukin 15 (IL15) and platelet-derived growth factor (PDGF).These signals are identified as nodes with only outgoing links (nodes with this property are called sources). The key output node of the network is Apoptosis, expressing programmed cell death. Note that this node has no outgoing links (a node with this property is called a sink). The nodes of the network include proteins, mRNAs, and concepts. The shape of the nodes indicates the cellular location: rectangles indicate intracellular components, ellipses indicate extracellular components, and diamonds indicate receptors. Conceptual nodes are represented by hexagons. The background of the non-conceptual nodes corresponds to the known status of these nodes in abnormally surviving T-LGL cells as compared to normal T cells: red (dark) indicates abnormally high expression or activity, green (lighter) means abnormally low expression or activity. The full names of the nodes can be found in [5, 6]. An arrowhead or a short perpendicular bar at the end of an edge indicates activation or inhibition, respectively. Figure reproduced from Ref. [6].

of signal transduction networks. In signaling networks it is possible that nodes have an auto-regulatory *loop*, an edge that both starts and ends at the node. Often it is beneficial to extend the definition of source and sink nodes to allow for the presence of a loop. For example, in Figure 4.2 the nodes Stimuli, IL15, and PDGF represent external signals acting on T cells, and indeed they are source nodes. The nodes Apoptosis, Proliferation, and Cytoskeleton signaling represent outcomes of the signal transduction process. Proliferation and Cytoskeleton signaling are sink nodes, and Apoptosis, which has a loop, can also be considered a sink node.

Centrality measures describe the importance of individual nodes in the network. The simplest of such measures is the node *degree*, which quantifies the number of edges connected to each node. For directed networks, the *in-* and *out-degree* of a node is defined as the number of edges coming into or going out of the node, respectively. The nodes whose combined in- and out-degrees are in the top 1% to 5% of all the nodes are termed *hubs*. These hub nodes often play an important role in the network. For example, the node representing the NFκB protein is a hub of the T-LGL network on Figure 4.2, having an out-degree of 11 and an in-degree of 5. This is not surprising, as NfκB is a transcription factor that is known to be important in cellular responses to various stimuli and in cell survival (see, e.g., Ref. [8]).

From a graph-theoretic standpoint, a *path* is a sequence of adjacent edges in the network. In networks that can have both positive and negative edges, the *sign of a path* is positive if there are no or an even number of negative edges in the path and is negative if there is an odd number of negative edges. A path containing two or more edges that begins and ends at the same node is called a *cycle*. The *length* of a path or a cycle is defined to be the number of its edges (loops can be considered as cycles of length one).

Network motifs are recurring patterns of interconnection with well-defined topologies [9]. Among these motifs are *feed-forward loops* (in which a pair of nodes is connected by both an edge or short path and a longer path) and *feedback loops* (directed cycles). An example of a feed-forward loop in Figure 4.2 is the subgraph formed by the nodes STAT3, P27, and Proliferation; this is an incoherent feed-forward loop, as the STAT3–Proliferation edge is positive and the path between them is negative. An example of a positive feedback loop on Figure 4.2 is the directed cycle between S1P, PDGFR, and SPHK1, while the cycle between TCR and CTLA4 is a negative feedback loop. Feed-forward loops are more abundant in the transcriptional regulatory and signaling networks of different organisms compared to randomized networks that keep each node's degree. They have been found to support several functions, such as filtering of noisy input signals, pulse generation, and response acceleration [9]. Positive feedback loops were found to support multistability while negative feedback loops could cause pulse generation or oscillations [10]. Examples that illustrate such behaviors will be presented in Section 4.7.

A signaling network, as all directed networks, is *strongly connected* if, for any two nodes in the network u and v, there is a directed path from u to v and another path from v to u. If a network is not strongly connected, it is informative to identify *strongly connected components* (or subgraphs) of the network. Having no strongly connected components (SCCs) indicates that the network has an acyclic structure (i.e., it does not contain feedback loops), while having a large SCC implies that the network has a central core. Signaling networks tend to have a strongly connected core of considerable size [11]. For example, the network in Figure 4.2 has a strongly connected component of 44 nodes, which represents 75% of all nodes. An SCC may have an in-component (nodes that can reach the SCC) and out-component (nodes that can be reached from the SCC). In biology, nodes in each of these subsets tend to have a common task. In signaling networks, the nodes of the in-component represent signals or their receptors and the nodes of the out-component are usually responsible for the transcription of target genes or for phenotypic changes [11]. The out-component of the T-LGL network in Figure 4.2 consists mainly of conceptual nodes that represent cell behaviors, such as apoptosis (the genetically determined process of cell destruction) or proliferation (the increase in the number of cells due to cell growth and division).

Software packages for network visualization and analysis include yEd Graph Editor, available from http://www.yworks.com/en/products/yfiles/yed/. Cytoscape[12], NetworkX [13], and Pajek [14].

Exercise 4.1. Consider the network depicted in Figure 4.3.

1. Is the network strongly connected? Explain your answer.
2. If the network is not strongly connected, identify its strongly connected components.
3. Does the network contain loops? If so, identify them.
4. Does the network contain cycles? If so, identify all cycles.
5. Are there any feed-forward loops? If so, identify them as positive, negative, or incoherent.
6. Are there any feedback loops? Identify them. Identify their sign as positive or negative. □

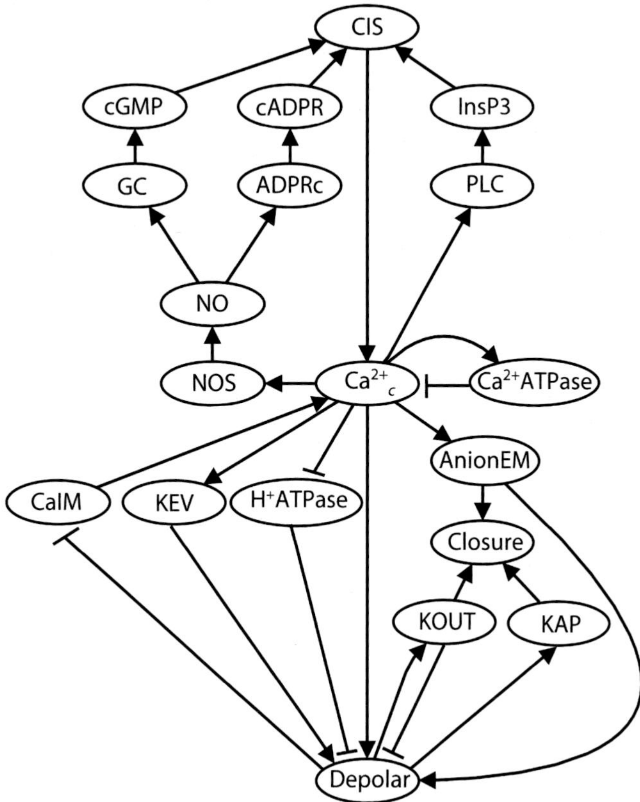

FIGURE 4.3 Figure for Exercise 4.1. The network is a part of a plant signal transduction network whose signal is the drought hormone abscisic acid and whose outcome is the closure of the stomata (microscopic pores on the leaves) [15, 16]. An arrowhead or a short perpendicular bar at the end of an edge indicates activation or inhibition, respectively. The full names of the nodes can be found in [15, 16]. Figure reprinted from Ref. [17] with permission from Elsevier.

4.3 DYNAMIC MODELING OF SIGNALING NETWORKS

Representation as a network of nodes connected pairwise by edges offers a coherent representation of a system of interacting biomolecules [7, 17, 18]. Going further, a *dynamic model* can describe how the abundances of the biomolecules in the network change over time due to their interactions. This is done by associating each node i of the signaling network with a variable x_i. Dynamic modeling approaches can be continuous or discrete according to the use of continuous or discrete variables. Continuous dynamic modeling [18, 19] describes the rate of change of each continuous variable x_i as a function of the other variables x_j in the signaling network. These models require the knowledge of mechanistic details for each interaction (e.g., the stoichiometric coefficients of the molecules that participate in a reaction and the kinetic rate functions) and their parameterization with rate constants. One also needs quantitative measurements of all the variables in the system in the initial condition and also in at least one stable state, to use for model validation. Continuous modeling is most feasible for well-characterized systems, where through decades of experimental work a sufficient amount of quantitative information has been gathered. Unfortunately, the state of the art for most systems is far from this mark: in many cases not all interactions have been mapped out, the detailed mechanisms are not known, there are no quantitative measurements of all the relevant variables, and kinetic parameter values are unknown and difficult to estimate. Continuous modeling is not practical for these types of systems.

As an alternative, discrete dynamic modeling such as Boolean network models [10, 20], multivalued logical models [21], and Petri nets [22] have been developed. These models use discrete variables that correspond to logic categories rather than quantitative values and describe the future value of a variable x_i (as opposed to its rate of change) as a function of the other variables x_j in the signaling network. Discrete dynamic models can be constructed from qualitative or relative measurements (e.g., whether a protein is more active in one condition compared to another), have no or very few kinetic parameters, and are able to provide a qualitative dynamic description of the system. They can be used to elucidate how perturbations may alter normal behavior and thus lead to testable predictions which are especially valuable in poorly understood large-scale systems. These approaches can be employed for systems with hundreds of components and have been used to model signal transduction networks in unicellular organisms, plants, animals, and humans (reviewed in Ref. [4]).

This chapter focuses on Boolean models. Construction of a Boolean model starts with a compilation of a list of components (nodes) and of the known interactions and regulatory relationships among these nodes, which will become the edges of the reconstructed interaction network. The model construction continues with determining Boolean functions which describe the regulation of each node based on the edges incident on the node and also using information from the literature. The collection of the resting or pre-stimulus states of the nodes will be used as an initial condition in the model. The model construction also includes a choice of how to represent the passing of time; as we'll see below, this choice has a subtle influence on certain outcomes of the model. Having chosen the transition functions, initial conditions, and the representation of time, running the model will provide a simulation of how the system evolves in time.

The model-indicated dynamic behaviors resulting from the simulations (e.g., long-term states) need to be compared with the available biological information on the behaviors of the system. If there are qualitative discrepancies that cast doubt on the model, the edges or Boolean functions of the model need to be rechecked and suitably revised. On the other hand, qualitative agreement between the model's results and biological knowledge increases our confidence in the model and allows its use to generate understanding and new predictions. For example, an often-used follow-up is a comprehensive analysis of the effects of node perturbations. We next describe the modeling process in detail.

4.4 THE REPRESENTATION OF NODE REGULATION IN BOOLEAN MODELS

The Boolean model of a signaling network associates each network node (i.e., gene, protein, molecule) i with a binary variable x_i which describes its expression level, concentration, or activity. The value $x_i = 1$ (ON) represents that component i is active or expressed, or has an above-threshold concentration; the value $x_i = 0$ (OFF) denotes that it is inactive or not expressed, or has a below-threshold concentration. The thresholds invoked in the definition of states do not need to be quantified, as long as it is known that a concentration level exists above which the component in question can effectively regulate its downstream targets. In Boolean models, the future state of node i, denoted by x_i^*, is determined based on a logic statement involving the current states of its regulators, i.e., $x_i^* = f_i$. This statement f_i, called a *Boolean transition function* (or a *Boolean rule*), represents the conditional dependency of the input (regulator) nodes in the regulation of the downstream (target) node. This function is usually expressed via the logic operators AND, OR, and NOT. For example, $f_4 = (x_1$ OR $x_2)$ AND (NOT x_3) is a Boolean function regulating the variable x_4. It indicates that x_4 will be ON when at least one of x_1 or x_2 is ON and simultaneously x_3 is OFF. When parentheses are used, as in this example, they determine the order of operations explicitly. Alternatively, the order of precedence of the logical operators may be used: NOT has the highest precedence, followed by AND, and then by OR, which has the lowest precedence. Any use of parentheses overrides the precedence rules. As an example, the rule for f_4 above can also be written as $f_4 = (x_1$ OR $x_2)$ AND NOT x_3.

A Boolean transition function can also be represented by a truth table, in which each row lists a possible combination of state values for the node's regulators and the associated output value of the function. Table 4.1 presents the truth tables for the Boolean functions corresponding to the operations NOT (third column), OR (fourth column), and AND (last column). The truth table of a Boolean function with k variables has 2^k rows and $k + 1$ columns (see Exercise 4.3).

Example 4.1. Consider the three-node signal transduction network depicted on Figure 4.4a. The source node A is the signal to the network, and both A and B positively regulate the sink node C. Let's identify transition functions compatible with this network.

TABLE 4.1 Truth Tables Illustrating the NOT, OR, and AND Operators

x_A	x_B	f_C = NOT x_A	$f_D = x_A$ OR x_B	$f_E = x_A$ AND x_B
0	0	1	0	0
0	1	1	1	0
1	0	0	1	0
1	1	0	1	1

The third column (f_C) indicates that the value of "NOT x_A" is the opposite (logic negation) of the value of x_A. The fourth column (f_D) indicates that "x_A OR x_B" is 1 whenever either input is 1. The last column (f_E) indicates that "x_A AND x_B" is 1 only when both inputs are 1.

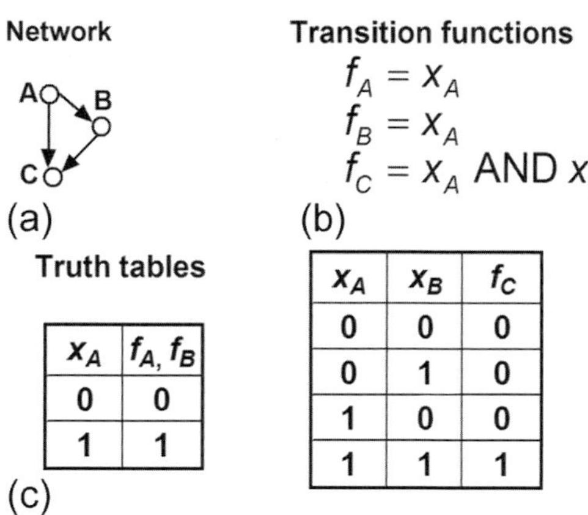

FIGURE 4.4 A Boolean model of a simple signal transduction network. (a) The signal transduction network. The edges with arrows represent positive effects. Note that the network does not uniquely determine the Boolean updating function for node *C*. (b) The Boolean transition functions in the model. The first transition function indicates that the state variable of node *A* does not change. The second transition function indicates that the state variable of node *B* follows the state of node *A* with a delay. The third transition function indicates that the condition for the ON state of node *C* is that both *A* and *B* are on. (c) The truth tables of the Boolean updating functions given in (b).

Because node *A* is the signal to the network, we are free to choose its transition function as long as it is independent of the other nodes. Let us assume that the state of node *A* stays constant, maintaining whatever value it started from. The corresponding transition function is $f_A = x_A$, and the equation governing the state of node *A* is $x_A^* = f_A = x_A$. Node *B* is positively regulated by node *A*, and is not regulated by anything else, thus its transition function is $f_B = x_A$. This indicates that the state of node *B* will follow the state of node *A* with a delay, $x_B^* = f_B = x_A$. Node *C* is positively regulated by node *A* and by node *B*. The network does not tell us how these two influences are cumulated. There are two choices: $f_C = x_A$ OR x_B, and $f_C = x_A$ AND x_B. The first function indicates that either *A* or *B* alone can successfully activate *C* (see column four of Table 4.1), while the second expresses the more stringent condition that both *A* and *B* need to be on simultaneously in order to activate *C* (see the last column of Table 4.1). Figure 4.4 indicates one of the two possible sets of transition functions, both as Boolean expressions (b) and as truth tables (c). □

Exercise 4.2. Construct the transition function and truth table for the networks in Figure 4.5. Consider both the AND and OR possibilities for the transition function of *C* for the network in Figure 4.5a and for the transition function of the node *B* for the network in Figure 4.5b. □

Exercise 4.3. Show that the truth table of a Boolean function with *k* variables has 2^k rows and $k + 1$ columns. *Hint*: Determine the number of different sequences of length *k* that can be formed from 0s and 1s. □

4.5 THE DYNAMICS OF BOOLEAN MODELS

In a Boolean model of a signal transduction network, time is usually discrete. This means that the model variables are updated only at fixed-time instances separated by a certain number of time steps and that no updates are being made between the time steps. As time is implicit in most Boolean models, one could think of a digital clock running in the background, with updates occurring only when the time on the clock changes. The time step between updates could vary from fractions of a second to hours, depending on the nature of the biological system [9]. Mathematically, the state of the system containing n nodes with associated state variables $x_1, x_2, \ldots, x_n$ at time t can be represented by a vector $(x_1(t), x_2(t), \ldots, x_n(t))$ with the ith element representing the state of node i at time t. By successively reevaluating each node's state while applying the corresponding transition function, the system's collective state evolves over time and eventually reaches a *steady state* (i.e., a state that remains unchanged over time) or a set of recurring states. These steady or recurring states are collectively referred to as *attractors*. Attractors that are not steady states are called *complex attractors*. For each attractor, its *basin of attraction* is comprised of all states that eventually lead to the attractor.

Exercise 4.4. Can you guess the attractor(s) of the Boolean model in Example 4.1? Consider the cases $x_A = 0$ and $x_A = 1$ separately. □

The transition functions of a Boolean model specify the rules for updating the network variables, but the order in which the updates are performed needs to be indicated separately. Various update schedules can be implemented via synchronous or asynchronous update algorithms. The *synchronous scheme* is the simplest update mode, wherein the states of all nodes are updated simultaneously according to the state of the system at the previous time step [20]. One significant disadvantage of this type of update is that it implicitly assumes that the timescales of all biological events in the system are similar and that the state transitions of components are synchronized. However, many systems include a mixture of biological events of different timescales (e.g., from fractions of seconds for protein-protein interactions to several minutes for transcription [9]), making the use of synchronous update inappropriate in those systems.

Asynchronous models aim to account for timescale diversity by updating the nodes in a nonsynchronous manner. There are *deterministic asynchronous* schemes with fixed individual timescales or fixed time delays. There also are *stochastic asynchronous* schemes wherein each node is updated with a certain probability, all nodes are updated according to a random sequence, or one randomly selected node is updated at a time step [23]. A parsimonious way to deal with diverse and unknown timescales is to use stochastic asynchronous update and do many simulations. We next present several examples that illustrate the different update algorithms.

Example 4.2. Assume that we have a network composed of nodes A and B. The edges do not matter in this example. Let's construct two deterministic and two stochastic updating schemes for this network.

In synchronous update, both nodes will be updated simultaneously at multiples of a time step, i.e., at time instances $1, 2, 3, \ldots, t$. If we are currently at time step t, the future state of a node means the state at time $t + 1$. Thus the state transitions of the two nodes will be $x_A{}^* = x_A(t+1) = f_A(t)$, $x_B{}^* = x_B(t+1) = f_B(t)$.

For another example of deterministic update, let us assume that node A can change state at multiples of a timescale t_A, while node B can change state at multiples of a timescale t_B. For simplicity let's designate the

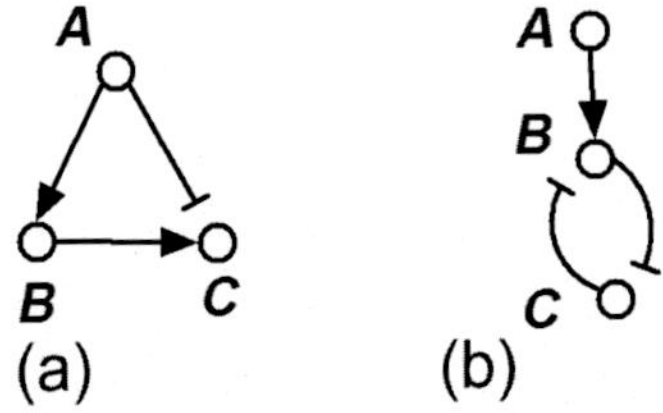

FIGURE 4.5 Figure for Exercise 4.2 and several of the follow-up exercises. Two simple signal transduction networks. For both networks, A is the source node (signal).

smaller timescale as the unit, and express the bigger timescale as an integer multiple of the smaller, for example, $t_A = 1$, $t_B = 2$. This means that node A will be updated at every time step while node B will be updated at even time steps. So the update scheme is A, A and B together, A, A and B, A, A and B....

A popular stochastic update, called *random order asynchronous update*, is performed as follows: (1) generate a permutation of the nodes and update them once in this order; (2) generate another permutation and update the nodes in the new order; (3) continue, by selecting a random permutation of the nodes at each step and updating the nodes in the order indicated by the permutation. The order of update in our example may look something like this: A, B; A, B; B, A; A, B; ...where the semicolons indicate the end of a time step, which here is interpreted as a round of update. Whenever node B is updated, it uses the state of node A obtained at its most recent update, which can be within the same round of update (such as in the first two rounds of our example). The same is true for node A. Notice the difference here in comparison with the synchronous update schedule where the update of B at time $t + 1$ always uses the value of A at time t.

Another frequently used stochastic update method is to update one randomly selected node at each time step. This method is called *general asynchronous update*. The order of update of our two nodes may look like this: B; A; A; A; B; B; ...When considering a long sequence of updates, it is possible that node A is updated more (or less) than node B using this method, while under random order asynchronous update they will be updated the same number of times. Nevertheless, because the probability of choosing each node is equal, on average they will be updated the same number of times. If we know that one node should be updated more frequently, we can choose unequal selection probabilities. □

A compact representation of all possible trajectories is visualized through the *state transition graph*, whose nodes are states of the system and whose edges denote the allowed transitions among the states according to the chosen updating scheme [23]. The attractors of the system can be determined from network analysis of the state transition graph. Fixed points will correspond to states that do not have any outgoing edges (transitions), only a loop. Each complex attractor forms a terminal strongly connected component of the state transition graph (i.e., a strongly connected component with an empty out-component).

Example 4.1. (continued). Consider again the signaling network and Boolean model in Figure 4.4. Let's represent the system's state as the triple $x_A\ x_B\ x_C$. Let us first determine the system's state transition graph in the absence of a signal ($x_A = 0$) when using synchronous update. We start from an initial state, let's say 011, and update each node's state using their corresponding transition functions. It is easiest to look up the function's output from the truth tables. Node A will keep its OFF ($x_A = 0$) state. The next state of node B is indicated by the first row of the truth table on the left, giving 0. The next state of node C can be looked up from the second row of the truth table on the right, yielding 0. Thus the next state of the system is 000. We have so far obtained the first edge of the state transition graph, from state 011 to state 000. Let's start from 000. Checking the first row of both truth tables, we find that the state remains 000. This state is thus a steady state (fixed point) of the system. Did you guess this state in Exercise 4.4?

We still have two states to consider as initial conditions. Let's start from 010. From the first row of the left truth table, the next state of node B is 0, and from the second row of the truth table on the right, the next state of node C is also 0. So the state 010 transitions to 000. Finally, starting from state 001, we find 000 as well. All initial conditions in which node A is OFF transition to the 000 steady state, thus the state transition graph has four nodes and four edges, starting from each state and all ending in 000, as shown in the left panel of Figure 4.6a. □

Exercise 4.5. Determine the state transition graph of the model in Figure 4.4 in the presence of a signal ($x_A = 1$) when using synchronous update. Compare with Figure 4.6a (right panel). □

Exercise 4.6. Determine the state transition graphs for the networks in Figure 4.5, assuming synchronous update. Consider both the AND and OR possibilities for the transition function of C for the network in panel (a) and for the transition function of B for the network in panel (b). Consider both the sustained absence ($x_A = 0$) and presence ($x_A = 1$) of node A. □

Example 4.1. (continued). Let us consider again the network from Figure 4.4 with the absence of signal ($x_A = 0$) but use general asynchronous update, when one node is updated at any given time step.

Because node A does not change state, its update does not need to be considered. But either node B or node C can be updated with equal probability, so the state transition graph will need to include both transitions. In general, the maximum number of transitions from any given state equals the number of nodes in the network.

Let's start with state 011 as we did for synchronous update. To update node B, we look up its next state from the first row of the truth table in Figure 4.4 on the left, and find that it is 0. The next state is thus 001, having the same state for node A (which does not change) and for node C (which was not updated). If we start from state 011 and update node C, its next state is 0, thus the next state of the system is 010. Thus state 011 has two successors, namely, 001 and 010, which is a markedly different result than the single successor, 000, which we found when using synchronous update (see Figure 4.6b, left panel, and compare with the state transition graph in Figure 4.6a, left panel). Indeed, the transition under synchronous update involved the state change of both node B and C, which is not possible under general asynchronous update.

The other two transitions we found for synchronous update involve a single state change, thus it is not a surprise to find that they are preserved. (Check this result.) When updating node B in state 001, or updating node C in state 010, the state remains unchanged, thus the transition is represented by a loop (Figure 4.6b, left panel). □

Exercise 4.7. Determine the state transition graph of the model in Figure 4.4 in the presence of signal ($x_A = 1$) when using general asynchronous update. Compare with Figure 4.6b (right panel). □

Figure 4.6 summarizes the state transition graph corresponding to synchronous update (a) and general asynchronous update (b). The asynchronous state transition graph has more edges, because there are up to twice as many distinct transitions when updating two nodes one by one instead of simultaneously. Most of the extra edges are loops and correspond to updates when a node's state is reevaluated but does not change. Nevertheless, the states that do not have any outgoing edges in addition to loops, i.e., the steady states 000 and 111, are identical for both types of update. Is this result generally true? Or do we expect any changes in the system's steady state if we switch from synchronous to asynchronous update? We will explore this question in the next section.

Exercise 4.8. Determine the state transition graphs for the networks in Figure 4.5 when using general asynchronous update. Consider both the AND and OR possibilities for the transition function of C for the

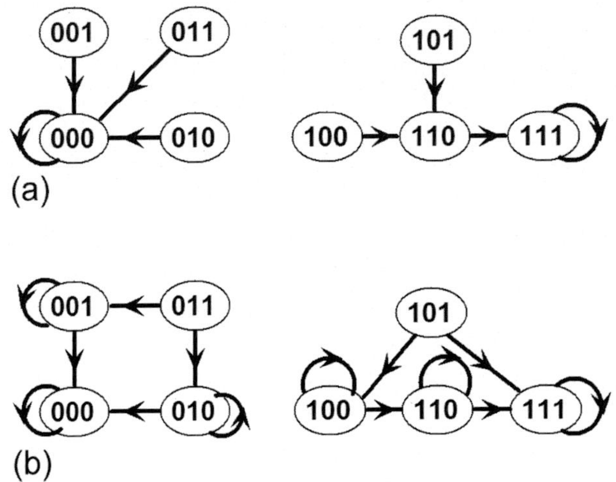

FIGURE 4.6 State transition graphs corresponding to the Boolean model presented in Figure 4.4. The symbols correspond to the states of the system, indicated in the order A, B, C; thus 000 represents $x_A = 0$, $x_B = 0$, $x_C = 0$. A directed edge between two states indicates the possibility of transition from the first state to the second by updating the nodes in the manner specified by the updating scheme. An edge that starts and ends at the same state (a loop) indicates that the state does not change during update. (a) The state transition graph corresponding to synchronous update, when all nodes are updated simultaneously. The two states that have loops are the fixed points of the system. (b) The state transition graph corresponding to updating one node at a time (general asynchronous update). While several states have loops, indicating that at least one of the nodes does not change state during update, only the two states that have no outgoing edges are fixed points of the system.

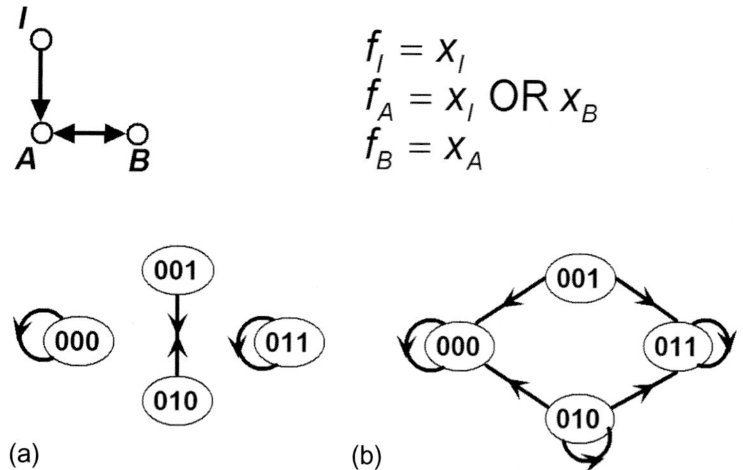

FIGURE 4.7 A simple signal transduction network composed of a source node I and two nodes, A and B, which form a mutual activation loop. It is assumed that the positive inputs from I and B are independently sufficient to activate node A. (a) The network's state transition graph corresponding to synchronous update, when the signal is OFF ($x_I = 0$). The states are specified in the node order I, A, B. (b) The state transition graph corresponding to general asynchronous update, when the signal is OFF ($x_I = 0$).

network in Figure 4.5a and for the transition function of B for the network in Figure 4.5b. Consider both the sustained absence ($x_A = 0$) and presence ($x_A = 1$) of node A. □

Exercise 4.9. For each of the cases considered in Exercises 4.6 and 4.8, compare the steady states obtained when using synchronous update and general asynchronous update. Are the steady states the same? □

Exercise 4.10. Consider the network in Figure 4.7. Determine the state transition graph of the network, using first synchronous update and then general asynchronous update. For the case $x_I = 0$, compare your graphs with those in Figure 4.7. □

4.6 ATTRACTOR ANALYSIS FOR STOCHASTIC ASYNCHRONOUS UPDATE

As we have already stated, attractors fall into two groups: fixed points (steady states), wherein the state of the system does not change, and complex attractors, wherein the system oscillates, regularly or irregularly, among a set of states. Fixed-point attractors usually correspond to the steady activation states of components or to cellular phenotypes in signaling networks. For example, the three fixed points of a Boolean model of a T-helper (Th) cell differentiation network [24] recapitulated the activation patterns of components observed in Th0, Th1, and Th2 cells, respectively. Complex attractors correspond to cyclic and oscillatory behaviors such as the cell cycle, circadian rhythms, or Ca^{2+} oscillations. The qualitative features of Boolean modeling make it suitable for analyzing the repertoire of behaviors in a large-scale system, such as its possible multistability (the existence of multiple stable steady states) [25], the initial conditions that lead to one attractor versus the other, and the activity changes of components following a perturbation. For example, the Boolean model of the T-cell apoptosis signaling network shown in Figure 4.2 indicated the existence of two fixed-point attractors, one corresponding to apoptosis and the other to a survival state which embodies the abnormal T-cell fate seen in the disease T-LGL leukemia [5, 6]. Analysis of the model also indicated the minimal perturbation that leads to the emergence of the abnormal survival state. Interestingly, this minimal perturbation involves only the overexpression of two external signals, IL15 and PDGF, suggesting that this disease does not necessarily have a genetic component. In this section we examine in what ways different update schedules may affect the attractors of a Boolean model. We consider fixed points (steady states) first.

Notice that arriving at a steady state in the state transition graph means that (after sufficiently many time steps) all of the system variables become constant with respect to time, i.e., $x_i^* = x_i$ for all nodes i. Thus, as the fixed points of a system are time independent, they are in fact the same for both synchronous and asynchronous updates. This also means that there is an alternative way to determine them. Because by definition $x_i^* = f_i$, the condition $x_i^* = x_i$ for all i leads to the set of equations $f_i = x_i$ for all nodes i, which we can solve. This is relatively easy if the network is small; for example, one can use elimination of variables. There are also advanced methodologies, such as transforming the Boolean equations into polynomial equations and solving them using Groebner bases (see Chapters 1 and 3 of [26]).

Exercise 4.11. Determine the steady states of Example 4.1 by solving the set of equations $f_A = x_A, f_B = x_B, f_C = x_C$. □

Example 4.3. Consider the hypothetical signal transduction network and its associated Boolean model in Figure 4.7. Let us compare the state transition graphs corresponding to synchronous update (a) and general asynchronous update (b) when $x_I = 0$. The determination of these state transition graphs was the subject of Exercise 4.10.

The synchronous state transition graph has two steady states (000 and 011) and a cyclic attractor formed by the states 001 and 010. The asynchronous state transition graph has the steady states 000 and 011, and no additional attractors. Indeed, synchronous models may exhibit limit cycles that are not present in stochastic asynchronous models. These limit cycles depend on two or more variables changing state at the same time. This synchronization among variables is not robust to stochasticity. The disappearance of the cyclic attractor also causes a change in the basins of attraction of the two steady states. Even if no attractors are lost when introducing asynchronicity, the choice of updating scheme can affect the attractors' basins. Because each state has a single successor under synchronous update, the attractors of a synchronous model have disjoint basins. But as stochastic asynchronous update allows several successors of a state, it is possible that a state is in the basin of two or more attractors. For example, in Figure 4.7b states 001 and 010 are in the basin of both steady states. □

Exercise 4.12. For each of the networks considered in Exercises 4.6 and 4.8,

1. Compare the complex attractors obtained when using synchronous update and general asynchronous update. Are they the same?
2. Find the basins of attraction for each of the steady states. □

Several software tools are available for Boolean dynamic modeling of biological systems. BooleanNet [27] can be used to simulate synchronous and random order asynchronous models and to determine their state transition graph. The R package BoolNet [28] provides attractor search and robustness analysis methods for synchronous, asynchronous, and probabilistic Boolean models. SimBoolNet, a plugin to the biological network analysis tool Cytoscape [12], determines state trajectories and attractors using sequential update (starting from the external signals). The software ADAM [29] performs analysis of synchronous Boolean models as examples of polynomial dynamical systems. For networks with less than 20 nodes ADAM can generate the full state transition graph. For larger networks it indicates the fixed points and limit cycles of user-specified length.

Exercise 4.13. Consider the network in Figure 4.8.

1. Determine the state transition graph and the attractors of the network using synchronous update. Find the basins of attraction for each attractor.
2. Repeat number 1, now using general asynchronous update of the node states. □

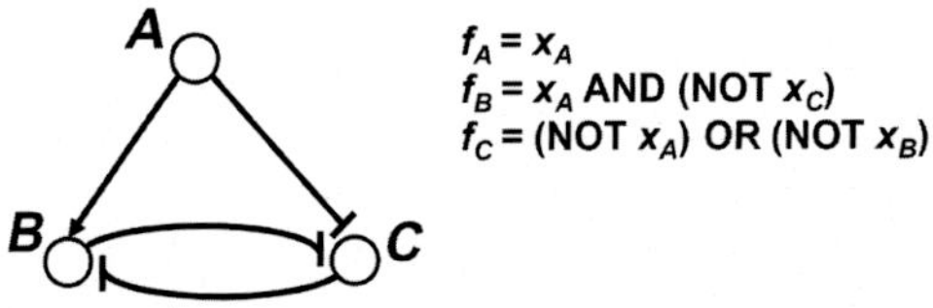

FIGURE 4.8 A simple three-node network for Exercise 4.13.

4.7 BOOLEAN MODELS CAPTURE CHARACTERISTIC DYNAMIC BEHAVIOR

Having now introduced the basic characteristics and properties of Boolean networks, it is important to ask if they can be used as realistic qualitative approximations of signal transduction dynamics in biology. In this section we give examples that demonstrate the ability of Boolean models to capture complex dynamic behavior such as excitation, adaptation, and multistability, which are common in signaling networks. For this section we assume that the reader has basic familiarity with continuous models described by ordinary differential equations. For a detailed introduction to the modeling of biochemical reactions with differential equations see Chapter 2, Section 3, of [26].

As we mentioned in Section 4.2, positive feedback loops support multistability [10], coherent feed-forward loops support the filtering of noisy input signals, and incoherent feed-forward loops support excitation-adaptation behavior [9, 25]. The review article [25] gives examples of continuous dynamic models exhibiting a transient excitation-adaptation behavior based on an incoherent feed-forward loop (Figure 4.9, top row) and a bistable response based on a positive feedback loop (Figure 4.9, bottom row). Let's see if Boolean models based on the same network motifs can qualitatively reproduce these behaviors.

Excitation-adaptation behavior is frequently observed in chemotaxis, which means cells' motion toward a chemical attractant, and can be based on a negative feedback loop [30] or an incoherent feed-forward loop [31].

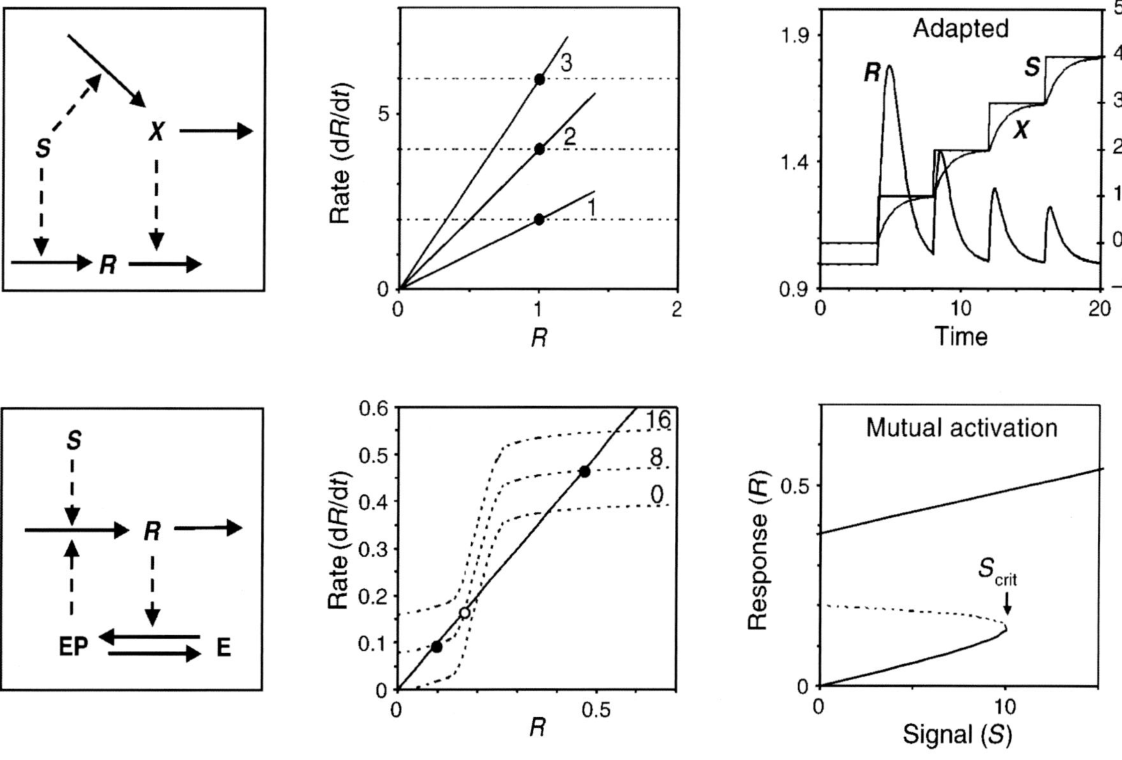

FIGURE 4.9 In the networks (left column), solid edges represent mass flow, such as synthesis or degradation of a protein, and dashed edges represent regulation of synthesis or of degradation. The source node of the networks represents the signal, and one node is designated as the response. The top row represents an example of feed-forward loop-based excitation-adaptation behavior, with S denoting the strength of the signal and X and R denoting the concentrations of two proteins. In the bottom network, the cycle between the EP and E nodes represents a phosphorylation-dephosphorylation cycle in which the total protein concentration is constant. This cycle is described by Michaelis-Menten kinetics, while the rest of the processes are assumed to follow mass action kinetics. The diagrams in the middle column indicate the absolute value of the rate of the synthesis and degradation of protein R as a function of the concentration of R. The diagram in the top right indicates the time course of the concentration of protein R when the value of the signal is increasing in a step-wise manner. The diagram in the bottom right indicates the steady state concentration of protein R as a function of the value of the signal. Figure reprinted from Ref. [26] with permission from Elsevier.

An example of feed-forward loop-based excitation-adaptation behavior is the model shown in the top row of Figure 4.9. In a continuous model, such processes are usually modeled by differential equations describing the dynamics of the concentrations of the biomolecules that make up the system. These models are continuous with respect to both time and the values of the model variables, and the model equations describe the rate of change for each of the variables as a function of all of them. The equations describing the continuous model for our example are

$$\frac{dR}{dt} = k_1 S - k_2 XR$$

$$\frac{dX}{dt} = k_3 S - k_4 X$$

where S is the value of the signal, X is the concentration of protein X, R is the concentration of the output protein R, and k_i, $i = 1 \ldots 4$ are rate constants. The analysis of this model, discussed in [25], yields that the steady-state (time-independent) concentration of R does not depend on the value of the signal. When the value of the signal undergoes step changes, the concentration of R increases transiently, peaks, then decreases to the signal-independent steady state (see top rightmost panel in Figure 4.9).

Example 4.4. Let's construct a Boolean model of the incoherent feed-forward loop of Figure 4.9 (top left). We start constructing the network of interactions by redefining the edges such that they represent regulatory relationships among the three nodes, S, X, and R. Because S catalyzes the synthesis of R, there is a positive relationship between S and R, indicated by a directed and positive edge starting from S and ending in R (Figure 4.10). Similarly, there needs to be a positive edge starting from S and ending in X. X catalyzes the degradation of R; thus the relationship between X and R is negative, indicated by a negative edge starting from X and ending in R. The uncatalyzed (free) degradation of X and R may each be represented as a negative loop at the respective node, but such degradation is usually left implicit when showing the networks that underlie Boolean models. The reason for this is that decay is implicitly incorporated as a default in all transition rules that depend only on the node's regulators. For example, $f_R = x_S$ already implements that x_R decays to OFF if x_S is OFF, even if x_R was ON before. The absence of decay, on the other hand, should be incorporated explicitly by including the node's own state variable in the transition rule, $f_R = x_S$ OR x_R. This case would be shown in the network by adding a positive loop to R.

Figure 4.10 illustrates the network underlying the Boolean model. We can assume that x_S stays constant ($f_S = x_S$), the Boolean rule of X is unambiguous ($f_X = x_S$), and the transition function of R closest to the continuous model is $f_R = x_S$ AND NOT x_X. The latter holds because S and X affect R simultaneously (that is, the concentration of R is the result of two simultaneously ongoing processes: the increase of the concentration due to the stimulus and the decrease of the concentration due to the catalyzed degradation of R by X). Let's assume that X and R have similar timescales, and use synchronous update. □

Exercise 4.14. Determine the state transition graph for Example 4.4 when using synchronous update. Compare with Figure 4.10b. Does the steady state of R depend on the state of the signal S? □

Now let's reproduce a step-wise increase in the signal variable x_S. The only such choice in a Boolean model is from $x_S = 0$ to $x_S = 1$. Starting with $x_S = 0$ and an arbitrary state for X and R, the system goes into the steady state 000 in one step. Let's now set $x_S = 1$, leading to the state 100 (see transition shown with dashed lines in Figure 4.10b). The state 100 is not an attractor, so the system's state will change into 111, wherein $R = 1$. Thus the step change in x_S drove x_R to change from 0 to 1 (excitation for R). The next state is 110, which is a steady state of the system. In this steady state, x_R is 0 (adaptation for R). Thus the Boolean model qualitatively reproduces the excitation-adaptation behavior: the change in x_S drove a transient excitation of x_R, but the steady state value of x_R was the same for both values of x_S. □

Exercise 4.15. Consider the steady state 110 in Figure 4.10b and implement a step change in S from 1 to 0. Is there an excitation (state change) in R? □

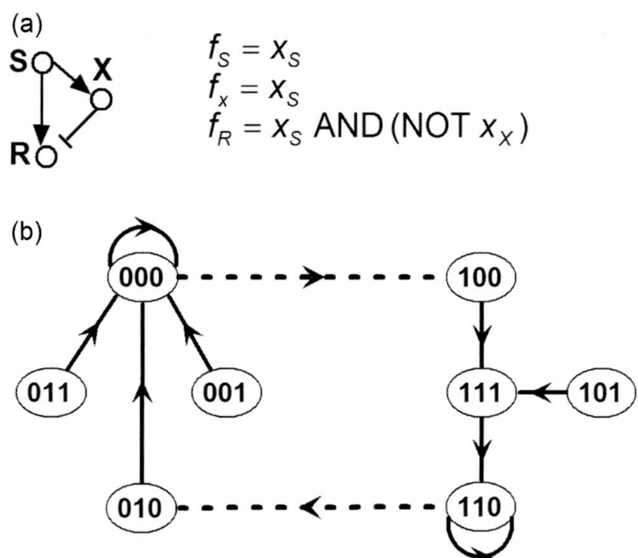

FIGURE 4.10 The Boolean correspondent of the model in the top row of Figure 4.9. (a) The network and Boolean transition functions. (b) The state transition graph of the model. The node states are shown in the order *S*, *X*, *R*. The transitions shown as dashed lines correspond to an externally set change in the value of the signal variable x_S. The path that starts from the 000 steady state and ends in the 110 steady state qualitatively reproduces the excitation-adaptation behavior of x_R seen in the continuous model.

Exercise 4.16. Consider the network in Figure 4.5a.

1. Does it have a commonality with the network in Figure 4.10? Explain.
2. Consider the transition function $f_C = x_B$ OR (NOT x_A) for node *C*. Using the synchronous state transition graphs calculated in Exercise 4.6, determine the trajectory of the system after x_A undergoes a step increase from state 001. Is there an excitation-adaptation behavior in x_C? □

Multistability is a phenomenon that arises often in physics, biology, and chemistry. Simply put, multistability is the ability of a system to achieve multiple steady states under the same external conditions. When there are two such states, we talk about *bistability*. In biology, bistability plays a key role in many fundamental processes such as cell division, differentiation, gene expression, cancer onset, and apoptosis. The example in [25] reproduced in the bottom row of Figure 4.9 illustrates that a signal-driven positive feedback loop can lead to bistability. The equation for the concentration of protein *R* in this model is

$$\frac{\mathrm{d}R}{\mathrm{d}t} = k_0 EP\,(R) + k_1 S - k_2 P$$

where *EP*(*R*) is a sigmoidal function shown as the lowest dashed curve in the bottom middle of Figure 4.9, and k_i, $i = 0 \ldots 2$ are rate constants. This model leads to an irreversible switch from the low-value steady state of *R* to the high-value one at a critical value of the signal (see bottom right panel of Figure 4.9). This leads to a history-dependent behavior (*hysteresis*): when the signal is gradually increased from zero, the steady state concentration of *R* increases on the curve corresponding to the lower-value stable steady state, then switches to the higher curve when *S* goes beyond the critical value. If we now gradually decrease the signal back to zero, the steady state value of *R* stays on the upper curve.

Example 4.5. Let's see if a Boolean model can recapitulate this memory-dependent behavior. First, note that the sigmoidal nature of *EP* as a function of *R* lends itself easily to a Boolean approximation. We could pick a value R_T (*T* for threshold) such that the value of the function $EP(R_T)$ is about one-half the size of the "jump." We'll consider *EP*(*R*) to be 0 for values of $R < R_T$ and 1 for values of $R \geq R_T$. With this, we can think of *EP* as a node in the Boolean network that is influenced by *R* and, in turn, it influences *R*. Because the function *EP*(*R*) is increasing, the node *R* influences the node *EP* in a positive way. In Figure 4.7 and associated Exercise 4.10 we have seen a Boolean model of a network having an input node (signal) and a positive feedback loop. Inspecting

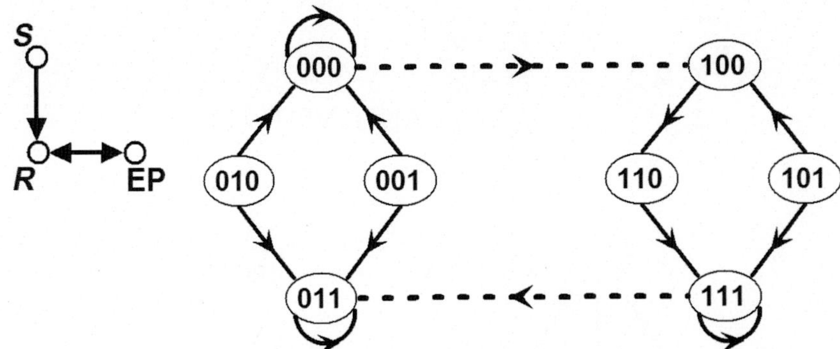

FIGURE 4.11 The Boolean correspondent of the model in the bottom row of Figure 4.9. The transitions shown as dashed lines correspond to an externally set change in the value of the signal variable x_S. For simplicity the loops that correspond to single node updates that do not change the node's state are not shown in the state transition graph, only loops that correspond to fixed points. The trajectory from the state 000 through 100, 110, 111, and 011 qualitatively reproduces the hysteresis of the continuous model. The order of variables is *S*, *R*, *EP*.

the regulation of *R* in the continuous model, we see that its synthesis is catalyzed independently by *S* and *EP*, thus an OR rule is the appropriate choice. Thus the model in Figure 4.7, with a suitable renaming of the node names, corresponds to the continuous model (see Figure 4.11). The transition functions are

$$f_S = x_S$$
$$f_R = x_S \quad \text{OR} \quad x_{EP}$$
$$f_{EP} = x_R$$

Let's consider general asynchronous update (the reader is encouraged to show as an exercise that using synchronous update gives similar results). As we have seen in Exercise 4.10, this model yields two fixed-point attractors when the signal is OFF, and a single attractor when the signal is ON. The continuous model also had two stable steady states for low values of the signal, and only one (the higher-value one) for signal levels above the critical level. Now consider that the system is initially in the steady state 000, and let's increase x_S to 1. The state 100 is not a steady state, and the system converges into the steady state 111. Thus the steady state value of *R* has switched from 0 to 1. Now let's decrease x_S to 0. The state 011 is a steady state, and the system remains there. Thus the steady state value of *R* did not go back to 0. We can conclude that the Boolean model qualitatively reproduces the hysteresis of the continuous model. □

Exercise 4.17. Consider the network in Figure 4.5b.

1. Does this have a commonality with the network in Figure 4.11? Explain.
2. Consider the transition function $f_B = x_A$ OR (NOT x_C) for node *B*. Using the general asynchronous state transition graphs calculated in Exercise 4.8, determine the trajectory of the system when x_A is switched to 1 from steady state 001, then switched to 0 in steady state 110. Does this system exhibit hysteresis? □

4.8 HOW TO DEAL WITH INCOMPLETE INFORMATION WHEN CONSTRUCTING THE MODEL

In the discussion so far, we have assumed that the entire signaling network is well known and well understood, with no knowledge gaps regarding its structure and mechanisms of interaction. In reality, however, there is often a need to deal with limited or incomplete experimental information while building a model. The following situations are typical: (1) not all links in network topology may be known with certainty; (2) when it is known that two or more nodes influence a node, the exact nature of the influence may still be unclear; (3) the initial condition for the system, which, ideally, should correspond to a relevant biological state, may not be known *a priori*; and (4) when detailed information on the biomolecular kinetics of the signal transduction is lacking, deciding on an update algorithm may be challenging. We now briefly address each of these cases and outline some possible remedies.

4.8.1 Dealing with Gaps in Network Construction

There are two major types of causal experimental evidence from which information on edges of a network can be extracted: physical or biochemical evidence indicating direct interaction between two components, and evidence of the effect of the genetic mutation or pharmacological inhibition of a particular component on another component. The latter evidence indicates a causal relationship between the two components, which may be due to a direct interaction or to a relationship mediated by other components. The integration of the indirect causal evidence is often challenging, as each such apparent pair-wise relationship may in fact reflect a set of adjacent edges (a path) in the network, and it may involve other known or unknown nodes. In some cases, evidence from multiple experiments yields composite causal relationships, which then need to be broken down to component-to-component relationships, depending on the concrete situation.

Example 4.6. Consider a hypothetical signal transduction network with input node I and output node O which includes two known mediators, $M1$ and $M2$, and an unknown number of so-far unidentified mediators. Assume that experimental inhibition of mediator $M1$ has led to the conclusion that $M1$ is a positive regulator of the signal transduction process. This can be schematically represented by drawing an edge between I and O (to stand for the whole signal transduction process) and also drawing an edge that starts from $M1$ and points to the edge between I and O. A similar experiment has led to the conclusion that $M2$ is also a positive regulator of the signal transduction process, thus we can draw an edge that starts from $M2$ and points to the edge between I and O. A third experiment has indicated that the up-regulation of I leads to the up-regulation of $M1$. This is represented by a positive edge directed from I to $M1$. A fourth experiment has led to an edge starting from $M1$ and ending in $M2$. Figure 4.12a shows the resulting network, which is not yet in the form of a graph, as there are edges pointing to edges.

Our goal now is to find the most parsimonious network graph consistent with the combined experimental evidence reflected in Figure 4.12a. To transform this into a graph, let us interpret the edge that starts at $M1$ and points to the $I \to O$ edge as $M1$ activating an unknown node situated between I and O. Let's represent this unknown node with a black dot. Similarly, we can interpret the edge that starts at $M2$ and points to the $I \to O$ edge as $M2$ activating a second unknown node situated between I and O (Figure 4.12b).

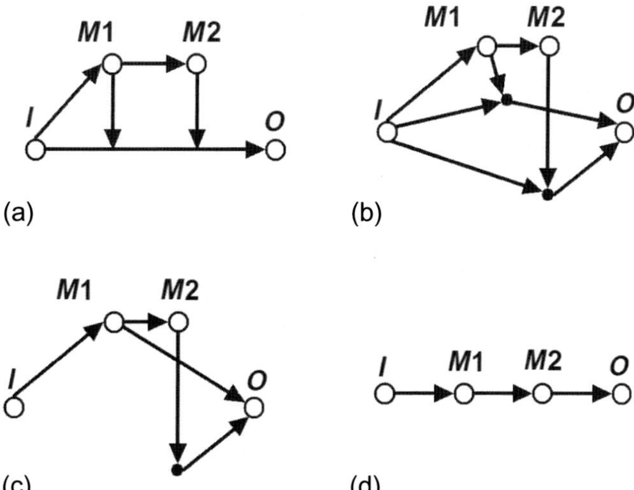

FIGURE 4.12 Illustration of the interpretation and simplification of causal information in order to find the most parsimonious signal transduction network. The interpretation involves the addition of unknown mediators, here shown as black dots. The simplification steps include transitive reduction and the collapsing of unknown mediator nodes.

Because the edges of this network correspond to causal effects but not to actual interactions, some of them are redundant as longer paths also express the same causal effect. For example, the edge between I and the upper unknown node (upper black dot) is explained by the two-edge path mediated by $M1$. Similarly, the edge between I and the lower black dot is explained by the three-edge path mediated by $M1$ and $M2$. These redundant edges can be deleted; this process is called *transitive reduction*. The upper black dot now has one incoming edge (from $M1$) and one outgoing edge (to O), thus it does not add any new causal information. For this reason, it can be eliminated by merging it with $M1$ (or alternatively with O), leading to a direct edge between $M1$ and O. Figure 4.12c shows the current incarnation of the network. Notice that we can now do another step of transitive reduction by deleting the edge between $M1$ and O, because it is explained by the $M1 \rightarrow M2 \rightarrow \bullet \rightarrow O$ path. Finally, we can compress the black dot between $M2$ and O, yielding the linear network shown in Figure 4.12c. In summary, the most parsimonious explanation of the four causal relationships is a linear path between I and O in which $M1$ is first and $M2$ is second. □

The approach illustrated in Example 4.6 was used to construct a model of drought-induced signal transduction in plants [15], specifically, the closure of stomata (microscopic pores on the leaves) in response to the drought hormone abscisic acid. We have seen a part of this network in Figure 4.3. Li et al. [15] collected more than 140 causal relationships derived from more than 50 literature citations on the regulation of stomatal closure by abscisic acid. The majority of these relationships were of type "C promotes the process (A promotes B)." The network resulting from the synthesis, interpretation, and simplification of these relationships had 54 nodes and 92 edges. The method was later formalized by DasGupta et al. [32, 33] and implemented in the software NET-SYNTHESIS. The software and its documentation can be downloaded from http://www2.cs.uic.edu/~dasgupta/network.synthesis/. The best use of this software is in iteration with additional literature search until the most appropriate network representation of the available experimental observations is found. This software can be used to construct the most parsimonious network consistent with a set of causal experimental evidence or to simplify an existing directed network in such a way that the causal relationships are preserved.

Exercise 4.18. Consider the list of causal evidence shown in Table 4.2. Use the software NET-SYNTHESIS to construct the most parsimonious signal transduction network. □

Exercise 4.19. Consider the network specified by the list of edges in Table 4.3. Use the software NET-SYNTHESIS to simplify this network by designating the nodes TCR, PDGFR, NFKB, and Caspase as pseudo-nodes and merging pseudo-nodes with regular nodes. An easy way to designate a node as pseudo-node in NET-SYNTHESIS is to precede its name by * (e.g., *TCR), either in the input file or by right-clicking on the node name in the displayed network. □

4.8.2 Dealing with Gaps in Transition Functions

We have already noted several times that for nodes with multiple regulators the knowledge of the incoming edges (positive and negative regulators) does not uniquely determine the dependency relationships among the node states. Thus, even complete knowledge of the networks does not by itself contain enough information to determine the transition rules for the nodes with multiple incoming edges. Assume, as an example, we know that two nodes A and B regulate a third node C. This could be an AND regulation (that is, both A and B would need to be ON to turn C ON) or it could be an OR relationship (when it would be enough for either one of A and B to be ON to turn C ON). Thus, additional information is needed.

One way to deal with this problem is to *knock out* one of the regulators, A or B, then examine the effect on C. In genetics, "gene knockout" is a technique used to make a gene inoperative. The term knockout here is used in the same sense: knockout of node A means setting and maintaining $x_A = 0$. If C remains permanently OFF after knockout of A, that would mean that both A and B are needed to turn C ON (thus, the transition function of C would be $f_C = x_A$ AND x_C); if not, we could conclude that the activation of C requires A or B, corresponding to the transition function $f_C = x_A$ OR x_C. In case of more than two regulators, the process also begins with examining the effect of knockout of one of the regulators on the state of the target node. If the information is still insufficient, several versions should be constructed and compared.

TABLE 4.2 A List of Causal Evidence Representative of What Could Be Synthesized from the Experimental Literature

Source Node	Causal Effect	Target Node or Edge	Direct Interaction?
ABA	Activates	InsPK	No
ABA	Activates	NO	No
ABA	Activates	PLD	No
NO	Activates	CIS	No
PA	Activates	ROS	No
ROS	Activates	CaIM	No
Ca^{2+}_{C}	Activates	AnionEM	No
Ca^{2+}_{C}	Activates	NO	No
PLD	Activates	PA	Yes
CIS	Activates	$Ca^{2+}{}_{C}$	Yes
CaIM	Activates	$Ca^{2+}{}_{C}$	Yes
AnionEM	Activates	Closure	Yes
KOUT	Activates	Closure	Yes
InsPK	Activates	ABA → CIS	No
Ca^{2+}_{ca}	Activates	NO → AnionEM	No
CaIM	Activates	ABA → KOUT	No

This list is derived from the work in [16] and has the same node names, but it is much simpler than the original.

Example 4.7. Consider the network in Figure 4.8, and assume the Boolean transition functions are unknown. Because A is the signal to the network, its transition function is clear: $f_A = x_A$. Inspecting the figure, the transition function for node B depends on x_A and on (NOT x_C), while the transition function of node C depends on (NOT x_A) and on (NOT x_B). Thus the possibilities are:

$$
\begin{aligned}
f_B &= x_A \text{ OR } (\text{NOT } x_C) \\
f'_B &= x_A \text{ AND } (\text{NOT } x_C)\,; \\
f_C &= (\text{NOT } x_A) \text{ OR } (\text{NOT } x_B) \\
f'_C &= (\text{NOT } x_A) \text{ AND } (\text{NOT } x_B)
\end{aligned}
$$

Let's now imagine that we search the literature and find evidence for a steady state in which $x_A = 1, x_B = 1$ and $x_C = 0$. Does this information eliminate any of the candidate transfer functions?

Plugging in those state values we obtain $f_B = 1, f'_B = 1, f_C = 0, f'_C = 0$. Both transfer function variants give the same result, which is in agreement with the node's steady states; thus, this information did not help limit the possibilities.

Imagine now that after more search we find that in the case when $x_A = 1$ and simultaneously node B is knocked out (i.e., x_B is set to 0), the steady state of C is 0. In this case $f_C = 1$ and $f'_C = 0$, and only the latter is consistent with the observed $x_C = 0$. Thus, we should conclude that the transfer function of node C is $f'_C =$ (NOT x_A) AND (NOT x_B).

Finally, let's assume that a third observation indicates that in the case when $x_A = 0$ and simultaneously node C is knocked out, the steady state of B is 1. In this case $f_B = 1$ and $f'_B = 0$, of which only the former is consistent with the observed steady state. Thus, we conclude that the transfer function of node B is $f_B = x_A$ OR (NOT x_C). □

TABLE 4.3 The List of Edges in a Two-Signal, One Output Signal Transduction Network Used in Exercise 4.19

Source Node	Causal Effect	Target Node
Stimuli	Activates	TCR
TCR	Activates	RAS
PDGF	Activates	PDGFR
S1P	Activates	PDGFR
PDGFR	Activates	S1P
RAS	Activates	FAS
S1P	Inhibits	FAS
FAS	Inhibits	S1P
FAS	Inhibits	NFKB
FAS	Activates	Caspase
NFKB	Inhibits	Caspase
Caspase	Activates	Apoptosis

This network is derived from the T cell apoptosis signaling network displayed in Figure 4.2, and has the same node names, but it is much simpler than the original.

The approach illustrated in Example 4.8 was used in [34] to construct and refine Boolean models from an initial signal transduction model by calibrating the model against measurements of protein abundance or activity. The initial signal transduction model was first simplified to collapse nodes for which no measurements were available, in a way similar to the collapsing of pseudo-nodes we saw in Example 4.6 and Exercise 4.19. Then an ensemble of models was generated from every possible transition function consistent with the network. Finally, those models were evaluated by comparing their steady states with the experimental observations and the most consistent and also most parsimonious model was selected. Application of this method to the signal transduction network that mediates early signaling downstream of seven cytokine and growth factor receptors in human liver cells, using measurements of sixteen proteins in this network, led to significant refinement (mostly edge deletion, but also a few additions) of an original database-derived network. The final network and Boolean model was validated by follow-up experimental measurements. This method is instantiated in the freely available software package CellNOpt [35].

Exercise 4.20. Consider the networks in Figure 4.5.

1. Assume that for the network in Figure 4.5a, an experimental observation is consistent with the steady state $x_A = x_B = x_C = 1$. What transition function does this imply for node C?
2. Assume that for the network in Figure 4.5b, an experimental observation is consistent with the steady state $x_A = x_C = 1$ and $x_B = 0$. What transition function does this imply for node B? □

4.8.3 Dealing with Gaps in Initial Condition

Ideally, the model's starting state should be the biologically relevant resting or pre-stimulus state, if it is known *a priori*. If the available information is insufficient, one can enumerate or, if that is difficult, sample a large number of initial conditions wherein certain nodes are in a known state while the state of others can vary. A large number of replicate simulations should be done, and the results need to be summarized over these replicate simulations. For example, one calculates the fraction of realizations of a certain attractor. We can think of these replicate simulations as a population of cells which differ in their pre-stimulus states, and the fraction of realizations of an attractor can be interpreted as the probability that the system attains the corresponding cellular phenotype. This approach was used, e.g., by Li et al. in [15] in the context of constructing a model of abscisic acid-induced closure. Because many of the nodes of this network are also involved in the response to other signals, for example, light and atmospheric CO_2, it is difficult to estimate their state prior to receiving the abscisic acid signal. Thus, the initial state of 38 intermediary (non-source, non-sink) nodes was randomly chosen, and 10,000 replicate simulations were performed. The authors studied the fraction of simulations in which the output node Closure was ON, and found that it reached 1 after eight updates of all the nodes (using random order asynchronous update). Thus, they were able to conclude that the initial state of the intermediary nodes did not affect the steady state of the node Closure.

Exercise 4.21. What initial conditions should be considered for Example 4.1 if we are interested in the system's response to a sustained signal? □

Exercise 4.22. How many initial states should be considered for an N-node network if we have no information on the actual initial state? □

4.8.4 Dealing with Gaps in Timing Information

We can choose an updating scheme that is most realistic for the biological system of interest, or compare different schemes with the same system. In cases where there is no information to guide the choice of update scheme, updating one node at a time (general asynchronous update) is the most parsimonious choice because its results are also representative of the random order update [16]. The biological system of interest for the study [16] was the signal transduction network by which plants respond to the drought hormone abscisic acid, first modeled in [15]. The study found that the state of the majority of the nodes, including Closure, stabilized regardless of the updating scheme used. A subset of nodes regulated by cytosolic Ca^{2+} had one or two different behaviors: fluctuations that eventually decayed, leading to a steady state in which cytosolic Ca^{2+} was OFF, and sustained oscillations if and only if strict relationships among the timing of Ca^{2+} production and decay were satisfied. There is evidence in the experimental literature for abscisic acid induced oscillations in Ca^{2+}, but not enough to establish whether they are sustained or not. Likewise, the timing or kinetics of Ca^{2+} production and decay are not known. However, the model suggested that a transient increase in Ca^{2+} was sufficient for a successful closure response and sustained oscillations were not necessary. We encourage the reader to examine this article as a way to develop further understanding of the effect that different updating schemes may have on the long-term behavior of Boolean models.

4.9 GENERATE NOVEL PREDICTIONS WITH THE MODEL

A Boolean model can be used to analyze the changes in the system's attractor repertoire in the case of system perturbations. Knockout of a component can be simulated by fixing the corresponding node in the OFF state; constitutive expression can be simulated by fixing the node's state as ON. Transient perturbations can also be studied by implementing temporary (reversible) changes to the node's states. The model can predict the changes in the attractors of the system and their basins of attraction and identify the perturbations that lead to dramatic changes. This way, perturbation analysis can identify key components that are essential to phenotype traits [5, 6, 15]. If the studied phenotype corresponds to a disease, the identified essential components are candidate targets for therapeutic interventions.

Exercise 4.23. Consider Example 4.1 (shown in Figure 4.4) when node B is knocked out (i.e., x_B is set to 0). What is the relevant state space now? Construct the state transition graph corresponding to synchronous update and general asynchronous update. Compare with Figure 4.13. □

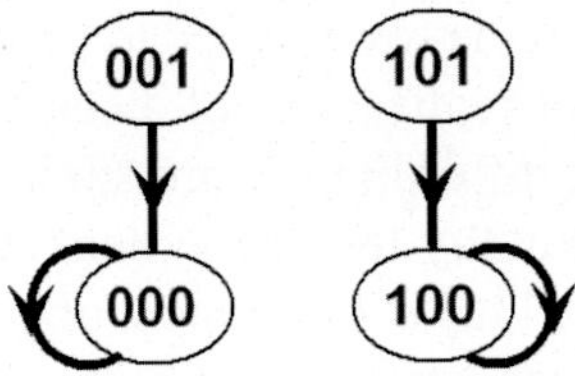

FIGURE 4.13 State transition graph for Example 4.1 (the network in Figure 4.4) when node *B* is knocked out. Because nodes *A* and *B* are not updated (*A* is a sustained signal and *B* is knocked out), synchronous and general asynchronous update are equivalent in this case.

Exercise 4.24. Compare Figure 4.6 with Figure 4.13. How did the steady states of the system change due to the knockout of node *B*? □

Exercise 4.25. Determine the state transition graph of the network in Example 4.1 (Figure 4.4) for synchronous update when node *C* is knocked out. □

Exercise 4.26. Determine the state transition graph of the network in Figure 4.4 for general asynchronous update when node *C* is knocked out. □

Exercise 4.27. Compare Figure 4.6 to your results in Exercises 4.25 and 4.26. How did the steady states of the system change due to the knockout of node *C*? □

An example of perturbation analysis is the study of nodes whose knockout can impair abscisic acid-induced closure [15, 16]. Using stochastic asynchronous update, the unperturbed system had a single fixed-point attractor which included the ON state of the node Closure. Systematic study of each intermediary node's knockout led to the identification of three perturbation categories: knockouts that led to the normal steady state (which represented 75% of all knockouts), knockouts leading to a steady state in which Closure was OFF (22.5% of all knockouts), and a single knockout, which represents the clamping of the cytosolic pH level, leading to a complex attractor in which Closure fluctuated between ON and OFF. Thus, one could conclude that the signal transduction process was robust to the large majority of perturbations, but also sensitive to the impairment of a few key nodes. These key nodes should be studied further to establish the ways in which their perturbations could be prevented. □

Exercise 4.28. Consider the model of Example 4.3. As shown in Figure 4.7b, the system has two steady states, 000 and 011, when the signal is OFF ($x_1 = 0$). Under general asynchronous update, both steady states are reachable from the initial conditions 001 and 010. Let's assume that steady state 000 is undesirable. Can you find a state manipulation (fixing the state of a node) such that state 000 becomes unreachable? □

As we have seen in Exercise 4.17, mutual inhibition among two nodes can lead to the same behavior as mutual activation between the same nodes. Mutual inhibition between two groups of nodes is in fact a key feature of the T-cell apoptosis signaling network (see Exercise 4.19). Indeed, as in Example 4.3, the Boolean model of this network has two steady states, one corresponding to apoptosis and one corresponding to an abnormal cell fate seen in T-LGL leukemia [5, 6]. A state manipulation that could potentially eliminate this latter steady state is to fix a node's state in the opposite state that it stabilizes in the T-LGL steady state. Considering this manipulation for each intermediary node in the network led to the identification of nineteen potential therapeutic targets for the disease [6]. More than half of these manipulations were supported by available experimental data or by follow-up experiments, and the rest can guide future experiments.

4.10 BOOLEAN RULE-BASED STRUCTURAL ANALYSIS OF CELLULAR NETWORKS

Analysis of all relevant dynamic trajectories of a system that is bigger than the simple examples we have considered here is complex and time consuming. The good news is that sometimes important conclusions can be drawn without dynamic simulations, based on graph-theoretic analysis alone. A first step in this direction is to resolve the ambiguity pertaining to nodes with multiple regulators by incorporating the Boolean rules into the

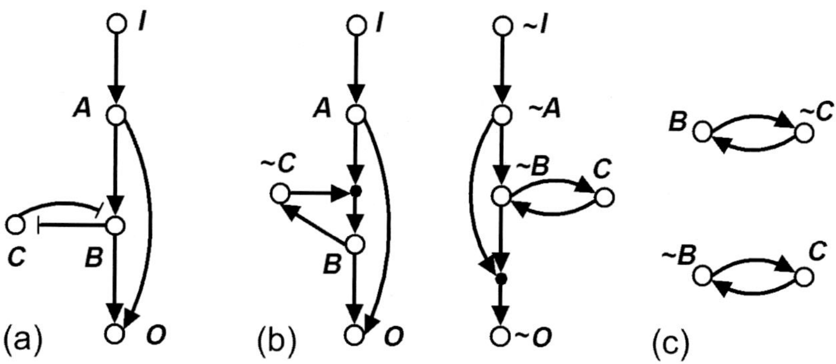

FIGURE 4.14 Illustration of methods that integrate the structure and logic of regulatory interactions. (a) A hypothetical signal transduction network. (b) The expanded representation of the network which integrates the Boolean transition functions $f_B = x_A$ AND NOT x_C, and $f_O = x_A$ OR x_B. The expanded network includes five complementary nodes, indicated by preceding the node name by ~, and two composite nodes, indicated by small black filled circles. (c) The stable motifs of the expanded network in the case of a sustained input signal ($x_I = 1$). The first stable motif corresponds to the state 11101 (in the order *I*, *A*, *B*, *C*, *O*), while the second stable motif corresponds to the state 11011.

network topology [36]. Specifically, one introduces a *complementary node* for each node of the network, and a *composite node* for each set of interactions with conditional dependency. We illustrate the process with the next example.

Example 4.8. Consider a signal transduction network composed of the input node *I*, intermediary nodes *A*, *B*, and *C* and the output node *O* (Figure 4.14a). This network shares some features of the core T-LGL network derived in [6]. The network does not completely specify the transition functions of nodes *B* and *O*. Let's specify the transition functions as $f_B = x_A$ AND (NOT x_C), and $f_O = x_A$ OR x_B. The complete set of transition functions now is:

$$f_A = x_I$$
$$f_B = x_A \text{ AND } (\text{NOT } x_C)$$
$$f_C = \text{NOT } x_B$$
$$f_O = x_A \text{ OR } x_B$$

Let's now construct the expanded network that integrates the transition functions. The expanded network features the addition of a complementary (negated) node for each real (original) node in the system, denoted by preceding the real node's name with ~. The state of this node is the negation of the state of the corresponding real node, and the transition function of the negated node is the logic negation of the transition function of the original node. For example, the transition function of the complementary node ~*A* is $f_{\sim A}$ = NOT $x_I = x_{\sim I}$, indicating that ~*A* is positively regulated by the complementary node ~*I*. The transition function of the complementary node ~*B* is $f_{\sim B}$ = NOT (x_A AND (NOT x_C)) = (NOT x_A) OR x_C. This means that ~*B* is positively regulated by ~*A* and *C*.[1] The transition function for ~*O* is $f_{\sim O}$ = NOT (x_A OR x_B) = (NOT x_A) AND (NOT x_B). The complete set of transition functions for complementary nodes is now

$$f_{\sim A} = \text{NOT } x_I$$
$$f_{\sim B} = (\text{NOT } x_A) \text{ OR } x_C$$
$$f_{\sim C} = x_B$$
$$f_{\sim O} = (\text{NOT } x_A) \text{ AND } (\text{NOT } x_B)$$

Next, we introduce composite nodes. For this the transition functions need to be specified in a *disjunctive normal format*, meaning that AND clauses are grouped and separated by OR's. For example, the expression (*A* AND *B*) OR (*C* AND *D*) is in a disjunctive normal form, while the expression *A* AND (*B* OR *C*) is not.

1. Recall the De Morgan laws for Boolean expressions: (1) NOT (*A* AND *B*) = (NOT *A*) OR (NOT *B*) and (2) NOT (*A* OR *B*) = (NOT *A*) AND (NOT *B*).

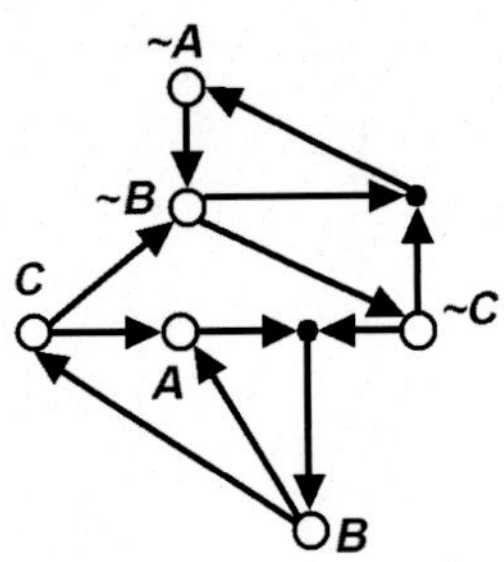

FIGURE 4.15 An expanded network for Exercise 4.29.

Inspecting the transition functions, we can verify that they are in the disjunctive normal format. We now add a composite node for each AND clause in the transition functions. Specifically, there will be a composite node for the expression x_A AND (NOT x_C), which regulates B, and another one for (NOT x_A) AND (NOT x_B), which regulates the complementary node $\sim O$.

The expanded network is shown on Figure 4.14b. Note that the update rules for all nodes with multiple inputs are now uniquely determined from the topology of the expanded network. All multiple inputs for a composite node are of type AND, while for the rest of the nodes multiple dependencies are of type OR. □

Exercise 4.29. Consider the expanded network in Figure 4.15. Construct the transition functions of the nodes A, B, C. Construct the transition functions of the complementary nodes $\sim A$, $\sim B$, $\sim C$. Verify that the transition function of each complementary node is the logic negation of the transition function of the respective original node. □

In Example 4.8 and Figure 4.14b, the expanded network is composed of two components that are disconnected from each other. The first component starts with the input node I, ends in the output node O, and contains A, B, $\sim C$ and a composite node shown as a black dot. The second component is made up by four complementary nodes, node C, and a second composite node. We can interpret the two subgraphs as the information transmission networks corresponding to the presence of the signal (left) and to the absence of the signal (right). The fact that the left subgraph contains both the input and output node indicates that there is at least one path that connects the input and output of the system (the signal and the system's response). The shortest such path is I, A, O. The next shortest is I, A, composite node, B, O. Is this path a context-independent conduit of information, or does its success depend on other nodes?

Because it involves a composite node, which stands for an AND clause embodying conditionality, this path is, in fact, dependent on the complementary node $\sim$C. Only a subgraph that contains all regulators of a composite node can serve as an independent information propagation conduit. This concept was termed an *elementary signaling mode* in [36] and is defined as the minimal set of components able to perform signal transduction independently. Thus, the network in Figure 4.14 contains two elementary signaling modes between input node I and output node O: the path I, A, O and the subgraph that contains I, A, the composite node, $\sim C$, B and O. Both of these are minimal because taking a node or edge away would obstruct the propagation of the signal.

The elementary signaling modes can be used to quantify the importance of nodes in mediating the signal. For example, in Figure 4.14b the loss of node A eliminates both elementary signaling modes between I and O, but the loss of node B leaves one of them intact. Application to several signaling networks, including those for abscisic acid-induced closure [15] and for T-cell apoptosis signaling [5], showed that nodes whose loss disrupts all elementary signaling modes were also essential to the model's dynamic attractor(s), in the sense that their knockout made this attractor unreachable [6, 36]. These results indicate that elementary signaling mode analysis, a method that involves Boolean logic and graph theory but no dynamic simulations, can be effectively used as a preliminary to or even as a substitute for dynamic perturbation analysis.

Exercise 4.30. Consider the model of Figure 4.14a in the case of a sustained signal ($x_I = 1$).

1. Determine the attractors of the system under general asynchronous update. Which of these attractors corresponds to a response to the signal?
2. Set node A to OFF. Determine the attractors of the system. Did at least one attractor remain that corresponds to a response to the signal? What is your conclusion? Is node A essential for the signal transduction process?
3. Set node B to OFF. Did at least one attractor remain that corresponds to a response to the signal? What is your conclusion? Is node B essential to the signal transduction process?
4. Let's assume that the ON state of the output node ($x_O = 1$) is undesirable. What node interventions would make this outcome impossible? □

The expanded network can also be used as a basis for network simplification. As shown in [37], a topological criterion can be used to identify network motifs (subgraphs) that stabilize in a fixed state regardless of the rest of the network. A *stable motif* in the expanded network is defined as the smallest strongly connected component (SCC) with the following properties: (1) the SCC cannot contain both a node and its complementary node and (2) if the SCC contains a composite node, it also contains all of its input nodes. For example, in Figure 4.14b the nodes $\sim B$ and C form a stable motif. The fixed state of the nodes in the stable motif can be directly read out from the expanded network: if the stable motif contains the node, the node stabilizes in the ON state, and if the stable motif contains a complementary node, the corresponding node stabilizes in the OFF state. These fixed states can be plugged into the transition functions of other nodes, leading to simpler functions, and consequently to a simpler expanded network. Iterative searching for stable motifs and network simplification leads to one of two possible outcomes: either there are no nodes with unknown states, in which case a fixed point of the system is identified, or no new stable motifs are found, in which case the remaining nodes are expected to oscillate. Thus stable motif analysis serves as a preliminary to or as a substitute for attractor analysis. For example, the T-cell apoptosis signaling network has four stable motifs. Iterative simplification of these stable motifs leads to the system's steady states, the same as those found from dynamic simulations. Interestingly, the stable motif formed by three nodes close to the PDGF signal can solely determine the steady state, regardless of the other motifs or the trajectory of the system leading up to the stabilization of the motif [37].

Overall, these integrated structural and logic methods are fruitful as exploratory analysis of large signaling networks where dynamic modeling is computationally impractical, or as a first step that guides follow-up targeted computational or experimental studies.

Exercise 4.31. Consider the sustained presence of the input signal ($x_I = 1$) in Figure 4.14a. Simplify the transition functions of the nodes and construct the expanded network corresponding to this case. Compare with Figure 4.14c. □

Exercise 4.32. Determine the stable motifs of the expanded network in Figure 4.14c. Compare with the steady states found in Exercise 4.30. □

Exercise 4.33. Consider again the network in Figure 4.14a, but this time consider the following update rules for nodes B and C: $f_B = x_A$ OR (NOT x_C), and $f_O = x_A$ AND x_B.

1. Construct the expanded network.
2. Determine the elementary signaling modes between input node I and output node O in the expanded network.
3. Determine the essential signal-mediating nodes based on the elementary signaling modes.
4. Consider the sustained absence of the signal $x_I = 0$. Determine the expanded network, its stable motifs, and the corresponding steady states. □

Exercise 4.34. Consider the network in Figure 4.16. Construct two sets of transition functions that are consistent with this network. For each set,

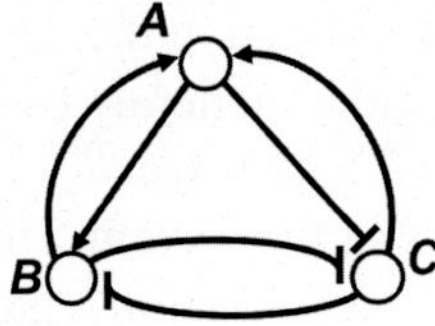

FIGURE 4.16 A simple three-node network for Exercise 4.34.

1. Construct the expanded network.
2. Determine the stable motifs in the expanded network and the corresponding steady states.
3. Verify your results for part 2 by determining the model's steady states analytically. □

4.11 CONCLUSIONS

Although Boolean network models have a limited capacity to describe the quantitative characteristics of dynamic systems, they do exhibit considerable dynamic richness and have proven effective in describing the qualitative behaviors of signal transduction networks, in predicting key components, and in proposing effective intervention strategies. The fact that Boolean models do not require the knowledge of kinetic parameters makes them a preferred choice for systems where these parameters have not been measured. Thus Boolean models pass the two key tests: they are useful, and they help us to better understand the systems for which they are formulated. The success of Boolean models illustrates that in at least a subset of biological systems the organization of network structure plays a more important role than the kinetic details of the individual interactions. Thus, Boolean networks can serve as a foundation for the modeling of signaling networks, upon which more detailed continuous models can be built as kinetic information and quantitative experimental data become available. The simpler Boolean model can be used for efficient exploratory analysis to fix the model's structure and to help develop a refined continuous model, which in turn can be further compared with quantitative biological observations.

REFERENCES

[1] Gomperts BD, Tatham PER, Kramer IM. Signal transduction. San Diego, CA: Academic Press; 2002.
[2] Gordon KJ, Blobe GC. Role of transforming growth factor-beta superfamily signaling pathways in human disease. Biochim Biophys Acta 2008;1782:197–228.
[3] Ikushima H, Miyazono K. TGFbeta signalling: a complex web in cancer progression. Nat Rev Cancer 2010;10:415–24.
[4] Albert R, Wang RS. Discrete dynamic modeling of cellular signaling networks. Methods Enzymol 2009;467:281–306.
[5] Zhang R, Shah MV, Yang J, Nyland SB, Liu X, Yun JK, et al. Network model of survival signaling in large granular lymphocyte leukemia. Proc Natl Acad Sci U S A 2008;105:16308–13.
[6] Saadatpour A, Wang RS, Liao A, Liu X, Loughran TP, Albert I, et al. Dynamical and structural analysis of a T cell survival network identifies novel candidate therapeutic targets for large granular lymphocyte leukemia. PLoS Comput Biol 2011;7:e1002267.
[7] Albert R, Barabasi AL. Statistical mechanics of complex networks. Rev Mod Phys 2002;74:47–97.
[8] Gilmore TD. Introduction to NF-kappaB: players, pathways, perspectives. Oncogene 2006;25:6680–4.
[9] Alon U. An introduction to systems biology: design principles of biological circuits. Boca Raton, FL: Chapman and Hall/CRC; 2006.
[10] Thomas R, d'Ari R. Biological feedback. Boca Raton, FL: CRC Press; 1990.
[11] Ma'ayan A, Jenkins SL, Neves S, Hasseldine A, Grace E, et al. Formation of regulatory patterns during signal propagation in a mammalian cellular network. Science 2005;309:1078–83.
[12] Smoot ME, Ono K, Ruscheinski J, Wang PL, Ideker T. Cytoscape 2.8: new features for data integration and network visualization. Bioinformatics 2011;27:431–2.
[13] Hagberg A, Schult D, Swart P. Exploring network structure, dynamics, and function using NetworkX. In: Varoquaux G, Vaught T, Millman J, editors. SciPy Conference, Pasadena, CA; 2008. p. 11–5.
[14] Batagelj V, Mrvar A. Pajek—analysis and visualization of large networks. In: Jünger M, Mutzel P, editors. Graph drawing software. Berlin: Springer; 2003. p. 77–103.

[15] Li S, Assmann SM, Albert R. Predicting essential components of signal transduction networks: a dynamic model of guard cell abscisic acid signaling. PLoS Biol 2006;4:e312.

[16] Saadatpour A, Albert I, Albert R. Attractor analysis of asynchronous Boolean models of signal transduction networks. J Theor Biol 2010;4:641–56.

[17] Kestler HA, Wawra C, Kracher B, Kuhl M. Network modeling of signal transduction: establishing the global view. Bioessays 2008;30:1110–25.

[18] Karlebach G, Shamir R. Modelling and analysis of gene regulatory networks. Nat Rev Mol Cell Biol 2008;9:770–80.

[19] Sobie EA, Lee YS, Jenkins SL, Iyengar R. Systems biology–biomedical modeling. Sci Signal 2011;4:tr2.

[20] Kauffman SA. The origins of order: self organization and selection in evolution. New York, NY: Oxford University Press; 1993.

[21] Aldridge BB, Saez-Rodriguez J, Muhlich JL, Sorger PK, Lauffenburger DA. Fuzzy logic analysis of kinase pathway crosstalk in TNF/EGF/insulin-induced signaling. PLoS Comput Biol 2009;5:e1000340.

[22] Chaouiya C. Petri net modelling of biological networks. Brief Bioinform 2007;8:210–9.

[23] Saadatpour A, Albert R. Boolean modeling of biological regulatory networks: a methodology tutorial. Methods 2013;62:3–12.

[24] Garg A, Di Cara A, Xenarios I, Mendoza L, De Micheli G. Synchronous versus asynchronous modeling of gene regulatory networks. Bioinformatics 2008;24:1917–25.

[25] Tyson JJ, Chen KC, Novak B. Sniffers, buzzers, toggles and blinkers: dynamics of regulatory and signaling pathways in the cell. Curr Opin Cell Biol 2003;15:221–31.

[26] Robeva RS, Hodge TL. Mathematical concepts and methods in modern biology: using modern discrete models. New York, NY: Academic Press; 2013.

[27] Albert I, Thakar J, Li S, Zhang R, Albert R. Boolean network simulations for life scientists. Source Code Biol Med 2008;3:16.

[28] Mussel C, Hopfensitz M, Kestler HA. BoolNet—an R package for generation, reconstruction and analysis of Boolean networks. Bioinformatics 2010;26:1378–80.

[29] Hinkelmann F, Brandon M, Guang B, McNeill R, Blekherman G, et al. ADAM: analysis of discrete models of biological systems using computer algebra. BMC Bioinform 2011;12:295.

[30] Yi TM, Huang Y, Simon MI, Doyle J. Robust perfect adaptation in bacterial chemotaxis through integral feedback control. Proc Natl Acad Sci U S A 2000;97:4649–53.

[31] Takeda K, Shao D, Adler M, Charest PG, Loomis WF, et al. Incoherent feedforward control governs adaptation of activated ras in a eukaryotic chemotaxis pathway. Sci Signal 2012;5:ra2.

[32] Dasgupta B, Albert R, Dondi R, Kachalo S, Sontag E, et al. A novel method for signal transduction network inference from indirect experimental evidence. J Comput Biol 2007;14:927–49.

[33] Kachalo S, Zhang R, Sontag E, Albert R, DasGupta B. NET-SYNTHESIS: a software for synthesis, inference and simplification of signal transduction networks. Bioinformatics 2008;24:293–5.

[34] Saez-Rodriguez J, Alexopoulos LG, Epperlein J, Samaga R, Lauffenburger DA, et al. Discrete logic modelling as a means to link protein signalling networks with functional analysis of mammalian signal transduction. Mol Syst Biol 2009;5:331.

[35] Terfve C, Cokelaer T, Henriques D, MacNamara A, Goncalves E, et al. CellNOptR: a flexible toolkit to train protein signaling networks to data using multiple logic formalisms. BMC Syst Biol 2012;6:133.

[36] Wang RS, Albert R. Elementary signaling modes predict the essentiality of signal transduction network components. BMC Syst Biol 2011;5:44.

[37] Zanudo JG, Albert R. An effective network reduction approach to find the dynamical repertoire of discrete dynamic networks. Chaos 2013;23:025111.

Chapter 5

Dynamics of Complex Boolean Networks: Canalization, Stability, and Criticality

Qijun He[1], Matthew Macauley[1] and Robin Davies[2]

[1]Department of Mathematical Sciences, Clemson University, Clemson, SC, USA, [2]Department of Biology, Sweet Briar College, Sweet Briar, VA, USA

5.1 INTRODUCTION

The *phenotype* of an organism consists of its observable traits, such as eye color, height, or wing type. One can also speak of phenotypes of a population or species, such as having tails, opposable thumbs, or body hair. The nature vs. nurture paradigm summarizes the two primary factors that determine phenotype, both at the individual and population level: (i) environment and (ii) genetic makeup, the latter of which is called *genotype*. On one hand, the phenotype of an organism (or population) must be robust enough to withstand changes to its environment and genotype. On the other hand, at the population level, it must be flexible enough to evolve and better adapt to these changes. Canalization is a measure of the stability of a phenotype with respect to outside changes.

The term canalization was coined by geneticist Conrad Hal Waddington in 1942 as an attempt to quantify the reduced sensitivity of a phenotype to genetic and environmental perturbations [1]. Other terms, such as robustness, developmental stability, and developmental homeostasis, are frequently used synonymously, though at other times subtle distinctions are made. From a big picture standpoint, there are many ways to define a quantitative stability measure, though these are all just models, and they depend on the framework one uses to study the problem. For example, a biostatistician might define canalization in terms of covariance. In this chapter, we will look at how this concept can be defined and quantified from a Boolean network model of a biological system. We will begin by introducing the concept of a gene regulatory network (GRN), which consists of a collection of genes, proteins, and enzymes that function together for a specific purpose. The first widely studied GRN was the lactose operon in *Escherichia coli*, mainly due to its simplicity. We will briefly introduce this network as a motivating example. In doing so, we will review the two main components of a GRN: the network topology (structure) and the regulatory functions. In our framework, these will be Boolean functions. We will pay particular attention to the class of canalizing Boolean functions, as these frequently arise in molecular networks. Along the way, we will examine some of the mathematical objects, ideas, and techniques that arise in the study of these Boolean network models.

Loosely speaking, complex systems such as Boolean network models fall into one of two dynamical regimes, characterized by whether small errors, or random external perturbations, tend to die out or propagate throughout the network [2]. As an analogy, consider a forest with thin underbrush in a wet climate. An isolated lightning strike is unlikely to spark a wildfire that spreads throughout the forest. In contrast, in a forest in an arid climate with thick underbrush, a lightning strike is much more likely to ignite a large fire that propagates throughout the forest. One can think of the lightning strike as an external perturbation to the system. In the first case, this

Algebraic and Discrete Mathematical Methods for Modern Biology. http://dx.doi.org/10.1016/B978-0-12-801213-0.00005-8
Copyright © 2015 Elsevier Inc. All rights reserved.

perturbation is most likely to die out; such a system is said to be stable, or *ordered*. In the second case, the perturbation is likely to propagate throughout the network (forest); such a system is said to be *chaotic*.

As another example, consider a social network model of a large population, where a few individuals are initially infected with an infectious, but very noncontagious, disease. Most likely, the disease will quickly die out and not become an epidemic. In contrast, suppose a few individuals are stricken with a highly contagious infectious disease, such as Ebola hemorrhagic fever or the Bubonic plague. Unless appropriate measures are taken, this infection will mostly likely spread throughout the population and become an epidemic. As will be described in Chapter 8, the first scenario occurs when the transmissibility of the disease is low enough that the expected number of secondary cases produced by an initially infected individual is $R_0 < 1$. The second scenario occurs when this constant R_0, called the *basic reproductive number*, is greater than 1; see also Ref. [3]. Thus, ordered epidemic networks are (generally) characterized by $R_0 < 1$, whereas chaotic ones are characterized by $R_0 > 1$. These two dynamical regimes are separated by the networks where $R_0 \approx 1$; these systems are said to lie in the *critical threshold*—on the brink of both order and chaos. Mathematically, they share certain salient characteristics of both regimes, and many real-world molecular networks are believed to lie in this threshold [4–6]. These networks are said to exhibit *critical dynamics* [7], and this has been observed in complex networks in a variety of areas—examples include but are not limited to climatology [8, 9], geology [10], neuroscience [11], and economics [12].

Lying on the phase transition between order and chaos, critical networks optimize the trade-off between robustness and adaptability [13]. Being on the brink of order endows critical networks with enough stability to be robust enough to withstand external perturbations, and being on the brink of chaos gives such networks the flexibility to express a wide range of dynamical behavior and to evolve. Because many real-world networks exhibit critical dynamics, it has been hypothesized that it is evolutionarily advantageous to lie in this phase transition, and there is evidence to support this theory [4]. Studies have shown that the most complex dynamics arise in networks near the critical threshold [2]. From an information theory standpoint, study structure-dynamics relationships have been shown to maximize diversity in the critical regime [14, 15]. The theory of self-organized criticality postulates that certain systems can self-organize to exhibit critical dynamics and emergent complex behavior [16]. Many scientists believe that criticality is a fundamental evolutionary mechanism present in a wide diversity of complex networks.

In a 2008 paper titled *Critical Dynamics in Genetic Regulatory Networks: Examples from Four Kingdoms* [4], the authors studied GRNs from thale cress (a weed), fruit flies, yeast, *E. coli*, and *Bacillus subtilis* (hay bacteria). They showed that all these networks exhibit critical dynamics. The thale cress, *Arabidopsis thaliana*, is a small flowering plant with a short genome consisting of approximately 135 Mb pairs. In the year 2000, it became the first plant to have its entire genome sequenced, and its flower morphogenesis network is the most well-studied plant GRN. The flower morphogenesis network controls the development of the four floral organs (sepals, petals, stamens, and pistil). This network has only 15 genes, and all of the regulatory functions are known. The segment polarity network of the fruit fly *Drosophila melanogaster* is another small well-studied GRN. The segment polarity network acts during the embryonic development of segmented organisms. Through integration of appropriate signaling, most notably Hedgehog and Wingless, this network ensures correct development of each segment in the organism. Though this GRN is larger, containing 60 genes, its network and regulatory functions are also known. The last three networks studied in Ref. [4] are much larger than the first two. As is true for many recently discovered networks, sheer size impedes the ability of scientists to determine the regulatory functions. For example, the gene transcription network in *E. coli* contains 1481 genes [17], the one in *Saccharomyces cerevisiae* (yeast) contains 3459 genes [18], and the *B. subtilis* (hay bacteria) network contains 830 genes [19]. Because the majority of the functions in these networks are unknown, the networks are analyzed by randomly assigning functions from a carefully constructed distribution. The details of how to select such functions are not important for this chapter, but roughly speaking, the physical network (or "wiring diagram") is constructed in a way so its main statistical properties match those in the observed GRNs. Next, the individual functions are randomly assigned. In Ref. [4], this was done with a bias (the proportion of 1s in its truth table) consistent with the actual microarray data.

The remainder of this chapter is organized as follows. In Section 5.2, we introduce Boolean models and establish some fundamentals of Boolean functions and the role of network topology. In Section 5.3, we show how the canalization concept can be captured in this setting and analyze some of the related mathematical concepts. This provides a background for the main topic of Section 5.4: the analysis of random Boolean networks (RBNs). We lay the groundwork for Boolean calculus, which is much simpler than standard calculus and does not require any prior calculus knowledge. This will help us analyze how small perturbations spread in RBNs and classify them into the ordered, chaotic, and critical dynamical regimes.

5.2 BOOLEAN NETWORK MODELS

5.2.1 Gene Regulatory Networks

A *GRN* is a collection of interacting genes, gene products (RNA, proteins, etc.), and enzymes that constitute a specific biological system or cell mechanism. The structure of a GRN can be represented as a directed graph, where the nodes represent components and the edges represent interactions. Sometimes the edges are "signed" to denote whether the interaction is an activation (positive) or repression/inhibition (negative). For example, a "positive edge" $A \longrightarrow B$ might mean that gene or enzyme A "activates," or "turns on" B, whereas a "negative edge" $A \longrightarrow\!\!\bullet\, B$ would be used to denote a relationship such as gene or enzyme A "inhibits," or "turns off" B. However, this graph is only part of the picture; it does not specify the interplay between multiple interactions. For example, the motif $A \longrightarrow B \longleftarrow C$ could describe a situation where genes A and C both activate gene B, but it does not describe how. For example, do A and C *both* need to be present (A AND C) for B to be transcribed? This would be the case if the cellular function needed to activate B requires both A and C to work together. However, it could also be the case that *either* A or C can activate B. Perhaps B plays such a vital cellular role that the cell needs to be robust enough to withstand the knockout of one of its activation genes.

As a second example, consider the motif $A \longrightarrow B \bullet\!\!\longleftarrow C$ which could describe a situation where gene A activates B but enzyme C inhibits it. What happens if A and C are both present? Perhaps B is transcribed ("on") if A is present *or* C is absent (A OR $\overline{C}$; where the bar denotes NOT). Or perhaps C takes precedence, like a "shut-down valve": B is transcribed if A is present *and* C is absent (A AND $\overline{C}$). Naturally, there are other possibilities besides these two, but the take-home message for now is that the signed graph does not completely determine the GRN; one also needs to know the regulatory functions. On the other hand, if all of the functions are known, the graph would easily be determined. Though this seems to render the graph redundant from a mathematical standpoint, the biological reality is that the network is usually discovered before the functions. For example, experiments involving gene knockouts or overexpressions can reveal dependencies between the nodes of a GRN, but further understanding is needed to determine the actual functions.

One needs to understand all the relationships in a GRN of a cellular process before proposing a mathematical model. Typically, the nodes are represented by variables and the interactions are represented by functions. In some models, the states of the nodes can take on values from an entire interval, e.g., $x_i \in [a, b]$. In other models, the states can only be discrete values, e.g., $x_i \in \{0, 1, 2\}$. Models of the former type are called *continuous-space*, and models of the latter type are called *discrete-space*. Time can also be continuous-valued, e.g., $t \in [0, \infty)$ or discretized, e.g., $t = 0, 1, 2, \ldots$, leading to continuous-time models and discrete-time models. Models that involve continuous variables and functions, such as differential equations, usually involve precise rate constants and hence are quantitative in nature. In contrast, discrete-space models, such as ones involving Boolean functions, are inherently qualitative. Though a discrete or Boolean approach may at first glance seem to be too simple, many studies have shown that these qualitative models encapsulate the essential features of the cellular dynamics and perform just as well as their continuous cousins when one factors in the uncertainty that comes with estimating rate constants.

As a simple motivating example, the *lac* operon in *E. coli* is responsible for bringing lactose into the cell and catalyzing the first step in its metabolism (breaking it down for energy), but it only does this if glucose is not present. The *lac* operon is one of the most well-understood networks, consisting of three structural genes, *LacZ*, *LacY*, and *LacA*, which encode three proteins. These proteins carry out specific functional duties, such as transportation of lactose, and the hydrolysis of lactose to glucose and galactose. All three genes are transcribed together, making a single mRNA, which, when translated, makes all three proteins. An adjacent

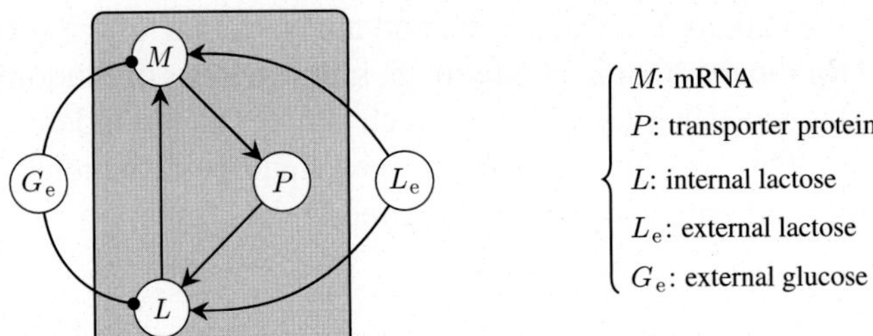

FIGURE 5.1 A simplified gene regulatory network of the *lac* operon, from a model proposed in Chapter 1 of Ref. [20].

regulatory gene, *LacI*, encodes the lac repressor, which binds to the operator region of the lac operon to turn the genes "off." There are a few other proteins involved in this process that help regulate the operon, though the details are unimportant to us here. A complete GRN consists of all genes, proteins, and enzymes that constitute the functional processes and regulation of this operon. However, one can consider simpler GRNs that aggregate several similar components into one or only focus on a particular part of the process. Figure 5.1 shows an example of a simplified GRN of a model proposed in Chapter 1 of Ref. [20] involving key features of the *lac* operon. The shaded region represents the cell. The three components inside the cell and the two components outside the cell are listed to the right of the network.

Recall that a GRN only gives partial information about the system. For example, in the GRN in Figure 5.1, the state of mRNA (M) is regulated by the nodes connected to incoming edges: L, L_e, and G_e. However, the network does not encode *how* these three components regulate M, only that it is a function involving L, L_e, and G_e, and that G_e inhibits M. In a continuous (e.g., calculus-based) model, where these variables[1] are continuous functions of t, this could mean that

$$\frac{\mathrm{dM}}{\mathrm{d}L} > 0, \quad \frac{\mathrm{dM}}{\mathrm{d}Le} > 0, \quad \frac{\mathrm{dM}}{\mathrm{d}Ge} < 0$$

In a Boolean model, where the relationships are described by Boolean logic, the value of M might be modeled by a Boolean function in variables L, L_e, and G_e, such as one of the following:

$$f_M(L, L_e, G_e) = (L \vee L_e) \wedge \overline{G_e} \quad \text{or} \quad f_M(L, L_e, G_e) = L \vee (L_e \wedge \overline{G_e}). \tag{5.1}$$

where $\wedge$ denotes logical AND, and $\vee$ denotes logical OR, and $\overline{X}$ denotes "NOT X." The preceding statements have different meanings, and one would need to understand the biology of the system before proposing which function to use in a model. For example, the first function would describe the scenario that "mRNA is transcribed if internal or external lactose is present, *and* if there is no external glucose." In contrast, the second function would describe the scenario that "mRNA is transcribed if internal lactose is present, or there is external lactose but not glucose." As with a calculus-based model, small changes in the functions can have big effects on the dynamics. It turns out that the first function in Eq. (5.1) was proposed as being the most realistic; see Chapter 1 of Ref. [20] for details.

The primary two features of Boolean network models are (i) the structure, or "topology," of the underlying network, and (ii) the actual Boolean functions at the nodes. In the remainder of the chapter, we will introduce the fundamental concepts of both of these.

5.2.2 Network Topology

A GRN is an example of a model over a graph (or network; we will use both interchangeably). The vertices, or nodes, represent components, and the edges represent interactions. In a GRN, these interactions are typically

---

1. In some models, external lactose L_e and external glucose G_e are treated as constant parameters. For our example, we will consider them as variables.

not symmetric: A affects B does not imply that B affects A. This results in the network being a directed graph. In addition, in this setting, one can attach additional information about whether this interaction is activating or repressing, leading to a signed directed graph like we saw previously.

In an epidemiological network model, as described in the beginning of this chapter, and in more detail in Chapter 8, vertices represent individuals and edges represent social contacts [21]. These interactions are naturally symmetric: if A comes in contact with B, then B comes in contact with A. In this setting, the underlying network is an undirected graph.

An edge that connects a node to itself is called a *loop*. For models over underdirected graphs, loops can usually be ignored. For example, in an epidemiological network, a loop would signify that an individual is in contact with themself. Because this is vacuously true, loops can be ignored in this setting. In some models, especially those where the graph is directed, the presence of loops is a vital part of the network topology. For example, in a gene network, a (directed) loop from a vertex A to itself might mean that gene A can regulate itself. Perhaps it is able to "turn itself off." This would be an example of a negative feedback loop and is a common motif in gene networks.

In all of these examples, there is an underlying network, or graph, whose *topology*, or structure, affects the model dynamics built on top of it. In different settings, this graph has different names: wiring diagram, dependency graph, underlying graph, or base graph, just to name a few. For small models, this graph may be known. However, for large models, this graph must be randomly generated. This leads to the question of how to algorithmically construct such a graph whose properties closely match the actual network. This question is just the tip of the iceberg of the field of random graph theory, and it is not the primary focus of this chapter. Rather, we will briefly introduce a few standard random graph models and characteristics that they possess. To make things simple, in this subsection we will usually consider the case of *undirected graphs without loops*.

One simple characteristic of a network is its degree distribution. A *neighbor* of a vertex is any other vertex connected to it by an edge, and the number of neighbors that a node has is its *degree*. Given a graph G, the average (mean) degree is a well-defined property of G, but the *degree distribution* gives more information. This is a function $P\colon \mathbb{N} \to [0, 1]$, where $P(k)$ is the fraction of nodes with degree k. For directed graphs, one needs to distinguish between out-neighbors and in-neighbors, so there are separate notions of in-degree and out-degree distributions.

There are other measures that can be computed for every node and made into a distribution function, but degree is the most common. The *clustering coefficient* of a node measures how much vertices tend to cluster together [22]. Specifically, given a vertex v, let N_v be its set of neighbors and $d_v = |N_v|$ its degree. A simple counting argument tells us that there are $\binom{d_v}{2} = d_v(d_v - 1)/2$ possible "triangles" that could exist involving v. That is, v and two neighbors $v', v'' \in N_v$ form a triangle if the edge $\{v', v''\}$ is also present. The clustering coefficient C_v of v represents the proportion of these triples that actually form triangles:

$$C_v = \frac{\text{number of triangles containing } v}{d_v(d_v - 1)/2} = \frac{2 \cdot |\{v', v''\} \in E(G) : v', v'' \in N_v|}{d_v(d_v - 1)}.$$

Figure 5.2 shows two graphs on six vertices and their degree distributions. Figure 5.3 shows a graph on seven vertices and the clustering coefficient of each vertex. Due to symmetry, there are only four "types" of vertices in this graph, and these can be classified by their degree.

$$P(k) = \begin{cases} 5/6, & k=1 \\ 1/6, & k=5 \\ 0, & \text{else} \end{cases} \qquad\qquad P(k) = \begin{cases} 1, & k=2 \\ 0, & \text{else.} \end{cases}$$

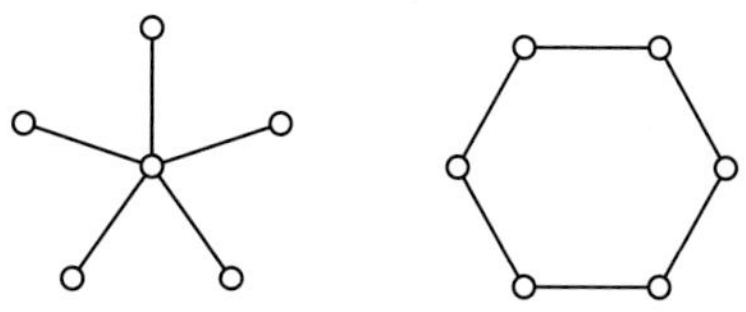

FIGURE 5.2 Two undirected graphs on six nodes and their degree distributions.

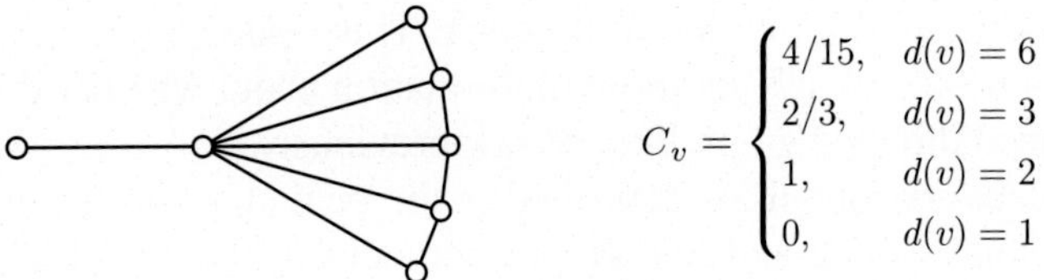

$$C_v = \begin{cases} 4/15, & d(v) = 6 \\ 2/3, & d(v) = 3 \\ 1, & d(v) = 2 \\ 0, & d(v) = 1 \end{cases}$$

FIGURE 5.3 A graph with has four "types" of vertices that can be described by their degree. The clustering coefficient of each is shown at right. Note that in general, two vertices could have the same degree but different clustering coefficients.

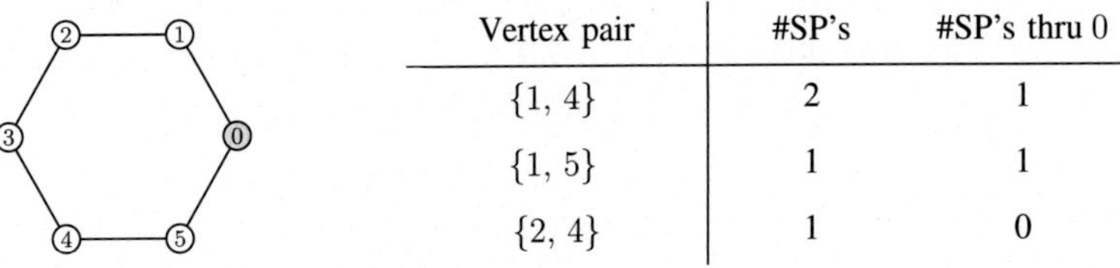

Vertex pair	#SP's	#SP's thru 0
$\{1, 4\}$	2	1
$\{1, 5\}$	1	1
$\{2, 4\}$	1	0

FIGURE 5.4 In the hexagon graph, there are $\binom{5}{2} = 10$ pairs of distinct vertices $\{s, t\}$ different from vertex 0. Eight of these pairs have a unique shortest path (SP) between them; the others have two shortest paths. Some examples are shown in the table on the right. Exactly 3 of these 12 shortest paths pass through 0, so the betweenness of vertex 0 (and by symmetry, any other vertex) is 1/4.

Like degree distribution, the clustering coefficient definition can be modified to handle directed graphs. However, it is arguably more natural a feature in underdirected graphs. For example, in a social network, the clustering coefficient measures the probability that two mutual friends also know each other.

The last feature of a network that we will introduce is "betweenness centrality." The *betweenness* of a node v is the proportion of all shortest paths in the network between pairs $\{u, w\} \subseteq V(G) - \{v\}$ of distinct nodes that pass through v. Note that $\{u, w\}$ may have multiple shortest paths between them. For example, in a star-graph, such as the one in Figure 5.2, the center vertex has betweenness 1, and all others have betweenness 0. For the hexagon graph, every vertex has betweenness 1/4. This calculation is derived in Figure 5.4. The betweenness measures the "centrality" of a node in a network. It has been studied in network-based models arising in sociology [23], ecology (food webs) [24], economics [25], and epidemiology [21], among others.

Exercise 5.1. Find two graphs that are structurally different but have the same degree distribution and clustering coefficients. □

Exercise 5.2. Find the clustering coefficient of each vertex of the two graphs in Figure 5.2. □

Exercise 5.3. Construct a graph that has two vertices of the same degree that have different clustering coefficients. □

Exercise 5.4. The clustering coefficients can be turned into a "distribution function," like what was done for the degree function, but with the minor difference that the domain must be the rational numbers, $\mathbb{Q}$. Namely, define $C \colon \mathbb{Q} \longrightarrow [0, 1]$, where $C(p/q)$ is the fraction of nodes with clustering coefficient p/q. Write out this function for the graph in Figure 5.3. □

Exercise 5.5. Calculate the betweenness of each node of the graph in Figure 5.3. □

Exercise 5.6. Calculate the betweenness of each node in the pentagon graph and in the heptagon (7-gon) graph. □

5.2.3 Network Topology and Random Networks

Quick! What does a "random graph" on 18 nodes look like? Is it tree-like, or does it have lots of cycles? Conjure up a mental image, and then take a peek at Figure 5.5. Which one of these three graphs is more like what you were picturing? Next question: What does a random graph on 1000 nodes looks like? Naturally, the answers to these questions depend on what the network is supposed to represent: a social network, protein network, wireless network, or something else. Once this is known, another question arises: How can one construct a "random" graph of a certain size with these salient features? This is what random graph models strive to accomplish.

Formally, a random graph model produces a probability distribution on the set of all graphs. One can view this as a way to algorithmically construct a "random" graph of a certain type. Each model has a few input parameters that must be specified. One of the first random graph models to be studied was proposed by Paul Erdős and Alfred Rényi and bears their name [26]. A random graph on n vertices using the Erdős-Rényi (ER) model $G(n,p)$ is constructed by including each of the $\binom{n}{2}$ possible edges with an independent probability of p, for some fixed $p \in [0,1]$. If $G = (V,E)$ is a graph on $|V| = n$ vertices and $|E| = M$ edges, then the probability that G is chosen by this process is

$$\Pr(G) = p^M(1-p)^{\binom{n}{2}-M}.$$

A variant of this model, denoted $G(n,M)$, chooses a graph uniformly from all n-vertex graphs with exactly M edges. ER graphs have been used in percolation models in physics, but they do not share key characteristics, such as high clustering, seen in many real-world graphs such as social networks.

One well-known feature of social networks is the "small-world" phenomenon, often referred to in popular culture as "six degrees of separation." In mathematical terms, this means that the social network describing personal acquaintances, a graph on roughly 7 billion nodes, has the curious property that the shortest path between any two nodes is at most 6. Because random graphs constructed by the ER model do not share this property (with probability almost 1), the ER model is not suitable for constructing social networks [22]. In the late 1990s, Watts and Strogatz proposed a "small-world network" model [22]. This was designed to preserve two key features of social networks that were notably absent from most existing random graph models: low shortest path and high clustering. The Watts-Strogatz model has three parameters: n (number of nodes), k (average degree), and p (rewiring probability). A graph is constructed by arranging n vertices $v_1, \ldots, v_n$ in a circle, with each one being connected to its k (an even integer) closest neighbors. Next, each edge is "rewired" with probability p. Specifically, for each vertex v_i in $v_1, \ldots, v_n$, take each edge (v_i, v_j) with $i < j$, and with probability p, replace v_j with a uniformly chosen vertex v_ℓ from the remaining vertices, excluding v_i and its neighbors. These rewirings represent "long-distance connections" common to most individuals in a social network: most individuals know a few select "random" people who live far away.

One of the major weaknesses of the Watts-Strogatz model is that its degree distribution is not realistic for many real-world networks. Assuming that the rewiring probability is small, most vertices will have roughly the same degree. A *scale-free network* is a network whose degree distribution is (approximately) a power law: $P(k) \sim k^{-\gamma}$, where $\gamma > 0$. We say "approximately" because this is usually taken to mean asymptotically (i.e.,

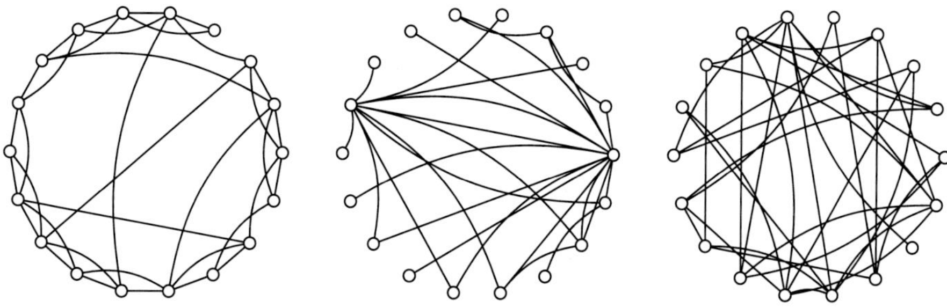

FIGURE 5.5 From left to right: A small-world, scale-free, and random network on 18 nodes and 36 edges. The small-world network has mostly "local" edges but also a few long-range connections. The scale-free network is characterized by a few "hubs." The random network has no clear distinguishing feature like the other two.

as the number of vertices $n \to \infty$). Scale-free networks are characterized by having a few large hubs, and many vertices of small degree. Many networks are believed to be roughly scale free. One such example is the World Wide Web: web sites such as Google, Facebook, and Amazon are highly connected, as they are linked from many web sites. In contrast, the vast majority of webpages have very few, if any links to them. Though the WWW network is directed—a link from i to j does not imply a link from j to i, the scale-free definition can be easily adapted to handle that. Another scale-free (directed) network is that of academic publications, where i is connected to j if i cites j. Most papers are rarely cited, whereas a few instrumental papers garner thousands of citations. Some social networks can also be scale-free. While a network where the edges represent physical contacts are less likely to have such large hubs, social networks where i is connected to j if i knows j, or if i follows j on Twitter, seem to be better candidates due to famous people being large hubs. Scale-free networks have also been observed in metabolic networks [27, 28] and protein networks [29].

Examples of a small-world, scale-free, and random network are shown in Figure 5.5. Given a network with complete data of vertices and edges, it is not difficult to determine whether it is scale free—just plot the degree distribution and see if it fits a power law. A harder question is how to create a scale-free network on n nodes, with power-law parameter γ. The most common method is the *Barabási-Albert* model [30], where nodes are added sequentially and connections to other nodes are added via a "preferential attachment" process. Roughly speaking, this means that as nodes are added, they are more likely to be connected up with high-degree nodes. This can be thought of as the common "rich get richer" saying, and the resulting algorithm generates a power-law degree distribution.

In contrast to many real-world networks, the out-degree distribution in some GRNs can be fairly close to uniform. Each gene, or node, is regulated by just a few other nodes. Stuart Kauffman introduced *NK-networks* [31] as a theoretical model of this phenomenon. In the literature, these are sometimes called *Kauffman networks* or *RBNs*. In this chapter, we will refer to the actual graph as the *NK*-network, and the model consisting of the graph and Boolean functions as an RBN. An *NK*-network has N vertices, each one with in-degree K. This means that every node is regulated by exactly K other nodes (i.e., the network has a uniform in-degree distribution). Kauffman introduced these networks as models of haploids with a single copy of each chromosome. The chromosome set has N distinct genes, and each gene occurs in exactly K alleles. The simplest case for higher eukaryotes is when each gene occurs in two alleles. This leads to an *NK*-network with $K = 2$, modeling what biologists call the *N-locus, two-allele additive fitness model.* These models from population genetics are similar to certain spin models arising in statistical physics [31]. One can argue about the merits of a theoretical model such as this, but in Kauffman's words: *If the NK-model were to serve no other purpose than to tune our intuitions about what such landscapes might look like, that alone would warrant our attention.*

Exercise 5.7. There are 11 distinct graphs on 4 vertices, which are shown below:

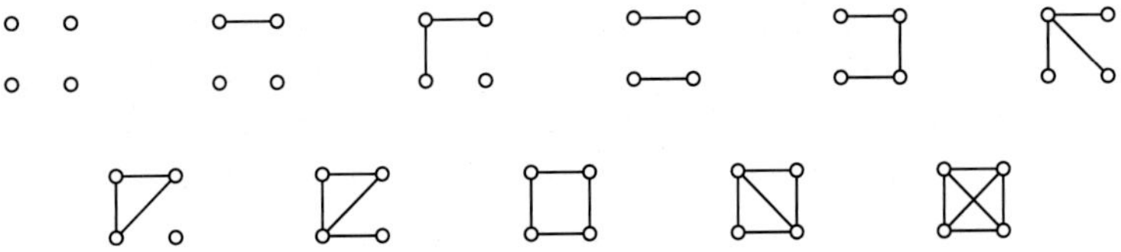

Compute the probability of each graph arising from the random graph model $G(4,p)$. Beware that most of these graphs can occur in multiple ways. For example, there are six different ways that the graph on 1 edge can arise. □

5.2.4 Boolean Functions

In addition to the network topology, the other main characteristic of a Boolean network is the underlying Boolean functions. Here, we will give a brief overview of Boolean variables, algebra, and functions, and then we will look at Boolean networks that utilize these. Throughout, $\mathbb{F}_2 = \{0, 1\}$ will denote the finite field of size 2. Because addition is always assumed to be modulo 2, for any $x \in \mathbb{F}_2$, the relation $x = -x$ holds. This makes subtraction signs unnecessary. However, sometimes they are used for notational convenience. For example,

TABLE 5.1 The Four Main Boolean Logic Operations: ∧ (AND), ∨ (OR), ⊕ (XOR), and – NOT

x	y	$x \wedge y$	$x \vee y$	$x \oplus y$	$\bar{x}$
0	0	0	0	0	1
0	1	0	1	1	1
1	0	0	1	1	0
1	1	1	1	0	0

$f(x) = (x - a)g(x)$ might be preferred over $f(x) = (x + a)g(x)$ if one is trying to emphasize that a is a root of $f(x)$. Because $x = -x$, we always have $x + x = 2x = 0$ as well.

When we speak of a Boolean function, it can safely be assumed to be either of the form $f\colon \mathbb{F}_2^n \to \mathbb{F}_2$ or of the form $f\colon \mathbb{F}_2^n \to \mathbb{F}_2^n$. Boolean functions can be written either in logical notation, or in polynomial notation. We will continue to let $\vee$ denote "logical OR," $\wedge$ denote "logical AND," and $\bar{x}$ denote "NOT x." Sometimes, the $\oplus$ symbol is used to denote "logical XOR," which returns 1 if x or y, but not both, have the value 1. Table 5.1 shows the output values of Boolean logical operations on two inputs. Note that $\oplus$ is just addition modulo 2. Henceforth, we will denote this as $+$ instead of $\oplus$. This also makes it clear how to generalize XOR to more than two variables. The other three Boolean operations can also be written using basic arithmetic; the following table shows how:

$$\begin{array}{ll} f(x,y) = x \wedge y & \quad f(x,y) = xy \\ g(x,y) = x \vee y & \quad g(x,y) = x + y + xy \\ h(x) = \bar{x} & \quad h(x) = 1 + x. \end{array}$$

As a result, we can take a function written as a logical expression and write it as a polynomial, and vice versa. We will use both notations, as sometimes one is more convenient than the other. Notice that just because a variable appears in a function does not mean that the function depends on it. For example, consider the following function, written in both logical and polynomial form:

$$f(x,y) = x + y(y+1), \quad f(x,y) = x \vee (y \wedge \bar{y}).$$

Because $y(y + 1) = 0$, this function can be simplified to $f(x, y) = x$. In this case, we say that y is a *fictitious variable* of f; it appears in the expression of $f(x, y)$ but does not affect the output.

In basic high school algebra, one multiplies out a polynomial to write it as a sum of monomials, and there is a unique way to do this. In a Boolean polynomial, the relation $x^2 = x$ holds regardless of x, so every exponent can be reduced down to 1. Although one could still multiply out every polynomial as is done in high school algebra, sometimes it is convenient to leave factors of $(1 + x)$, or $\bar{x}$, in a monomial—we call this an "extended monomial." Because $x(x + 1) = 0$ for all $x \in \mathbb{F}_2$, in every extended monomial, it can be assumed that each variable x either appears as x, as $x + 1$ (equivalently, $\bar{x}$), or does not appear at all.

Definition 5.1. A Boolean function $M(x_1, \ldots, x_n)$ is an *extended monomial* if

$$M(x_1, \ldots, x_n) = \prod_{i=1}^{n} y_i,$$

where y_i is either x_i, or $1 + x_i$, or 1. □

Clearly, every polynomial can be written as a sum of extended monomials. However, there is no longer necessarily a unique way to do this. For example, the following polynomial is a sum of four extended monomials, but has six monomials (which are also extended monomials) when fully multiplied out:

$$\begin{aligned} f(x,y,z) &= xy + (x+1)yz + x + (y+1)(z+1) = xy + \bar{x}yz + x + \bar{y}\bar{z} \\ &= x + y + z + xy + xyz + 1. \end{aligned}$$

TABLE 5.2 The Truth Tables of All 16 Boolean Functions on Two Inputs

(x_1, x_2)																
$(0,0)$	0	1														
$(0,1)$	0	1														
$(1,0)$	0	1														
$(1,1)$	0	1														

The first column represents the input $x = (x_1, x_2)$, and each of the other 16 columns represents the output of a Boolean function. For example, the first two of these columns describe the two constant functions. Filling out the remainder of this table is Exercise 5.8.

Exercise 5.8. There are 16 Boolean functions on two variables. Each one is determined by where it sends the four elements of $\mathbb{F}_2^2$: $(0,0)$, $(0,1)$, $(1,0)$, and $(1,1)$, and so these 16 functions can all be arranged in a table, as in Table 5.2.

Fill out the remaining 14 functions, and write each one in both its polynomial form and its Boolean form. Given a Boolean function $f(x)$, the two-column table consisting of the input $x = (x_1, x_2)$ and the output $f(x)$ is called the *truth table* of the function. Thus, a filled-out Table 5.2 would represent the truth tables of all 16 Boolean functions on two variables. □

Exercise 5.9. A Boolean function $f\colon \mathbb{F}_2^n \to \mathbb{F}_2$ is *symmetric* if $f(x_1, \ldots, x_n) = f(x_{\pi_1}, \ldots, x_{\pi_n})$ for all permutations π of $\{1, \ldots, n\}$. In other words, the output is only determined by how many bits of x are equal to 1.

(a) Determine which of the 16 Boolean functions on two variables (see previous exercise) are symmetric.
(b) How many symmetric Boolean functions are there on n variables?
(c) Find all symmetric Boolean functions on three variables. For each one, either identify it as a commonly known function (e.g., constant, AND, OR) or give it a simple name that accurately describes it. □

5.2.5 Boolean Networks

Given a GRN, one can create a *Boolean network* model by assigning Boolean variables to each node and representing the interactions as Boolean functions. Boolean networks are discrete-time, discrete-space dynamical systems first proposed by Stuart Kauffman in 1969 as models of GRNs [32]. There is a finite set X of nodes. Sometimes, it is convenient to take $X = \{1, 2, \ldots, n\}$, and other times, letters are used for the nodes, as in the example in Figure 5.1. Each node $i \in X$ is said to have a *state* $x_i \in \{0, 1\}$, which can be a quantity or a qualitative measure (ON vs. OFF, present vs. absent, or high vs. low). A directed edge (i,j) means that the state of j depends on the state of i. Thus, the state of j should be a function of the states of nodes i connected to incoming edges of j. Unlike a GRN, the edges in a Boolean network are directed but usually unsigned. That is, edge $i \longrightarrow j$, which we sometimes write as (i,j), means that i "affects" the state of j, but it does not encode whether this interaction is an activation, an inhibition, or something more complicated. The directed graph of all nodes and edges in Boolean network is called the *wiring diagram*. If i affects its own state, then we include the self-loop (i, i).

If we take the vertex set of a Boolean network to be $X = \{1, 2, \ldots, n\}$, then the state of a node i is a Boolean variable $x_i \in \mathbb{F}_2 = \{0, 1\}$, and the vector

$$x(t) = (x_1(t), \ldots, x_n(t)) \in \mathbb{F}_2^n$$

is called the *system state*. Time is also discretized into steps $t = 0, 1, 2, \ldots$. Each node j has an *update function* $f_j\colon \mathbb{F}_2^n \to \mathbb{F}_2$ that determines the value of x_j for the next time step. Though the domain of f_j is $\mathbb{F}_2^n$, this function can only depend on the states of the nodes i such that (i,j) is an edge in the wiring diagram. Sometimes, these functions are written using Boolean logic variables, and other times they are written in polynomial form. For example, if we want to say that "gene C is on if gene A is on and enzyme B is not present," we may write

$$f_C(t+1) = f_C(A(t), B(t)) = A(t) \wedge \overline{B(t)}. \tag{5.2}$$

Because it is understood that Boolean variables are functions of time, we will usually just write the preceding example as

$$f_C(A,B) = A \wedge \overline{B}.$$

At each time step t, the states of each node are recomputed via the *global update function* $f\colon \mathbb{F}_2^n \to \mathbb{F}_2^n$ to get a new system state, $x(t+1)$. The most commonly used global update function in Boolean models simply updates the nodes synchronously:

$$x(t+1) = f(x(t)) = (f_1(x(t)), \ldots, f_n(x(t))). \tag{5.3}$$

Some models use an asynchronous update [33], but this raises the question of which of the possible $n!$ update orders to use. Thus, we will henceforth assume that a synchronous global update is used.

Definition 5.2. A *Boolean network* is a pair $(X, \mathcal{F})$ consisting of a finite set of nodes X and a set $\mathcal{F} = \{f_i\}_{i \in X}$ of update functions, where each $f_i\colon \mathbb{F}_2^n \to \mathbb{F}_2$. □

Given a Boolean network $(X, \mathcal{F})$ as defined earlier, one can construct its global update map, its wiring diagram, and a directed graph, called its *phase space*, that completely encodes the dynamics.

Definition 5.3. The *phase space* of a Boolean network is the directed graph whose nodes are the 2^n system states and whose edge set is

$$E = \{(x, f(x)) \mid x \in \mathbb{F}_2^n\}.$$

□

Each node in the phase space has exactly one outgoing edge. Consequently, there are two types of nodes: those that lie on a directed cycle, called *periodic states*, and those that do not, called *transient states*. Every periodic state lies on a *cycle* of length $k \geq 1$. States on length-1 cycles are called *fixed points*. Transient states lie on chains that lead into periodic cycles. A transient state that has no predecessor is called a *garden-of-Eden state*.

The wiring diagram of a Boolean network is also easy to construct—it is a directed graph with vertex set X and an edge (i,j) for each x_i that appears in the equation for f_j and is not fictitious.

As an example, consider a Boolean network on three nodes: $X = \{1, 2, 3\}$, and with update functions $\mathcal{F} = \{f_i\}_{i=1}^3$ as shown in Figure 5.6. These functions are shown in *polynomial form*, rather than Boolean logic

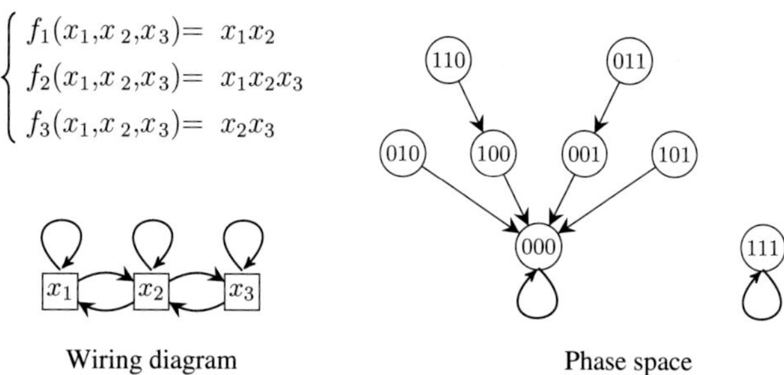

FIGURE 5.6 A simple Boolean network $(X, \mathcal{F})$ on three nodes.

form. For example, the function $f_2(x_1,x_2,x_3) = x_1x_2x_3$ as Boolean expression can be written $f_2(x_1,x_2,x_3) = x_1 \wedge x_2 \wedge x_3$. The phase space from Figure 5.6 was constructed with the freely available Web-based software Analysis of Dynamic Algebraic Models (ADAM). As of the writing of this book, this is currently located at http://adam.plantsimlab.org/, though this is likely not its permanent home. Instead of spending time in this chapter explaining how to use it, we encourage the interested reader to check it out; it is very straightforward to use.

Though Boolean networks are widely used as models of biological networks, one must remember the old adage that "all models are wrong, but some models are useful." Like any mathematical model, Boolean networks have several artifacts that have drawn criticism. One of these is the synchronous update: biological networks do not have a universal "central clock." Another issue is that the network is assumed to be static, whereas in reality, edges are continually added, removed, and changed. For example, consider the model of a disease network where edges represent social contacts. However, these social contacts are usually temporary. You might be in contact with person A in the morning, and then person B in the evening. In this case, you cannot infect person A with person B's disease, despite the fact that you're connected to both in a static social network. The interdisciplinary field of "evolving networks" is very popular due to these issues and more, but it is still in its infancy [34].

Exercise 5.10. Consider the Boolean network on three nodes whose update functions are

$$f_1(x_1,x_2,x_3) = x_1 \vee x_2, \quad f_2(x_1,x_2,x_3) = x_1 \vee x_3, \quad f_3(x_1,x_2,x_3) = x_2.$$

Write the functions in polynomial form. Then draw the wiring diagram and the phase space of this Boolean network. Identify the limit points and the transient points. Using the ADAM software (http://adam.plantsimlab.org/) will save considerable time and accuracy. □

Exercise 5.11. Let $(X, \mathcal{F})$ be a Boolean network with global update function $f\colon \mathbb{F}_2^n \to \mathbb{F}_2^n$. The set of transient states of f can be expressed formally as the set

$$\text{Trans}(f) := \{x \in \mathbb{F}_2^n \mid f^{(k)}(x) \neq x \text{ for all } k \geq 1\},$$

where $f^{(k)}(x)$ denotes f applied to x exactly k times. Complete the following statements with formal mathematical definitions.

(a) The set of periodic points of f is:

$$\text{Per}(f) := \left\{x \in \mathbb{F}_2^n \mid \qquad\qquad \right\}.$$

(b) The fixed point set of f is:

$$\text{Fix}(f) : \left\{x \in \mathbb{F}_2^n \mid \qquad\qquad \right\}.$$

(c) The garden-of-Eden states of f are the set:

$$\text{GoE}(f) := \left\{x \in \mathbb{F}_2^n \mid \qquad\qquad \right\}.$$

□

Exercise 5.12 (Exploratory). Can you create a Boolean network on three nodes whose phase space consists of a single length-8 cycle? Can you create one whose phase space consists of a single chain leading into a fixed point? □

5.3 CANALIZATION

5.3.1 Canalizing Boolean Functions

Although Boolean networks can be fun to play around with in their own right, there is particular interest in identifying and studying networks that arise from actual biological systems. One can ask what types of directed graphs arise as wiring diagrams, or what types of functions are more likely to model such systems. Some functions tend to be more "biologically meaningful" than others. One such class of functions are the *canalizing functions*, which were introduced by Stuart Kauffman, nearly 25 years after he invented Boolean networks [31]. Loosely speaking, a function is canalizing if one of its variables can completely determine the output of the function by taking on a certain value. This phenomenon is common in biology because it describes instances where, e.g., a certain cellular process will not begin if a particular substrate is present, or a reaction needs a particular enzyme to catalyze it. In this section, we will see how this property is characterized in a Boolean model.

For example, in the simple model of the *lac operon* from Figure 5.1, mRNA (M) will be transcribed (produced) in the next time step if lactose is available in the cell or externally and in addition if external glucose (G_e) is present. This can be modeled by the following Boolean function:

$$f_M\colon \mathbb{F}_2^3 \longrightarrow \mathbb{F}_2, \qquad f_M(L, L_\mathrm{e}, G_\mathrm{e}) = (L \vee L_\mathrm{e}) \wedge \overline{G_\mathrm{e}}. \tag{5.4}$$

Here, if $G_\mathrm{e} = 1$, then $f_M = 0$ regardless of the values of L and L_e. We say that G_e is a *canalizing* variable. In contrast, L is not canalizing because for either value of L, the output of f_M can still be either 0 or 1. Similarly, L_e is not a canalizing variable.

Definition 5.4. A Boolean function $f\colon \mathbb{F}_2^n \to \mathbb{F}_2^n$ is *canalizing* if there is a Boolean function $g(x_1, \ldots, x_{i-1}, x_{i+1}, \ldots, x_n)$ and $a, b \in \mathbb{F}_2$ such that

$$f(x_1, \ldots, x_n) = \begin{cases} b & x_i = a \\ g \not\equiv b & x_i \neq a. \end{cases}$$

In this case, x_i is a *canalizing variable*, the input a is its *canalizing value*, and the output value b when $x_i = a$ is the corresponding *canalized value*. □

The $g \not\equiv b$ condition ensures that $g(x)$ is not a constant function. Having this in the definition ensures that constant functions are not defined to be canalizing, even though they trivially have a variable (any x_i will do) such that if it takes on a particular value (any), the output $f(x)$ is fixed. It is worth noting, however, that this convention is not entirely universal—some books and papers use the convention that constant functions are indeed canalizing.

Returning to the example above, the function f_M is canalizing because it can be written as

$$f_M(L, L_\mathrm{e}, G_\mathrm{e}) = \begin{cases} 0 & G_\mathrm{e} = 1 \\ L \vee L_\mathrm{e} & G_\mathrm{e} \neq 1. \end{cases}$$

Here, G_e is the canalizing variable, and if it takes its canalizing value of 1, then the function f_M outputs the canalized value of 0. If G_e does not take its canalizing value of 1, then $f_M = L \vee L_\mathrm{e}$, which is not a constant function.

As another example, the functions $f_1(x, y) = x \wedge y$ and $f_2(x, y) = x \vee y$ are canalizing. For f_1, both x and y are canalizing variables because if either one is 0 (the "canalizing value"), this forces the output to be 0 (the "canalized value"). Similarly, x and y are both canalizing variables of $f_2(x, y)$ with canalizing values 1 and canalized value 1.

Exercise 5.13. For each of the 16 Boolean functions from Exercise 5.8, determine if it is canalizing. □

5.3.2 Nested Canalizing Functions

The concept of a *nested canalizing* function was introduced by Stuart Kauffman and collaborators in 2003 in a paper on modeling the yeast transcription network [35]. As with canalizing functions, it was introduced because of the observation that many functions in Boolean network models have this particular property. In short, nested canalizing means "recursively canalizing" in the following sense: If a function is canalizing, then there is some variable x_j that can determine the output of f if it gets the right input, a_j. If $x_j \neq a_j$, then f is a Boolean function

g on 1 fewer variable. One may ask if this function is canalizing, and repeat this process. A function is nested canalizing if canalizing variables can be recursively "picked off" in this manner until there are none left.

Definition 5.5. A Boolean function $f\colon \mathbb{F}_2^n \to \mathbb{F}_2$ is a *nested canalizing function* (NCF) if there is some reordering of the variables $x_1, \ldots, x_n$ so that f can be written as

$$f(x) = \begin{cases} b_1 & x_1 = a_1 \\ b_2 & x_1 \neq a_1,\ x_2 = a_2 \\ b_3 & x_1 \neq a_1,\ x_2 \neq a_2,\ x_3 = a_3 \\ \vdots & \vdots \\ b_n & x_1 \neq a_1, \ldots,\ x_{n-1} \neq a_{n-1},\ x_n = a_n \\ \overline{b_n} & x_1 \neq a_1, \ldots,\ x_n \neq a_n \end{cases} \tag{5.5}$$

for some choice of constants $a_1, \ldots, a_n \in \mathbb{F}_2$ and $b_1, \ldots, b_n \in \mathbb{F}_2$. □

The definition of NCF requires that at least one (of all possible $n!$) orderings of the variables yields an equation of the form in Eq. (5.5). Sometimes, any ordering will do, but other times, there are only a few. We will say that a permutation of the variables that yields an NCF is a *nested canalizing sequence*. For example, all six orderings of the variables $\{x, y, z\}$ in the function $f(x, y, z) = x \vee y \vee z$ are nested canalizing sequences. An NCF written as in Eq. (5.5) is said to be in *standard form*. Naturally, this need not be unique—every canalizing sequence gives rise to a way to write the function in standard form.

In 2007, Jarrah et al. [36] discovered that what Kauffman had been calling NCFs had been independently studied in the electrical engineering (EE) community, under the name of *unate cascade functions*. Because these functions had been widely studied, theorems about them from EE immediately gave insight to those in mathematical biology studying NCFs. For example, a function is nested canalizing if and only if it can be written as a composition of AND and OR functions, as described by the following proposition [36].

Proposition 5.1. *A Boolean function $f\colon \mathbb{F}_2^n \to \mathbb{F}_2$ is nested canalizing if and only if it there is some reordering of the variables $x_1, \ldots, x_n$ so that f has the following form:*

$$f(x_1, \ldots, x_n) = y_1 \Diamond_1 (y_2 \Diamond_2 (\cdots (y_{n-1} \Diamond_{n-1} y_n)) \cdots)$$

where

- *each y_i is either x_i or $1 + x_i$;*
- *each $\Diamond_i$ is either $\wedge$ or $\vee$.*

The interest in unate cascade functions (i.e., NCFs) by electrical engineers arose because this is the precise class of functions whose "binary decision diagrams" (BDD) have minimal average path length (APL) [37]. A *BDD* is a simple data structure that can efficiently represent a Boolean function, and it additionally serves as a convenient visual aid and as a quick evaluation tool. The basic idea is to minimize the memory needed to store the function and the time steps needed to evaluate the function with respect to a fixed variable order.

A Boolean function $f(x_1, \ldots, x_n)$ can be also represented (inefficiently) as a *binary decision tree*. To do this, one needs to first specify a fixed *variable order* on $\{x_1, \ldots, x_n\}$. The binary decision tree gives a quick way to compute the function output when the values of the variables are plugged in according to this order. This is best seen by an example: consider the function

$$f(x_1, x_2, x_3) = x_1 \wedge (x_2 \vee x_3) = x_1x_2x_3 + x_1x_2 + x_1x_3$$

with variable order $x_1 < x_2 < x_3$, which means "x_1 comes first, then x_2, then x_3." The binary decision tree of this function is shown on the left in Figure 5.7. The root vertex is labeled with the first variable (in this case, x_1). There are two outgoing edges—one corresponding to setting $x_1 = 0$ (dashed, and to the left), and the other to $x_1 = 1$ (solid, and to the right). Both of the children are labeled with the next variable in the fixed order (in this case, x_2). This process is repeated: each nonleaf node is labeled with a variable x_i and has exactly two children (nodes directly "below" it). The nodes are labeled by level, from top to bottom. Finally, each node labeled with

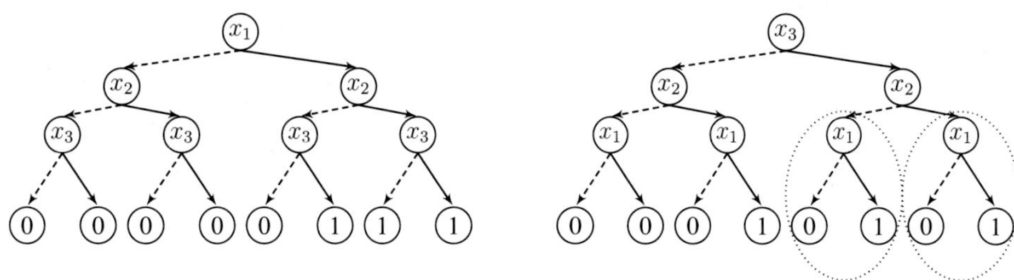

FIGURE 5.7 Two binary decision trees of $f = x_1x_2x_3 + x_1x_2 + x_1x_3$, with respect to variable orders $x_1 < x_2 < x_3$ (left) and $x_3 < x_2 < x_1$ (right). A dashed edge out of node x_i means that $x_i = 0$, and a solid edge out of x_i means $x_i = 1$. Each evaluation of $x_1 = a_1$, $x_2 = a_2$, and $x_3 = a_3$ corresponds to a unique leaf which is labeled by $f(a_1, a_2, a_3)$.

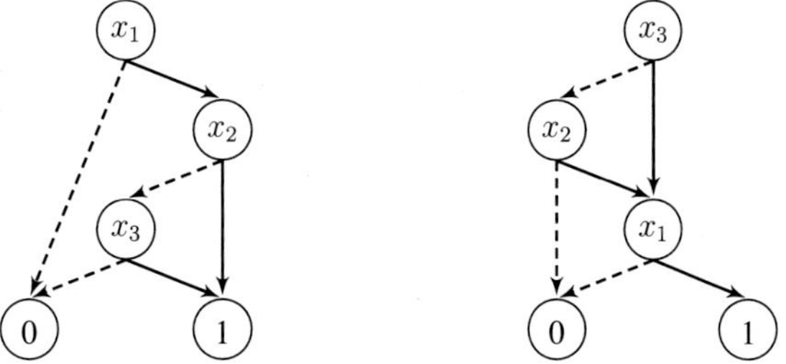

FIGURE 5.8 Two BDDs of $f = x_1x_2x_3 + x_1x_2 + x_1x_3$. The one on the left arises from the variable order $x_1 < x_2 < x_3$, and the one on the right arises from $x_3 < x_2 < x_1$.

the last variable (in this case, x_3) also has two children, but these are leaves. Notice that in general, there are 2^n leaves, each one having a unique length-n path back to the root. This path uniquely describes an evaluation of all n variables. Label each leaf with either 0 or 1—the value of the function $f(x_1, \ldots, x_n)$ corresponding to this particular evaluation.

A binary decision tree is an extremely inefficient way to represent a Boolean function—it requires 2^n leaf nodes and $2^n - 1$ interior nodes, which is prohibitively large for functions of more than just a few variables. It has other drawbacks as well: it carries a lot of redundant information, and the trees of two n-variable functions look structurally identical. A *BDD* can be thought of as a "reduced" version of a binary decision tree, in that it carries the same information but without the redundancies. For example, the diagram on the left of Figure 5.8 is the BDD for the function $f = x_1x_2x_3 + x_1x_2 + x_1x_3$, with order $x_1 < x_2 < x_3$. Notice how any evaluation $f(a_1, a_2, a_3)$ can be computed in the same manner as for the binary decision tree—start at the root, and follow the paths corresponding to $x_1 = a_1$, $x_2 = a_2$, and $x_3 = a_3$ until a node labeled with 0 or 1 is reached. Also notice how the fact that f is an NCF with the variable order $x_1 < x_2 < x_3$ can be visualized from the BDD: Starting at the node labeled x_1, there is an edge ($x_1 = 0$) down to a leaf. However, the other edge ($x_1 \neq 0$) leads to a node labeled x_2, which also has a direct edge ($x_2 = 1$) to a leaf. To summarize, there is a unique node at every level, and each node has a unique path to a leaf node. On the other hand, the BDD on the right of Figure 5.8 for the same function f but with respect to variable order $x_3 < x_2 < x_1$ does not have that property. Specifically, there is no edge from x_3 to a leaf. Even stronger: upon following either edge from x_3, one can still reach both 0 and 1 nodes. Therefore, x_3 is not a canalizing variable.

In any rooted tree, every node has a canonical *subtree* consisting of itself and all its descendants. However, the BDDs in Figure 5.8 are not trees, but acyclic directed graphs. In such a structure, every node still has an analogue of a subtree that we called a *substructure*. Specifically, the substructure of v is the directed graph consisting of all of the nodes and edges that can be reached via a directed path from v.

In general, the problem of how to construct a BDD given a Boolean function and a fixed variable order is difficult. If a binary decision tree has already been constructed, then it can be easily reduced to a BDD by repeatedly applying the following operations:

(i) Merge identical substructures that have the same parent node, and then eliminate that node.
(ii) Merge identical substructures that have different parents.

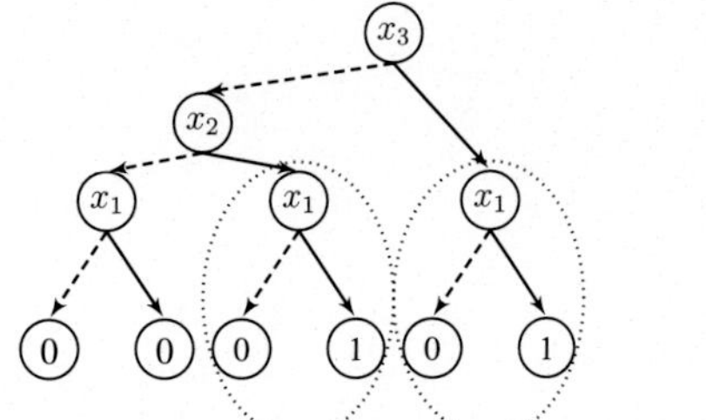

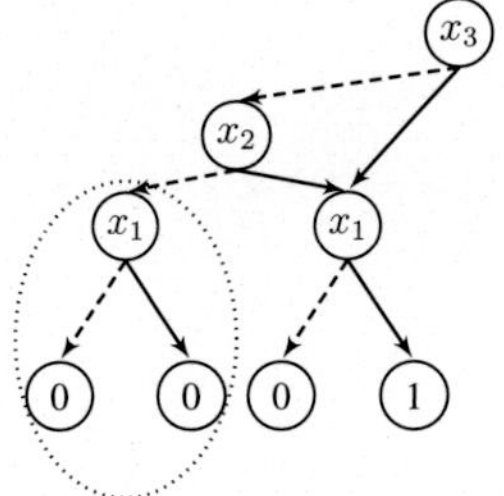

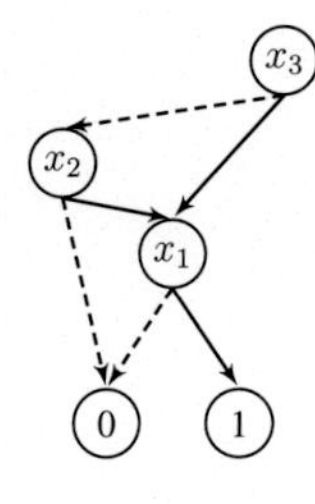

FIGURE 5.9 The merging process applied to the binary decision tree of $f = x_1x_2x_3 + x_1x_2 + x_1x_3$, with order $x_3 < x_2 < x_1$. The tree on the left is obtained by merging the two circled subtrees in Figure 5.7 and removing the parent vertex. The middle tree is formed by merging the two circled subtrees on the leftmost tree, and the tree on the right is formed by eliminating the x_1-labeled vertex.

These two operations are applied repeatedly as long as they are applicable and the resulting graph will be a BDD. As an example of this, the binary decision tree of $f = x_1x_2x_3 + x_1x_2 + x_1x_3$ with respect to order $x_3 < x_2 < x_1$ is shown in Figure 5.7 on the right. Notice that for three of the four nodes labeled x_1, the substructures (subtrees) rooted at those nodes are identical. When the two circled subtrees are merged and their parent node (labeled x_2) is eliminated, we obtain the diagram on the left of Figure 5.9. This diagram also has two identical subtrees rooted at x_1-nodes, but with different parents. Merging these subtrees gives the diagram in the middle of Figure 5.9, which is no longer a tree. Finally, the x_1-node that has two 0-children can be eliminated, yielding the BDD which is shown on the right in Figure 5.9.

It is not obvious, but every Boolean function has, for a given ordering of variables, a unique BDD [37]. On the other hand, different variable orders of the same function generally have BDDs that are structurally different. For example, both diagrams in Figure 5.8 are BDDs of the same function, but with respect to different variable orderings.

One of the primary utilities of BDDs is that they represent a concise representation of a Boolean function. Every evaluation of a function's variables corresponds to a path from the root to the leaves. The APL of a BDD, taken over all 2^n evaluations, is in some sense a measure of the function's complexity. Because this depends on the variable order, we say that the *APL* of a Boolean function is the minimal APL of one of its BDDs, taken over all possible orderings. We denote this as $\text{APL}_{f(x)}$, and it describes how quickly f can be evaluated on average. An NCF has the property that at every interior vertex, there is a direct edge to a leaf, which lowers the APL considerably. Obviously, constant functions have the simplest BDDs—they would consist of just a single vertex and no edges. The number of levels of a BDD describes how many variables the function depends on. By definition, an NCF depends on all variables, and so its diagram must have n levels (or $n + 1$, if the leaves are included). The next result, from the EE community, says that NCFs are the "quickest" functions to evaluate. It may be no coincidence that these functions often arise in biological networks—they may possess some sort of evolutionary advantage.

Theorem 5.1 (Jarrah et al. [36]). *The n-variable Boolean functions with no fictitious variables that have minimal APL are precisely the NCFs.*

Exercise 5.14. The function $f(x, y, z) = x \vee y \vee z$ is nested canalizing, and all six orderings of the variables are nested canalizing sequences. Find a 3-variable NCF with fewer than six nested canalizing sequences. Can you find an NCF that has only one? □

Exercise 5.15. Find a 3-variable function $f\colon \mathbb{F}_2^3 \to \mathbb{F}_2$ that satisfies the definition of being nested canalizing from Definition 5.5 except for the last line, requiring $f(x) = \overline{b_3}$. That is,

$$f(x) = \begin{cases} b_1 & x_1 = a_1 \\ b_2 & x_1 \neq a_1,\ x_2 = a_2 \\ b_3 & x_1 \neq a_1,\ x_2 \neq a_2,\ x_3 = a_3 \end{cases}$$

What property must f have for this to hold? Draw its BDD with respect to the variable order $x_1 < x_2 < x_3$. Explain why it is reasonable to not call such a function nested canalizing. □

Exercise 5.16. Determine the Boolean functions described by the following diagrams by writing out the truth table. Are these functions nested canalizing? Simplify the diagrams by merging identical substructures until you wind up with a BDD.

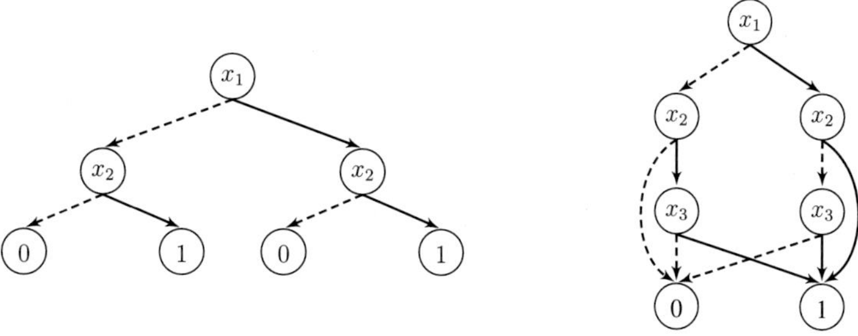

□

Exercise 5.17. Consider the Boolean function $f(x_1, x_2) = x_1 + x_2$. What is APL_f? □

Exercise 5.18. What is the largest APL that a Boolean function of n variables can have? Which functions achieve this? □

Exercise 5.19. Compute the APL of the BDDs shown in Figure 5.8. What is the APL of the Boolean function $f(x_1, x_2, x_3) = x_1x_2x_3 + x_1x_2 + x_1x_3$? □

Exercise 5.20. For NCFs, any canalizing order will yield a BDD with minimal APL. Can you derive a general formula for the APL of an NCF on n variables? □

5.3.3 Canalizing Depth

Though most Boolean functions that arise in gene network models are nested canalizing, there are exceptions. For example, in a model of the *lac* operon found in Ref. [20], the update function for internal lactose is canalizing but not nested canalizing:

$$f_L(E, L, G_e, L_e) = \overline{G_e} \wedge [(E \wedge L_e) \vee (L \wedge \overline{E})] \tag{5.6}$$

where E is a term indicating the presence of β-galactosidase and lactose permease, both of which are produced simultaneously if mRNA is present. In plain English, this says that there will be lactose (L) present in the next time step if there is no external glucose ($\overline{G_e}$), *and* either there is external lactose (L_e) and the necessary transporter protein (E), or there already is internal lactose (L) and no β-galactosidase (E). One may argue that in such a model, external lactose is more of a parameter than a variable. If L_e is treated as a parameter (i.e., a constant), then the function in Eq. (5.6) is indeed an NCF. Despite this, the preceding example shows that some functions can be "partially" nested canalizing.

In the example from Eq. (5.6), the variable G_e is canalizing, and f_L can be written as

$$f_L(E, L, G_e, L_e) = \begin{cases} 0 & \text{if } G_e = 1 \\ (E \wedge L_e) \vee (L \wedge \overline{E}) & \text{if } G_e \neq 1 \end{cases}$$

and the function $g(E, L, L_e) = (E \wedge L_e) \vee (L \wedge \overline{E})$ is not canalizing.

Definition 5.6. Suppose a Boolean function $f\colon \mathbb{F}_2^n \to \mathbb{F}_2$, under some reordering of the variables $x_1, \ldots, x_n$ can be written as

$$f(x) = \begin{cases} b_1 & x_1 = a_1 \\ b_2 & x_1 \neq a_1,\ x_2 = a_2 \\ b_3 & x_1 \neq a_1,\ x_2 \neq a_2,\ x_3 = a_3 \\ \vdots & \vdots \\ b_d & x_1 \neq a_1, \ldots,\ x_{d-1} \neq a_{d-1},\ x_d = a_d \\ g \not\equiv b_d & x_1 \neq a_1, \ldots,\ x_d \neq a_d \end{cases} \tag{5.7}$$

for some choice of constants $a_1, \ldots, a_d \in \mathbb{F}_2$ and $b_1, \ldots, b_d \in \mathbb{F}_2$, and noncanalizing function g. Then we say that f has *canalizing depth d*. □

For completeness, we may say that a noncanalizing function has canalizing depth 0. Notice that the NCFs are precisely the functions of canalizing depth n. Sometimes, canalizing functions of depth d, where $1 \leq d \leq n$, are called *partially nested canalizing*. These functions are quite new in the literature, and they are still being studied [38]. Our motivation to introducing them is twofold: (i) to show the reader how the canalization and nested canalization concepts can be generalized, and (ii) these functions make for a great example of a Derrida plot to measure the stability of RBNs, which will be done in the next section.

5.3.4 Dominant Variables of NCFs

In a NCF, some variables are more "dominant" than others. For example, in the function $f(x, y, z) = xyz + x$, which has two nested canalizing sequences: x, y, z and x, z, y, variable x is the most dominant variable because it comes first in both nested canalizing sequences. On the other hand, y and z seem to have the same dominance. To formalize this concept, we need to understand how to decompose an arbitrary NCF into "extended monomial layers."

Recall that an extended monomial is a product of distinct variables from $\{x_1, \ldots, x_n, \bar{x}_1, \ldots, \bar{x}_n\}$ such that for each i, at most one of x_i and $\bar{x}_i$ can appear. An extended monomial $M = y_1 y_2 \cdots y_n$ containing every variable is an NCF, and *any* ordering of the variables is a nested canalizing sequence. In other words, all variables in an extended monomial have the same dominance. For a general NCF, we will describe how to obtain a unique algebraic form using extended monomials. In this form, all the variables will be partitioned into layers, where each layer contains an extended monomial of the corresponding variables. The variables in outer layers will be more dominant than those in inner layers. Variables in the same layer will have the same dominance.

Theorem 5.2 (Li et al. [39]). *Given $n \geq 2$, the function $f(x_1, \ldots, x_n)$ is an NCF if and only if it can be uniquely written as*

$$f(x_1, \ldots, x_n) = M_1(M_2(\cdots(M_{r-1}(M_r + 1) + 1)\cdots) + 1) + b,$$

where each M_j is an extended monomial, and M_r contains at least two variables. Moreover, for each $i = 1, \ldots, n$, the variable y_i (either x_i or $1 + x_i$) appears in exactly one M_j. The variables in each M_i are said to be in the ith monomial layer.

The peculiar requirement that the innermost layer M_r contains at least two variables is needed because if it only had one variable, then it could be "absorbed" into the M_{r-1} extended monomial. This will be explored in Exercises 5.21 and 5.22.

To illustrate Theorem 5.2 with an example, consider two elementary Boolean functions on three variables: the AND and OR functions. It is clear right away that AND has a one-monomial layer, but less clear that OR does as well. It helps to write OR as the negation of NOR, which is the product of the negation of the variables:

$$\text{and}(x, y, z) = xyz, \qquad \text{or}(x, y, z) = \text{nor}(x, y, z) + 1 = \bar{x}\bar{y}\bar{z} + 1.$$

For a more complicated example, consider the following function:

$$\begin{aligned} f(x_1,\ldots,x_7) &= x_1(x_2+1)(x_3x_4(x_5+1)(x_6x_7+1)+1) \\ &= x_1\overline{x_2}(x_3x_4\overline{x_5}(x_6x_7+1)+1). \end{aligned}$$

By Theorem 5.2, this function is an NCF with three monomial layers: the most dominant variables are x_1 and x_2 because they appear in M_1. The next layer contains the variables that appear in M_2: x_3, x_4, and x_5. Finally, the last layer consists of the variables x_6 and x_7 because they appear in M_3. The layer structure can be seen from the standard NCF form of f:

$$f = \begin{cases} 0 & x_1 = 0 \\ 0 & x_1 \neq 0,\ x_2 = 1 \\ 1 & x_1 \neq 0,\ x_2 \neq 1,\ x_3 = 0 \\ 1 & x_1 \neq 0,\ \ldots,\ x_3 \neq 0,\ x_4 = 0 \\ 1 & x_1 \neq 0,\ \ldots,\ x_4 \neq 0,\ x_5 = 1 \\ 0 & x_1 \neq 0,\ \ldots,\ x_5 \neq 1,\ x_6 = 0 \\ 0 & x_1 \neq 0,\ \ldots,\ x_6 \neq 0,\ x_7 = 0 \\ 1 & x_1 \neq 0,\ \ldots,\ x_6 \neq 0,\ x_7 \neq 0 \end{cases}$$

Each layer appears as a block of variables with the same canalized output. Note that for the last variable, we can choose either 0 (next-to-last line) or 1 (last line) as the canalized output, thus the last layer has at least two variables.

As mentioned before, there might be more than one nested canalizing sequence for a given NCF. We can now find all these sequences using monomial layers [39].

Proposition 5.2. *Given $n \geq 2$, let the function $f(x_1,\ldots,x_n)$ be an NCF of the form shown in Theorem 5.2. Rearranging the order of variables in any layer yields a nested canalizing sequence. Moreover, every nested canalizing sequence can be realized in this manner.*

If layer M_i contains k_i variables, then there are $k_i!$ orderings of the variables within it. Thus there are $\prod_{i=1}^{r}(k_i!)$ different nested canalizing sequences in all. For example, the 7-variable 3-layer NCF given earlier has $2!3!2! = 24$ nested canalizing sequences. Note that because M_r must contain at least two variables, then every NCF on more than two variables must have at least two nested canalizing sequences.

Exercise 5.21. The NCF $f(x_1,x_2,x_3) = x_1(x_2(x_3+1)+1)$ has only two monomial layers, because of the requirement that the innermost layer must have at least two variables. Find these two layers. That is, write this function as

$$f(x_1,x_2,x_3) = x_1(x_2(x_3+1)+1) = M_1(M_2+1)+1)+b$$

where M_1 and M_2 are extended monomials and $b = 0$ or 1. Find all nested canalizing sequences for f. □

Exercise 5.22. Consider the following standard form of an NCF f:

$$f = \begin{cases} 1 & x_1 = 0 \\ 0 & x_1 \neq 0,\ x_2 = 1 \\ 1 & x_1 \neq 0,\ x_2 \neq 1,\ x_3 = 0 \\ 0 & x_1 \neq 0,\ x_2 \neq 1,\ x_3 \neq 0,\ x_4 = 1 \\ 1 & x_1 \neq 0,\ x_2 \neq 1,\ x_3 \neq 0,\ x_4 \neq 1 \end{cases}$$

(a) How many extended monomial layers does f have? How many different nested canalizing sequences does f have in total?
(b) Write down a different nested canalizing sequence of f other than x_1, x_2, x_3, x_4. What is the standard form of f with respect to the sequence you wrote down? □

5.4 DYNAMICS OVER COMPLEX NETWORKS

As we have seen, many real-world phenomena can be modeled as dynamical processes over large complex networks. Examples include GRNs, protein-protein interaction networks (see Figure 5.10), the spread of a disease or a rumor through a social network, a virus or worm spreading through a computer network, or a posting that has gone "viral" through a social media network. For a more thorough study of these topics, some great references include the books *Networks: an introduction* [40] by Mark Newman, *The structure and dynamics of networks* [41] by Newman, Barabási, and Watts, as well as *Dynamical processes on complex networks* [42], by Barrat et al., and the 134-page survey article *Complex networks: structure and dynamics* [43] by Boccaletti et al. In this final section, we will focus on one specific aspect: stability of dynamics from a mean-field perspective from statistical physics, and we will return to the theme of critical dynamics that was introduced in the beginning.

Recall that an *NK*-network is a directed graph on N (a large number of) vertices, each having in-degree K (a small number). Kauffman proposed this as a theoretical model for a chromosome with N genes, each one appearing in K alleles [31]. A common existing model in population genetics was the *N-locus, two-allele additive fitness model*, which is the case of an *NK*-network with $K = 2$. Though the in-degree of the nodes in such a network is uniform, the out-degree need not be. Some nodes will not regulate any other nodes, whereas others will regulate many. In general, the out-degree distribution of an *NK*-network could vary widely. It could be the case that each node has out-degree equal to K, but more likely, it will follow a power-law or binomial distribution (approximately normal, for large N). The only requirement is that the average in-degree be

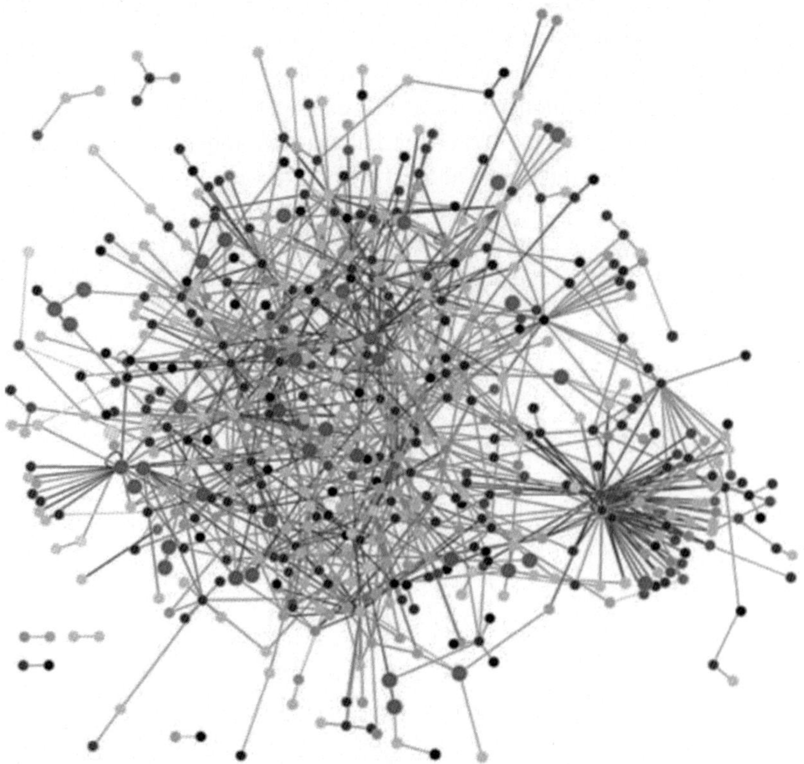

FIGURE 5.10 The protein interaction network of *Treponema pallidum*, from Ref. [46].

exactly K. It turns out that the two most influential parameters in an NK-network are N and K themselves, rather than the network topology. It has also been shown that NK-networks with $K = 2$ have dynamics in the critical phase, between ordered and chaotic [44].

In this section, well will describe how to quantify this phase transition using the mean-field theory from statistical physics. The main technique is using a simple plot called a *Derrida plot*, introduced in 1986 by Derrida and Pomeau [45]. We will also examine the role of the functions in the dynamics of Boolean networks. Though the regulatory functions in RBN models are often constructed by sampling from all 2^{2^K} functions on K inputs uniformly, not all of these functions arise with the same frequency in actual gene networks. Kauffman defined canalizing, and then nested canalizing Boolean functions because most observed regulatory functions share these properties. Mathematical biologists wish to understand if networks built with these functions exhibit any special features. For example, are they more or less stable than a "random" network? To do this, we will introduce the *activity* of variables, and the *sensitivity* of the function. These can be defined using basic Boolean calculus. In a nutshell, (traditional) calculus is the study of the rate of changes of functions, and Boolean calculus measures how Boolean functions change—the output either stays the same or flips ($0 \leftrightarrow 1$) as the input Boolean vector changes. Ordinary calculus is not a prerequisite, though knowing some certainly would not hurt; the primary benefit would be that it would aid in recognizing parallels and further motivating the basic definitions.

5.4.1 Boolean Calculus

In vector calculus, the partial derivative $\partial f/\partial x_j$ of a real-valued function measures how the function changes if the input vector changes by a small amount in the x_j-direction. There is a natural analogue of this for Boolean functions. The *Boolean partial derivative* of $f\colon \mathbb{F}_2^n \to \mathbb{F}_2$ with respect to x_j is a function defined as

$$\frac{\partial f}{\partial x_j}\colon \mathbb{F}_2^n \longrightarrow \mathbb{F}_2, \qquad \frac{\partial f(x)}{\partial x_j} = f(x + e_j) + f(x),$$

where e_j is the vector with a 1 in the jth coordinate and 0s elsewhere. Note that $\partial f(x)/\partial x_j$ is 1 if flipping the jth bit of the vector x changes the output, and 0 otherwise. This doesn't measure how much the function changes (that no longer applies), just whether it changes: "yes" (1) or "no" (0).

The *activity* of the variable x_j in f is a number between 0 and 1 defined as

$$\alpha_j^f = \frac{1}{2^n} \sum_{x \in \mathbb{F}_2^n} \frac{\partial f(x)}{\partial x_j}. \tag{5.8}$$

The activity of x_j is precisely the expected value of $\frac{\partial f(x)}{\partial x_j}$ over all 2^n input vectors, and it measures how often toggling the jth bit of the input vector toggles the output of f. There is no obvious analogue of this concept in classical calculus.

The *sensitivity* of f at $x \in \mathbb{F}_2^n$ is an integer between 0 and n defined by

$$s^f(x) = \sum_{i=1}^{n} \chi(f(x + e_i) \neq f(x)),$$

where χ is an "indicator function" that is 1 if $f(x + e_i) \neq f(x)$, and 0 otherwise. The sensitivity $s^f(x)$ measures the number of ways that toggling a bit of x toggles the output of f. The average sensitivity of f is the expected value of $s^f(x)$ taken uniformly over all $x \in \{0, 1\}^n$, i.e.,

$$s^f = E\left[s^f(x)\right] = \sum_{i=1}^{n} \alpha_i^f. \tag{5.9}$$

In some sense, the average sensitivity of a function is a measure of its stability. As with the activity, there is no clear analogue of this concept in standard calculus. For an example, constant functions have average sensitivity 0: flipping an input bit will never change the output of the function. In contrast, the *parity* function

$$f(x_1,\ldots,x_n) = x_1 + \cdots + x_n$$

has average sensitivity 1, because given any input vector, flipping any single bit will change the output. Intuitively, the variables in canalizing functions should have lower activities than those in arbitrary functions, and the actual functions should have a lower than average sensitivity. To motivate this, suppose f is canalizing, with $f = b$ if $x_i = a$. For half of the input vectors (that is, 2^{n-1} out of 2^n), flipping one of the $n-1$ bits (excluding x_i) will not change the output. One would expect that for a random Boolean function, flipping one of the bits would change the output roughly 50% of the time. Indeed, this loss of sensitivity can be quantified.

Theorem 5.3 (Shmulevich and Kauffman [47]). *Consider a random canalizing function $f(x_1,\ldots,x_n)$ with canalizing variable x_1. Then the activity vector of the variables is $E[\alpha^f] = (\frac{1}{2}, \frac{1}{4}, \ldots, \frac{1}{4})$, and hence the average sensitivity is*

$$s^f = \frac{1}{2} + (n-1)\frac{1}{4} = \frac{n+1}{4}.$$

Because canalizing functions have lower activities and sensitivities than "random" Boolean functions, one would expect that Boolean network models built with canalizing functions would be more stable than those built with arbitrary functions. This is indeed the case and is the main focus of the remainder of this chapter. In addition, we will see how the canalizing depth affects the stability.

Exercise 5.23. Compute the partial derivatives $\partial f/\partial x_j$ of each of the following functions. Describe the resulting function; sometimes it will be a commonly known Boolean function.

(i) The logical OR function: $f(x_1,\ldots,x_n) = x_1 \vee \cdots \vee x_n$
(ii) The logical AND function: $f(x_1,\ldots,x_n) = x_1 \wedge \cdots \wedge x_n$
(iii) The parity function: $f(x_1,\ldots,x_n) = x_1 + \cdots + x_n$, where the sum is taken modulo 2
(iv) The *k-threshold* function on n variables, defined by

$$f\colon \mathbb{F}_2^n \longrightarrow \mathbb{F}_2^n, \qquad f(x) = \begin{cases} 1 & x_i = 1, \text{ for at least } k \text{ entries of } x \\ 0 & \text{otherwise} \end{cases}$$

All these functions are *symmetric*, which means that $f(x_1,\ldots,x_n) = f(x_{\pi_1},\ldots,x_{\pi_n})$ for all permutations π of $\{1,\ldots,n\}$. Therefore, $\partial f/\partial x_i = \partial f/\partial x_j$ for any i and j. □

Exercise 5.24. For each of the following symmetric functions on three variables, compute the activities of a variable (it does not matter which one, due to symmetry) and the compute the sensitivity of the function.

(i) A constant function
(ii) The logical OR function: $f(x_1,x_2,x_3) = x_1 \vee x_2 \vee x_3$
(iii) The logical AND function: $f(x_1,x_2,x_3) = x_1 \wedge x_2 \wedge x_3$
(iv) The parity function: $f(x_1,x_2,x_3) = x_1 + x_2 + x_3 \pmod 2$
(v) The k-threshold function for $k = 1, 2$, and 3 (See the previous exercise. Note that the 0-threshold and 4-threshold functions are the constant functions.) □

Exercise 5.25. Generalize the previous exercise from 3 to n-variable symmetric functions. □

Exercise 5.26. Consider the following two NCFs:

$$f_1(x_1,x_2,x_3,x_4) = x_1x_2x_3x_4, \qquad f_3(x_1,x_2,x_3) = x_1(x_2(x_3x_4+1)+1),$$

which have one and three extended monomial layers, respectively. Write out the truth table of each function. Then, compute the activity of each variable and the average sensitivity of the function. Make a conjecture as to which layers tend to have the variables with the highest activities. Do you see a trend between the layer number and average sensitivity (and hence stability of Boolean network models that use these functions)? Check Ref. [39] for the answer and more on this. □

5.4.2 Derrida Plots and the Three Dynamical Regimes

Consider a large RBN on N nodes. The state space likely has some fixed points and some large limit cycles. But which of these is prevalent? What does the state space look like, and what can we say about the dynamics of the system? Here are some general questions that, although we cannot answer exactly, we would like to gain a better qualitative understanding of:

- What are the lengths of the limit cycles?
- How many limit cycles are there?
- What are the lengths of the transient chains?
- Are small perturbations in a Boolean network likely to die out or spread throughout the network?

Because the state space is a directed graph on 2^N nodes, it is much too large to compute for most real models. In mathematics, when problems like this arise, a common approach is to run numerical simulations. In Ref. [44], Kauffman did this using NK-networks for various small values of K. He noticed distinct differences in the dynamics in the cases of $K \leq 2$ and $K > 2$, with $K = 2$ being the "critical" threshold. For example, the limit cycles of networks with $K \leq 2$ tend to be small, whereas if $K \geq 3$, there tend to be many fewer limit cycles but they are exponentially large. These questions can be analytically studied using an *annealed approximation* from statistical physics [45]. In this section, we will focus on the question about whether perturbations die out or spread, how to measure this experimentally, and what it tells us about stability. The main tool will be *Derrida plots* [45].

The *Hamming distance* between two Boolean vectors $x, x' \in \mathbb{F}_2^N$ is the number of bits in which they differ. Because this is an integer between 0 and N, it is usually convenient to divide this quantity by N to get the *normalized Hamming distance*. Henceforth, when we speak of Hamming distance, we will assume it is normalized, and so a "small" Hamming distance is close to 0, and a "large" Hamming distance is close to 1. Denote the normalized Hamming distance between x and x' by

$$\overline{\mathrm{H}}(x, x') := \frac{1}{N} \sum_{i=1}^{N} |x_i - x_i'|.$$

A Derrida plot is a graph that measures the expected size of a perturbation at time $t + 1$ vs. its size at time t. This is usually determined computationally through repeated experiments. Consider a fixed Boolean network on N nodes. Choose a random Boolean vector x, uniformly from $\mathbb{F}_2^N$. For each $M = 1, 2, \ldots, N$, perturb this vector by randomly flipping M of its bits, yielding a new vector $x^{(M)}$. If $f \colon \mathbb{F}_2^N \to \mathbb{F}_2^N$ is the global update function of the Boolean network, then define

$$\rho(t, M) = \overline{\mathrm{H}}(x - x^{(M)}) = \frac{M}{N}, \quad \rho(t + 1, M) = \overline{\mathrm{H}}(f(x) - f(x^{(M)})).$$

If M is understood or fixed, then we frequently write $\rho(t)$ instead of $\rho(t, M)$. One can think of $\rho(t)$ as the size of a random size-M perturbation introduced at time t, and $\rho(t + 1)$ as the size of the perturbation at time $t + 1$. If $\rho(t + 1) > \rho(t)$, then the perturbation has grown. If $\rho(t + 1) < \rho(t)$, then the perturbation has shrunk. Naturally, one could have a very stable network but by chance, choose x to be very sensitive to perturbations, or vice versa.

Therefore, this experiment must be carried out repeatedly to shed any real insight. Fortunately, computing this for a single vector can be done extremely quickly by a computer, even for large N. This experiment can then be repeated for thousands of random vectors $x \in \mathbb{F}_2^N$. As the number of experiments grows, the value of $\rho(t+1, M)$ approaches the actual expected value $E[\rho(t+1, M)]$, which is defined to be the Derrida plot of the Boolean network.

Definition 5.7. The *Derrida plot* of an ensemble of networks is the graph of the function

$$\lambda\colon \left\{0, \frac{1}{N}, \frac{2}{N}, \ldots, \frac{N-1}{N}, 1\right\} \longrightarrow [0,1], \quad \lambda\left(\frac{M}{N}\right) = E[\rho(t+1, M)].$$

□

The stability of a Boolean network depends on whether a *small* perturbation is likely to die out or spread. Therefore, the most important aspect of a Derrida plot is whether it lies above, below, or roughly on the main diagonal $y = x$ in the interval $(0, \varepsilon)$ for small ε (e.g., $\varepsilon = \frac{1}{10}$, or $\frac{1}{5}$).

Definition 5.8. A Boolean network on N nodes is said to be
(1) *chaotic* if the graph of λ on $(0, \varepsilon)$ lies above the line $y = x$ for some small $\varepsilon > 0$;
(2) *ordered* if the graph of λ on $(0, \varepsilon)$ lies below the line $y = x$ for some small $\varepsilon > 0$.
It is extremely unlikely that the graph of λ would ever lie exactly on the line $y = x$ for an interval $(0, \varepsilon)$. However, if it lies very close to this line, then the Boolean network is said to be *critical*. Naturally, there is a gray area as to what "very close" means. □

Sometimes it is convenient to speak of the chaotic and critical *regimes* of the space of Boolean networks, which are separated by the narrow critical threshold. Critical networks have salient characteristics shared by both ordered and chaotic networks. For example, they must be robust enough to withstand random perturbations, yet flexible enough to adapt and evolve when necessary. In Ref. [48], the authors argue that networks from living organisms possess a robustness that puts them in either the ordered or critical regimes, and their experimental results and analysis involving networks found in eukaryotic cells support this theory.

5.4.3 Ensembles of RBNs

In the previous section, we considered a single instance of a Boolean network. In many scenarios, one wants to understand the dynamics of a typical network given certain parameters. For example, what dynamical regime do *NK*-networks tend to operate in, and how does this depend on K? To answer this, constructing a single Boolean network is not enough. Further experiments must be done to generate a large number of *NK*-networks and randomly assign Boolean functions to each network. This large collection is called an *ensemble of RBNs*. The Derrida plots of each representative are then averaged to yield the Derrida plot of the entire ensemble. An example of this is shown in Figure 5.11, from Ref. [31]. Notice how for $K = 3, 4, 5$, the Derrida plots lie above the line $y = x$ for small input values; these networks lie in the chaotic regime. In contrast, the Derrida plot for $K = 2$ lies below the line, but "close enough" to be reasonably considered critical.

Sometimes, the regulatory functions are chosen via some distribution. For example, the functions might be chosen uniformly from all 2^{2^K} Boolean functions on K variables. Alternatively the functions could be chosen uniformly from all canalizing functions. In other experiments, the functions are chosen with a *bias* p, based on the number of 1s in its truth table. Specifically, suppose a function $f\colon \mathbb{F}_2^N \to \mathbb{F}_2$ has exactly M inputs x such that $f(x) = 1$, and $2^N - M$ for which $f(x) = 0$. This function will be chosen with probability $p^M(1-p)^{2^N - M}$. As an example of this, consider the following four Boolean functions on two variables, shown in Table 5.3.

The probabilities of each function occurring under a bias of p is

$$P(x \wedge y) = p^1(1-p)^3, \quad P(x \vee y) = p^3(1-p)^1, \quad P(x \oplus y) = p^2(1-p)^2, \quad P(1) = p^4(1-p)^0.$$

In the special case of bias $p = 1/2$, every Boolean function is chosen uniformly.

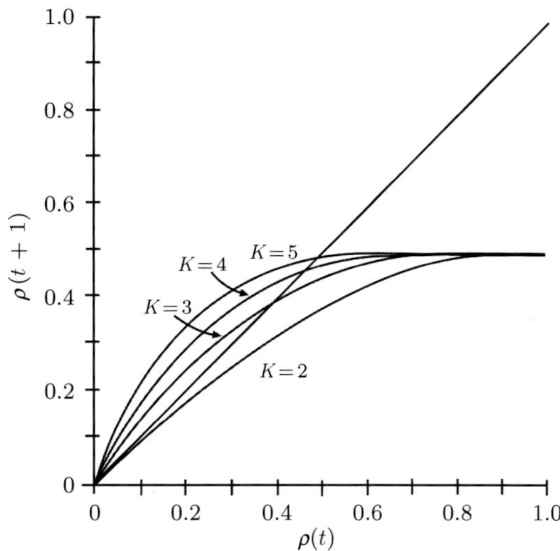

FIGURE 5.11 Derrida plots for random *NK*-networks for various values of *K*, where the Boolean functions are sampled uniformly, as derived in Ref. [31].

TABLE 5.3 Four Boolean Functions on Two Variables: AND, OR, XOR, and a Constant Function

x	y	$x \wedge y$	$x \vee y$	$x \oplus y$	1
0	0	0	0	0	1
0	1	0	1	1	1
1	0	0	1	1	1
1	1	1	1	0	1

The dynamical regime of an RBN is closely related to its average sensitivity. Recall that this is an integer s between 0 and K that describes the number of ways that toggling a bit will change the output. Ordered networks are characterized by $s < 1$, and chaotic networks by $s > 1$. Though the average sensitivity of a single RBN is difficult to compute, the calculations often become easier for an ensemble. For example, the average sensitivity of an ensemble of random *NK*-networks with bias p is $s = 2Kp(1 - p)$; see Ref. [45, 47]. Thus, if the functions are assigned uniformly ($p = 1/2$), an RBN is *chaotic* if $K > 2$, *ordered* if $K < 2$, and *critical* if $K = 2$.

To tie together this section with the previous one, we will end with an example of a Derrida plot from an ensemble of networks based on their canalizing depth. These are the results of experiments done in Ref. [38]. Consider an ensemble of *NK*-networks, where $N = 100$ and $K = 12$. Obviously, this is a large K, and a randomly selected function will have an expected sensitivity of $s = 6$, and thus an RBN built with arbitrary functions should be chaotic. It is natural to ask how the stability changes if we sample from canalizing functions. Say that a function is *d-canalizing* if it is has canalizing depth *at least d*. For completeness, say that the 0-canalizing functions are precisely the noncanalizing functions. For each $d = 0, 2, 4, \ldots, 12$, an ensemble of 1000 RBNs was created by sampling from the d-canalizing functions uniformly. For each of these networks, a random state $x \in \mathbb{F}_2^{100}$ was chosen and randomly perturbed, and the Derrida plot was created by averaging over the entire ensemble. The results of this are shown in Figure 5.12. When sampling uniformly over all functions ($d = 0$), the Derrida plot rises sharply above the diagonal $x = y$—this indicates chaotic dynamics. In contrast, sampling from the two-canalizing functions gives a Derrida plot that also lies above the $x = y$ line, but much closer to it. These networks are chaotic, but close to the critical threshold. Next, consider the case where the functions are

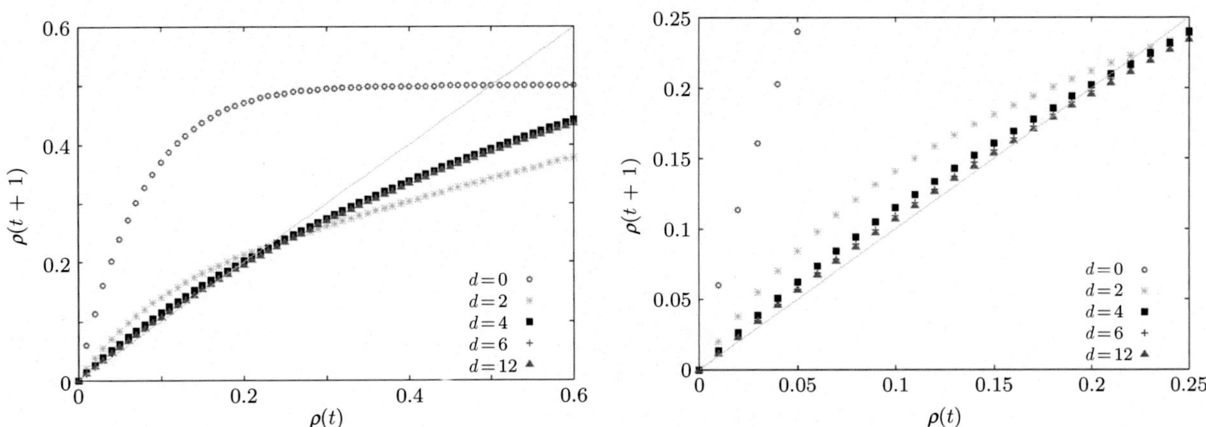

FIGURE 5.12 An example Derrida plot measuring the stability of functions based on their canalizing depth. On the x-axis is the (normalized) size of a perturbation, and on the y-axis is the size of that perturbation in the next time step. The plot on the right is a zoomed in version of the one on the left.

4-canalizing. From the Derrida plot, we see that this also yields chaotic networks, but these are even closer to being critical. In fact, one could make the argument that this Derrida curve is close enough to the diagonal for the ensemble to be classified as lying in the critical regime. Finally, notice how the dynamics of the networks using NCFs ($d = 12$) are almost the same as the four-canalizing and six-canalizing networks. Overall, we take from these simple experiments the intuition that networks built with canalizing functions are significantly more stable than plain random networks, and that stability increases as canalizing depth increases. However, it does so with diminishing returns, in that there is not much difference between networks with functions of small canalizing depth vs. those with full canalizing depth (i.e., NCFs).

Exercise 5.27. Explain why a Derrida curve should always lie below the line $y = x$ for values of ρ close to 1. □

ACKNOWLEDGMENTS

The authors thank Catherine Gurri for help on making some of the figures. The first and second authors were supported by National Science Foundation Grant DMS-#1211691.

REFERENCES

[1] Waddington CH. Canalisation of development and the inheritance of acquired characters. Nature 1942;150:563-4.
[2] Drossel B. Random Boolean networks. Weinheim, Germany: Wiley-VCH Verlag GmbH & Co.; 2009. p. 69-110 [chapter 3].
[3] Diekmann O, Heesterbeek JAP. Mathematical epidemiology of infectious diseases. Chichester: Wiley; 2000. p. 146.
[4] Balleza E, Alvarez-Buylla E, Chaos A, Kauffman SA, Shmulevich I, Aldana M. Critical dynamics in genetic regulatory networks: examples from four kingdoms. PLoS One 2008;3:e2456.
[5] Nykter M, Price ND, Aldana M, Ramsey SA, Kauffman SA, Hood LE, et al. Shmulevich I Gene expression dynamics in the macrophage exhibit criticality. Proc Natl Acad Sci USA 2008;105:1897-900.
[6] Rämö P, Kesseli J, Yli-Harja O. Perturbation avalanches and criticality in gene regulatory networks. J Theor Biol 2006;242:164-70.
[7] Hohenberg PC, Halperin BI. Theory of dynamic critical phenomena. Rev Mod Phys 1977;49:435-79.
[8] Arakawa A. Scaling tropical rain. Nat Phys 2006;2:373-4.
[9] Peters O, Neelin JD. Critical phenomena in atmospheric precipitation, Nat Phys 2006;2:393-6.
[10] Heimpel M. Critical behavior and the evolution of fault strength during earthquake cycles. Nature 1997;388:865-8.
[11] Werner G. Metastability, criticality and phase transitions in brain and its models. Biosystems 2007;90:496-508.
[12] Lux T, Marchesi M. Scaling and criticality in a stochastic multi-agent model of a financial market. Nature 1999;397:498-500.

[13] Aldana M, Balleza E, Kauffman S, Resendis O. Robustness and evolvability in gene regulatory networks. J Theor Biol 2007;245: 433-48.
[14] Sethna JP, Dahmen KA, Myers CR. Crackling noise. Nature 2001;410:242-50.
[15] Nykter M, Price ND, Larjo A, Aho T, Kauffman SA, Yli-Harja O, et al. Critical networks exhibit maximal information diversity in structure-dynamics relationships. Phys Rev Lett 2008;100:058702.
[16] Jensen HJ. Self-organized criticality: emergent complex behavior in physical and biological systems. New York: Cambridge University Press; 1998. p. 10.
[17] Salgado H, Gama-Castro S, Peralta-Gil M, Díaz-Peredo E, Sánchez-Solano F, Santos-Zavaleta A, et al. RegulonDB (version 5.0): *Escherichia coli* K-12 transcriptional regulatory network, operon organization, and growth conditions. Nucleic Acids Res 2006;34:D394-7.
[18] Luscombe NM, Babu MM, Yu H, Snyder M, Teichmann SA, Gerstein M. Genomic analysis of regulatory network dynamics reveals large topological changes. Nature 2004;431:308-12.
[19] Makita Y, Nakao M, Ogasawara N, Nakai K. DBTBS: database of transcriptional regulation in *Bacillus subtilis* and its contribution to comparative genomics. Nucleic Acids Res 2004;32:D75-7.
[20] Robeva R, Hodge T. Mathematical concepts and methods in modern biology: using modern discrete models. Academic Press; 2013.
[21] Eubank S. Network based models of infectious disease spread. Jpn J Infect Dis 2005;58:S9-13.
[22] Watts DJ, Strogatz SH. Collective dynamics of 'small-world' networks. Nature 1998;393:440-2.
[23] Freeman L. A set of measures of centrality based on betweenness. Sociometry 1977;40:35-41.
[24] Martín González MA, Dalsgaard B, Olesen B. Centrality measures and the importance of generalist species in pollination networks. Ecol Complex 2010;7:36-43.
[25] Zhao Z, Pei G, Huang F, Liu X. Equity importance modeling with financial network and betweenness centrality. In: Proceedings of 16th international conference on computing in economics and finance; 2010.
[26] Erdős P, Rényi A. On the evolution of random graphs. Publ Math Inst Hung Acad Sci 1960;5:17-61.
[27] Jeong H, Tombor B, Albert R, Oltvai ZN, Barabási A-L. The large-scale organization of metabolic networks. Nature 2000;407:651-4.
[28] Ravasz E, Somera AL, Mongru DA, Oltvai ZN, Barabási A-L. Hierarchical organization of modularity in metabolic networks. Science 2002;297:1551-5.
[29] Jeong H, Mason S, Barabási A-L, Oltvai ZN. Lethality and centrality in protein networks. Nature 2001;411:41-2.
[30] Barabási A-L, Albert R. Emergence of scaling in random networks. Science 1999;286:509-12.
[31] Kauffman SA. The origins of order: self-organization and selection in evolution. New York: Oxford University Press; 1993.
[32] Kauffman SA. Metabolic stability and epigenesis in randomly constructed genetic nets. J Theor Biol 1969;22:437-67.
[33] Mortveit Henning S, Reidys Christian M. An introduction to sequential dynamical systems. Universitext, Springer Verlag; 2007.
[34] Aggarwal C, Subbian K. Evolutionary network analysis: a survey. ACM Comput Surv (CSUR) 2014;47:10.
[35] Kauffman SA, Peterson C, Samuelsson B, Troein C. Random Boolean network models and the yeast transcriptional network. Proc Natl Acad Sci USA 2003;100:14796-9.
[36] Jarrah AS, Raposa B, Laubenbacher R. Nested canalyzing, unate cascade, and polynomial functions. Phys D 2007;233:167-74.
[37] Butler JT, Sasao T, Matsuura M. Average path length of binary decision diagrams. IEEE Trans Comput 2005;54:1041-53.
[38] Layne L, Dimitrova ES, Macauley M. Nested canalyzing depth and network stability. Bull Math Biol 2012;74:422-33.
[39] Li Y, Adeyeye JO, Murrugarra D, Aguilar B, Laubenbacher R. Boolean nested canalizing functions: a comprehensive analysis. Theor Comput Sci 2013;481:24-36.
[40] Newman MEJ. Networks: an introduction. Oxford, UK: Oxford University Press; 2010.
[41] Newman M, Barabási AL, Watts DJ. The structure and dynamics of networks. Princeton, NJ: Princeton University Press; 2006.
[42] Barrat A, Barthélemy M, Vespignani A. Dynamical processes on complex networks. Cambridge: Cambridge University Press; 2008. p. 1.
[43] Boccaletti S, Latora V, Moreno Y, Chavez M, Hwang DU. Complex networks: structure and dynamics. Phys Rep 2006;424:175-308.
[44] Kauffman SA. Emergent properties in random complex automata . Phys D 1984;10:145-56.
[45] Derrida B, Pomeau Y. Random networks of automata: a simple annealed approximation Europhys Lett 1986;1:45-9.
[46] Titz B, Rajagopala SV, Goll J, Häuser R, McKevitt MT, Palzkill T, et al. The binary protein interactome of *Treponema pallidum*—the syphilis spirochete. PLoS One 2008;3:e2292.
[47] Shmulevich I, Kauffman SA. Activities and sensitivities in Boolean network models. Phys Rev Lett 2004;93:048701.
[48] Shmulevich I, Kauffman SA, Aldana M. Eukaryotic cells are dynamically ordered or critical but not chaotic. Proc Natl Acad Sci USA 2005;102:13439-44.

Chapter 6

Steady State Analysis of Boolean Models: A Dimension Reduction Approach

Alan Veliz-Cuba[1,2] and David Murrugarra[3]

[1]*Department of Mathematics, University of Houston, Houston, TX, USA,* [2]*Department of BioSciences, Rice University, Houston, TX, USA,* [3]*Department of Mathematics, University of Kentucky, Lexington, KY, USA*

6.1 INTRODUCTION

Boolean models, an especial class of discrete models, have a long tradition in mathematical modeling. Discrete models, such as finite state machines, Boolean networks, Petri nets, or agent-based models, play an important role in modeling processes that can be viewed as evolving in discrete time, in which state variables have only finitely many possible states. Decision processes, electrical switching networks, or intracellular molecular networks represent examples. Boolean networks, as computational models for biological systems, were introduced by Kauffmann [1] and Thomas and D'Ari [2]. Boolean network models have been efficiently used to model biological systems such as the yeast cell cycle network [3], the Th regulatory network [4], the *lac* operon [5], the p53-mdm2 complex [6, 7], and many others [8–13] and have also been used for theoretical analysis [14–20].

A Boolean network (BN) can be defined as a dynamical system that is discrete in time as well as in variable states. More formally, consider a collection $x_1, \dots, x_n$ of variables, each of which can take on values in the binary set $\{0, 1\}$. A Boolean network in the variables $x_1, \dots, x_n$ is a function

$$f = (f_1, \dots, f_n) : \{0, 1\}^n \to \{0, 1\}^n$$

where each coordinate function f_i is a Boolean function on a subset of $\{x_1, \dots, x_n\}$, which represents how the future value of the ith variable depends on the present values of the variables.

Coordinate functions can be represented using the AND ($\wedge$), OR ($\vee$), and NOT ($\neg$) operators. For example, the Boolean function $f_1(x_2, x_3)$ that satisfies $f_1(0,0) = 0$, $f_1(0,1) = 1$, $f_1(1,0) = 0$, and $f_1(1,1) = 0$ can be written as $f_1(x_2, x_3) =$ NOT x_2 AND x_3, or as $f_1(x_2, x_3) = \neg x_2 \wedge x_3$. The AND operator usually represents regulation that requires the presence of all species, whereas the OR operator usually represents regulation where one species is sufficient. The NOT operator usually describes negative regulation.

Given a Boolean network $f = (f_1, \dots, f_n)$, a directed graph G with the n nodes $x_1, \dots, x_n$ can be associated with f. There is a directed edge from x_j to x_i if x_j appears in f_i. In the context of a molecular network model, this graph represents the wiring diagram of the network. Graphs can be represented by their adjacency matrices. Let A be the $n \times n$ adjacency matrix of this graph, G. That is, $A = (a_{ij})$ is defined as follows:

$$a_{ij} = \begin{cases} 1 & \text{if } f_i \text{ depends on } x_j, \\ 0 & \text{otherwise.} \end{cases}$$

Algebraic and Discrete Mathematical Methods for Modern Biology. http://dx.doi.org/10.1016/B978-0-12-801213-0.00006-X
Copyright © 2015 Elsevier Inc. All rights reserved.

An edge $x_i \to x_j$ in the wiring diagram is said to be positive if

$$f_j(x_1, \ldots, x_{i-1}, 0, x_{i+1}, \ldots, x_n) \quad \leq \quad f_j(x_1, \ldots, x_{i-1}, 1, x_{i+1}, \ldots, x_n);$$

the edge is negative if the inequality is reversed. Negative edges are usually denoted by circles or blunt arrows. If all Boolean functions f_i are unate, then all edges can be given a sign assignment [21].

The dynamical properties of a Boolean network are given by the difference equation $x(t+1) = f(x(t))$; that is, the dynamics is generated by iteration of f. More precisely, the dynamics of f is given by the state space graph S, defined as the graph with vertices in $\mathbb{K}^n = \{0,1\}^n$, which has an edge from $x \in \{0,1\}^n$ to $y \in \{0,1\}^n$ if and only if $y = f(x)$. In this context, the problem of finding the states $x \in \{0,1\}^n$ where the system will get stabilized is of particular importance. These special points of the state space are called attractors of a Boolean network, and these may include steady states, where $f(x) = x$, and cycles, where $f^r(x) = x$ for an integer $r > 1$. Notice that the state space has 2^n states, that is, the state space of Boolean networks grows exponentially with the number of nodes. This fact presents a serious challenge to the computational analysis of the dynamical properties of large Boolean networks. For instance, the problem of finding the steady states of a Boolean network has been shown to be NP-complete [22, 23].

This chapter presents reduction approaches for steady state analysis of Boolean networks [24–27]. The methods consist of "steady state approximations" that allow the removal of nodes while preserving important dynamical properties and thus reducing the size of the Boolean network. More specifically, the algorithm generates a reduced Boolean network that has a smaller number of nodes (and therefore a smaller state space) than the original network; the reduced network will retain the same number of steady states and topological features of the original network so that the analysis of the original network can be performed by focusing on the reduced network.

This chapter is organized as follows: In Section 6.2, we illustrate the ideas of the reduction method using a toy example of the *lac* operon and show how the steady states of the reduced network can be easily used to find the steady states of the original network. In Section 6.3, we formally present the method and show its theoretical properties. In Sections 6.4 and 6.5, we show implementations of the reduction algorithm using the algebra software Mathematica and Macaulay2, respectively. In Section 6.6, we show applications of the reduction method and how conclusions about the reduced networks can be used to predict behavior of the original systems. In Section 6.7, we present a reduction method tailored specifically to AND networks. A conclusion follows in Section 6.8.

6.2 AN EXAMPLE: TOY MODEL OF THE *lac* OPERON

In this section we illustrate the ideas of the reduction method on a toy model of the *lac* operon (Table 6.1). The *lac* operon is a gene network in charge of metabolizing lactose in the cell. The variables that we include in the model are mRNA of the lac genes (M); the repressor protein (R); the proteins permease (P) and beta-galactosidase (B); and intracellular lactose (L) and allolactose (A). The wiring diagram of this model is in Figure 6.1. For simplicity we consider the case in which no glucose is in the environment and extracellular lactose is available.

TABLE 6.1 Toy Model of the *lac* Operon

Boolean Function	Justification
$f_M = \neg R$	R represses mRNA production
$f_P = M$	P is produced by translation of M
$f_B = M$	B is produced by translation of M
$f_R = \neg A$	A inactivates the repressor protein
$f_A = L \wedge B$	A is produced from lactose by the action of beta-galactosidase
$f_L = P$	Permease brings extracellular lactose into the cell

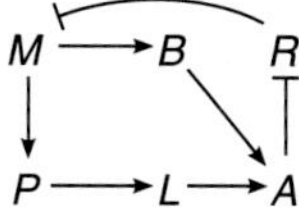

FIGURE 6.1 Wiring diagram of toy model of the *lac* operon. Blunt arrows denote negative edges.

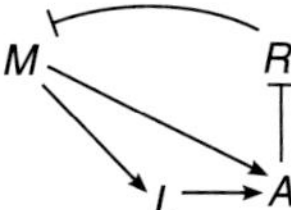

FIGURE 6.2 Wiring diagram of the reduced network of the toy model of the *lac* operon.

Thus, we have the Boolean model $f = (f_M, f_P, f_B, f_R, f_A, f_L) : \{0,1\}^6 \to \{0,1\}^6$, where f_x gives the value of $x(t+1)$ as function of the state of the system at time t. Now we are interested in finding the steady states of the model; that is, all values of $x = (M, P, B, R, A, L)$ such that $f(x) = x$. By an exhaustive search, it can be shown that $x = 000100$ and $x = 111011$ are steady states (parentheses have been omitted for simplicity). Note that when using the exhaustive search approach, one needs to compute $f(x)$ for each $x \in \{0,1\}^6$, and thus, this approach is unfeasible for large models.

Exercise 6.1. Check that $x = 000100$ and $x = 111011$ are in fact steady states; that is, check that $f(x) = x$. □

Exercise 6.2. Consider $x^0 = 000000$ and compute $x^{t+1} := f(x^t)$ for $t = 0, \ldots, 10$. Repeat the process for $x^0 = 111111$. □

We now illustrate the idea of the reduction. First, we consider the variable P. When the system reaches equilibrium, we will have $P(t) = P(t+1) = f_P(x(t)) = M(t)$, and thus we can make the assumption that $P = M$ from the beginning and not change the steady state behavior. Alternatively, if we assume that the reactions involved in the Boolean function $f_P = M$ are fast compared to other reactions, we can consider that P reaches its future value instantly and consider the steady state approximation $P(t) = M(t)$. Similarly, we may assume that $B(t) = M(t)$. Note that knowledge of P and B is not necessary to simulate the system, and thus, we obtain the reduced network

$$\begin{aligned} f_M &= \neg R, \\ f_R &= \neg A, \\ f_A &= L \wedge M, \\ f_L &= M. \end{aligned}$$

Exercise 6.3. By exhaustive enumeration, find all steady states of $f(M, R, A, L) = (\neg R, \neg A, L \wedge M, M)$. □

The wiring diagram of this reduced network is given in Figure 6.2. Thus, by using steady state approximations we have obtained a smaller network.

We now will see how the steady states of the original network are related to the reduced network. By an exhaustive search, we determine that the steady states of the reduced network are $y_1 = 0100$ and $y_2 = 1011$. For an easier comparison, we included the steady states of the original and the reduced network in Table 6.2.

We can see that the steady states of the original and reduced network are essentially the same. More precisely, the function $\pi(M, P, B, R, A, L) = (M, R, A, L)$ defines a bijection between the steady states. Furthermore, the steady states of the reduced network and the equations $B = M$, $P = M$ are sufficient to recover the steady states of the original network.

Now we take the reduction further. We continue from the reduced network above and make the steady state approximation $A = f_A = L \wedge M$. In this case we replace every occurrence of A in the network by $L \wedge M$; thus, $f_R = \neg A$ becomes $f_R = \neg(L \wedge M) = \neg L \vee \neg M$, and we obtain the network

TABLE 6.2 Steady States of Original and Reduced Models

Steady States									
Original Network						Reduced Network			
M	*P*	*B*	*R*	*A*	*L*	*M*	*R*	*A*	*L*
0	0	0	1	0	0	0	1	0	0
1	1	1	0	1	1	1	0	1	1

$$f_M = \neg R,$$

$$f_R = \neg L \vee \neg M,$$

$$f_L = M.$$

Exercise 6.4. By an exhaustive enumeration, find all steady states of $f(M, R, L) = (\neg R, \neg L \vee \neg M, M)$ by exhaustive enumeration and compare with Exercise 6.3. □

This idea can be carried on further to remove node R. The reduction process is summarized in Table 6.3. The equations under the reduced networks are the steady state approximations, which can later be used to recover the steady state of a network using the steady states of the reduced networks.

The last network is a two-variable network, $(f_M, f_R) : \{0,1\} \to \{0,1\}$ given by $f_M = \neg R, f_R = \neg M$ and wiring diagram shown in Figure 6.3. The steady states are easily found by inspection, and they are 01 and 10. Now, using the steady states and the equations $L = M$, $A = L \wedge M$, $P = M$, $B = M$, we easily find the steady states of the original network. For example, using $MR = 01$ we compute $L = 0, A = 0 \wedge 0, P = 0$, and $B = 0$; that is, $x = 000100$.

TABLE 6.3 Sequence of Reductions

Original BN	Remove *P*, *M*	Remove *A*	Remove *L*
$f_M = \neg R$	$f_M = \neg R$	$f_M = \neg R$	$f_M = \neg R$
$f_P = M$	$f_R = \neg A$	$f_R = \neg L \vee \neg M$	$f_R = \neg M$
$f_B = M$	$f_A = L \wedge M$	$f_L = M$	$L = M$
$f_R = \neg A$	$f_L = M$	$A = L \wedge M$	$A = L \wedge M$
$f_A = L \wedge B$	$P = M$	$P = M$	$P = M$
$f_L = P$	$B = M$	$B = M$	$B = M$

FIGURE 6.3 Wiring diagram for the last reduced network of the toy model of the *lac* operon.

6.3 GENERAL REDUCTION

In this section, we will formally define the reduction algorithm.

6.3.1 Definition

We now provide the reduction steps to reduce a Boolean network and its corresponding wiring diagram. The idea behind the reduction method is simple: the wiring diagram and Boolean functions should reflect actual regulation and hence nonfunctional edges ($x_i \to x_j$ in the wiring diagram such that f_j does not actually depend on x_i), and variables should be removed; on the other hand, vertices in the wiring diagram can be deleted, without losing some of the important information, by allowing its functionality to be "passed on" to other variables. This reduction method has been shown to preserve the number of steady states [24].

1. We simplify the Boolean functions and wiring diagram:
   (a) Reduce Boolean expressions using Boolean algebra. This will delete variables that are not functional.
   (b) Delete edges that do not correspond to Boolean expressions. That is, we delete edges that are nonfunctional.
2. We delete vertices with no self-loop, that is, vertices whose Boolean function does not depend on itself. Let x_i be a vertex such that f_{x_i} does not depend on x_i.
   (a) For all vertices y such that there is an edge $x_i \to y$, that is, for all vertices whose Boolean function depends on x_i, replace the Boolean function for y, $f_y(x_1, \ldots, x_i, \ldots, x_k)$, by $f_y(x_1, \ldots, f_{x_i}, \ldots, x_k)$.
   (b) Replace edges $y \to x_i \to z$ by $y \to z$ and delete x_i (and edges from/to x_i).

Exercise 6.5. Consider $f_1 = f_1(x_1,x_2,x_3) = x_2 \wedge x_3$, $f_2 = f_2(x_1,x_2,x_3) = x_1 \vee \neg x_3$, $f_3 = f_3(x_1,x_2,x_3) = x_1 \vee x_2$. Compute $f_1(x_1,x_2,f_3)$ and $f_2(x_2,x_2,f_3)$. □

Exercise 6.6. Reduce the following Boolean expressions using Boolean algebra or truth tables.

1. $x_1 \wedge \neg x_1$,
2. $(x_1 \vee x_2) \wedge \neg x_1$,
3. $(x_1 \vee x_2) \wedge x_1$,
4. $(x_1 \wedge x_2) \vee x_1$,
5. $(x_1 \vee x_2) \wedge x_2$,
6. $(x_1 \vee \neg x_2) \wedge (x_1 \vee x_2)$,
7. $(x_1 \vee x_2) \wedge x_1 \wedge x_2$,
8. $x_1 \wedge x_2 \wedge x_3 \wedge \neg x_1$. □

Exercise 6.7. Consider $f_1 = f_1(x_1,x_2,x_3) = x_1 \wedge x_2 \wedge \neg x_3$, $f_2 = f_2(x_1,x_2,x_3) = \neg x_3$, $f_3 = f_3(x_1,x_2,x_3) = x_1 \vee x_2$. Compute and simplify $f_1(x_1,x_2,f_3)$ and $f_2(x_2,x_2,f_3)$. □

6.3.2 Examples

Example 6.1. We consider the Boolean network $f(x) = (x_2, (x_1 \wedge x_3) \vee \neg x_2, \neg x_1)$ with wiring diagram shown in Figure 6.4 (left) and remove the last variable, x_3. Then, the new Boolean functions for the first two variables will be

$$h_1(x_1,x_2) = f_1(x_1,x_2,f_3) = f_1(x_1,x_2,\neg x_1) = x_2,$$
$$h_2(x_1,x_2) = f_2(x_1,x_2,f_3) = f_2(x_1,x_2,\neg x_1) = (x_1 \wedge \neg x_1) \vee \neg x_2.$$

FIGURE 6.4 Wiring diagram of the Boolean network from Example 6.1. Left: original network; right: reduced network after removing x_3. Circles denote negative edges.

Now, we need to check if all variables that appear in the Boolean expressions actually affect the function. For the first Boolean function of the reduced system, $h_1(x_1,x_2) = x_2$, there is nothing to simplify; however, for the second Boolean function of the reduced system, $h_2(x_1,x_2) = (x_1 \wedge \neg x_1) \vee \neg x_2$, we see that $(x_1 \wedge \neg x_1)$ can be simplified to 0, so h_2 simplifies to $0 \vee x_2 = x_2$. Thus, the (simplified) reduced system is

$$h_1(x_1,x_2) = x_2,$$
$$h_2(x_1,x_2) = \neg x_2,$$

with wiring diagram shown in Figure 6.4 (right). □

Example 6.1 illustrates why it is important to simplify the Boolean expressions. Sometimes the reduced system has "redundant" variables, and we need to identify them to obtain a better representation and an accurate wiring diagram.

Exercise 6.8. Find the steady states of f and h, if any. □

Exercise 6.9. Consider the Boolean network $f(x) = (x_2, (x_1 \wedge x_3) \vee x_2, \neg x_1)$. Reduce the network by removing variable x_3 (make sure you simplify the Boolean expressions in the reduced system). Find the steady states of the original and reduced networks, and compare their first and second coordinates. Also, using the steady states of the reduced network, compute f_3. □

Exercise 6.10. Consider the Boolean network $f(x) = (x_2, (x_1 \wedge x_3) \wedge x_2, \neg x_1)$. Reduce the network by removing variable x_3 (make sure you simplify the Boolean expressions in the reduced system). Find the steady states of the original and reduced networks, and compare their first and second coordinates. Also, using the steady states of the reduced network, compute f_3. □

It turns out that removing variables using this method preserves the number of steady states. Furthermore, the steady states of the reduced network are simply the projection of the steady states of the original system onto the variables that were not reduced. In fact, if we know the steady states of the reduced network, we can reconstruct the steady states of the original network by using the equations from the reduction steps. We explain this in detail in the next example.

Example 6.2. We consider the Boolean network $f = f(x_1,x_2,x_3,x_4,x_5) = (x_5 \vee \neg x_2 \vee x_4, \neg x_1 \wedge \neg x_3, \neg x_2, \neg x_2, x_1 \vee x_4)$ with wiring diagram shown in Figure 6.5.

We first remove variable x_5 by replacing it with $x_1 \vee x_4$ and obtain

$$((x_1 \vee x_4) \vee \neg x_2 \vee x_4, \neg x_1 \wedge \neg x_3, \neg x_2, \neg x_2),$$

which simplifies to

$$(x_1 \vee \neg x_2 \vee x_4, \neg x_1 \wedge \neg x_3, \neg x_2, \neg x_2).$$

We also save the equation used for the reduction, $x_5 = x_1 \vee x_4$.

We now remove variable x_4 by replacing it with $\neg x_2$ and obtain (after simplifying)

$$(x_1 \vee \neg x_2, \neg x_1 \wedge \neg x_3, \neg x_2),$$

and the additional equations $x_4 = \neg x_2$ $x_5 = x_1 \vee x_4$.

Finally, we remove variable x_3 by replacing it with $\neg x_2$ and obtain (after simplifying)

$$(x_1 \vee \neg x_2, \neg x_1 \wedge x_2),$$

and the additional equations $x_3 = \neg x_2$, $x_4 = \neg x_2$ $x_5 = x_1 \vee x_4$.

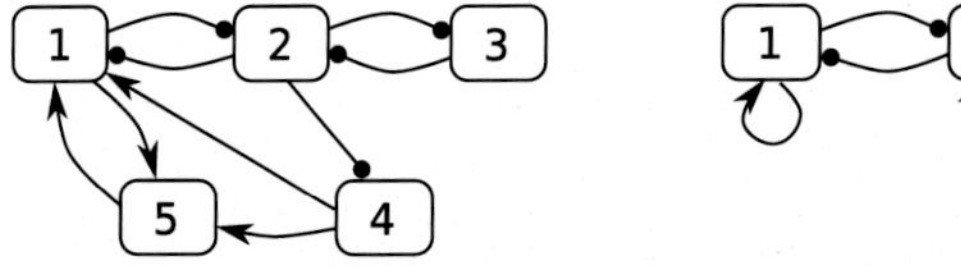

FIGURE 6.5 Wiring diagram of the Boolean network from Example 6.2. Left: original network; right: reduced network after removing x_5, x_4, x_3 (in that order).

That is, we have the following reduced system and equations.

$$h_1(x_1,x_2) = x_1 \vee \neg x_2,$$
$$h_2(x_1,x_2) = \neg x_1 \wedge x_2,$$
$$x_3 = \neg x_2,$$
$$x_4 = \neg x_2,$$
$$x_5 = x_1 \vee x_4.$$

Now the steady states of the reduced network are $(x_1,x_2) = (0,1)$, $(1,0)$ (Exercise 6.11). How can we recover the steady states of the original system? For each steady state of the reduced network, we simply substitute into the equations we obtained during the reduction. For example, for $(x_1,x_2) = (1,0)$ we obtain: $x_3 = \neg x_2 = \neg 0 = 1$, $x_4 = \neg x_2 = \neg 0 = 1$, $x_5 = x_1 \vee x_4 = 1 \vee 1 = 1$; that is $x = 10111$ will be a steady state of the original Boolean network. Similarly, using $(x_1,x_2) = (0,1)$, we obtain that $x = 01000$ will be a steady state of the Boolean network. □

Example 6.2 shows the power of the reduction algorithm: not only does it allow us to reduce the network without changing the number of steady states, but the algorithm also gives a systematic way to recover the steady states of the original network. This last point is important if we want to be able to implement the algorithm in a computer. In this sense, the reduction algorithm gives a smaller system such that the steady states of the original and reduced network are in an algorithmic one-to-one correspondence.

Exercise 6.11. Find the steady states of the Boolean network $h(x_1,x_2) = (x_1 \vee \neg x_2, \neg x_1 \wedge x_2)$. Also, draw the wiring diagram of h. □

Exercise 6.12. Consider the Boolean network $f = f(x_1,x_2,x_3,x_4,x_5) = (x_5 \vee \neg x_2 \vee x_4, \neg x_1 \wedge \neg x_3, \neg x_2, \neg x_2, x_1 \vee x_4)$ from Example 6.2, and reduce the network by removing variables x_3, x_4, and x_5. Compare this reduced network with the reduced network from Example 6.2. Are the reduced networks equal? Are the equations for x_3, x_4, x_5 the same? Use the reduced network and the equations for x_3, x_4, x_5 to find the steady states of f. Compare your answers with the steady states found in Example 6.2. □

Exercise 6.13. Consider the Boolean network $f = f(x_1,x_2,x_3,x_4,x_5) = (x_5 \vee \neg x_2 \vee x_4, \neg x_1 \wedge \neg x_3, \neg x_2, \neg x_2, x_1 \vee x_4)$ from Example 6.2 and reduce the network by removing variables x_1, x_3, and x_4. Compare this reduced network with the reduced network from Example 6.2. Are the reduced networks equal? Are the equations for x_1, x_3, x_4 the same? Use the reduced network and the equations for x_1, x_3, x_4 to find the steady states of f. Compare your answers with the steady states found in Example 6.2. □

Exercise 6.14. Consider the Boolean network $f = f(x_1,x_2,x_3,x_4,x_5) = (x_5 \vee \neg x_2 \vee x_4, \neg x_1 \wedge \neg x_3, \neg x_2, \neg x_2, x_1 \vee x_4)$ from Example 6.2 and reduce the network by removing variables x_5, x_2, and x_4. Compare this reduced network with the reduced network from Example 6.2. Are the reduced networks equal? Are the equations for x_5, x_2, x_4 the same? Use the reduced network and the equations for x_5, x_2, x_4 to find the steady states of f. Compare your answers with the steady states found in Example 6.2. □

It is important to mention that one can only remove variables that do not have a self-loop in their wiring diagram. That is, one can only remove x_i if and only if f_i does not depend on x_i. Also, when one is reducing a network, some loops may appear, so the set of removable variables can change depending on the order selected for reduction.

Exercise 6.15. Reduce the network $f = (x_3, x_1 \vee x_3, x_1 \vee x_2)$ using all possible orders (remember that variables with self-loops cannot be removed), and draw the wiring diagrams of the reduced networks (and the intermediate wiring diagrams). Do all reduced networks have the same size? Compute the steady states of the reduced networks, and use them to find the steady states of the original network. □

Exercise 6.16. Reduce the network $f = (x_4, x_4, x_4, (x_1 \wedge x_2) \vee x_3)$ using all possible orders (remember that variables with self-loops cannot be removed) and draw the wiring diagrams of the reduced networks (and the intermediate wiring diagrams). Do all reduced networks have the same size? Compute the steady states of the reduced networks, and use them to find the steady states of the original network. □

6.4 IMPLEMENTING THE REDUCTION ALGORITHM USING BOOLEAN ALGEBRA

We have implemented the reduction algorithm in Mathematica (ReductionBN.nb) to aid in the reduction of larger models. The file ReductionBN.nb is available for download from the book's companion Web site. The main ingredients in implementing it are a way to encode functions (Boolean expressions in Mathematica) and a way to simplify (FullSimplify and BooleanMinimize functions in Mathematica). Once you open the file in Mathematica, run the first cell to make the functions available.

The code is designed to accept variables that start with letters and only contain alphanumeric characters. It only accepts the Boolean operators $\wedge$, $\vee$, and $\neg$ (i.e., no *XOR*, *NAND*, etc.), which are encoded as &&, ||, and !, respectively. To define a Boolean network, one has to "declare" the variables and functions in quotations, as follows.

Example 6.3. Consider $f = (x_5 \vee \neg x_2 \vee x_4, \neg x_1 \wedge \neg x_3, \neg x_2, \neg x_2, x_1 \vee x_4)$. Then, we declare the variables and functions by typing and running the following in a Mathematica cell.

```
 In[1]:=
X={"x1", "x2", "x3", "x4", "x5"};
F={"(x5||!x2)||x4","!x1&&!x3","!x2","!x2","x1||x4"};
```

To remove a variable, say x_5, we type and run

```
 In[2]:=
REMOVEVERTEX[{X, F}, "x5"]
Out[2]:=
{{x1,x2,x3,x4},{x1||!x2||x4,!x1&&!x3,!x2,!x2}}
```

To remove variables recursively (as in Example 6.2), we use

```
 In[3]:= {Xr,Fr} = {X,F};
         {Xr,Fr} = REMOVEVERTEX[{Xr, Fr}, "x5"];
         {Xr,Fr} = REMOVEVERTEX[{Xr, Fr}, "x4"];
         {Xr,Fr} = REMOVEVERTEX[{Xr, Fr}, "x3"];
         Print[Xr]; Print[Fr];
Out[3]:= {x1,x2}
         {x1 || !x2,!x1 && x2}
```

which is the reduced network we obtained by hand in Example 6.2.

When we simply want to reduce as many variables as possible and do not have preference in the order, we can use the functions REDUCEALLF and REDUCEALLB to reduce the network. REDUCEALLF and REDUCEALLB remove variables recursively starting at x_1 and x_n, respectively.

```
 In[4]:= REDUCEALLB[{X, F}]
Out[4]:= {{x1, x2}, {x1 || !x2, !x1 && x2}}
 In[5]:= REDUCEALLF[{X, F}]
Out[5]:= {{x2, x5}, {x2 && !x5, !x2 || x5}}
```

□

Exercise 6.17. Repeat Exercises 6.12–6.14 using the Mathematica code and compare with your answers. □

Exercise 6.18. Reduce the network $f = (x_5 \vee \neg x_2 \vee x_4, \neg x_1 \wedge \neg x_3, \neg x_2, \neg x_2, x_1 \vee x_4)$ using all possible orders (preferably writing a piece of code that goes over all possible reductions). Do all reduced networks have the same size? Is there a variable that always appears in the reduced networks? Is there a variable that never appears in the reduced networks? □

Exercise 6.19. Reduce the network $f = (x_6, x_6, x_6, x_6, x_6, x_1 \vee (x_2 \wedge x_3) \vee (x_4 \wedge x_5))$ using REDUCEALLF and REDUCEALLB. Do the reduced networks have the same size? □

Exercise 6.20. Create examples of networks such that REDUCEALLF and REDUCEALLB give reduced networks of very different sizes. □

6.5 IMPLEMENTING THE REDUCTION ALGORITHM USING POLYNOMIAL ALGEBRA

An alternative approach for implementing reduction algorithms is to use polynomial algebra. This approach has the advantages that the Boolean representation will be unique, that polynomial algebra has been studied extensively, and that there are many software tools available for computational algebra. In this section, we show how tools from computational algebra can be used to reduce Boolean models. We will give examples using the algebra software Macaulay2 [28] (this software is available online at http://habanero.math.cornell.edu:3690/).

6.5.1 Background

The main observation that makes possible the use of computational algebra tools is that any Boolean function can be written as a polynomial. In fact, the polynomial can be chosen so that the variables are *square-free*; that is, no factors of the form x_i^2 are necessary.

The following formulas will be useful in the conversion of a Boolean representation into a polynomial representation:

$$
\begin{aligned}
a \vee b &= a + b + a * b,\\
a \wedge b &= a * b,\\
\neg a &= 1 + a,\\
a^2 &= a,\\
2a &= 0
\end{aligned}
\tag{6.1}
$$

where the "+" and "*" on the right-hand side are addition and multiplication modulus 2 (as is customary, the "*" is usually omitted). The last two equations correspond to the fact that we are only interested in values of the variables in $\{0, 1\}$.

Example 6.4. Using Eq. (6.1) we can determine the polynomial representation of $x_1 \wedge (x_2 \vee x_3)$:

$$
\begin{aligned}
x_1 \wedge (x_2 \vee (\neg x_3)) &= x_1 \wedge (x_2 \vee (1 + x_3)),\\
&= x_1 \wedge (x_2 + (1 + x_3) + x_2(1 + x_3)),\\
&= x_1 \wedge (x_2 + (1 + x_3) + x_2(1 + x_3)),\\
&= x_1(x_2 + (1 + x_3) + x_2(1 + x_3)),\\
&= x_1(x_2 + 1 + x_3 + x_2 + x_2x_3),\\
&= x_1(1 + x_3 + x_2x_3),\\
&= x_1 + x_1x_3 + x_1x_2x_3.
\end{aligned}
$$
□

Exercise 6.21. Use formulas from Eq. (6.1) to find the polynomial representation of the following Boolean functions:

1. $x_1 \wedge \neg x_1$,
2. $(x_1 \vee x_2) \wedge x_1$,
3. $(x_1 \vee x_2) \wedge \neg x_1$,
4. $(x_1 \wedge x_2) \vee x_1$,
5. $(x_1 \vee x_2) \wedge x_2$,
6. $(x_1 \vee \neg x_2) \wedge (x_1 \vee x_2)$,
7. $(x_1 \vee x_2) \wedge x_1 \wedge x_2$,
8. $x_1 \wedge x_2 \wedge x_3 \wedge \neg x_1$. □

Exercise 6.22. Consider $f_1 = x_1 \wedge x_2 \wedge \neg x_3, f_2 = \neg x_3, f_3 = x_1 \vee x_2$. Compute and simplify $f_1(x_1, x_2, f_3)$ and $f_2(x_1, x_2, f_3)$ using polynomial algebra. □

6.5.2 Using Polynomial Algebra Software to Reduce Boolean Networks

We now use Macaulay2 for our calculations.

Example 6.5. We will use the Boolean network from Example 6.2, $f = f(x_1, x_2, x_3, x_4, x_5) = (x_5 \vee \neg x_2 \vee x_4, \neg x_1 \wedge \neg x_3, \neg x_2, \neg x_2, x_1 \vee x_4)$. First declare that we want to work with five variables

```
i1:R=ZZ/2[x1,x2,x3,x4,x5]/ideal(x1^2-x1,x2^2-x2,x3^2-x3,x4^2-x4,x5^2-x5);
```

`ZZ/2` means that we want the calculations to be done modulus 2; `[x1,x2,x3,x4,x5]` means that we are working with polynomials in the variables x_1, x_2, x_3, x_4, x_5; and `/ideal(x1^2-x1,x2^2-x2,x3^2-x3,x4^2-x4,x5^2-x5)` means that we want to force the equality $a^2 = a$.

To make the calculations easier, we define "$a|b = a + b + ab$" and "$a\&b = a * b$" in Macaulay2 as follows:

```
i2 : RingElement | RingElement :=(x,y)->x+y+x*y;
i3 : RingElement & RingElement :=(x,y)->x*y;
```

Now we can find the polynomial representation of the Boolean network.

```
i4 : f1=x5 | (1+x2) | x4 ;
i5 : f2=(1+x1) & (1+x3);
i6 : f3=1+x2;
i7 : f4=1+x2;
i8 : f5=x1 | x4;
```

To see the polynomial representation of a function we simply type `f1`

```
i9 : f1
```

Macaulay2 returns:

```
o9 : x2*x4*x5 + x2*x4 + x2*x5 + x2 + 1
```

As in Example 6.2, we will remove variables x_5, x_4, x_3. To remove variable x_5, we evaluate the polynomial setting $x_5 = f_5$:

```
i10 : f1=sub(f1,{x5=>f5});
i11 : f2=sub(f2,{x5=>f5});
i12 : f3=sub(f3,{x5=>f5});
i13 : f4=sub(f4,{x5=>f5});
```

We next remove x_4 and then x_3

```
i14 : f1=sub(f1,{x4=>f4});
i15 : f2=sub(f2,{x4=>f4});
i16 : f3=sub(f3,{x4=>f4});
i17 : f1=sub(f1,{x3=>f3});
i18 : f2=sub(f2,{x3=>f3});
```

To see the resulting reduced network, we simply type `(f1,f2)`. □

```
i19 : (f1,f2)
```

Macaulay2 returns:

```
o19 = (x1*x2 + x2 + 1, x1*x2 + x2)
o19 : Sequence
```

Exercise 6.23. Compute the polynomial representation of the reduced network from Example 6.2, and verify that it is equal to the reduced network from Example 6.5. □

Exercise 6.24. Use computational algebra to reduce the Boolean network from Example 6.5 using the order x_3, x_4, x_5. □

Exercise 6.25. Use computational algebra to reduce the Boolean network from Example 6.5 using the order x_1, x_3, x_4. □

Exercise 6.26. Use computational algebra to reduce the Boolean network from Example 6.5 using the order x_5, x_2, x_4. □

Exercise 6.27. Use computational algebra to reduce the network defined by $f = (x_6, x_6, x_6, x_6, x_6, x_1 \vee (x_2 \wedge x_3) \vee (x_4 \wedge x_5))$. First start with x_1 (then x_2, and so on); then repeat the reduction staring with x_6 (then x_5, and so on). Before removing a variable x_i, make sure that f_i does not depend on x_i; if it does, then continue with the next variable. □

6.6 APPLICATIONS

6.6.1 The *lac* Operon

The *lac* operon is a gene network in charge of metabolizing lactose in the cell. The variables that we include in the model are mRNA of the lac genes (M); the repressor protein (R) that inhibits mRNA production; the protein permease (P), which brings lactose inside the cell; the protein beta-galactosidase (B), which transforms lactose into allolactose; intracellular lactose (L); and allolactose (A). The *lac* operon is regulated by the level of extracellular lactose (L_e) and extracellular glucose (L_e). See Ref. [29] for an introduction to the *lac* operon.

In Ref. [5], the authors created the first Boolean model capable of explaining bistability in the *lac* operon. Bistability happens when the *lac* operon can be ON or OFF under the same environmental conditions. That is, in a population of cells sharing the same environment, some cells will have their *lac* operon ON and others will have it off. Bistability was shown to occur for medium levels of extracellular lactose only. If the level of extracellular lactose is low, then the *lac* operon will be off, and if the level of extracellular lactose is high, then the *lac* operon will be on. Because we have three regions where the dependence on extracellular lactose is very different, extracellular lactose, intracellular lactose, allolactose, and the repressor protein were modeled as having three possible states (low/medium/high). To model this system as a Boolean network (two states only), extra variables had to be introduced. For example, the level of extracellular lactose was modeled as 00 for the low state, 01 for the medium state, and 11 for the high state. The *lac* operon model is given in Table 6.4, and its wiring diagram is given in Figure 6.6.

In this model x_{11}, x_{12}, and x_{13} are high, medium extracellular lactose (L_e), and extracellular glucose (G_e), respectively; these variables are parameters for the model and can take any value in $\{0, 1\}$.

Exercise 6.28. By hand or using software, reduce the model of the *lac* operon starting from the last variable. □

Using the reduction algorithm, we find the reduced Boolean network, which is given in Table 6.5. Note that this is a Boolean network in one variable only and three parameters.

Now let us analyze the model under different environmental conditions. The most interesting case is when there is no extracellular glucose ($x_{13} = 0$) and there is a medium level of extracellular lactose ($x_{11} = 0, x_{12} = 1$). For these values of the external parameters, we obtain $h(x_1) = x_1$, which has two steady states $x_1 = 0, x_1 = 1$. Thus, we conclude that the original Boolean network has two steady states. That is, the system exhibits bistability, which is in agreement with biological data [30].

For other parameters bistability is not observed, which is also in agreement with biological data [30]. For example, for $x_{11} = 1$, $x_{12} = 1$, and $x_{13} = 1$, we obtain $h(x_1) = 0$. Then, the only steady state of the reduced network is $x_1 = 0$. Thus, the original network has a unique steady state in this case.

TABLE 6.4 Boolean Functions for the Boolean Model for the *lac* Operon

Variable	Boolean Function
x_1 = *lac* mRNA (*M*)	$f_1 = x_4 \wedge \neg x_5 \wedge x_6$
x_2 = *lac* permease (*P*)	$f_2 = x_1$
x_3 = *lac* β-galactosidase (*P*)	$f_3 = x_1$
x_4 = CAP (*C*)	$f_4 = \neg x_{13}$
x_5 = high LACI (*R*)	$f_5 = \neg x_7 \wedge \neg x_8$
x_6 = medium LACI (*R*)	$f_6 = (\neg x_7 \wedge \neg x_8) \vee x_5$
x_7 = high allolactose (*A*)	$f_7 = x_9 \wedge x_3$
x_8 = medium allolactose (*A*)	$f_8 = x_9 \vee x_{10}$
x_9 = high lactose (*L*)	$f_9 = x_2 \wedge x_{11} \wedge \neg x_{13}$
x_{10} = medium lactose (*L*)	$f_{10} = ((x_{12} \wedge x_2) \vee x_{11}) \wedge \neg x_{13}$

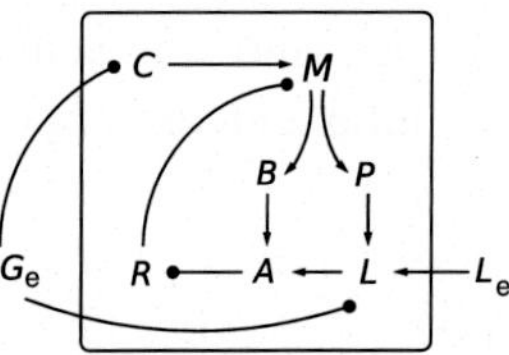

FIGURE 6.6 Wiring diagram of the Boolean model of the *lac* operon. Molecular components that are represented by more than a single variable have been collapsed into a single node (e.g., L_e represents x_{11} and x_{12}).

TABLE 6.5 Boolean Functions for the Reduced Version of the Boolean Model for the *lac* Operon

Variable	Boolean Function
x_1 = *lac* mRNA	$h(x_1) = ((x_1 \wedge x_{12}) \vee x_{11}) \wedge \neg x_{13}$
Variable	**Polynomial Function**
x_1 = *lac* mRNA	$h(x_1) = x_1x_{11}x_{12}x_{13} + x_1x_{11}x_{12} + x_1x_{12}x_{13} + x_1x_{12} + x_{11}x_{13} + x_{11}$

Exercise 6.29. Analyze $h(x_1) = ((x_1 \wedge x_{12}) \vee x_{11}) \wedge \neg x_{13}$ for low extracellular lactose ($x_{11} = 0$, $x_{12} = 0$) and low extracellular glucose ($x_{13} = 0$). How many steady states does the reduced network have? How many steady states does the original network have? □

Exercise 6.30. Analyze $h(x_1) = ((x_1 \wedge x_{12}) \vee x_{11}) \wedge \neg x_{13}$ for low extracellular lactose ($x_{11} = 0$, $x_{12} = 0$) and high extracellular glucose ($x_{13} = 1$). How many steady states does the reduced network have? How many steady states does the original network have? □

Exercise 6.31. Analyze $h(x_1) = ((x_1 \wedge x_{12}) \vee x_{11}) \wedge \neg x_{13}$ for high extracellular lactose ($x_{11} = 1, x_{12} = 1$) and low extracellular glucose ($x_{13} = 0$). How many steady states does the reduced network have? How many steady states does the original network have? □

6.6.2 Th-Cell Differentiation

Th cells are cells in the immune system in charge of different responses. Some of the most important types of cells are Th1 and Th2 cells, which can be classified as proinflamatory and anti-inflammatory, respectively. These two cells differentiate from a precursor type, Th0 (naive Th cell). See Ref. [31] for an introduction to Th-cell differentiation.

In Ref. [4], the authors proposed a Boolean model to explain the mechanism for differentiation of Th0 cells into Th1 or Th2. The Th-cell differentiation model is given in Table 6.6, and its wiring diagram is given in Figure 6.7.

TABLE 6.6 Boolean Functions for the Boolean Model of Th-Cell Differentiation

Variable	Boolean Function
x_1 = GATA3	$f_1 = (x_1 \vee x_{21}) \wedge \neg x_{22}$
x_2 = IFN-β	$f_2 = 0$
x_3 = IFN-βR	$f_3 = x_2$
x_4 = IFN-γ	$f_4 = (x_{14} \vee x_{16} \vee x_{20} \vee x_{22}) \wedge \neg x_{19}$
x_5 = IFN-γ R	$f_5 = x_4$
x_6 = IL-10	$f_6 = x_1$
x_7 = IL-10R	$f_7 = x_6$
x_8 = IL-12	$f_8 = 0$
x_9 = IL-12R	$f_9 = x_8 \wedge \neg x_{21}$
x_{10} = IL-18	$f_{10} = 0$
x_{11} = IL-18R	$f_{11} = x_{10} \wedge \neg x_{21}$
x_{12} = IL-4	$f_{12} = x_1 \wedge \neg x_{18}$
x_{13} = IL-4R	$f_{13} = x_{12} \wedge \neg x_{17}$
x_{14} = IRAK	$f_{14} = x_{11}$
x_{15} = JAK1	$f_{15} = x_5 \wedge \neg x_{17}$
x_{16} = NFAT	$f_{16} = x_{23}$
x_{17} = SOCS1	$f_{17} = x_{18} \vee x_{22}$
x_{18} = STAT1	$f_{18} = x_3 \vee x_{15}$
x_{19} = STAT3	$f_{19} = x_7$
x_{20} = STAT4	$f_{20} = x_9 \wedge \neg x_1$
x_{21} = STAT6	$f_{21} = x_{13}$
x_{22} = T-bet	$f_{22} = (x_{18} \vee x_{22}) \wedge \neg x_1$
x_{23} = TCR	$f_{23} = 0$

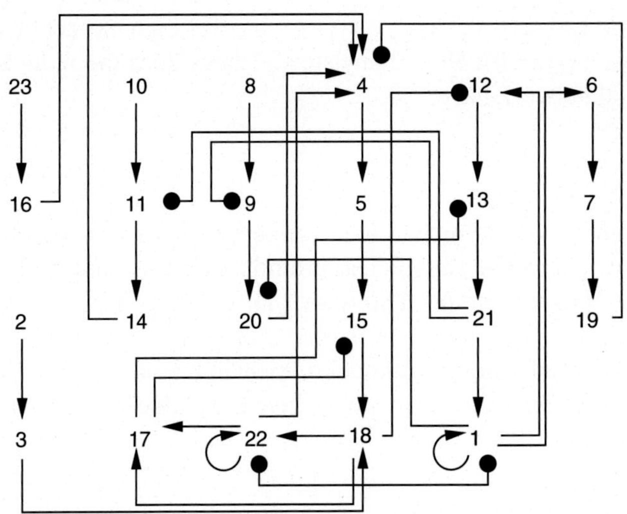

FIGURE 6.7 Wiring diagram of the Boolean model of Th-cell differentiation.

TABLE 6.7 Boolean Functions for the Reduced Version of the Boolean Model of Th-Cell Differentiation

Variable	Boolean Function
$x_1 =$ GATA3	$h_1(x_1, x_{22}) = x_1 \wedge \neg x_{22}$
$x_{22} =$ T-bet	$h_{22}(x_1, x_{22}) = \neg x_1 \wedge x_{22}$
Variable	**Polynomial Function**
$x_1 =$ GATA3	$h_1(x_1, x_{22}) = x_1 x_{12} + x_1$
$x_{22} =$ T-bet	$h_{22}(x_1, x_{22}) = x_1 x_{22} + x_{22}$

Exercise 6.32. By hand or using a software package, reduce the model of Th-cell differentiation starting from the last variable. □

Using the reduction algorithm, we find the reduced Boolean network, which is given in Table 6.7. Note that this is a Boolean network in two variables.

Now let us analyze the model. The steady states of the reduced network are $(x_1, x_{22}) = (0,0)$, $(0,1)$, $(1,0)$ (Exercise 6.33). The steady state $(0,0)$ corresponds to GATA3 and T-bet being inactive, which is the "signature" of Th0 cells. The steady state $(0,1)$ corresponds to T-bet being active only, which is the signature of Th1 cells. The steady state $(1,0)$ corresponds to GATA3 being active only, which is the signature of Th2 cells. In summary, the reduced network, and therefore the original Boolean network, is able to replicate the three main types of Th cells in agreement with biological data.

Exercise 6.33. Find the steady states of the reduced Boolean network $h_1 = x_1 x_{12} + x_1$, $h_{22} = x_1 x_{22} + x_{22}$. □

6.7 AND BOOLEAN MODELS

So far the reduction method allows us to reduce any network regardless of the Boolean operators. However, interactions between biological components are not arbitrary, and there are classes of Boolean functions that appear more frequently than others [21]. Hence, it is important to study the reduction of these classes of networks.

In this section, we present a reduction method tailored specifically to AND Boolean models [32]. These networks are of the form $f = (f_1, \ldots, f_n) : \{0, 1\}^n \to \{0, 1\}^n$, where $f_i = x_{j_1} \wedge x_{j_2} \wedge \ldots \wedge x_{j_k}$.

Example 6.6. The following Boolean network is an AND network.

In Boolean form	or in polynomial form
$f_1 = x_2 \wedge x_3,$	$f_1 = x_2x_3,$
$f_2 = x_1 \wedge x_2 \wedge x_6,$	$f_2 = x_1x_2x_6,$
$f_3 = x_2,$	$f_3 = x_2,$
$f_4 = x_1 \wedge x_5,$	$f_4 = x_5x_1,$
$f_5 = x_5 \wedge x_6,$	$f_5 = x_5x_6,$
$f_6 = x_1 \wedge x_5 \wedge x_6,$	$f_6 = x_1x_5x_6.$

□

Exercise 6.34. Find the wiring diagram of the following Boolean networks.

1. $f(x) = (x_2x_3, x_1x_2x_6, x_2, x_1x_5, x_5x_6, x_1x_5x_6)$
2. $f(x) = (x_2, x_3, x_4, x_1)$
3. $f(x) = (x_4, x_1, x_2, x_3)$
4. $f(x) = (x_6, x_6, x_6, x_6, x_6, x_1x_2x_3x_4x_5)$
5. $f(x) = (x_2x_3, x_1x_3, x_1x_2)$
6. $f(x) = (x_2, x_1, x_4, x_5, x_1x_3)$ □

An important observation for AND networks is that they are completely determined by the wiring diagram. For example, $f_1 = x_2 \wedge x_3$ will be represented by arrows from x_2 and x_3 to x_1 (and no other arrows to x_1), and having arrows from x_2 and x_3 to x_1 (and no other arrows to x_1) implies that f_1 must be $x_2 \wedge x_3$ because only the $\wedge$ operator is allowed.

Exercise 6.35. Find the AND networks associated to the wiring diagrams shown in Figure 6.8. □

6.7.1 Background

We now introduce terminology that we need for the reduction of AND networks.

A subgraph G_i of a wiring diagram is *strongly connected* if there is a directed path from any vertex of G_i to any other vertex in G_i. In Figure 6.9 we see a graph and some strongly connected subgraphs.

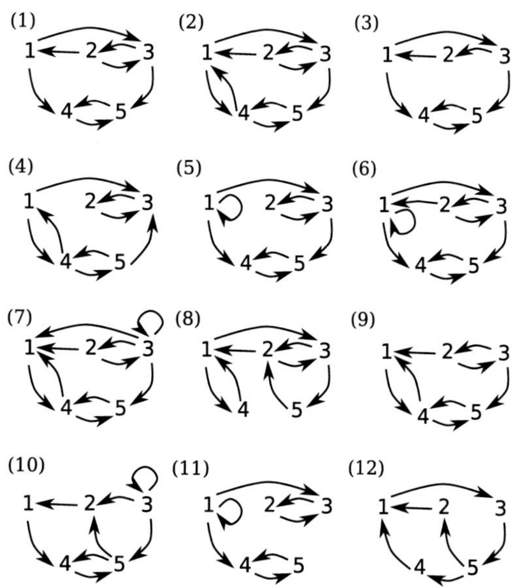

FIGURE 6.8 Wiring diagrams for Exercises 6.35, 6.36 and 6.38.

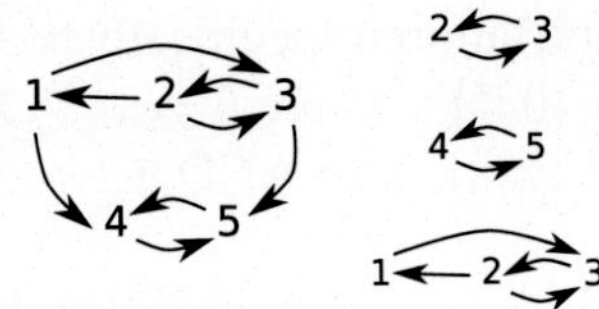

FIGURE 6.9 Wiring diagram (left) and some strongly connected subgraphs (right).

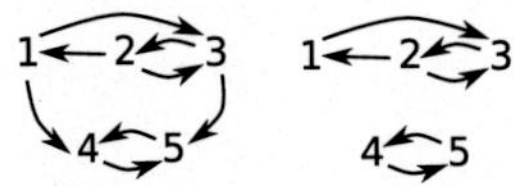

FIGURE 6.10 Wiring diagram (left) and strongly connected components (right).

A subgraph G_i of a wiring diagram is a *strongly connected component* (scc) if it is strongly connected and is the largest strongly connected subgraph that contains G_i, that is, G_i cannot be made larger and still be strongly connected. If a scc has only one node and it does not have a self-loop, then we call that scc *trivial*. In Figure 6.10, we see a wiring diagram and all its strongly connected components.

Exercise 6.36. Find the strongly connected components of the wiring diagrams in Figure 6.8. □

Exercise 6.37. Find all steady states of the AND networks corresponding to the wiring diagrams in Figure 6.8. □

The main result in the reduction of AND networks is that strongly connected components can be replaced by a single node with a self-loop.[1] This follows from the fact that for a steady state, all variables in the same scc will have the same value.

Example 6.7. Consider the Boolean network $f = (x_2x_3, x_1x_3, x_2, x_1x_5, x_6, x_3x_4)$ with wiring diagram shown in Figure 6.11 (left). To reduce this network we replace the scc by single nodes with self-loops. For example, we can use x_1 for the first ssc and x_4 for the second scc. Also, we collapse all edges from any vertex in a scc to another scc as a single edge. In this example, the edges $x_1 \to x_4$ and $x_3 \to x_6$ are collapsed into the edge $x_1 \to x_4$ in the reduced network. Thus, we obtain the reduced network given in Figure 6.11 (right), which corresponds to the Boolean network $h = h(x_1, x_4) = (x_1, x_1x_4)$ and has steady states $(x_1, x_4) = (0, 0), (1, 0), (1, 1)$. Then, the steady states of the original AND network are found using the equations $x_2 = x_3 = x_1$ and $x_6 = x_5 = x_4$: $x = 000000$, 111000, 111111. □

Exercise 6.38. Find all steady states of the AND networks corresponding to the wiring diagrams in Figure 6.8 using reduction of AND networks. □

Exercise 6.39. Use reduction of AND networks to find the steady states of the following Boolean networks.

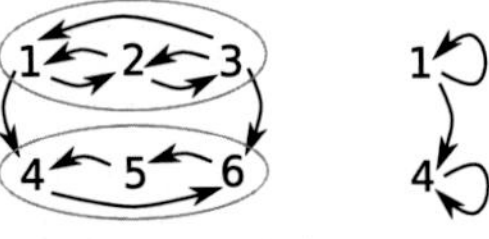

FIGURE 6.11 Wiring diagram of AND network from Example 6.7 (left) with its strongly connected components highlighted in circles and the reduced wiring diagram (right).

1. This is not true for trivial scc, but for simplicity, all scc in this section are nontrivial.

FIGURE 6.12 Wiring diagrams for Exercise 6.40.

1.	2.	3.	4.	5.	6.
$f_1 = x_4x_5$	$f_1 = x_4x_5$	$f_1 = x_4x_5$	$f_1 = x_4x_5$	$f_1 = x_4x_5$	$f_1 = x_4x_5$
$f_2 = x_3x_5$	$f_2 = x_3x_5$	$f_2 = x_3$	$f_2 = x_3x_9$	$f_2 = x_3$	$f_2 = x_3$
$f_3 = x_2x_1x_6$	$f_3 = x_2x_1x_6$	$f_3 = x_2x_6$	$f_3 = x_2x_9x_6$	$f_3 = x_2x_8$	$f_3 = x_2x_9$
$f_4 = x_5$	$f_4 = x_5$	$f_4 = x_5$	$f_4 = x_5$	$f_4 = x_5$	$f_4 = x_5$
$f_5 = x_1x_8$	$f_5 = x_1$	$f_5 = x_1$	$f_5 = x_1x_2$	$f_5 = x_1$	$f_5 = x_1$
$f_6 = x_9x_5x_3$	$f_6 = x_9x_3x_7$	$f_6 = x_9x_3x_7$	$f_6 = x_9x_7$	$f_6 = x_9x_7$	$f_6 = x_9x_3$
$f_7 = x_8x_1$	$f_7 = x_8x_1$	$f_7 = x_8x_1$	$f_7 = x_8x_1$	$f_7 = x_8x_1x_2$	$f_7 = x_8x_1x_2$
$f_8 = x_7x_4$	$f_8 = x_7x_4$	$f_8 = x_7x_4$	$f_8 = x_7x_4$	$f_8 = x_7x_4x_3$	$f_8 = x_7x_4x_3$
$f_9 = x_6x_4$	$f_9 = x_6x_3x_8$	$f_9 = x_6x_3x_8$	$f_9 = x_6x_8$	$f_9 = x_6x_8$	$f_9 = x_6x_3$

□

Exercise 6.40. Find all steady states of the AND networks corresponding to the wiring diagrams in Figure 6.12 using reduction of AND networks. □

6.8 CONCLUSION

For many biological systems the precise regulatory mechanism are unknown, and Boolean networks provide a strong theoretical framework to analyze them. Although the analysis of small Boolean models can easily be done by exhaustive enumeration, the analysis of large models is still an open problem. Addressing this problem is important because the dimension of published models is growing to over 100 [12, 13].

Here we presented a method for steady state analysis based on dimension reduction. In essence, the method works by removing nodes in the network and passing forward its regulatory information to downstream nodes. Computing the steady states of the reduced Boolean network is easier, and the algorithm also provides a way to reconstruct the steady states of the original network.

Because biological systems are not arbitrary, it is important to understand how reduction methods perform or can be optimized for certain classes of Boolean networks. For example, for AND networks there exists a very efficient reduction method that allows the computation of steady states of large AND networks (Section 6.7). Thus, although typical reduction methods can work on any Boolean network, one can obtain better algorithms for networks that are biologically relevant. Similar approaches were used in Ref. [33] to design a reduction algorithm optimized for *AND-NOT networks*. AND-NOT networks are Boolean networks of the form $f : \{0, 1\}^n \to \{0, 1\}^n$, where $f_i(x) = y_{i_1} \wedge y_{i_2} \wedge \ldots \wedge y_{i_k}$ such that y_j is either x_j or $\neg x_j$; that is, each f_i is a conjunction of positive (x_j) or negative ($\neg x_j$) variables.

One disadvantage of reduction methods is that they do not provide a way to compute the steady states of the reduced network. However, other steady state analysis tools can be used on the reduced network. Indeed, it has been shown recently that the combination of different methods with network reduction is much more powerful than any particular method [34]. Namely, starting from an arbitrary Boolean network, one can construct its *AND-NOT representation* [35], then apply a network reduction algorithm optimized for AND-NOT networks [33], and then use tools from computational algebra to find the steady states of the reduced Boolean network [36, 37], which can be used to compute the steady states of the original network. Thus, the best way for steady state analysis of Boolean models will probably be a synergistic method where network reduction plays a key role.

REFERENCES

[1] Kauffman S. Metabolic stability and epigenesis in randomly constructed genetic nets. J Theor Biol 1969;22(3):437-67.

[2] Thomas R, D'Ari R. Biological feedback. 1st ed. Boca Raton, FL: CRC Press; 1989.

[3] Li F, Long T, Lu Y, Ouyang Q, Tang C. The yeast cell-cycle network is robustly designed. Proc Natl Acad Sci USA 2004;101(14):4781-6.

[4] Mendoza L. A network model for the control of the differentiation process in Th cells. Biosystems 2006; 84(2):101-14 [Dynamical Modeling of Biological Regulatory Networks].

[5] Veliz-Cuba A, Stigler B. Boolean models can explain bistability in the *lac* operon. J Comput Biol 2011;18(6):783-94.

[6] Murrugarra D, Veliz-Cuba A, Aguilar B, Arat S, Laubenbacher R. Modeling stochasticity and variability in gene regulatory networks. EURASIP J Bioinf Syst Biol 2012;2012:5.

[7] Abou-Jaoudé W, Ouattara D, Kaufman M. From structure to dynamics: frequency tuning in the p53-mdm2 network. I. logical approach. J Theor Biol 2009;258(4):561-77.

[8] Davidich M, Bornholdt S. Boolean network model predicts cell cycle sequence of fission yeast. PLoS One 2008;3(2):1672.

[9] Albert R, Othmer H. The topology of the regulatory interactions predicts the expression pattern of the segment polarity genes in *Drosophila melanogaster*. J Theor Biol 2003;223:1-18.

[10] Saadatpour A, Wang R, Liao A, Liu X, Loughran T, Albert I, et al. Dynamical and structural analysis of a T-cell survival network identifies novel candidate therapeutic targets for large granular lymphocyte leukemia. PLoS Comput Biol 2011;7(11):1002267.

[11] Zhang R, Shah M, Yang J, Nyland S, Liu X, Yun J, et al. Network model of survival signaling in large granular lymphocyte leukemia. PNAS 2008;105(42):16308-13.

[12] Helikar T, Konvalina J, Heidel J, Rogers J. Emergent decision-making in biological signal transduction networks. PNAS 2008;105(6):1913-8.

[13] Helikar T, Kochi N, Kowal B, Dimri M, Naramura M, Raja S, et al. A comprehensive, multi-scale dynamical model of ErbB receptor signal transduction in human mammary epithelial cells. PLoS One 2013;8(4):61757.

[14] Sontag E, Veliz-Cuba A, Laubenbacher R, Jarrah A. The effect of negative feedback loops on the dynamics of Boolean networks. Biophys J 2008;95:518-26.

[15] Veliz-Cuba A, Laubenbacher R. On the computation of fixed points in Boolean networks. J Appl Math Comput 2011;39(1-2):145-53.

[16] Xu W, Ching W, Zhang S, Li W, Chen X. A matrix perturbation method for computing the steady-state probability distributions of probabilistic Boolean networks with gene perturbations. J Comput Appl Math 2011;235(8):2242-51.

[17] Veliz-Cuba A. An algebraic approach to reverse engineering finite dynamical systems arising from biology. SIAM J Appl Dyn Syst 2012;11(1):31-48.

[18] Li W, Cui L, Ng M. On computation of the steady-state probability distribution of probabilistic Boolean networks with gene perturbation. J Comput Appl Math 2012;236(16):4067-81.

[19] Veliz-Cuba A, Murrugarra D, Laubenbacher R. Structure and dynamics of acyclic networks. Discrete Event Dyn Syst. 2014;24(4):647-658.

[20] Laubenbacher R, Hinkelmann F, Murrugarra D, Veliz-Cuba A. Algebraic models and their use in systems biology. In: Jonoska N, Saito M, editors. Discrete and topological models in molecular biology. Berlin, Heidelberg: Springer; 2013. p. 443-74.

[21] Raeymaekers L. Dynamics of Boolean networks controlled by biologically meaningful functions. J Theor Biol 2002;218(3):331-41.

[22] Akutsu T, Kuhara S, Maruyama O, Miyano S. A system for identifying genetic networks from gene expression patterns produced by gene disruptions and overexpressions. Genome Inform 1998;9:151-60.

[23] Zhao Q. A remark on "scalar equations for synchronous Boolean networks with biological applications" by C. Farrow, J. Heidel, J. Maloney, and J. Rogers. IEEE Trans Neural Netw 2005;16(6):1715-6.

[24] Veliz-Cuba A. Reduction of Boolean network models. J Theor Biol 2011;289:167-72.

[25] Saadatpour A, Albert R, Reluga T. A reduction method for Boolean network models proven to conserve attractors. SIAM J Appl Dyn Syst 2013;12(4):1997-2011.

[26] Saadatpour A, Albert I, Albert R. Attractor analysis of asynchronous Boolean models of signal transduction networks. J Theor Biol 2010;266(4):641-56.

[27] Naldi A, Remy E, Thieffry D, Chaouiya C. A reduction of logical regulatory graphs preserving essential dynamical properties. In: Degano P, Gorrieri R, editors. Computational methods in systems biology, vol. 5688. Lecture notes in computer science. Berlin, Heidelberg: Springer; 2009. p. 266-80.

[28] Grayson D, Stillman M. Macaulay2, a software system for research in algebraic geometry. Available at: http://www.math.uiuc.edu/Macaulay2/.

[29] Robeva R, Kirkwood B, Davies R. Mechanisms of gene regulation: Boolean network models of the lactose operon in *Escherichia coli*. In: Robeva R, Hodge T, editors. Mathematical concepts and methods in modern biology. Boston: Academic Press; 2013. p. 1-35.

[30] Ozbudak E, Thattai M, Lim H, Shraiman B, van Oudenaarden A. Multistability in the lactose utilization network of *Escherichia coli*. Nature 2004;427(6976):737-40.

[31] Mendoza L. A network model for the control of the differentiation process in Th cells. Biosystems 2006;84(2):101-14.

[32] Jarrah A, Laubenbacher R, Veliz-Cuba A. The dynamics of conjunctive and disjunctive Boolean network models. Bull Math Biol 2010;72(6):1425-47.

[33] Veliz-Cuba A, Aguilar B, Laubenbacher R. Dimension Reduction of large sparse AND-NOT network models. Electronic Notes in Theoretical Computer Science, 2014;accepted.

[34] Veliz-Cuba A, Aguilar B, Hinkelmann F, Laubenbacher R. Steady state analysis of Boolean molecular network models via model reduction and computational algebra. BMC Bioinf 2014;15:221.

[35] Veliz-Cuba A, Buschur K, Hamershock R, Kniss A, Wolff E, Laubenbacher R. AND-NOT logic framework for steady state analysis of Boolean network models. Appl Math Inf Sci 2013;7(4):1263-74.

[36] Veliz-Cuba A, Jarrah A, Laubenbacher R. The polynomial algebra of discrete models in systems biology. Bioinformatics 2010;26:1637-43.

[37] Hinkelmann F, Brandon M, Guang B, McNeill R, Blekherman G, Veliz-Cuba A, et al. ADAM: analysis of discrete models of biological systems using computer algebra. BMC Bioinf 2011;12:295.

Chapter 7

BioModel Engineering with Petri Nets

Mary Ann Blätke[1], Monika Heiner[2] and Wolfgang Marwan[1]
[1]Otto-von-Guericke University, Magdeburg, Germany, [2]Brandenburg University of Technology, Cottbus, Germany

7.1 INTRODUCTION

BioModel Engineering

Systematically constructing, maintaining, and deploying artifacts are typical attributes of sound engineering, independent of the application field. Along these lines, BioModel Engineering is the science of engineering computational models of biochemical processes in an efficient, sustainable, and trustworthy manner.

In *Systems Biology*, models are used to describe our abstract understanding of biochemical processes and to predict their behavior, for example, in response to perturbations like mutations, chemical interventions, or changes in the environment. In *Synthetic Biology*, models support the reliable design and redesign of molecular networks and may serve as design templates for novel synthetic biological systems. Ideally, computational models have high explanatory and predictive power.

In this chapter, we will explore how this ambitious aim might become achievable by the support of the unifying power of Petri nets, serving as umbrella language for modeling in the qualitative, stochastic, continuous, and hybrid paradigms. An introduction to BioModel Engineering illustrated by means of signal transduction pathways can be found in [1]; a more rigorous discussion of how to go from structure to behavior is given in [2].

Petri Nets

The basic ideas of Petri nets, as we understand them today, were introduced by Carl Adam Petri in his PhD thesis in 1962 [3]. Petri nets are basically a formal language with a graphical representation and operational (execution) semantics. They permit the modeling of relationships between two types of nodes, which are called places and transitions in the Petri net terminology and typically refer to objects and events. The objects can be biologically interpreted as any (bio-)chemical species, such as genes, mRNAs, proteins, metabolites, and other biomolecules, whereas events model (bio-)chemical reactions at variable resolution of details, such as association, dissociation, regulatory activation and deactivation, transport, transcription, and so on. Quantitative information such as the stoichiometry of (bio-)chemical reactions or the amount of species involved can be described as well. The graph structure of a Petri net is unambiguous and explicitly distinguishes between situations involving alternatives or concurrency.

Enhancing these key modeling principles by the powerful concept of discrete data types, as we know them from programming languages, yields colored Petri nets. Here, "color" can just be seen as a mere synonym for "discrete data type." They are particularly strong, as we will later see, in dealing with larger networks comprising similar processes in similar components.

Petri nets may represent species on different abstraction levels, reaching from molecules via single cells to multicellular organisms and populations. This allows us to model different types of biological networks, such as

Algebraic and Discrete Mathematical Methods for Modern Biology. http://dx.doi.org/10.1016/B978-0-12-801213-0.00007-1
Copyright © 2015 Elsevier Inc. All rights reserved.

metabolic networks, signaling networks, or gene regulatory networks, and to put them into a larger context of biological structure and function. Different abstraction levels, as well as different network types, can be integrated in a single Petri net to yield a coherent model.

The basic modeling concept of (qualitative) Petri nets can be further enriched by kinetic information. Then, Petri nets can be interpreted in the stochastic, continuous, or hybrid modeling paradigms, opening the door to explore one and the same model in different settings.

The operational semantics that is defined in the formal language of Petri nets allows us to execute a model, given by its graph, its initial state, and its kinetics, if any, and thus to explore its behavior, either in a merely qualitative way or in a continuous, stochastic, or hybrid way. There are numerous tools supporting a wealth of analysis techniques to qualitatively and quantitatively investigate the structure and behavior of a given Petri net, and thus to examine a biological reaction network from different, complementary angles.

As we will demonstrate, Petri nets offer a straightforward and intuitive approach for a unifying framework to construct and deploy computational models. Their application to biological processes facilitates the integration of different types of qualitative and quantitative information, of different types of biomolecular reactions as well as complex functional interactions, and of small- to large-scale or even multilevel networks.

Framework

Our unifying Petri net framework (see Figure 7.1) can be divided into two levels: uncolored [4] and colored [5]. Each level comprises a family of related Petri net classes, sharing structure, but being specialized by their kinetic information. Specifically, the uncolored level contains qualitative, that is, time-free Petri nets ($\mathcal{PN}$), as well as quantitative, that is, time-dependent Petri nets such as stochastic Petri nets ($\mathcal{SPN}$), continuous Petri nets ($\mathcal{CPN}$), and hybrid Petri nets ($\mathcal{HPN}$). The colored level consists of the colored counterparts of the uncolored level, thus containing colored qualitative Petri nets ($\mathcal{PN}^{\mathcal{C}}$), colored stochastic Petri nets ($\mathcal{SPN}^{\mathcal{C}}$), colored continuous Petri nets ($\mathcal{CPN}^{\mathcal{C}}$), and colored hybrid Petri nets ($\mathcal{HPN}^{\mathcal{C}}$).

A model given in a specific net class can be converted into any other net class. This conversion may involve a loss or enrichment of information, if we move between the qualitative and quantitative paradigms (see arrows

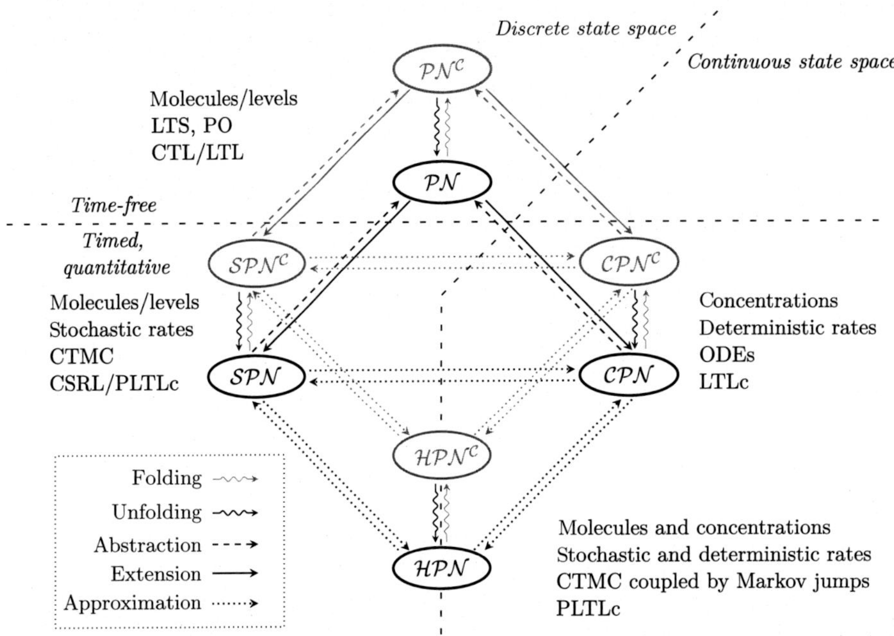

FIGURE 7.1 The unifying Petri net framework, adapted from [5], with kind permission from Springer Science + Business Media B.V., Figure 1, p. 399. See Table 7.1 for abbreviations.

TABLE 7.1 Abbreviations Used in Figure 7.1

CSRL	Continuous stochastic reward logic
CTL	Computational tree logic
CTMC	Continuous-time Markov chain
LTL	Linear-time logic
LTLc	Linear-time logic with constraints
LTS	Labeled transition system
ODEs	System of ordinary differential equations
PO	Partial order semantics
PLTLc	Probabilistic linear-time logic with constraints

abstraction and *extension*). The conversion may also change the graphical representation of a model (without changing the actual network) if we move between the uncolored and colored levels (see arrows *folding* and *unfolding*). The conversion may also just mean a different interpretation of the kinetic information (rates), without changing anything in the model specification, if we move between the stochastic, continuous, and hybrid paradigms (see arrows *approximation*).

In this chapter, we will illustrate the complementary power of these Petri net classes with a running example—a (simplified) model of a gene regulatory network of a circadian clock mechanism. When running through the different modeling paradigms and learning their specific analysis features, we will explain all technical terms in more detail; see Table 7.1 for the abbreviations used in Figure 7.1. We deliberately avoid any formal definitions, which can be found in the references we provide under *Further Reading*.

Tools

A crucial point of BioModel Engineering is reproducibility, and reproducible computational research [6] in Systems Biology requires reliable software tools. In this chapter, we will use the following public domain tools.

- *Snoopy* [4, 5] is a platform supporting the construction and animation and/or simulation of various types of Petri nets, including all Petri net classes used in this chapter, with an automatic conversion between them. Snoopy supports several data exchange formats, among them the Systems Biology Markup Language (SBML, level 1 and 2), which opens the door to many tools popular in Systems and Synthetic Biology.
- *Charlie* [7] applies standard analysis techniques of Petri net theory to determine structural and behavioral properties of Petri nets, complemented by explicit model checking of computational tree logic (CTL) and linear-time logic (LTL). Charlie reads all uncolored Petri nets designed with Snoopy.
- *Marcie* [8–10] combines exact analysis techniques gaining their efficiency by symbolic data structures (Interval Decision Diagrams) with approximative analysis techniques building on fast adaptive uniformization (FAU) and parallelized stochastic simulation (Gillespie, tau leaping). It supports model checking of CTL, continuous stochastic logic (CSL), and probabilistic linear-time logic with constraints (PLTLc). Marcie reads $\mathcal{PN}$ and $\mathcal{SPN}$ given in the Abstract Net Description Language (ANDL), one of the export features of Snoopy.
- *MC2(PLTLc)* [11] is a Monte Carlo Model Checker for linear-time logic with constraints (LTLc) and PLTLc, operating on stochastic, deterministic, and hybrid simulation traces; deterministic parameter scans; or even wetlab data. We use MC2 to analyze deterministic and hybrid traces generated by Snoopy.

These tools are recommended for solving the exercises in this chapter. The first three can be downloaded from http://www-dssz.informatik.tu-cottbus.de, the last one from http://www.brunel.ac.uk/research/centres/cssb/software-systems-and-databases/mc2.

All Petri nets used in this chapter are available at http://www-dssz.informatik.tu-cottbus.de/examples. They can also be found as Supplementary Material on the volume's website.

7.2 RUNNING CASE STUDY

A broad variety of organisms, from cyanobacteria to mammals, use circadian clocks to regulate processes with daily periodicity, keeping track of the day/night rhythm, and, thus, to adapt their behavior accordingly. Most known circadian clocks are believed to be driven by gene regulatory networks involving interconnected positive and negative regulators as common features [12]; see Figure 7.2. The positive regulator (activator) enhances its own expression and that of the negative regulator (repressor). The repressor diminishes (inhibits) the activator activity. Degradation of the repressor and re-expression of the activator close the loop. The interplay between activator and repressor permits stable oscillation with constant period over a wide range of internal and external fluctuations [13].

To elucidate circadian oscillator mechanisms is one of the challenges in Systems Biology; there are numerous computational models around. The simplified model, see Figure 7.2, which we use in this chapter as a running example to illustrate the Petri net approach, has been published in [14], where it has been encoded as a system of ordinary differential equations (ODEs). The variables with their initial conditions are summarized in Table 7.2, and the chemical reactions in Table 7.3. The kinetics of all reactions are governed by the mass-action law; the kinetic parameters are given in Table 7.4.

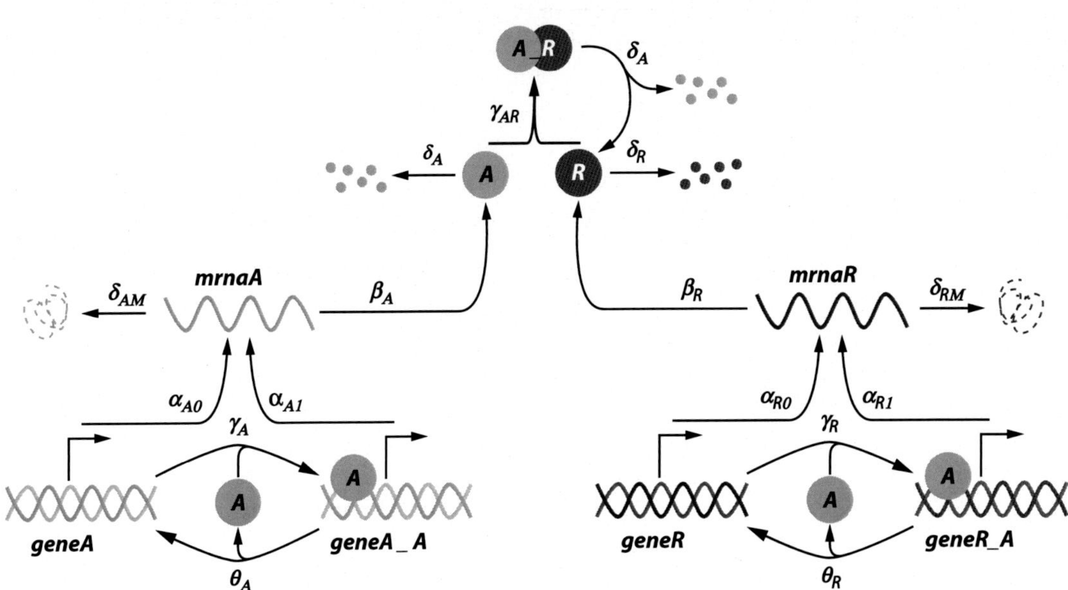

FIGURE 7.2 General scheme of a gene regulatory network of circadian oscillation (inspired by and redrawn after [14]), comprising four levels: (1) gene regulation, (2) mRNA synthesis by transcription of the DNA, (3) protein synthesis by translation of the RNA, and (4) protein-protein interaction. (1) The activator protein A binds to the promoter of its encoding gene as well as to the promoter of the gene encoding the repressor protein R. (2) According to the model, there is a basal as well as an increased transcription of the activator mRNA and the repressor mRNA due to binding of the activator to the corresponding promoters. (3) The mRNAs are translated into the respective proteins. (4) The activator protein and repressor protein form a complex with each other. Degradation of the activator protein occurs upon dissociation of the complex. The mRNAs and the unbound proteins are degraded as well.

TABLE 7.2 Variables of the Circadian Clock Model

Variable[a]	Description	Initial Value[b] (mol)
A	Activator protein	0
geneA	Gene encoding the activator protein (activator gene, for short)	1
geneA_A	Complex of activator gene and activator protein	0
mrnaA	mRNA of activator protein	0
R	Repressor protein	0
geneR	Gene encoding the repressor protein (repressor gene, for short)	1
geneR_A	Complex of repressor gene and activator protein	0
mrnaR	mRNA of repressor protein	0
A_R	Complex of activator protein and repressor protein	0

[a]The underscore in a variable name indicates a compound.
[b]The cellular volume is assumed to be the unity, thus concentrations equal number of molecules.

TABLE 7.3 Chemical Reaction Equations[a] of the Circadian Clock Model

r1:	$geneA + A$	$\xrightarrow{\gamma_A}$	$geneA\_A$	r9:	$geneR + A$	$\xrightarrow{\gamma_R}$	$geneR\_A$
r2:	$geneA\_A$	$\xrightarrow{\theta_A}$	$geneA + A$	r10:	$geneR\_A$	$\xrightarrow{\theta_R}$	$geneR + A$
r3:	$geneA$	$\xrightarrow{\alpha_{A0}}$	$geneA + mrnaA$	r11:	$geneR$	$\xrightarrow{\alpha_{R0}}$	$geneR + mrnaR$
r4:	$geneA\_A$	$\xrightarrow{\alpha_{A1}}$	$geneA\_A + mrnaA$	r12:	$geneR\_A$	$\xrightarrow{\alpha_{R1}}$	$geneR\_A + mrnaR$
r5:	$mrnaA$	$\xrightarrow{\beta_A}$	$mrnaA + A$	r13:	$mrnaR$	$\xrightarrow{\beta_R}$	$mrnaR + R$
r6:	$mrnaA$	$\xrightarrow{\delta_{AM}}$	$\emptyset$	r14:	$mrnaR$	$\xrightarrow{\delta_{RM}}$	$\emptyset$
r7:	A	$\xrightarrow{\delta_A}$	$\emptyset$	r15:	R	$\xrightarrow{\delta_R}$	$\emptyset$
r8:	$A + R$	$\xrightarrow{\gamma_{AR}}$	A_R	r16:	A_R	$\xrightarrow{\delta_A}$	R

[a]These equations have been extracted from the ODEs given in [14]. The derived equations are unique, as the ODEs fulfill the three conditions established in [66].

TABLE 7.4 Parameter Values of the Circadian Clock Model

Parameter	Value	Parameter	Value
α_{A0}	50 h^{-1}	α_{R0}	0.01 h^{-1}
α_{A1}	500 h^{-1}	α_{R1}	50 h^{-1}
β_A	50 h^{-1}	β_R	5 h^{-1}
γ_A	1 mol^{-1} h^{-1}	γ_R	1 mol^{-1} h^{-1}
γ_{AR}	2 mol^{-1} h^{-1}		
δ_{AM}	10 h^{-1}	δ_{RM}	0.5 h^{-1}
δ_A	1 h^{-1}	δ_R	0.2 h^{-1}
θ_A	50 h^{-1}	θ_R	100 h^{-1}

7.3 PETRI NETS ($\mathcal{PN}$)

7.3.1 Modeling

Basics

Petri nets belong to the graph formalisms, that is, their basic ingredients are nodes and arcs describing the relationship between the nodes. More specifically and technically speaking, Petri nets are bipartite, directed multigraphs; see Figure 7.3.

- *Bipartite*: There are two types of nodes, called *places* and *transitions*, which form disjunctive node sets. Places are typically represented as circles and transitions as squares. There are no limits for their interpretation; see Table 7.5 for a few examples.
- *Directed*: Directed arcs, represented as arrows, connect places with transitions and vice versa, thereby specifying which biomolecules serve as precursors (making the pre-places) or products (making the post-places) for each reaction.

  Likewise, we speak of the pre-transitions of a place, which refers to the set of transitions (reactions) producing the species modeled by this place, and the post-transitions of a place, which refers to the set of transitions (reactions) consuming the species modeled by this place.

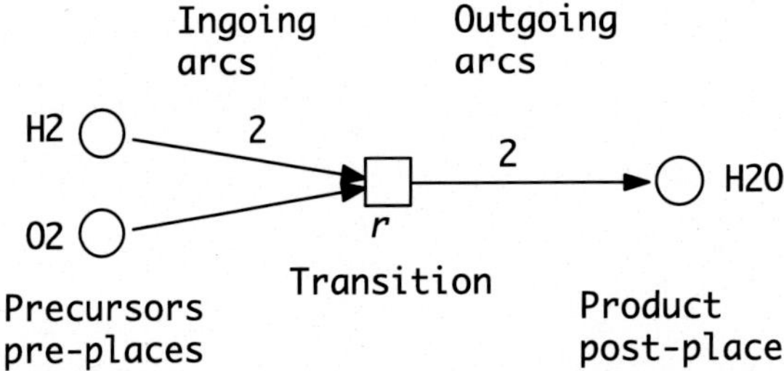

FIGURE 7.3 Petri net for the well-known chemical reaction $r: 2H_2 + O_2 \rightarrow 2H_2O$. *Hint*: Node names have to obey the same constraints as known from most programming languages for identifiers. The first character has to be a letter or underscore, followed by any combination of letters, numbers, and underscores; no other special characters are allowed, neither subscripts nor superscripts.

TABLE 7.5 Some Interpretations of Petri Net Nodes

Places	Transitions
Object	Event
(Bio-)chemical species	(Bio-)chemical reaction
Gene	Regulation, transcription
mRNA	Regulation, translation, splicing, degradation
Protein	(Un-)binding, covalent modification, conformational change
Cell	Differentiation, growth, migration
Tissue	Muscular contraction, absorption of water and nutrients, elimination of waste products
Organ	Filtration, respiration, digestion
Organism	Reproduction, metamorphosis, death
Population	Interaction, communication, locomotion

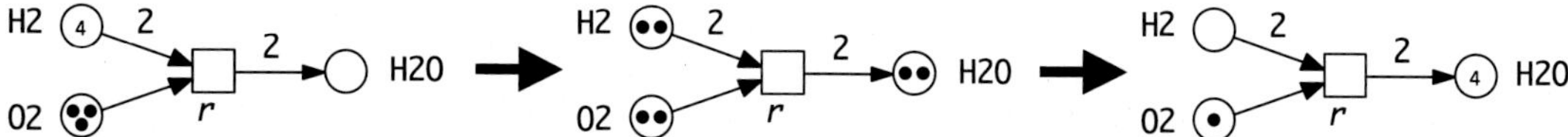

FIGURE 7.4 Petri net for $r : 2H_2 + O_2 \rightarrow 2H_2O$ and three possible states, each connected by a firing of the transition r. Initially, there are four tokens on place H_2 and three tokens on place O_2, while place H_2O is clean (zero tokens). The transition r is not enabled anymore in the marking reached after these two single firing steps.

The bipartite property precludes arcs between nodes of the same type.

- *Multigraph*: Two given nodes may be connected by multiple arcs, typically abbreviated to one weighted arc. The weight is shown as a natural number next to the arc. The default value is 1, and usually not explicitly given. Arc weights permit us to conveniently specify the stoichiometry of (bio-)chemical reactions.

In addition to the network structure and in contrast with standard graph formalisms, Petri nets enjoy an execution semantics. For this purpose, they also embody movable objects; see Figure 7.4.

- *Tokens*: The (discrete) quantitative amounts of the involved biomolecules are represented by tokens residing on places. The token numbers are given by black dots or natural numbers. The number zero is the default value, and usually not explicitly given. Tokens may be understood as the number of molecules, the concentration level of a species, the activity level of a gene, or simply a gene in its active state or a protein in a certain conformation.
- *Marking (state)*: A specific distribution of tokens over all places is called marking; it gives the current state of the model. We will use "marking" and "state" as pure synonyms.

Having the network structure and the initial state, let's bring the Petri net to life by moving the tokens through the net. The given arc weights define how many of these tokens on a certain place are consumed or produced by a transition.

- *Enabledness*: An action that is encoded by a transition can only take place if the corresponding pre-places host sufficient amounts of tokens according to the weights of the transition's ingoing arcs. If this condition is fulfilled, then the transition is *enabled* and *may fire* (occur), that is, it can execute the encoded action. An enabled transition is never forced to fire.
- *Firing*: Upon firing, a transition consumes tokens from its precursors (pre-places) according to the arc weight of the ingoing arcs, and produces new tokens on its products (post-places) according to the arc weights of the outgoing arcs. The firing happens atomically (i.e., there are no states in between) and does note consume any time.

  Firing generally changes the current distribution of tokens, and the system reaches a new state.
- *Behavior*: We obtain the dynamic behavior of a Petri net by repeating these steps of looking for enabled transitions and randomly choosing one single transition among the enabled ones to let it fire.

Typical Scenarios

Having these principles in mind, it is straightforward to design Petri net models for a biochemical system in a bottom-up manner. Each reaction yields a single transition, and all species involved are represented as places. Ingoing arcs of a transition connect precursors (substrates) and outgoing arcs connect products; see Figure 7.3. Let's consider some special cases.

- *Side condition*: A species that is required for a reaction to take place, but is not consumed upon firing, establishes a side condition. A place representing a species that is only a side condition for a reaction, such as enzymes, kinases, and transporters, is connected with the corresponding transition in both directions, that is, via two opposed arcs. That way, the tokens on the side condition are not used up and remain available for further use. This situation is often graphically represented by a special arrow with a black dot as head, called *read arc* (also *test arc*); see Figure 7.5b, g, and i-k. This abstract representation of catalytic activity neglects the formation of enzyme-substrate complexes, and thus effects like substrate saturation; see Figure 7.7 for a refined representation of enzymatic reactions.

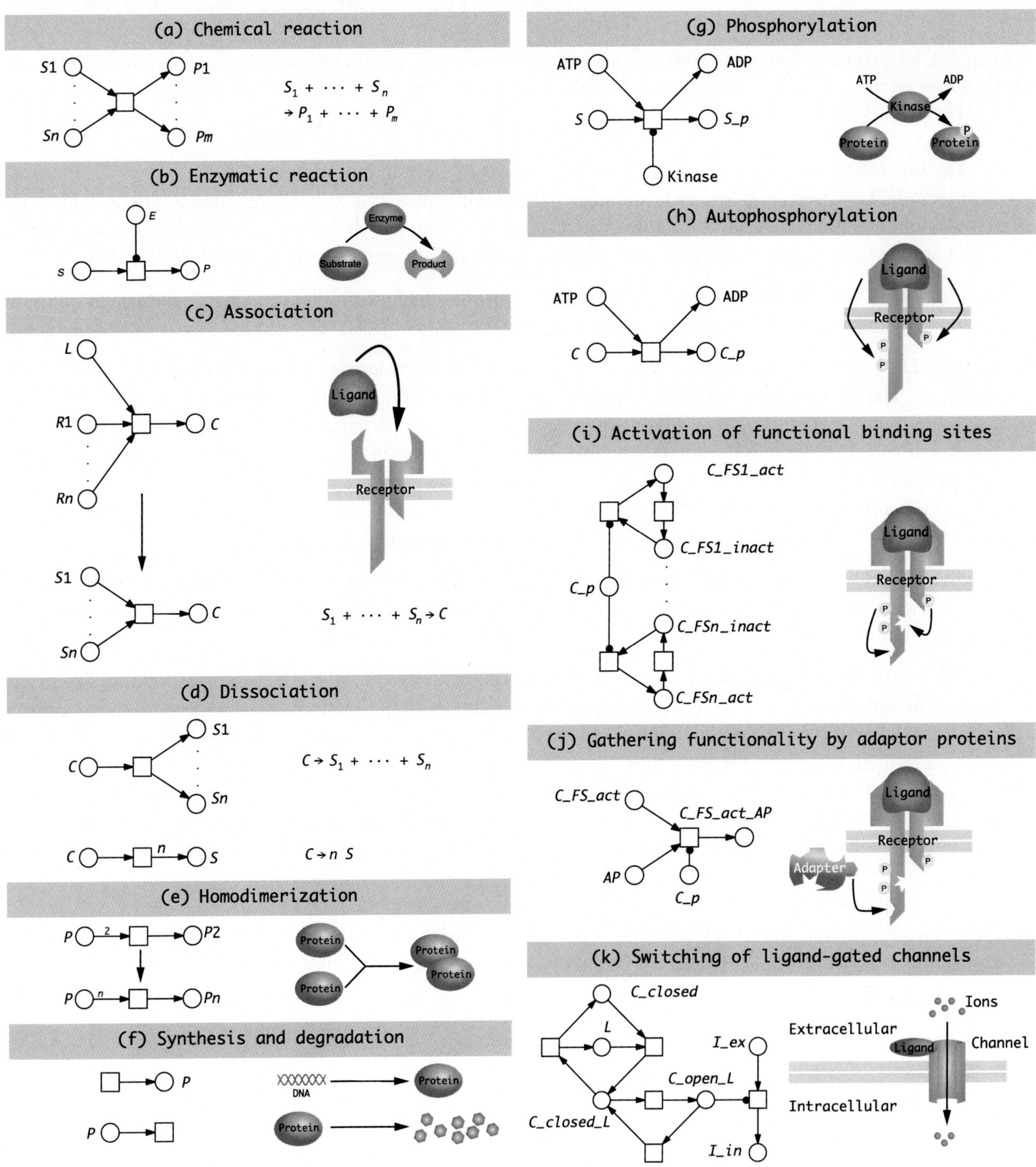

FIGURE 7.5 Petri net components of typical biochemical mechanisms; for abbreviations see Table 7.6. Please note the abstraction by the use of side conditions.

TABLE 7.6 Abbreviations Used in Figures 7.5 and 7.6

Abbreviation	Meaning	Abbreviation	Meaning
C	Complex	in	Intracellular
CA	Carrier	L	Ligand
Ch	Channel	p	Phosphorylated
E	Enzyme	P	Product
ex	Extracellular	R	Receptor
FS	Functional binding site	S	Substrate

- *Homodimerization*: When two identical molecules form a complex, then the reactants are modeled by only one place and an arc weight is used to indicate the stoichiometry of the reaction, meaning the number of reactants needed to form the complex; see Figure 7.5e.
- *Synthesis and degradation*: The network under consideration is often embedded into an implicitly assumed environment, from which it receives tokens by a transition without precursors (input transition), or to which it delivers tokens by a transition without products (output transition). Input transitions can fire as often as they want and describe synthesis events, while output transitions reflect degradation events; see Figure 7.5f.
- *Activation of functional binding sites*: The activation of functional binding sites is an example of a sidecondition, described above. The receptor complex must be phosphorylated to switch thes state of its binding sites from inactive to active; in turn the bindings sites can be (spontaneously) inactivated; see Figure 7.5i.
- *Switching of ligand-gated channels*: A ligand binds to the closed channel, and the channel can switch into an open state after the complex formation. The open state of the channel is a side condition for the flow of ions through the channel. Both the binding of the ligand to the channel and the conformal change of the channel are reversible processes; see Figure 7.5k.
- *Reversible reaction*: The forward and the backward reaction are modeled separately by two opposite transitions, that is, the precursors of one direction are the products of the opposite direction, and vice versa; see Figure 7.5k.
- *Carrier-mediated transport*: A carrier protein mediates the transport of large molecules through the membrane. In the example shown, the extracellular substrates bind to the empty carrier when it is open to the extracellular side. After binding of the extracellular substrate, the carrier first switches to a closed state and then opens to the intracellular side, which allows the release of the previously extracellular substrate into the intracellular space. In antiporter types of transporters, as shown in Figure 7.6a3, the carrier can transport an intracellular substrate to the extracellular side during its catalytic cycle, thus kinetically coupling two transport processes.

Figures 7.5 and 7.6 provide more modeling examples of some typical mechanisms that can be found in biochemical systems. Exercise 7.1 may help you draw on your own your very first Petri nets with Snoopy.

Enhanced Modeling Comfort

Biochemical networks tend to become huge. Thus, in addition to a basic toolkit for drawing Petri nets, Snoopy supports two distinguished features for the design and systematic construction of larger Petri nets; see Figure 7.7 and Exercise 7.2.

- *Macro nodes*: There are two types of macro nodes. Macro transitions (drawn as two centric squares) help hide transition-bordered subnets (i.e., subnets having only transitions as interface to the supernet). Likewise,

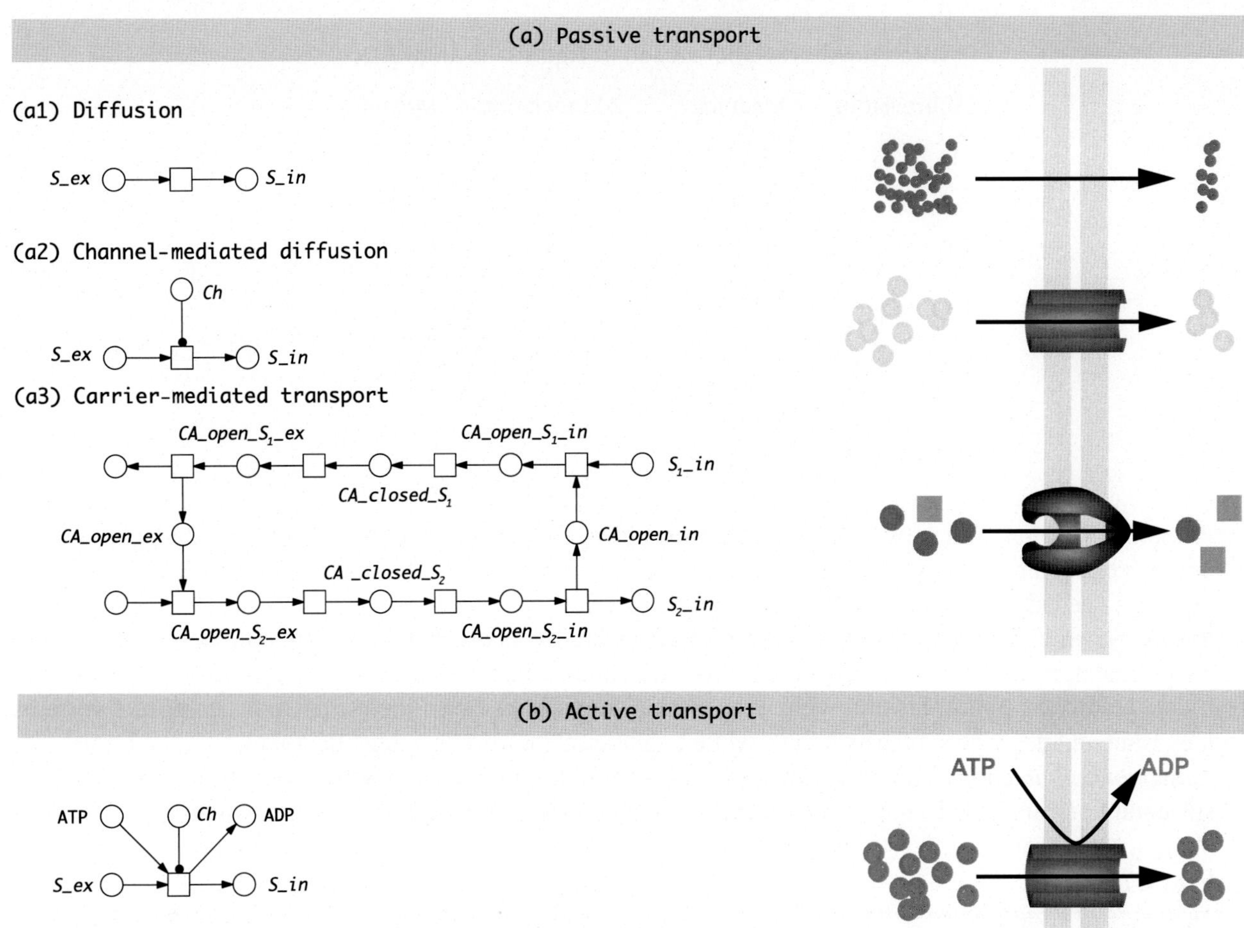

FIGURE 7.6 Petri net components of typical examples of transport mechanisms over biological membranes; for abbreviations see Table 7.6.

macro places (drawn as two centric circles) permit us to hide place-bordered subnets (i.e., subnets having only places as interface to the supernet). Both types of macro nodes allow the hierarchical structuring of Petri nets. The left column in Figure 7.7 demonstrates the nested use of macro transitions.

- *Logical nodes*: Places and transitions can be specified as logical nodes (also called fusion nodes). They are automatically colored gray. Logical nodes with the same name are logically identical, that is, graphical copies of a given node. They are often used for species involved in several reactions or reactions involving species spread over the network. See Figure 7.7 for the use of logical places (middle column) and logical transitions (right column).

Macro nodes and logical nodes help us deal with larger networks. They may contribute to a net's readability and, thus, are crucial for nontrivial networks. They do not change the actual network structure, just its representation. Likewise, they do not extend the expressiveness of the modeling language.

Running Example

Now we are ready to turn our running example into a Petri net. This could be done by modeling every reaction in Table 7.3 separately, and declaring all places to be logical ones. This list of Petri net components would directly correspond to the list of reactions.

However, to take advantage of a graphical representation and in order to highlight the structure of the network, it would be better to merge all places with identical names. This is what we do, except for the activator protein

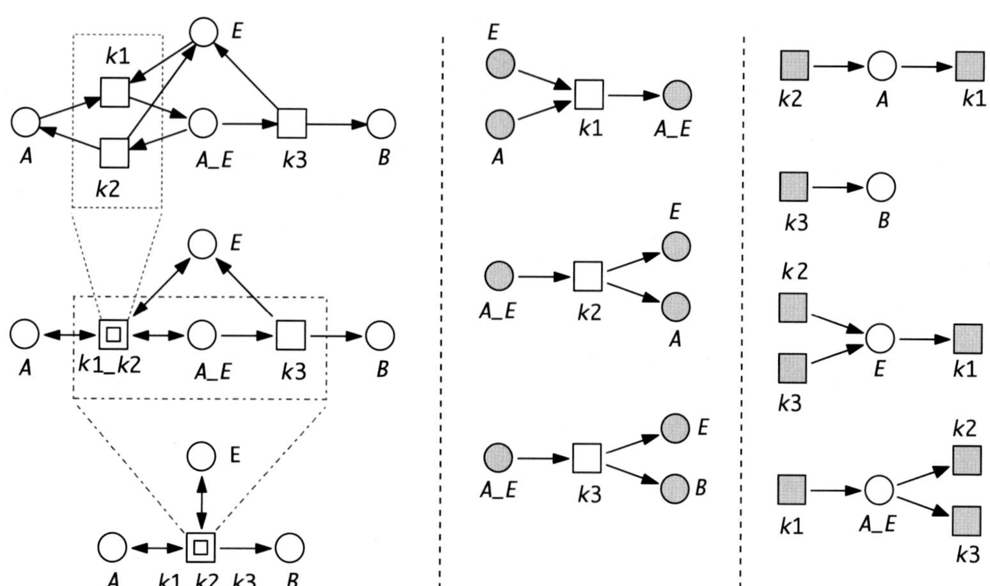

FIGURE 7.7 Different, but equivalent graphical representations of one and the same $\mathcal{PN}$ modeling the enzymatic reaction $A + E \leftrightarrow A\_E \rightarrow B + E$, with A_E being the enzyme-substrate complex. (Left) Macro transitions yield hierarchically structured models; (middle) reaction-centric representation by the use of logical places; (right) species-centric representation by the use of logical transitions.

A, which we choose to represent by three logical places to get the Petri net structure nice and tidy. The result, see Figure 7.8, closely resembles the general scheme in Figure 7.2. The Petri net is made of 9 places (variables), 16 transitions (reactions), 27 (standard) arcs, and 6 read arcs (which corresponds to 39 standard arcs in total). Initially, there is a token each on place *geneA* and *geneR*, and all other places are clean (compare initial values in Table 7.2).

Exploring Model Behavior

Before continuing reading, we suggest playing with the Petri nets introduced so far; see Exercise 7.3. Each execution exemplifies some possible net behavior. This behavior may include the following scenarios (compare running example).

- *Sequential reactions*: reflect causality (e.g., reaction *r*7 cannot occur before *r*5 occurs, which in turn first requires the occurrence of either *r*3 or *r*4).
- *Alternative reactions*: compete for tokens on shared pre-places and therefore branch into alternative behavior; in Petri net terminology, the transitions are said to be in *conflict* (e.g., reactions *r*7 and *r*8 share the pre-place *A*; a token on *A* can either be consumed by *r*7 or by *r*8).
- *Concurrent reactions*: are neither in causal nor conflict relation; thus, they are independent and can fire in any order or even concurrently (e.g., reactions *r*3 and *r*11).

State Space

By playing the token game, we can produce and observe any reachable state. All states that can be reached from a given state by any firing sequence of arbitrary length constitute the set of reachable states. The set of states reachable from the initial state is said to be the *state space* of a given Petri net. The state space can be finite or infinite.

Each execution run of a Petri net corresponds to a walk through its state space; see Exercise 7.4. The basic principle is given as pseudocode in Algorithm 1; Exercise 7.5 deals with an improvement.

Executing a Petri net generally involves decisions between alternative behavior. Encountered alternatives (conflicts) are taken nondeterministically (automatic mode) or user-guided (manual mode). To get an exhaustive

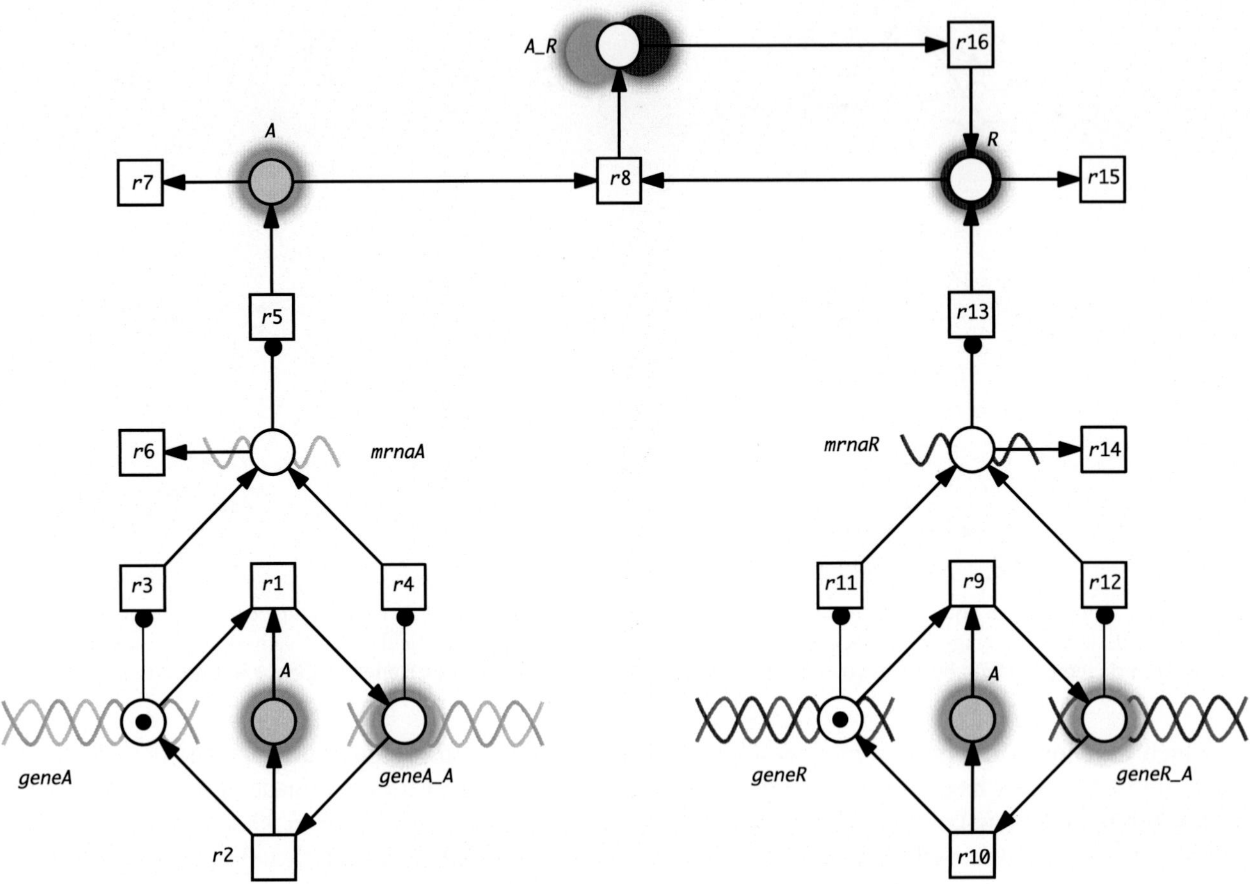

FIGURE 7.8 Petri net for the circadian clock model, according to [14], enhanced by some illustrative icons; compare general scheme in Figure 7.2. See Table 7.2 for a description of the places (variables), and Table 7.3 for a list of the transitions (reactions).

ALGORITHM 1 Basic $\mathcal{PN}$ simulation (animation) algorithm

Require: $\mathcal{PN}$ with initial state s_0;
number of steps N;

integer count ← 0
state s ← s_0 ▹ make initial state to current state
write(*count*, *s*) ▹ add s_0 to trace

while *count* < *N* **do**
  count ← *count* + 1
  pick one transition t enabled at s ▹ single-step firing
  s ← *fire*(*s*, *t*) ▹ compute new state *s* by firing of *t*
  write(*count*, *s*) ▹ add *s* to trace
end while

---

picture, we have to consider all possible execution sequences. Obviously, that is impossible for systems with infinite behavior, which can be caused by cyclic behavior and/or infinite state space. Even if the state space is finite, the number of all cycle-free execution sequences may be far too huge to be exhaustively explored by animation. In the next section, we will learn more about the options we have to systematically analyze the behavior a given Petri net may exhibit.

7.3.2 Analysis

$\mathcal{PN}$ are time-free models, however, the qualitative analysis considers all possible behavior under any timing. Thus, a Petri net itself implicitly contains all possible time-dependent behavior. However, to learn about analysis techniques, we need to know first how to express the properties of interest in terms of the Petri net terminology.

Behavioral Properties

It is common sense to distinguish between *general* behavioral properties, which can be applied to any network without considering its special functionality, and *special* behavioral properties, which do reflect the intended functionality. We introduce here the basic ideas and start with the first category.

There are three orthogonal *general behavioral properties* that are usually explored first to gain some insights into the behavior of a Petri net; see Exercise 7.6.

- *Boundedness*: A place is said to be k-bounded (bounded for short) if the maximal number of tokens on this place is bounded by a constant k in all reachable states. A Petri net is k-bounded (bounded for short) if all its places are k-bounded.
  Boundedness precludes overflow by unlimited increase of tokens and ensures a finite state space.
- *Liveness*: A transition is said to be live if, whatever happens, it is always possible to reach a state where this transition is enabled. A Petri net is live, if all transitions are live.
  In a live net, all transitions are able to contribute to the net behavior forever.
- *Reversibility*: A Petri net is said to be reversible if the initial state can be reached again from each reachable state. This obviously involves the reversibility of every reachable state.
  A reversible net has the capability of self-reinitialization.

Further behavioral properties of general interest are:

- *Dead state*: States where no transition is enabled are called dead states. These are terminal states of the system behavior; no further changes are possible and all activities are switched off. A live Petri net does not have dead states, but freedom of dead states does not ensure liveness.
- *Dynamic conflict*: It may happen that two transitions are enabled, but the firing of one transition disables the other one. Such a situation is called a dynamic conflict. The occurrence of dynamic conflicts indicates alternative (branching) system behavior; decisions between alternatives are taken nondeterministically.

Special behavioral properties can often be expressed in terms of *reachability/nonreachability of states*. More subtle properties require advanced query languages, for which we will deploy temporal logic.

Running Example

By playing the token game for our running example (see Exercise 7.4), we can easily figure out the 1-boundedness of the places *geneA*, *geneA_A*, *geneR*, and *geneR_A*. But the entire net is unbounded, because all other places are unbounded. If the transcription of mRNA or the generation of proteins occur faster than their respective degradations, infinite many tokens (molecules) will be accumulated.

Furthermore, we can argue that this Petri net is apparently live and reversible, which precludes dead states. It is easy to find a dynamic conflict, as the place A is shared by four transitions ($r1$, $r7$, $r8$, $r9$). Consequently, if there is just one token on A, then all four transitions are enabled, but only one of them can fire, which involves the consumption of the token, which in turn disables the other three transitions. The decision, which transition to fire is taken nondeterministically.

Analysis Techniques

When models become more complicated, it might not be obvious anymore to decide behavioral properties by reasoning only. Then we need mathematically sound analysis techniques. Petri net theory offers a rich body of such analysis techniques; most of them are implemented in our analysis tool Charlie. Some of them have their counterparts in other frameworks, but some are unique features of Petri net theory.

Basically, two different ways to explore the behavior can be distinguished: (1) static analyses consider the whole state space without constructing it, and (2) dynamic analyses do construct the full or partial state space for deciding the properties of interest. We sketch here only a few of them to give a flavor of the kind of analysis techniques we have.

Static Analyses

Static analysis techniques aim at structural properties and permit us, if they hold, to deduce behavioral properties of Petri nets from their structure without constructing any state space. Most structural properties depend on the graph structure only. If such a property holds, it holds in any initial state. Some properties also take into account the initial state. The most important structural properties can be classified into elementary graph properties, siphons/traps, and place/transition invariants.

Elementary Graph Properties

Elementary graph properties simply explore the net structure and can be easily checked. As their computational load can be neglected, they are automatically decided when calling Charlie. They usually reflect the modeling approach, so they may be used for preliminary consistency checks to preclude production faults when constructing a Petri net. Occasionally they permit on their own conclusions on behavioral properties. The following list obeys the order as applied in Charlie's result vector.

- *PUR*: A Petri net is *pure* if there are no two nodes connected in both directions. This precludes read arcs. In this case, the net structure is fully represented by the incidence matrix, which is used for the computation of invariants.
- *ORD*: A Petri net is *ordinary* if all arc weights are equal to 1. This includes homogeneity. A nonordinary Petri net cannot be live and 1-bounded at the same time.
- *HOM*: A Petri net is *homogeneous* if all outgoing arcs of a given place have the same multiplicity.
- *CSV*: A Petri net is *conservative* if all transitions add in total exactly as many tokens to their post-places as they subtract from their pre-places, or briefly, all transitions fire token-preservingly. A conservative Petri net is structurally bounded, that is, bounded for any initial state.
- *SCF*: A Petri net is *static conflict free* if there are no two transitions sharing a pre-place. Transitions involved in a static conflict may compete for the tokens on shared pre-places. Thus, static conflicts indicate situations where dynamic conflicts, that is, nondeterministic choices, may occur in the system behavior. However, it depends on the token situation whether a conflict does actually occur dynamically. There is no nondeterminism in SCF nets.
- *CON*: A Petri net is *connected* if it holds for every two nodes a and b: there is an undirected path between a and b. Disconnected parts of a Petri net cannot influence each other, so it is advisable to analyze them separately.
- *SC*: A Petri net is strongly connected if it holds for every two nodes a and b: there is a directed path from a to b. Strong connectedness involves connectedness and the absence of boundary nodes. It is a necessary condition for a Petri net to be bounded & live.
- *FT0, TF0, FP0, PF0*: These properties ask for *boundary nodes*, which exist in four types.
  - *Input transition*: a transition without pre-places (FT0),
  - *Output transition*: a transition without post-places (TF0),
  - *Input place*: a place without pre-transitions (FP0),
  - *Output place*: a place without post-transitions (PF0).

  Boundary nodes typically model interconnections of an open system with its environment. A net without boundary nodes is self-contained, that is, a closed system. It needs a nonclean initial marking to become live.

  A Petri net with input transitions is unbounded, because the firing of input transitions does not depend on any precondition. A Petri net with input places is not live because the tokens on an input place are sooner or later used up. Actually, a net with boundary nodes cannot be strongly connected, and thus not bounded & live.

- *Net structure class*: We distinguish the following classes of net structures: state machine (SM), synchronization graph (SG), free choice (FC), extended free choice (EFC), and extended simple (ES).

Siphons/Traps

These notions are unique features of Petri net theory.

- *Siphon*: A nonempty set S of places of a Petri net is called a siphon if every transition that fires tokens onto a place in S also has a pre-place in this set; that is, the set of pre-transitions of S is contained in the set of post-transitions of S.

  Pre-transitions of a siphon cannot fire if the place set is clean, that is, none of the places carries a token. Therefore, a siphon cannot get tokens again as soon as it is clean, and then all its post-transitions are dead.
- *Trap*: On the contrary, a nonempty set Q of places of a Petri net is called a trap if every transition that subtracts tokens from a place of the trap set also has a post-place in this set; that is, the set of post-transitions of Q is contained in the set of pre-transitions of Q.

  Post-transitions of a trap always return tokens to the place set. Therefore, once a trap contains tokens, it cannot become clean again.
- *Siphon Trap Property*: Siphon and trap are closely related but contrasting notions. When they come on their own, we usually get deficient behavior. However, both notions have the power to perfectly complement each other. A Petri net satisfies the siphon trap property (STP) if every siphon includes an initially marked trap (this property obviously depends on the initial state).

  For certain combinations of structural properties, we can derive behavioral properties.
  - If a $\mathcal{PN}$ is ORD and does not have any siphons, then it is live.
  - If a $\mathcal{PN}$ is ORD, then STP precludes dead states.
  - If a $\mathcal{PN}$ is ORD, ES, and the STP holds, then the net is live.
  - If a $\mathcal{PN}$ is ORD and EFC, then it is live iff the STP holds.
- *Bad siphons*: A siphon is called bad if it does not include a trap. Then, there is no initial state to satisfy the STP. Without appropriate timing constraints, a bad siphon may always become clean, thus precluding the liveness of its post-transitions.

Place/Transition Invariants

Place and transition invariants (P- and T-invariants for short) play a crucial role in analyzing biological systems, thanks to their biological interpretations. Both of them are computed by solving a linear equation system that describes the Petri net structure and that is independent of the initial state; see also Exercise 7.7.

- *P-invariant*: A P-invariant specifies a set of places over which the weighted token count keeps constant whatever happens in the Petri net. So, a place belonging to a P-invariant is k-bounded. We get the upper bound k by multiplying the invariant with the initial state.

  A net where each place belongs to a P-invariant is called *covered with P-invariants (CPI)*, which is a sufficient criterion for boundedness.

  In metabolic networks, P-invariants often correspond to conservation laws in chemistry, reflecting substrate conservations, while in signal-transduction networks, P-invariants often correspond to proteins and their possible states.
- *T-invariant*: A T-invariant specifies a multiset of transitions; it can be interpreted in two different ways. The multiset either says how often a transition has to fire to obtain a cycle in the system behavior, or the multiset gives relative firing rates, keeping the Petri net in a steady state.

  A net where each transition belongs to a T-invariant is called *covered with T-invariants (CTI)*. A bounded & live Petri net is CTI.

  T-invariants are often identified with elementary functional modules; the composition of their behavior yields the total system behavior.

Technically, invariants are vectors. Any linear combination of P-invariants (T-invariants) yields again a P-invariant (T-invariant). Therefore, one is usually interested in minimal invariants; that is, invariants that cannot be described by a linear combination.

Invariants induce subnets, comprising the nodes belonging to the invariant and all pre- and post-nodes. These mass-preserving subnets (P-invariants) or state-reproducing subnets (T-invariants) may be used for modularization, validation, or reduction of the model; see Exercise 7.8.

Running Example

Let's see what we can learn about the net behavior by just looking at the net structure and the initial state (compare Charlie's analysis protocol in the Supplementary Material).

- The net is connected, but not strongly connected, as there are four output transitions: $r6$, $r7$, $r14$, $r15$. Consequently, the net cannot be bounded & live.
- The net is not pure (because there are read arcs), but ordinary (because all arc weights are 1), and consequently it is also homogeneous.
- The net is not conservative; that is, there are transitions changing the number of tokens. All transitions triggered by read arcs increase the token number by their firing; all output transitions ($r6$, $r7$, $r14$, $r15$) and the three reactions forming compounds ($r1, r8, r9$) decrease the token number. Thus, the number of tokens in the net varies and the current total number depends on the behavior.
- There is one structural conflict set, comprising $r1, r7, r8, r9, r15$. The transitions $r1, r7, r8, r9$ share the pre-place A, thus they compete for the tokens on A. $r15$ gets involved, as it shares a pre-place with $r8$; thus, firing of $r15$ may have an influence on the enabledness of $r8$. This structure goes beyond the ES property.
- There are siphons, and the STP holds, which precludes bad siphons and dead states.
- There are two P-invariants (given in a simplified notation),
  $pi1 = \{geneA, geneA\_A\}, pi2 = \{geneR, geneR\_A\}$,
  covering four places, but not the entire net; so the net is not CPI. For each P-invariant, the constant token sum is 1, which confirms our expectations: Whatever happens in the net, each gene either exists in its free form or is bound to an activator protein; the gene will neither disappear nor be multiplied.

  All other places are unbounded for the given initial state, as we can easily deduce by playing with the net.
- There are nine T-invariants (given in a simplified notation),
  $ti1 = \{r1, r2\}, ti2 = \{r3, r6\}, ti3 = \{r4, r6\}, ti4 = \{r5, r7\}$,
  $ti5 = \{r5, r8, r16\}, ti6 = \{r9, r10\}, ti7 = \{r11, r14\}, ti8 = \{r12, r14\}$,
  $ti9 = \{r13, r15\}$,
  covering the net (CTI). Most of the T-invariants comprise just two transitions with complementary effects on the state, for example, $r5$ produces one token on A, and $r7$ consumes one token on A. The only exception is $ti5$, which corresponds to the following cycle: the production of the activator protein A ($r5$) is counterbalanced by A and R forming a complex ($r8$), and dissociating the complex in turn degrades A while releasing the temporarily used repressor protein R ($r16$).

  A balanced firing of the (multi-)sets of transitions defined by the T-invariants reproduces a given state, and thus makes the Petri net bounded under specific timing constraints.

Sets or multisets of nodes, such as P/T-invariants, siphons, or traps that have been computed with Charlie and written to a result file, can be read by Snoopy to visualize the subnetworks induced by them; see Exercise 7.8.

Dynamic Analyses

If the static analyses do not suffice, we have to compute a data structure describing the whole system behavior. The easiest way to do this is by constructing a dedicated graph—the reachability graph; see Algorithm 2. The nodes of a reachability graph represent all reachable states; the edges in between are labeled by single transitions, the firing of which causes the state change. The given algorithm makes a breadth-first construction, which is specifically appropriate if one looks for shortest paths. Alternatively, a depth-first construction keeps extending a path until an already known state is encountered, which then triggers backtracking. Generally, it does not matter in which order unprocessed states are picked.

ALGORITHM 2 Reachability graph construction

Require: $\mathcal{PN}$ with initial state s_0;

```
edgeSet E ← ∅                                   ▷ E—reachability graph's edges
stateSet S ← {s0}                               ▷ S—reachability graph's nodes
stateSet U ← {s0}                               ▷ U—set of unprocessed states
while U ≠ ∅ do
  pick one state s out of U
  U ← U \ {s}
  for all transitions t enabled at s do
    s' ← fire(s, t)                             ▷ s'—successor state by firing of t
    if s' ∉ S then
      S ← S ∪ {s'}                              ▷ new node
      U ← U ∪ {s'}
    end if ;
    E ← E ∪ {(s, t, s')}                        ▷ new edge
  end for
end while
```

The reachability graph gives a concise representation of all possible single-step firing sequences. Consequently, concurrent system behavior is described by enumerating all permutations of concurrent transitions, called interleaving firing sequences for short. Thus, the reachability graph represents the *interleaving semantics*. The much more challenging partial order (true concurrency) semantics (PO) goes beyond the scope of this introductory chapter.

Obviously, the reachability graph is only finite for bounded Petri nets. Even for bounded models, the reachability graph may exceed the available memory. If we succeed in constructing the complete reachability graph, we are able to decide behavioral Petri net properties.

- *Boundedness*: A Petri net is k-bounded iff there is no node in the reachability graph with a token number larger than k for any place.
- *Reversibility*: A Petri net is reversible iff the reachability graph is strongly connected.
- *Dead states*: A Petri net has dead states iff the reachability graph contains terminal nodes, that is, nodes without outgoing edges.
- *Liveness*: In order to decide liveness, we partition the reachability graph into strongly connected components (SCC), that is, maximal sets of strongly connected nodes. A SCC is called terminal if no other SCC is reachable in the partitioned graph. A transition is live iff it is included in all terminal SCCs of the partitioned reachability graph. A Petri net is live iff this holds for all transitions.
- *Dynamic conflicts*: The occurrence of dynamic conflicts is checked at best during the construction of the reachability graph, because branching nodes do not necessarily mean alternative system behavior.
- *Model checking*: The behavioral properties of interest have to be expressed in temporal logic. A logic that perfectly fits the purpose of exploring a reachability graph is CTL. CTL extends standard propositional logics by path quantifiers (**A**—for all paths, **E**—there is one path) and temporal operators (including **G**—globally, **F**—finally). CTL provides a flexible query language and is specifically appropriate for special behavioral properties.

Below we give some examples to illustrate the power of this analysis approach; a more comprehensive discussion goes beyond the scope of this introductory chapter.

Running Example

The net is structurally unbounded, yielding an infinite state space, which in turn precludes the exhaustive construction of the reachability graph. However, the static analysis reveals CTI, which suggests that the net may become bounded under certain timing constraints. To mimic this fact in our qualitative model, we introduce a bounded version for our running example; see Figure 7.9.

We deploy *inhibitory arcs* with arc weights to limit token generation. An inhibitory arc goes always from a place to a transition and puts an additional constraint on the enabledness. The transition can only be enabled if the token number on the tested place is smaller than the arc weight. Please note that inhibitory arcs bring the Turing power and reduce the analysis techniques to the dynamic ones. Their use for our running example does not reflect biomolecular mechanisms.

By the use of inhibitory arcs, the state space becomes finite but explosively grows, as illustrated in Table 7.7. In any case, we are able to decide the three orthogonal behavioral properties: the net is C-bounded, with $C = \max(C1, C2)$; reversible; and live. There are dynamic conflicts, as expected, and we can check behavioral properties expressed as CTL properties. As an illustration, we give three examples of *special behavioral properties*. CTL can also be used to query general behavioral properties; see Exercise 7.9.

- Forever it holds that gene A is either free or bound to the activator protein.
  AG$[(geneA = 1\,\&\,geneA\_A = 0)|(geneA = 0\,\&\,geneA\_A = 1)]$
- It is forever possible that there are at least k molecules of protein A; that is, there will be new proteins A forever, which involves liveness of transitions $r7$. This formula yields true for any $k \leq C2$.
  AG EF$[A \geq k]$
- It is possible that there are at the same time at least k molecules of each protein species (activator and repressor). This formula yields true for any $k \leq C2$.
  EF$[A \geq k\ \&\ R \geq k]$

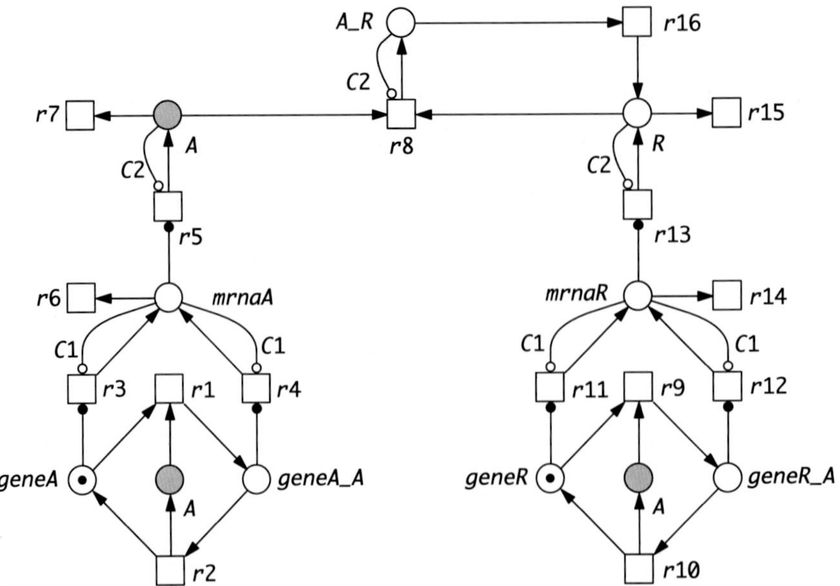

FIGURE 7.9 The running example in a bounded version—maximal $C1$ tokens can be accumulated on the mRNA places, and maximal $C2$ tokens by mRNA translation on the protein places. Inhibitory arcs (hollow circle as arc head) with arc weight $C1$ ($C2$) limit the generation of mRNA (proteins). $C1$ and $C2$ are constants to conveniently parameterize the model; compare Table 7.7. These inhibitory arcs do not reflect biomolecular mechanisms.

TABLE 7.7 State-Space Growth for Increasing *C1*, *C2*; Compare Figure 7.9, Computed with Marcie's Symbolic Engine Encoding the State Space by Interval Decision Diagrams

C1	C2 = 1	C2 = 10	C2 = 100	C2 = 1000
1	240	33,792 (10^4)	24,889,632 (10^7)	2,408,809,602,021 (10^{10})
10	7260 (10^3)	1,022,208 (10^6)	752,911,368 (10^8)	728,664,904,707,172 (10^{11})
100	612,060 (10^5)	86,178,048 (10^7)	6,347,478,406,061 (10^{10})	61,430,666,905,608 (10^{13})
1000	60,120,060 (10^7)	8,464,904,448 (10^9)	62,348,590,360, 616,263 (10^{12})	60,340,740,760,616, 263,646,566 (10^{15})

7.3.3 Further Reading

- *A bit of history*: $\mathcal{PN}$ have been initiated by concepts proposed by Carl Adam Petri in his PhD thesis in 1962 [3]. The first substantial results making up the still growing body of Petri net theory appeared around 1970. Initial textbooks devoted to Petri nets were issued in the beginning of the eighties. See [15] for an early survey paper, and [16] for a recent insider view on Petri net history.
- *$\mathcal{PN}$ in Systems Biology*: The idea to represent biochemical networks by Petri nets is rather intuitive, and was mentioned by Carl Adam Petri himself in the seventies in one of his internal research reports on interpretations of net theory; see also [17]. The deployment in Systems Biology was first published in [18–20]; for a review paper, see [21].
- *More background*: Many more details and related formal definitions can be found in [38]. See [23] for more hints about the tool handling of Snoopy and Charlie, and [24] for a detailed discussion of bad siphons.
- *P/T-invariant*: Invariants have belonged to the standard body of Petri net theory from the very beginning [25]. An approachable introduction can be found in [26].

  Notions closely related to T-invariants are *elementary modes* [27], *extreme pathways* [28], and *generic pathways* [29]. All four notions coincide for networks without reversible reactions. Likewise, P-invariants are elsewhere known as chemical moieties or conservation relations, see, for example, [30].

  Model validation by means of P/T-invariant analysis is demonstrated in [31] for three case studies: apoptosis, carbon metabolism in potato tuber, and the glycolysis and pentose phosphate metabolism. T-invariants have been used in [26] to derive coarse network structures highlighting the structural principles inherent in the functional modules, and in [32] to identify the core and fragile node of a Hypoxia response network; see also Section 7.5.3 (ODEs reduction by T-invariant analysis).
- *Model checking*: is an established and very powerful analysis technique for behavioral properties, which, however, requires some expertise in temporal logic. See [33] for an introduction to CTL, and [34] for LTL. Typical patterns for biologically relevant properties can be found in [35].
- *Beyond standard Petri nets*: $\mathcal{PN}$ have been enhanced to extended Petri nets ($\mathcal{XPN}$) by adding *special arcs* such as inhibitor arcs, equal arcs, and reset arcs. They can be further enriched to include *state-dependent arcs* (also known as self-modifying arcs); that is, arc multiplicities may be state-dependent expressions in terms of a transition's pre-places. Each of these features extends the expressiveness of the modeling language and brings the Turing power. However, the gain of Turing power comes with a loss of analysis power, which is restricted to the dynamic analysis techniques, and thus to the analysis of models with finite state space of manageable size.
- *Further examples*: The application of $\mathcal{PN}$ and their detailed stepwise qualitative analysis, including model checking, has been outlined in [36] for a case study in model-driven Synthetic Biology, and in several book chapters using different running examples: glycolysis and pentose phosphate pathway [37], Raf kinase

inhibitor protein (RKIP)-inhibited extracellular signal regulated kinase (ERK) pathway [38], mitogen-activated protein kinase (MAPK) cascade [22], and genes in a regulatory cycle (repressilator) [39, 40].

7.3.4 Exercises

Exercise 7.1. Drawing a Petri Net. To draw your first Petri net, repeat the following steps.

1. Create a new net (*File* → *New*) and choose the appropriate net class. You will get a new drawing window with the name "*unnamed*."
2. Select the graph element *Place* in the menu panel on the left-hand side, and left-click on the drawing window at those positions where you want to get a place. Each mouse click creates a new place.
3. Similarly, select the graph element *Transition* in the menu panel, and create transitions.
4. To connect two nodes, select *Edge* (which is just a synonym for *arc*) in the menu panel, click on the source node and move the pointer, while keeping the mouse button pressed, to the target node, where the mouse button is released.
5. A double click on a graph element (place, transition, edge) opens an attribute window, which allows you to edit the element-specific properties.
6. Finally, do not forget to save your work under a new name (*File* → *Save as*).
   To familiarize yourself with the basic editing features, draw the $\mathcal{PN}$ in Figure 7.3. □

Exercise 7.2. Drawing a Hierarchical Petri Net. To hierarchically structure a Petri net, repeat the following steps.

1. Draw your flat net (or portion of it) as you wish to have it.
2. Select the subgraph that you want to be abstracted by a macro node, and select *Hierarchy* → *Coarse*. Choose the appropriate coarse element (macro place, macro transition) and hit the OK button.
3. A double click on the macro node opens its attribute window and allows you to assign a suitable name; a click on the entry with this name in the hierarchy panel on the left-hand side opens the sub-graph in a separate window. The blue net parts have been automatically generated and represent the connection of the subgraph (macro node) with the neighboring nodes on the next higher hierarchy level.
4. Finally, do not forget to save your work under a new name (*File* → *Save as*).
   To familiarize yourself with this feature, construct the hierarchical Petri net in Figure 7.7, left column. □

Exercise 7.3. Petri Net Simulation. Snoopy visualizes the token flow, so you can conveniently explore the net behavior. To animate a Petri net, follow these steps.

1. Open the net, and open the animate window (*View* → *Animation Mode*).
2. A left click on a place increases the token number by 1; a right click decreases the token number by 1.
3. *Manual mode*: A left click on a transition triggers the firing of this transition, if it is enabled.
4. *Automatic mode*: Clicking directly on the control panel in the animation window leaves the decision of which transition to fire to the simulation/animation algorithm.

The set of all enabled transitions is determined in each step of the automatic mode. There are three strategies to choose the transition(s), among all enabled transitions, to be fired in the next execution step.

(a) *Single step*: one single transition is randomly chosen.
(b) *Intermediate step*: an arbitrary subset of concurrent transitions is randomly chosen.
(c) *Maximal step*: a maximal set of concurrent transitions is randomly chosen.

Explore these three strategies for the running example and find for each strategy a firing sequence where each transition fires at least once. To make it a challenge, try to find short sequences (in terms of number of steps).

Hint. Steps just done can be played backward up to a depth of 10. This default value can be changed in the *Global Preferences* dialog. □

Exercise 7.4. Walking Through the State Space. Explore the state space of the running example by playing the token game. Try to answer the following questions.

(a) Is it possible to reach a state with exactly one token on every place? Prove your answer by a witness firing sequence, or explain the nonreachability.
(b) Is the state space finite or infinite? Try to figure out the maximal token numbers you can get on each place. Try to characterize the set of reachable states by a regular expression.
(c) Having played with the net for a while, is it always possible to reach again the given initial state? Explain your answer.
(d) A state, in which no transition is enabled is called a *dead state*. Are there reachable dead states for the given initial state? Explain your answer. □

Exercise 7.5. Improved Petri Net Simulation. Algorithm 1 gives the very basic procedure for simulating a Petri net. This naive algorithm assumes that there is at least one enabled transition in each step, otherwise the algorithm would encounter a dead state—a state where none of the transitions is enabled. Improve this pseudocode with the following features.
(a) The algorithm terminates upon encountering a dead state.
(b) The algorithm allows you to choose between single/intermediate/maximal step firing; see Exercise 7.3. □

Exercise 7.6. Orthogonality of Behavioral Properties. The basic Petri net properties of boundedness, liveness, and reversibility are orthogonal, that is, independent from each other.
(a) Prove this statement by providing a net as small as possible for each possible combination (obviously, there are eight of them). Argue for every net why the properties hold as you claim.

If you find this too easy, repeat the exercise with the additional constraints that the nets should be
(b) ordinary
(c) ordinary and pure
(d) ordinary, pure, and strongly connected. □

Exercise 7.7. Incidence Matrix. The computation of P/T-invariants builds on a matrix, which is called incidence matrix in Petri net theory, and stoichiometric matrix in chemical network theory. The incidence matrix encodes the structure of a Petri net by having as many rows as there are places and as many columns as there are transitions. The integer matrix entry (p, t) gives the token change on place p by the firing of transition t.
(a) Show that a Petri net with read arcs cannot be uniquely described by the incidence matrix. Which other situations are not fully reflected by the incidence matrix? Give examples.
(b) A well-defined matrix operation is the transposition, which exchanges rows and columns. If we apply the matrix transposition to the incidence matrix of a Petri net, what happens with the Petri net and its invariants? □

Exercise 7.8. Visualizing Invariants. Invariants are computed with Charlie, but can be visualized with Snoopy using the following steps.
1. Open the net to which the invariants belong.
2. Go to *Extras* → *Load Node Set file*, and select the text file written by Charlie (its default extension is *inv*).
3. A new window pops up, which allows you to run through all invariants and highlight the induced subnets in your favorite color. Single invariants or sets of invariants can be chosen. Invariants generally overlap; thus, when selecting a set of invariants, one can choose between *intersection* and *union*.
4. If you want to save the highlighting, choose *keep coloring* before closing the window.

To familiarize yourself with these features, generate for the running example a $\mathcal{PN}$, showing the subnets induced by the two P-invariants in color. □

Exercise 7.9. CTL Model Checking. While CTL model checking fits particularly for the decision of special properties, it can also be used to decide general properties. Express for the bounded version of the running example the following properties in CTL and check them with Marcie.
(a) Check the liveness of reactions $r1$ and $r8$. What had to be done to decide the liveness of the net?
(b) Check the reversibility of the net.

(c) Check the token conservation of the two P-invariants.
(d) Checking the upper bounds for A, R, and A_R by the queries
EF[$A > C2$], **EF**[$R > C2$], **EF**[$A\_R > C2$]
yields in the last case FALSE (as expected), but for the other two cases TRUE. Why? What are the correct upper bounds? □

Exercise 7.10. Behavior of Open vs. Closed Systems. Consider the following toy example for a system of chemical equations.

$$\begin{aligned} 2C + O_2 &\rightarrow 2CO \\ C + O_2 &\rightarrow CO_2 \\ C + CO_2 &\rightleftarrows 2CO \end{aligned}$$

Derive a corresponding Petri net model and explore its behavior by qualitative analysis techniques.
(a) Which structural properties hold?
(b) Determine and interpret the P- and T-invariants.
(c) The $\mathcal{PN}$ is bounded, but neither live nor reversible (for any initial state). Construct a live and reversible $\mathcal{PN}$ by assuming appropriate interconnections with an environment. This could be done as
(c.1) Open system—using input/output transitions, and
(c.2) Closed system—avoiding input/output transitions.
Do not forget to define appropriate but minimal initial states.
(d) Determine and interpret the P- and T-invariants for the open and closed systems. Which structural properties are different?
(e) Determine the behavioral properties for the open and closed systems.
(f) Try to construct the reachability graph for the open system. What is the problem? Construct the reachability graph for the closed system. Decide liveness, reversibility, and existence of dynamic conflicts in terms of the reachability graph. Are there concurrent reactions? Check the feasibility of the minimal T-invariants.
(g) The infinitely repeated occurrence of the first chemical reaction $2C + O_2 \rightarrow 2CO$ could be translated into
(g.1) **AG**[$(C \geq 2$ & $O_2 \geq 1) \rightarrow$ **AX**$(CO \geq 2)$]
(g.2) **AG**[$(C \geq 2$ & $O_2 \geq 1) \rightarrow$ **EX**$(CO \geq 2)$]
(g.3) **AG**[$(C \geq 2$ & $O_2 \geq 1) \rightarrow$ **AF**$(CO \geq 2)$]
(g.4) **AG**[$(C \geq 2$ & $O_2 \geq 1) \rightarrow$ **EF**$(CO \geq 2)$].
What is the difference? Which of them hold?
You should be able to solve all subtasks of this exercise without tool support, but you can check most of your answers with the help of Snoopy, Charlie, and Marcie. □

7.4 STOCHASTIC PETRI NETS ($\mathcal{SPN}$)

In this section, we introduce time and we learn how specific kinetic assumptions will typically restrict the qualitatively infinite state space to a quantitatively finite state space, if we forget about the states with a probability below a certain threshold. We suggest starting with Exercise 7.11.

7.4.1 Modeling

Basics

Stochastic Petri nets build on standard (i.e., qualitative, time-free) Petri nets. As in the qualitative case, a stochastic Petri net maintains a discrete number of tokens on its places. But contrary to the time-free case, a stochastic firing rate (compare Figure 7.1) is associated with each transition. The firing rate determines a stochastic waiting time before an enabled transition actually fires, provided it did not lose its license to fire in between. The waiting times are random variables following an exponential probability distribution (with the parameter λ). Therefore, all transition firing sequences of the qualitative Petri net can still occur, but the probability of taking a certain path depends on the individual firing rates of transitions in conflict, and the probability of reaching a certain state depends on the rates of all transitions involved. Snoopy provides the following features to specify state-dependent firing rates (i.e., of the parameter λ of the exponential probability distribution).

1. *Rate functions* are, technically speaking, arbitrary mathematical functions. To keep a close relation to the net structure, only a transition's pre-places are allowed as variables in its rate function. Popular kinetics such as mass-action kinetics (see examples below), Michaelis-Menten kinetics, and level semantics are supported by predefined function patterns. Of course, each transition gets its own rate function, making up together a list of rate functions. Several of such rate function lists can be maintained, allowing for flexible models.
2. *Parameters* are real-valued constants used in rate functions. Several sets of parameters can be maintained.
3. *Modifiers* are special arcs, graphically represented by dashed lines, which always go from a place to a transition. Pre-places connected with a transition by a modifier arc may modify the transition's firing rate, but do not have an influence on the transition's enabledness (contrary to places connected by standard or special arcs).

The firing itself of a stochastic transition does not consume time and follows the standard firing rule of qualitative Petri nets. To get a feeling for stochasticity and its influence on the net behavior, let's consider two simple examples, where tokens could be read as, for example, molecules.

- $\mathcal{SPN}_1$: Figure 7.10 shows an irreversible reaction, and Figure 7.11 a few related computational experiments by varying the number of tokens N. Single stochastic runs look general differently (a-c), while stochasticity is leveled out by increasing the number of tokens (d), or averaging over an increasing number of runs (e), which generally approaches the deterministic behavior of $\mathcal{CPN}$ (f); see Section 7.5.
- $\mathcal{SPN}_2$: Figure 7.12 shows a reversible reaction, of which we explore the steady state in Figure 7.13 by varying the number of tokens N. Single stochastic simulation runs reveal that higher numbers of tokens decrease the size of fluctuations (a).

  This observation is approved by the probability distribution in the steady state (b), which is a discrete function showing for every possible (discrete) token number $0, \ldots, N$ its probability in the steady state. The bell-shape distribution becomes sharper with increasing numbers of tokens, but the curves' amplitudes decrease as the probability mass of 1 has to be distributed over increasingly more possible (discrete) values. Thanks to the simple net structure, there is a closed-form solution for the probability of having j tokens on A in the steady state: $p(j) = \binom{N}{j} \frac{k_1^j \cdot k_2^{N-j}}{(k_1+k_2)^N}$. To support the comparison for different N, the diagram is normalized and shows j/N, the probability distribution for the fraction of tokens on A in the steady state.

Running Example

Turning our $\mathcal{PN}$, which we have designed and explored in the previous section, into an $\mathcal{SPN}$ simply requires us to assign rate functions to all transitions. The default rate function is *MassAction*(1), that is, mass-action kinetics with parameter 1. It is automatically assigned to every stochastic transition, such as when doing the $\mathcal{PN} \to \mathcal{SPN}$ export. All reactions in our running example apply mass-action kinetics, so we just have to define the kinetic parameters given in Table 7.4 and use them in the rate functions. The resulting $\mathcal{SPN}$ is provided as Supplementary Material; see also Exercise 7.11.

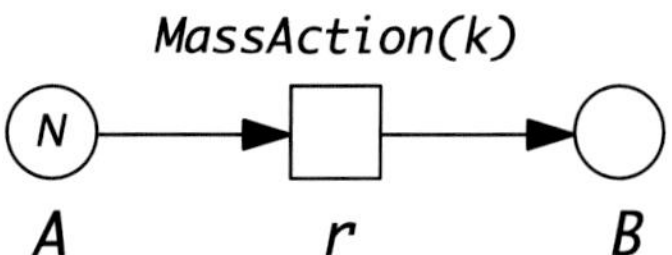

FIGURE 7.10 $\mathcal{SPN}_1$ for the irreversible reaction $A \xrightarrow{k} B$, with initially N tokens on A; see Figure 7.11 for some simulation results with $k = 1$ and $N = 10, 1000$.

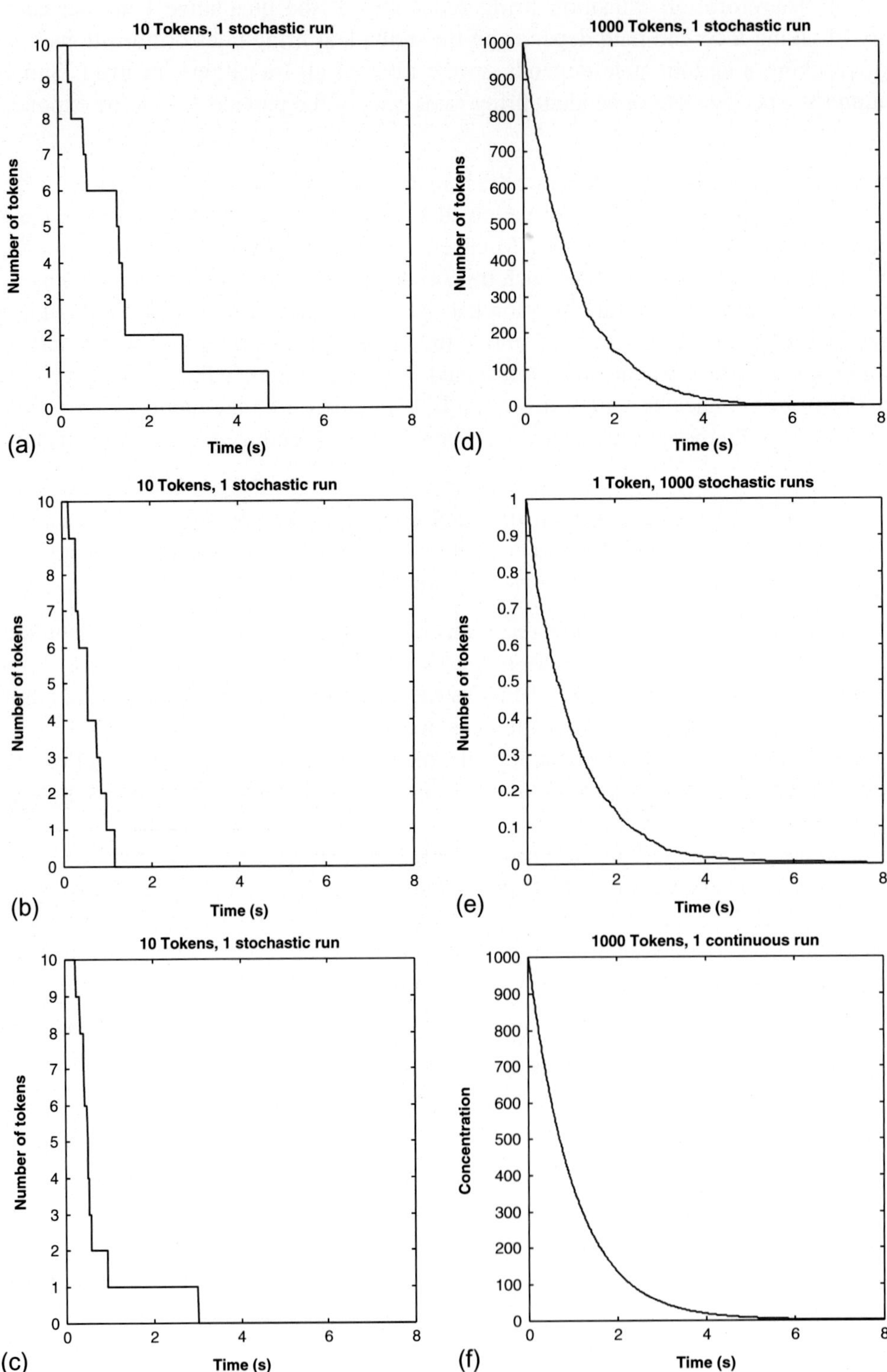

FIGURE 7.11 Stochastic simulations of $\mathcal{SPN}_1$ in Figure 7.10 in different settings. (a)-(c) Three single runs for $N = 10$; each run is different due to stochasticity. Stochasticity is leveled out by increasing N, or averaging over an increasing number of runs: (d) $N = 1000$, single run; (e) $N = 1$ and 1000 runs; (f) continuous run. This discussion is adapted from [41].

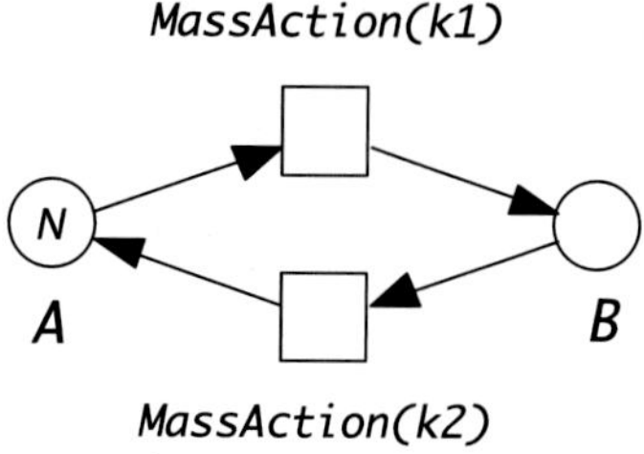

FIGURE 7.12 $\mathcal{SPN}_2$ for the reversible reaction $A \underset{k2}{\overset{k1}{\rightleftharpoons}} B$, with initially N tokens on A; see Figure 7.13 for some simulation results with $k_1 = 1, k_2 = 1$, and $N = 10, 100, 500$.

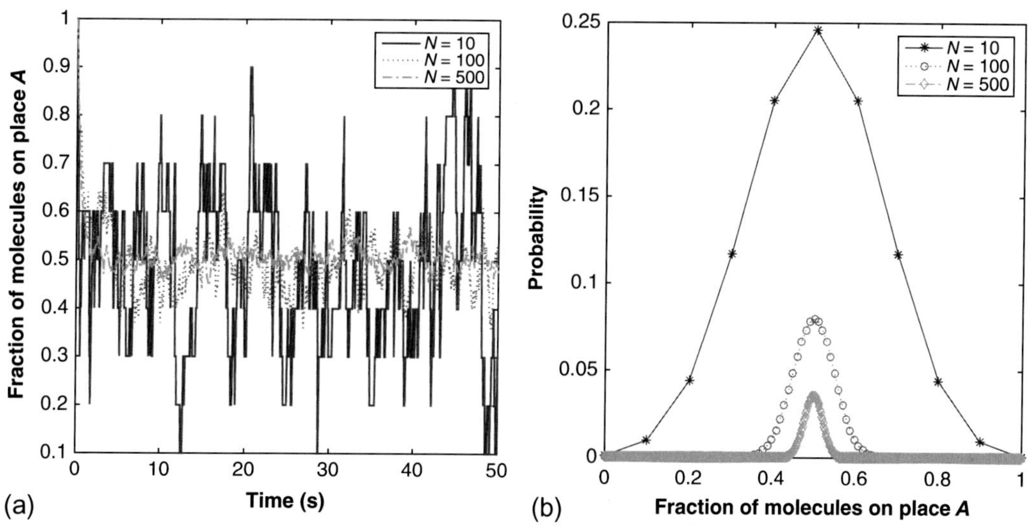

FIGURE 7.13 Exploring $\mathcal{SPN}_2$ in Figure 7.12 with $N = 10$, $N = 100$, and $N = 500$ tokens (molecules). (a) Single stochastic simulation runs; fluctuations decrease with increasing numbers of tokens. (b) Probability distribution of the fractions of tokens on A in the steady state. Please note that this is actually a discrete function. This discussion is adapted from [42].

7.4.2 Analysis

Basics

Stochastic Petri nets with exponentially distributed firing delays for all transitions fulfill the Markov property; thus their semantics are described by a continuous time Markov chain (CTMC). The CTMC for a given $\mathcal{SPN}$ is basically isomorphic to the reachability graph of its corresponding $\mathcal{PN}$, but edges are labeled with the generally state-dependent transition rates. Parallel transitions (transitions sharing pre- and post-states) are an exception; there is only one arc in the CTMC labeled with the sum of their rates. Thus, all $\mathcal{PN}$ analysis techniques can still be applied, and all behavioral properties that hold for the $\mathcal{PN}$ are still valid for the corresponding $\mathcal{SPN}$. Additionally, we have the following techniques to explore stochasticity.

Stochastic Simulation

Stochastic simulation algorithms (SSA) generate random walks through the CTMC. The most famous one is Gillespie's SSA, also known as the *direct method*; Algorithm 3 shows the basic idea. An SSA basically consists of two steps. (1) Determine when the next transition will fire (line 5). (2) Determine which transition will fire (line 7). There are two different ways to compute (1). (1a) Determine duration until next firing, which is applied in the direct method. (1b) Determine time of next firing, which is applied in the *first-reaction* and *next-reaction* methods. All three algorithms are exact in the sense that no event is skipped. A popular improvement is the *tau leaping method*, which aims at efficient simulation of stiff systems. It deploys time jumps for fast-forward steps, and thus generally skips events.

We obtain approximated traces by averaging over a number of simulation runs, as done in Figure 7.11e. See Exercise 7.12 for the options we have to specify simulation experiments.

ALGORITHM 3 Basic $\mathcal{SPN}$ simulation algorithm

Require: $\mathcal{SPN}$ with initial state s_0;
simulation interval $[\tau_0, \tau_{end}]$;

time $\tau \leftarrow \tau_0$
state $s \leftarrow s_0$ ▷ make initial state to current state
write(τ, s) ▷ add s_0 to trace
while $\tau < \tau_{end}$ **do**
  determine duration $\Delta\tau$ *until next firing*
  $\tau \leftarrow \tau + \Delta\tau$
  determine transition t enabled at s and firing at time τ
  $s \leftarrow$ *fire*(s, t) ▷ compute new state s by firing of t
  write(τ, s) ▷ add s to trace
end while

Simulative Model Checking

For a systematic exploration of simulation traces we use PLTLc, a probabilistic extension of LTL with constraints, to express behavioral properties of interest. Simulative model checking considers a finite set of finite outputs, in other words, just a finite subset of the state space, generated by an (at best) exact SSA. This permits us to explore huge or even infinite state spaces in reasonable time, or just to obtain a first rough estimate. Each trace is evaluated to a Boolean truth value, and the probability of a behavioral property holding true is approximated by the number of traces with true values over the whole sample set. One has to consider a sufficient number of simulation traces to obtain trustworthy approximations. The number of traces required increases with the expected confidence in the numerical results. Rare events may dramatically increase the required size of the sample set.

Numerical Methods

As long as the underlying semantics of a stochastic Petri net is described by a finite CTMC of manageable size, it can be analyzed using standard stochastic analysis techniques such as transient analysis, steady-state analysis, or (exact) model checking. Transient analysis means to compute the transient probabilities to be in a certain state at a specific time point using, for example, the uniformization method. Steady-state analysis computes the steady-state probabilities using, for example, Jacobi iteration or Gaussian-Seidel iteration. In numerical model checking, behavioral properties can be checked, which have been expressed in, for example, CSL—a stochastic counterpart of CTL, where the path quantifiers are replaced by probabilities.

For illustration, we perform transient analysis for the two introductory examples $\mathcal{SPN}_1$ and $\mathcal{SPN}_2$; see Figures 7.14 and 7.15.

Running Example

Simulating the $\mathcal{SPN}$ for our running examples generates sustained oscillation for all unbounded places, with each single run being different; see Figure 7.16 for two traces showing the proteins A and R.

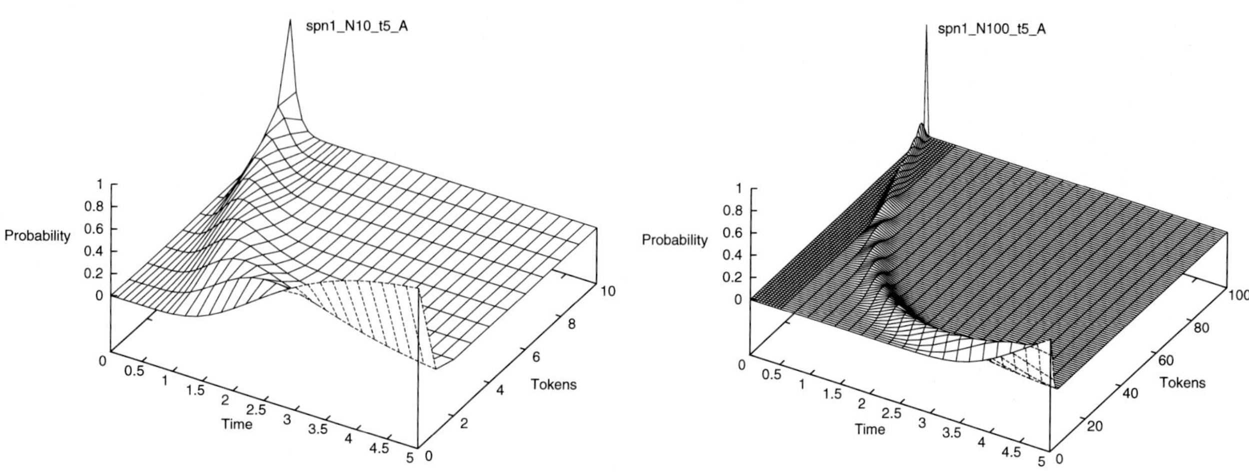

FIGURE 7.14 Transient analysis for $\mathcal{SPN}_1$ in Figure 7.10 for $N = 10$ (left) and $N = 100$ (right). The 3D plots show the evolution of the probability distribution of the token number on A up to time point 5.

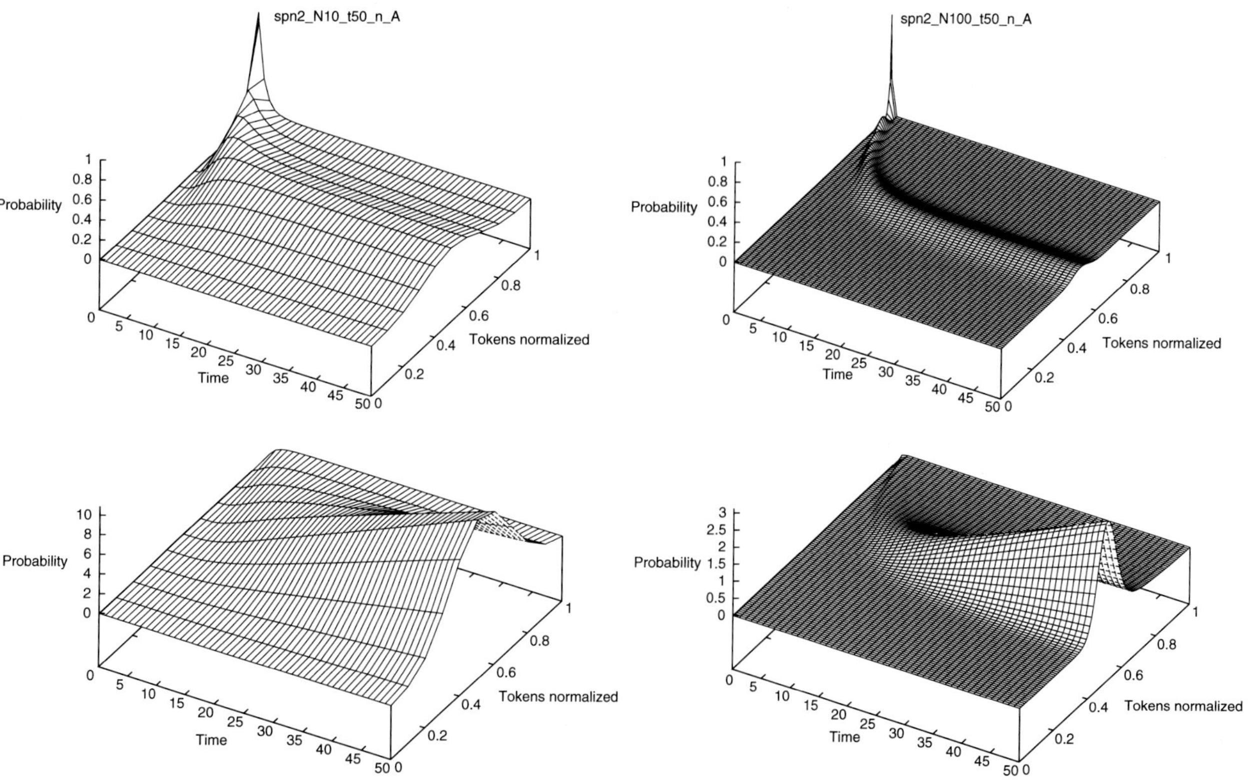

FIGURE 7.15 Transient analysis for $\mathcal{SPN}_2$ in Figure 7.12 for $N = 10$ (left) and $N = 100$ (right). The 3D plots in the first row show the evolution of the probability distribution of the fraction of tokens on A up to time point 50. The second row gives the cumulative probability distributions, sharpening the effect; compare the closed-form solution in Figure 7.13b.

Next, we use a PLTLc-specific feature to explore the value range for all structurally unbounded places—the *free variables*, which are specified by a leading $.

- What is the probability that up to time point τ one of the mRNAs rises above v? We do not exactly know which mRNA will start rising, so we use the disjunction
  $\mathbf{P}_{=?}[\mathbf{F}_{[0,\tau]}\ mrnaA > \$v \mid mrnaR > \$v]$.

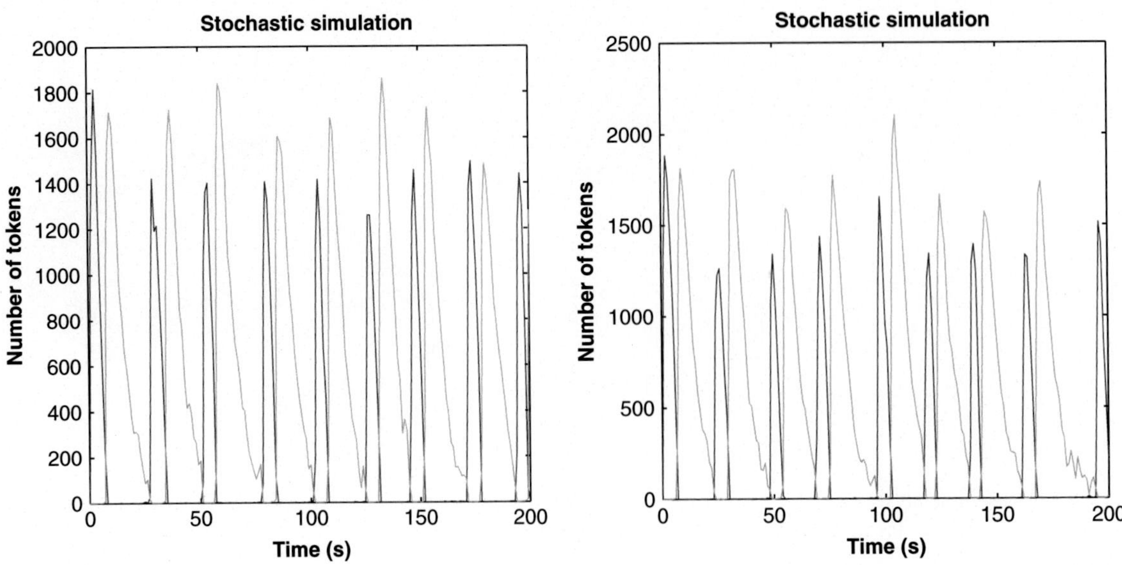

FIGURE 7.16 Two plots of single stochastic simulation runs (showing A and R) when reading the running example as $\mathcal{SPN}$. Each single run looks different in terms of oscillation.

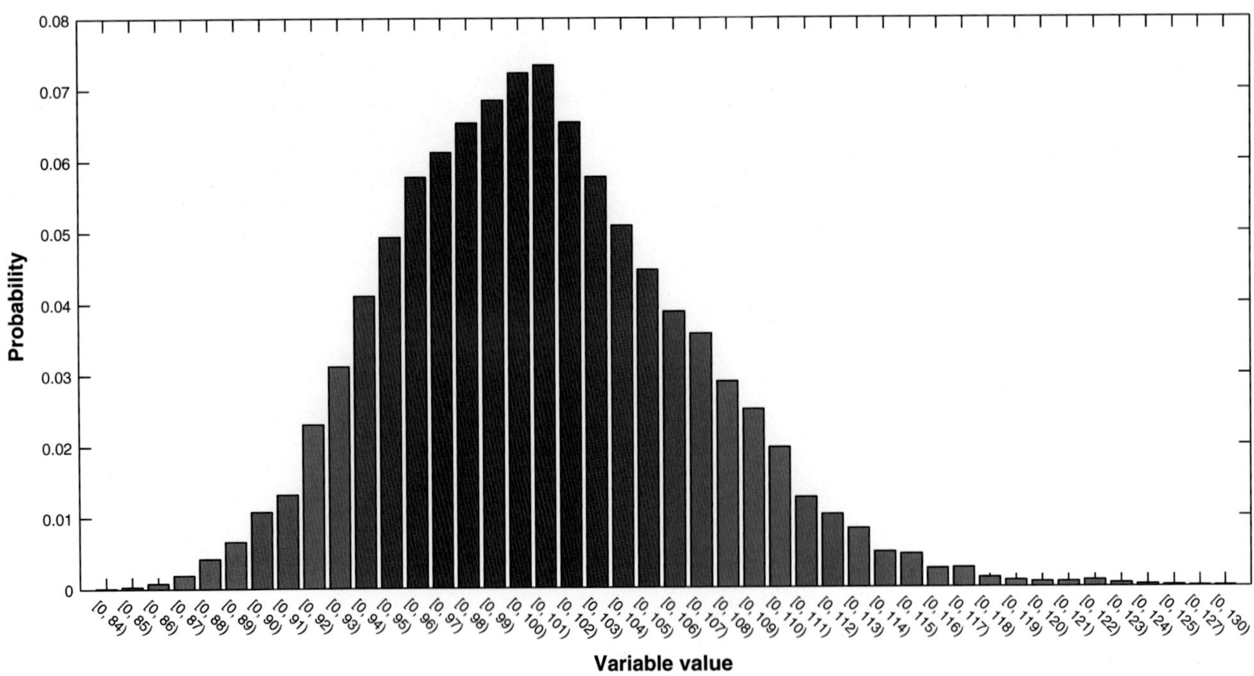

FIGURE 7.17 Probability distribution of the value range for the two mRNA places, determined for the $\mathcal{SPN}$ model by 10,000 stochastic simulation runs with $\tau = 100$. Increasing the number of runs smooths the bell shape, but does not shift the value range. Values beyond 120 are most unlikely.

Simulative model checking yields the domain of the free variable v and the probability of each interval; see Figure 7.17. We observe that values beyond 120 become increasingly unlikely (less than 0.001). Thus, we take $C1 = 120$ as upper bound (see Figure 7.9). Likewise, we determine the probability distributions for A, R, and A_R (see Supplementary Material), yielding the upper bound $C2 = 2100$, which cuts the infinite state space down to 814,963,013,843,528 (10^{14}) states, while stochastic simulation runs still exhibit oscillation.

However, this state-space size still exceeds by several magnitudes the resources of current computers as required for numerical algorithms; see Exercise 7.14. Thus, we confine ourselves to simulative model checking to explore the PLTLc counterparts to our CTL properties, previously considered in Section 7.3.2.

- What is the probability that forever (generally up to time point τ), gene A is either free or bound to the activator protein?
  $\mathbf{P}_{=?}[\mathbf{G}_{[0,\tau]}\ (geneA = 1\ \&\ geneA\_A = 0)\ |\ (geneA = 0\ \&\ geneA\_A = 1)]$
  As this property is structurally ensured, we expect probability 1 as answer. Please note that in terms of simulative model checking, *forever* usually means in the entire (finite) trace as far as generated. However, there are advanced techniques of simulative model checking with steady-state and time-unbounded temporal operators, which permit us to compute, for example,
  $\mathbf{P}_{=?}[\mathbf{G}(geneA = 1\ \&\ geneA\_A = 0)\ |\ (geneA = 0\ \&\ geneA\_A = 1)]$.
  Here, there is no time interval specified, but the trace generation is terminated as soon as the steady state has been reached.
- What is the probability of having k molecules of protein A at time point τ?
  $\mathbf{P}_{=?}[\mathbf{F}_{[\tau,\tau]}(A = k)]$
  Increasing τ will finally approach the steady-state probabilities. Then, varying k yields the steady-state probability distribution. For any $k > C2$, we expect neglectable probabilities for the unbounded model version, and zero probabilities for the bounded model.
- What is the probability of having sometimes (finally) at least k molecules of each protein (activator and repressor) at the same time, when considering traces up to time point τ?
  $\mathbf{P}_{=?}[\mathbf{F}_{[0,\tau]}(A \geq k\ \&\ R \geq k)]$
  The probability is technically larger than 0 for any k (with $k \leq C2$ for the bounded model), as the property holds for the $\mathcal{PN}$, but drops dramatically with increasing k, and reaches very fast insignificant values that are practically 0.

By means of simulative model checking, these queries can be equally applied to the $\mathcal{SPN}$ model in its unbounded and bounded versions, and thus can be deployed to check the behavioral equivalence of both models.

7.4.3 Further Reading

- *A bit of history*: $\mathcal{SPN}$ were introduced in 1982 for performance evaluation of computer systems [43]. Comprehensive textbooks [44, 45] are out of print but available as pdfs on the authors' websites. Do not miss the preface in [44] if you are into $\mathcal{SPN}$ history.
- *$\mathcal{SPN}$ in Systems Biology*: The use of $\mathcal{SPN}$ to explore stochastic systems in molecular biology was first demonstrated in [46]; further examples can be found in, for example, [47, 48]. The level concept to discretize and approximate $\mathcal{CPN}$ by $\mathcal{SPN}$ has been discussed in [38]. For a gentle introduction to stochastic modeling in Systems Biology, see [49].
- *More background*: Formal definitions of (biochemically interpreted) $\mathcal{SPN}$ can be found in [38, 50].
  SSAs with state-dependent rates go back to [51], followed by many improvements; a well-founded discussion of related issues can be found in [52].
  A comprehensive textbook of stochastic analysis techniques and relate algorithms is [53]. See [8, 54] to learn more about Marcie's symbolic approach for the numerical methods, and [55] for a comparative study of stochastic analysis techniques, also providing user guidelines for when to use which kind of technique.
- *Model checking*: A gentle introduction to the use of simulative model checking can be found in [50]. Consult Marcie's manual [10] for more examples of CSL and PLTLc properties and specific questions regarding tool handling, for example, how to generate the 3D plots with Marcie, assisted by Gnuplot.
  A more recent extension of stochastic temporal logic is rewards, yielding continuous stochastic reward logic (CSRL); see [56] for details of symbolic CSRL model checking. Recent advances in simulative model checking with steady-state and time-unbounded temporal operators are presented in [57].

- *Extensions*: $\mathcal{SPN}$ have been enriched by special arcs and special transitions, yielding in summary $\mathcal{XSPN}$. They come along with four special arc types and state-dependent arcs as available for $\mathcal{XPN}$, and furthermore three special transition types:
  - *Immediate* transitions (zero waiting time, always highest priority), which will be used in Section 7.6.1; see, for example, $\mathcal{HPN}_2$.
  - *Deterministic* transitions (deterministic waiting time, relative to the time point where the transition gets enabled), and
  - *Scheduled* transitions (scheduled to fire, if any, at single or equidistant, absolute points of the simulation time).

    The priority of the latter two lies between the immediate and stochastic transitions.

  In Snoopy, we do not differentiate between different classes of stochastic Petri nets. Thus, we usually call our extended stochastic Petri nets simply $\mathcal{SPN}$ if confusion is precluded; for details, see [50].

  The unrestricted use of special (immediate, deterministic, scheduled) transitions destroys the Markov property. But the adaptation of Gillespie's SSA is rather straightforward and supported in Snoopy.
- *Further examples*: To learn from case studies, here are some suggestions:
  - $\mathcal{SPN}$ were used in [36] for a case study in model-driven Synthetic Biology, in [22, 38] to model and analyze signal transduction pathways, and in [39, 40] for genes in a regulatory cycle (repressilator); the detailed analyses deploy analytical and simulative model checking, if possible.
  - $\mathcal{XSPN}$ are applied in [50] to a classical example of prokaryotic gene regulation, the lac operon, to demonstrate their power for model-based design of wetlab experiments, and in [58] to investigate phosphate regulation in enteric bacteria.

7.4.4 Exercises

Exercise 7.11. Changing the Net Class. Snoopy supports the conversion of a (qualitative) Petri net into a quantitative Petri net (or vice versa), which allows us to reuse the structure. Only the additional attributes have to be set.

1. Open the $\mathcal{PN}$; go to the export Window (*File* → *Export*); choose the appropriate target(s), among them $\mathcal{SPN}$; and hit the OK button.
2. Open the $\mathcal{SPN}$ just generated. It looks exactly the same as the $\mathcal{PN}$ (when *show rate functions* is switched off). It also has the same constants and functions; see *Declarations* panel.
3. The default rate function automatically set for stochastic transitions is *MassAction*(1).
4. Now we are ready to specify $\mathcal{SPN}$-specific attributes.
   - Add the kinetic parameters as constants in the *Declarations* panel.
   - Double-clicking on a transition opens the *Edit Properties* window. Go to *Function* in order to change the rate functions.

Generate for the running example the $\mathcal{SPN}$ out of its $\mathcal{PN}$. □

Exercise 7.12. Repeated Stochastic Simulation. Take the $\mathcal{SPN}$ of the running example and familiarize yourself with Snoopy's simulator. The simulation can be parameterized in four ways.

- *Model configuration*: There are different sets of initial states (markings), rate functions and parameters can be defined, and one of each has to be chosen for a given experiment.
- *Simulation time*: The simulation starts always at time 0 and goes until the specified *Interval end*, if no dead state is encountered. But, the simulation data are only recorded for the simulation interval given by *interval start* and *interval end*.
- *Output step count*: The number of output steps in the specified simulation interval defines the output grid of the recorded simulation data.
- *Number of runs and threads*: The number of simulation runs to be averaged and the number of parallel threads to be deployed have to be specified.

Having started the simulation, the progress bar indicates how far the simulation has been going until now, and the time consumed by the simulation is displayed too.

There are three export options:

- *Direct export*: The averaged result of the selected places or transitions is exported.
- *Single trace export*: The result of every simulation run is exported separately.
- *Exact trace export*: Every change of the marking of the selected places or the rates of the selected transitions at any time point is exported.

The simulation results can be shown as

- *Table*: Each column represents a selected place (transition) and each row shows the averaged marking (firing times) at one output step point.
- *Plot*: Each curve stands for one selected place (transition); the x-axis holds the time and the y-axis holds the number of tokens (firing times).

Different tables and plots can be created to switch conveniently between different views of the simulation results. Each view is characterized by a set of selected places (transitions). Simulation plots can also be saved. For more details, consult Snoopy's FAQ. □

Exercise 7.13. Exploring Stochasticity. Take the $\mathcal{SPN}_1$ in Figure 7.10 and the $\mathcal{SPN}_2$ in Figure 7.12 to explore the effect of stochastic simulations by varying the initial token number N on A, the kinetic parameters k (k_1, k_2), and the number of runs. Find suitable simulation options for *interval end* and *output step count*. □

Exercise 7.14. Estimating Memory Consumption. Stochastic analyses are often quite memory-expensive. To get the idea, calculate the following estimates. We assume that the memory consumption for the Petri net itself and the algorithms can be neglected. All considered data structures store double values (8 byte).

(a) *Stochastic simulation*: Two types of data structures are required.
   - One matrix serves as the result table and stores a value for each place and each time point in the output grid.
   - Two vectors in the size of the number of places keep the initial and current state.

   Estimate the required memory for Petri nets with 10, 100, 1000, 10,000, and 100,000 places, and for output grids of 10, 100, 1000, 10,000, and 100,000 time points.

   Assuming you have a computer with 1/4/8/16/32 GB (free) memory, which kind of simulation experiments can you handle?

(b) *Stochastic numerical methods*: Two types of data structures are required.
   - One square matrix in the size of the number of states encodes the CTMC.

     Marcie does not explicitly hold the CTMC; instead it recomputes, when required, the matrix on the fly.
   - Two to four vectors in the size of the state space keep the results and auxiliary data, depending on the engine and the used algorithms.

   Let's consider the algorithms requiring two vectors. Estimate the required memory for the state spaces given in Table 7.7 for (b1) Marcie (CTMC memory consumption can be neglected), and (b2) a hypothetical tool explicitly encoding the CTMC matrix.

   Assuming you have a computer with 1/4/8/16/32 GB (free) memory, which problem size can you handle? □

Exercise 7.15. Exploring the Running Example. Take the running example (unbounded version) and perform the following computational experiments.

(a) Change δ_R to 0.08. How does the dynamic behavior change? Can oscillations still be obtained after this perturbation?
(b) Consider averaged stochastic simulations with increasing number of runs. How does the observable behavior change? Why? □

Exercise 7.16. Exploring the Bounded Version of the Running Example. Take the $\mathcal{XPN}$ in Figure 7.9, which is a bounded version of the running case study; read it as an $\mathcal{SPN}$ with the kinetic parameters of Table 7.4;

and explore its behavior by stochastic simulation. You will soon realize that it is behaviorally equivalent to the structurally unbounded version for $C1 \geq 120$, $C2 \geq 2100$.

(a) To simplify life, two constants $(C1, C2)$ have been used in this bounded-model version. However, a more detailed analysis reveals more precise different bounds for *mrnaA*, *mrnaR*, *A*, *R*, and *A_R*. Adjust the $\mathcal{SPN}$ appropriately to distinguish between five constants imposing individual upper bounds.

(b) Vary the constants and observe the differences in the results (single runs), specifically for decreasing values. Explain your observations.

Hint. The insights gained in Exercise 7.9d might be worth taking into consideration. □

7.5 CONTINUOUS PETRI NETS ($\mathcal{CPN}$)

To explore time-dependent behavior, $\mathcal{SPN}$ seem to be the natural choice, as biochemical networks are inherently governed by stochastic laws. However, if molecules are in high numbers, and stochastic effects can be neglected, one can equally take a deterministic approach to explore how the averaged concentrations of species evolve over time. We suggest starting with Exercise 7.17.

7.5.1 Modeling

Basics

Continuous Petri nets are another quantified version of the standard notion of qualitative Petri nets. Like their ancestor, they are bipartite, directed multigraphs, however, arc weights and the numbers assigned to places are now nonnegative real numbers. Thus, the number of tokens on each place is replaced by a token value, which we interpret, for example, as the concentration of a given species. The instantaneous firing of a continuous transition is carried out like a continuous flow, whereby the strength of the flow is determined by a deterministic firing rate function, which each transition gets assigned (compare Figure 7.1). Deterministic rate functions are typically state-dependent and supported by the same features outlined in Section 7.4.1 for stochastic rate functions.

A continuous transition is enabled if the token value of all its preplaces is larger than zero. This coincides for mass-action kinetics with rates larger than zero. Due to the influence of time, a continuous transition is forced to fire as soon as possible. Altogether, the semantics (behavior) of a continuous Petri net is defined by an ODEs. Each place subject to changes gets its own equation, describing the continuous change over time of its token value by the continuous increase of its pretransitions' flow and the continuous decrease by its post-transitions' flow. To simplify the notation in the generated ODEs, places are usually interpreted as (nonnegative) real variables. Each equation corresponds basically to a line in the incidence matrix (stoichiometric matrix). A transition that is pre- and post-transition yields two terms, which can be reduced by algebraically transforming the right-hand side of the equation.

For an introduction, let's consider three simple examples.

- $\mathcal{CPN}_1$: Reading $\mathcal{SPN}_1$ in Figure 7.10 as $\mathcal{CPN}$ induces the ODEs

$$\frac{dA}{dt} = -(k \cdot A), \quad \frac{dB}{dt} = +(k \cdot A)$$

  a deterministic simulation run is shown in Figure 7.11f.
- $\mathcal{CPN}_2$: Reading $\mathcal{SPN}_2$ in Figure 7.12 as $\mathcal{CPN}$ induces the ODEs

$$\frac{dA}{dt} = (k2 \cdot B) - (k1 \cdot A), \quad \frac{dB}{dt} = (k1 \cdot A) - (k2 \cdot B).$$

- $\mathcal{CPN}_3$: Figure 7.18a illustrates a couple of ways to keep a model flexible by the use of constants. It also illustrates a crucial difference in how $\mathcal{CPN}$ behave—there are no conflicts. The induced ODEs are given in Table 7.8, and simulation results in Figure 7.18b.

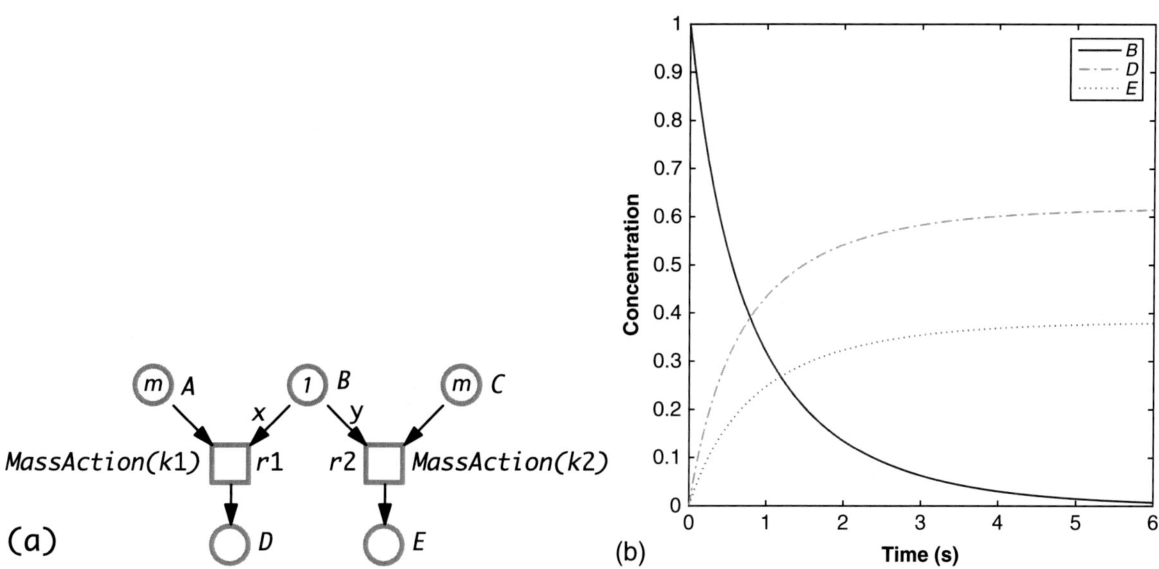

FIGURE 7.18 $\mathcal{CPN}$ do not know conflicts. (a) $\mathcal{CPN}_3$—the two transitions $r1$ and $r2$ share the token value of B. It depends on the parameters $k1, k2$, and the stoichiometric factors x, y how much each transition gets. (b) Deterministic simulation of $\mathcal{CPN}_3$ with $k1 = 1$, $k2 = 0.5$, and all other constants set to 1.

TABLE 7.8 The ODEs Induced by $\mathcal{CPN}_3$ Shown in Figure 7.18a and Generating the Deterministic Simulation Trace Shown in Figure 7.18b

$\frac{dA}{dt} = -(k1 \cdot A \cdot B^x)$

$\frac{dB}{dt} = -x \cdot (k1 \cdot A \cdot B^x) - y \cdot (k2 \cdot B^y \cdot C)$

$\frac{dC}{dt} = -(k2 \cdot B^y \cdot C)$

$\frac{dD}{dt} = +(k1 \cdot A \cdot B^x)$

$\frac{dE}{dt} = +(k2 \cdot B^y \cdot C)$

In summary, $\mathcal{CPN}$ offer a graphical way to specify ODEs. Due to the explicit structure, we expect to get descriptions that are less error-prone compared to the ones created manually in a textual notation from scratch. $\mathcal{SPN}$ and $\mathcal{CPN}$ have the power to approximate each other, as shown in Figure 7.1.

Running Example

If we read tokens on each place as (real-valued) concentrations of species and associate a deterministic rate function with each transition, for example, the same mathematical functions used as stochastic rates for the $\mathcal{SPN}$, we can immediately consider it a $\mathcal{CPN}$. The underlying ODEs of the continuous model, as generated by Snoopy when replacing read arcs by two opposite arcs, are given in Table 7.9; see also Exercise 7.18b.

7.5.2 Analysis

Basics

As a $\mathcal{CPN}$ can be read as a structured description of ODEs, all standard ODE analysis techniques can be deployed for its exploration, for example, bifurcation analysis, sensitivity analysis, or parameter scanning. Currently,

TABLE 7.9 The Unreduced ODEs Induced by Reading the Running Example as $\mathcal{CPN}$ with Mass-Action Kinetics, as Generated by Snoopy

$$\frac{dgeneA}{dt} = (ta \cdot geneA\_A) + (aa0 \cdot geneA) - (ga \cdot geneA \cdot A) - (aa0 \cdot geneA)$$

$$\frac{dA}{dt} = (ta \cdot geneA\_A) + (ba \cdot mrnaA) + (tr \cdot geneR\_A)$$
$$-(ga \cdot geneA \cdot A) - (da \cdot A) - (gr \cdot A \cdot geneR) - (gar \cdot A \cdot R)$$

$$\frac{dgeneA\_A}{dt} = (ga \cdot geneA \cdot A) + (aa1 \cdot geneA\_A) - (ta \cdot geneA\_A) - (aa1 \cdot geneA\_A)$$

$$\frac{dmrnaA}{dt} = (aa0 \cdot geneA) + (aa1 \cdot geneA\_A) + (ba \cdot mrnaA) - (dam \cdot mrnaA) - (ba \cdot mrnaA)$$

$$\frac{dmrnaR}{dt} = (ar1 \cdot geneR\_A) + (ar0 \cdot geneR) + (br \cdot mrnaR) - (drm \cdot mrnaR) - (br \cdot mrnaR)$$

$$\frac{dgeneR\_A}{dt} = (gr \cdot A \cdot geneR) + (ar1 \cdot geneR\_A) - (tr \cdot geneR\_A) - (ar1 \cdot geneR\_A)$$

$$\frac{dgeneR}{dt} = (tr \cdot geneR\_A) + (ar0 \cdot geneR) - (gr \cdot A \cdot geneR) - (ar0 \cdot geneR)$$

$$\frac{dR}{dt} = (br \cdot mrnaR) + (da \cdot A\_R) - (dr \cdot R) - (gar \cdot A \cdot R)$$

$$\frac{dA\_R}{dt} = (gar \cdot A \cdot R) - (da \cdot A\_R)$$

Note: We deliberately give the unreduced ODEs to better illustrate the relation to the generating $\mathcal{CPN}$.

this requires exporting to suitable tools, for example, Matlab, which is supported by Snoopy. Additionally, the following two analysis techniques are directly available within our toolbox.

Deterministic Simulation

As soon as there are transitions with more than one preplace, we get typically a nonlinear system, which calls for a numerical treatment of the system on hand; see Algorithm 4 for the basic principle. In more sophisticated algorithms, h may be a variable step size to cope with stiff systems. For efficiency reasons, Snoopy does not check if a continuous transition is enabled, which may result in negative concentration values, as it may also be caused by an inappropriate step size. In the case of mass-action kinetics, nonenabledness and a rate of zero coincide.

Snoopy supports 14 different stiff/unstiff ODE integrators to numerically solve the ODEs. These ODE solvers range from simple fixed-step-size solvers (e.g., Euler), which are suitable for unstiff $\mathcal{CPN}$ models, to more

---

ALGORITHM 4 Basic $\mathcal{CPN}$ simulation algorithm

Require: $\mathcal{CPN}$ with initial state s_0;
simulation interval $[\tau_0, \tau_{end}]$;
step size h *with* $h < (\tau_{end} - \tau_0)$;

define function f by constructing the ODEs induced by $\mathcal{CPN}$
time $\tau \leftarrow \tau_0$
state $s \leftarrow s_0$ ▷ initialize ODE solver with s_0
$write(\tau, s)$ ▷ add s_0 to trace

while $\tau < \tau_{end}$ **do**
  $\tau \leftarrow \tau + h$
  $s \leftarrow s + h \cdot f(s)$
  $write(\tau, s)$ ▷ add s to trace
end while

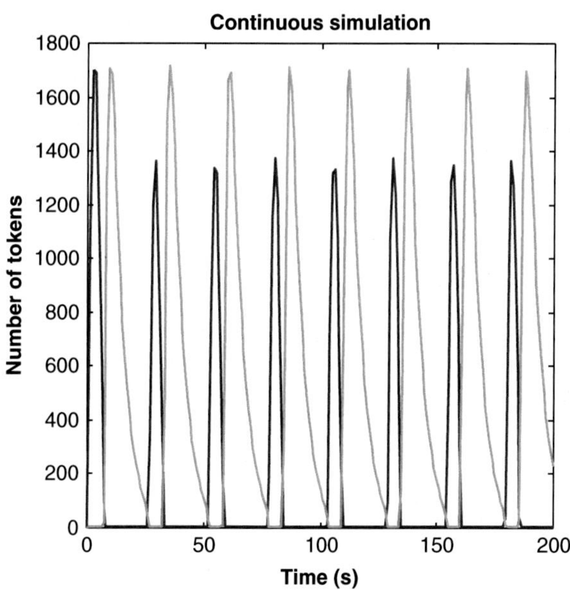

FIGURE 7.19 Plot of a deterministic simulation run (showing A and R) when reading the running example as $\mathcal{CPN}$.

sophisticated variable-order, variable-step, multistep solvers (e.g., Backward Differentiation Formulas (BDFs)), which are advisable for stiff $\mathcal{CPN}$ models.

Simulative Model Checking

The behavior of a $\mathcal{CPN}$ model is deterministic, that is, each run with the same parameters yields the same results. Thus, the state space can be considered continuous and linear. It can be explored by using, for example, continuous linear temporal logic with constraints (LTLc) or PLTLc in a deterministic setting. Both are interpreted over the continuous simulation trace generated by numerically integrating ODEs.

Running Example

Running continuous simulation for the $\mathcal{CPN}$ model of the running example yields plots as illustrated in Figure 7.19. In contrast to the stochastic runs, the oscillations vary less in their amplitude, and each run looks exactly the same.

For illustration, we specify the following PLTLc property.

- Does the value of protein A first rise and then fall; in other words, does there exists a peak in the trace?

  $\mathbf{P}_{=?}[\mathbf{F}[(\mathrm{d}(A) > 0)\ \&\ \mathbf{F}[(\mathrm{d}(A) < 0)]]]$

  The function d(*species*) returns the derivative of the concentration of the species at each time point. The probability of this query is 1, that is, there is a peak; see Figure 7.19.

7.5.3 Further Reading

- *A bit of history*: $\mathcal{CPN}$, also known as fluid Petri nets, were originally proposed as a $\mathcal{PN}$ relaxation technique in order to overcome the state explosion problem. The foundations of fluidization or continuization (i.e., getting a continuous-state approximation) were laid around 1990 [59]. $\mathcal{CPN}$ come along as autonomous (time-free) or nonautonomous (timed) versions with constant or variable rates; see [60] for a comprehensive, but not always easy-to-read textbook, and [16] for a recent survey.
- *$\mathcal{CPN}$ in Systems Biology*: The use of continuous Petri nets for the structured design of ODEs describing biochemical processes was introduced in [61]. Along these lines, a novel methodology for the engineering

of biochemical network models is proposed in [62] and illustrated for signaling pathways by the step-wise composition of hierarchically structured $\mathcal{CPN}$s.

- *More background*: See [38] for formal definitions of $\mathcal{CPN}$ as used in this chapter, and [63] for more details about the $\mathcal{CPN}$ features supported by Snoopy; deterministic simulation algorithms in general; and Snoopy's implementation of them, which builds on the library SUNDIALS CVODE [64].
- *Model checking*: Please see [65] for the use of LTLc, and [11] for details on how to use MC2 for simulative model checking and a couple of biological examples.
- *$\mathcal{CPN}$ versus ODEs descriptions*: A $\mathcal{CPN}$ uniquely defines an ODEs, but generally not vice versa. Three sufficient criteria to uniquely extract the structure hidden in a given ODEs are discussed in [66]. The fact that different representation styles are not just a matter of taste is also argued in [67].
- *$\mathcal{SPN}$ versus $\mathcal{CPN}$ behavior*: The relationship between the strongly connected state space of the $\mathcal{PN}$ (and thus $\mathcal{SPN}$) and the steady-state behavior of the corresponding $\mathcal{CPN}$ is explored by means of a signaling network in [61].

  However, moving between the stochastic and continuous (deterministic) modeling paradigms may generally involve counterintuitive effects in the observable behavior. For example, a trap—a set of places that can never become empty in the discrete case as soon as it has obtained a token—can be emptied in the continuous case [68]. Consequently, a continuous counterpart of the STP was introduced in [69]. See Exercise 7.22 and [16, 70] for other surprising effects.
- *ODEs reduction by T-invariant analysis*: Static analysis of a hypoxia control network, previously introduced as ODEs, helped in [32] to reveal three functional modules (T-invariants) responsible for degrading the hypoxia-inducible factor (HIF), and to identify superfluous network parts not required for the essential steady-state behavior. The core network is represented as hierarchical $\mathcal{CPN}$, highlighting the structural principles inherent in the three functional modules and identifying the fragile node in the network, without which the switch-like ODEs behavior is shown to be completely absent.
- *Further examples*: $\mathcal{CPN}$ have been used in [37] for a case study in model-driven Synthetic Biology, in [23, 39] to model and analyse signal transduction pathways, and in [40, 41] for genes in a regulatory cycle (repressilator); the detailed analyses deploy analytical and simulative model checking.

7.5.4 Exercises

Exercise 7.17. Changing the Net Class. Snoopy supports the conversion between quantitative Petri nets, which allows us, for example, to read an $\mathcal{SPN}$ as $\mathcal{CPN}$ (or vice versa) without having to do any further changes.

1. Open the $\mathcal{SPN}$; go to the export Window (*File* → *Export*); choose the appropriate target(s), among them $\mathcal{CPN}$; and hit the OK button.
2. Open the $\mathcal{CPN}$ just generated. It looks exactly the same as the $\mathcal{SPN}$. It also has
   - The same constants and functions; see *Declarations* panel.
   - The same rate functions. Double-clicking on a transition opens the *Edit Properties* window. Go to *Function* in order to check the rate functions.
3. Use constants if you want to scale the kinetic parameters; see [49].
4. Moving to the continuous paradigm may require you to adjust the type of some arcs; see also Exercise 7.18b. To change the type of an arc, select the arc in question, and go to *Edit* → *Convert To*. □

Take the $\mathcal{SPN}$ of the running example and produce its $\mathcal{CPN}$ in two versions: (a) keeping all read arcs, and (b) replacing all read arcs with two opposite arcs. The latter in turn should generate the ODEs given in Table 7.9.

Exercise 7.18. Deterministic Simulation. Familiarize yourself with Snoopy's deterministic simulator. Like the stochastic simulation of $\mathcal{SPN}$, the deterministic simulation of $\mathcal{CPN}$ can be parameterized in several ways; see Exercise 7.12. For more details, see also Snoopy's FAQ webpage.

(a) Additionally, one can choose between stiff and unstiff solvers. Explore their differences by playing with the kinetic parameters and initial states of $\mathcal{CPN}_1$, $\mathcal{CPN}_2$, and $\mathcal{CPN}_3$.

(b) In contrast to $\mathcal{SPN}$, read arcs and two opposite arcs usually cause different behavior in $\mathcal{CPN}$. Take the running example and compare its behavior for the $\mathcal{CPN}$ keeping the read arcs and the $\mathcal{CPN}$ where read arcs are replaced by two opposite arcs; see Exercise 7.17. Play also with the weights of the read arcs, which can now be nonnegative real numbers.

(c) Take the $\mathcal{CPN}$ of the running example and change δ_R to 0.08. How does the dynamic behavior change? Can oscillations still be obtained after this perturbation? □

Exercise 7.19. Transforming ODEs into $\mathcal{CPN}$. Given is the following ODEs:

$$\frac{dA_1}{dt} = k_2 \cdot A_2 - k_1 \cdot A_1 \qquad \frac{dB_1}{dt} = -k_3 \cdot A_2 \cdot B_1 + k_6 \cdot B_2$$

$$\frac{dA_2}{dt} = k_1 \cdot A_1 - k_2 \cdot A_2 \qquad \frac{dB_2}{dt} = k_3 \cdot A_2 \cdot B_1 - k_4 \cdot A_2 \cdot B_2 + k_5 \cdot B_3 - k_6 \cdot B_2$$

$$\frac{dB_3}{dt} = k_4 \cdot A_2 \cdot B_2 - k_5 \cdot B_3$$

(a) Reveal the structure hidden in this ODEs by deriving a corresponding $\mathcal{CPN}$. You can check your solution by drawing the $\mathcal{CPN}$ with Snoopy, and doing *export* $\rightarrow$ *ODEs to text (LaTeX)*.
(b) Is the solution unique?
(c) Which structural and behavioral properties hold for the corresponding discrete Petri net?
(d) Construct a suitable initial state and explore the discrete and continuous behavior. □

Exercise 7.20. Transforming $\mathcal{CPN}$ into ODEs. Find at least three examples of two different $\mathcal{CPN}$s defining the same ODEs, up to some algebraic transformations. □

Exercise 7.21. Exploring the Limits. We claim that any ODEs (no matter what it looks like) can be generated by a $\mathcal{CPN}$. Give a simple procedure to define such a $\mathcal{CPN}$, assuming the ODEs are given. □

Exercise 7.22. Comparing Stochastic and Deterministic Behavior. Find at least three Petri nets that, when read as $\mathcal{SPN}$ and $\mathcal{CPN}$ with mass-action kinetics, reveal substantially different behavior with respect to their general behavioral properties. □

Exercise 7.23. Absolute Concentration Robustness. A place of a $\mathcal{CPN}$ (a variable of an ODEs) is said to have absolute concentration robustness (ACR) if its concentration is the same in all positive steady states; that is, it does not depend on the initial state, but only on the kinetic constants. For illustration, let's explore the following toy two-species mass-action system [71]:

$A + B \xrightarrow{\alpha} 2B\,,\, B \xrightarrow{\beta} A.$

(a) Consider this system as $\mathcal{PN}$. Determine the structural properties, and the general behavioral properties. *Hint.* there is a bad siphon.
(b) Consider this system as $\mathcal{CPN}$ and compare its behavior with the $\mathcal{PN}$, while varying the initial state. □

7.6 HYBRID PETRI NETS ($\mathcal{HPN}$)

If you cannot make up your mind whether to choose $\mathcal{SPN}$ or $\mathcal{CPN}$, then $\mathcal{HPN}$ might be the right choice. This net class combines under one roof everything $\mathcal{SPN}$ and $\mathcal{CPN}$ have to offer. We suggest starting with Exercise 7.24.

7.6.1 Modeling

Basics

Biochemical systems may involve reactions from more than one type of biological network, such as gene regulation, signal transduction, or metabolic pathways. Combining reactions belonging to distinct network types tends to result in stiff systems. For example, species in gene regulatory networks may contain few numbers of molecules, while species in metabolic networks often involve large numbers of molecules.

To support this kind of scenario, Snoopy integrates all functionalities of its stochastic Petri nets (see Section 7.4) and continuous Petri nets (see Section 7.5) into one net class, yielding generalized hybrid Petri nets ($\mathcal{GHPN}$), or hybrid Petri nets for short ($\mathcal{HPN}$). $\mathcal{HPN}$ are specifically tailored to models that require an interplay between stochastic and continuous behavior. They provide a trade-off between accuracy and runtime of model simulation by adjusting the number of stochastic transitions appropriately, which can be done either statically (by the user) or dynamically (by the simulation algorithms).

A typical $\mathcal{HPN}$ application is the hybrid representation of biochemical reactions at different scales, where slow reactions are represented by stochastic transitions and fast reactions by continuous transitions. Likewise, species occurring in small numbers are modeled by discrete places, and species occurring in larger numbers by continuous places. Stochastic and continuous network parts may influence each other by arcs connecting discrete and continuous nodes. The restrictions are kept to a minimum: discrete places cannot be connected with continuous transitions by standard arcs, as the continuous flow of the continuous transition contradicts the discrete number of tokens allowed to be on discrete places.

Often, discrete and continuous net parts are interconnected by read or inhibitory arcs, controlling the enabledness by specifying thresholds. Figure 7.20 shows a simple example in two versions. A discrete control cycle (the outer cycle) regulates the behavior of the inner continuous net part to keep the value of the continuous place *A* between two thresholds, specified by the two constants *LB* (lower bound) and *UB* (upper bound). The continuous in/outflows are alternately switched on/off by discrete transitions, which are either stochastic or immediate transitions.

- $\mathcal{HPN}_1$: *stochastic transitions*—there is a stochastic time delay between enabling and firing. The values of *A* go beyond *LB* and *UB*; repeated single runs are different.
- $\mathcal{HPN}_2$: *immediate transitions*—there is no time delay between enabling and firing. The values of *A* vary exactly between *LB* and *UB*; repeated single runs are identical.

Initially, the place *A* is clean, and the continuous inflow is on. The inflow is switched off when the threshold *UB* is reached (read arcs), while the continuous outflow is switched off when the value of the place *A* falls below the threshold *LB* (inhibitory arcs). See Figure 7.21 for hybrid simulation runs.

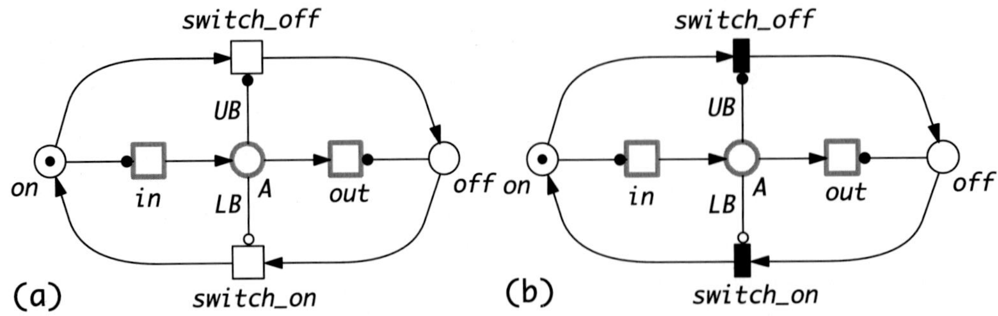

FIGURE 7.20 Two hybrid Petri nets regulating the continuous place *A* to keep its token value in the interval [*LB*, *UB*]. We apply the drawing convention to show continuous nodes in thicker lines to visually distinguish them from the discrete nodes. The continuous in/outflows are alternately switched on/off by discrete transitions, (a) $\mathcal{HPN}_1$: stochastic transitions and (b) $\mathcal{HPN}_2$: immediate transitions. All (stochastic/continuous) rates apply mass-action kinetics. See Figure 7.21 for hybrid simulation runs.

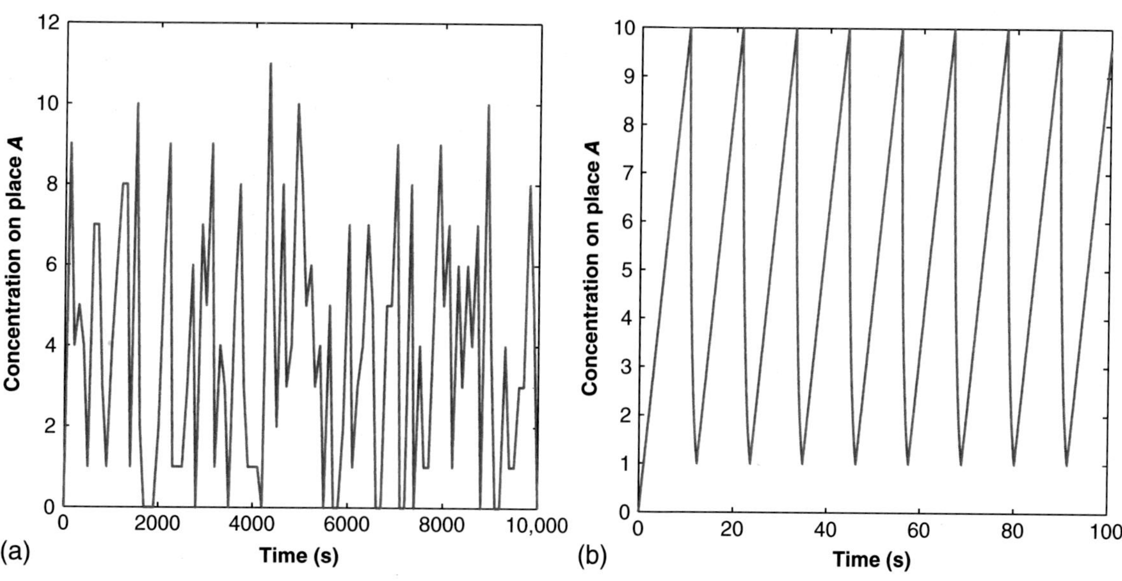

FIGURE 7.21 Hybrid simulation runs for $\mathcal{HPN}_1$ and $\mathcal{HPN}_2$ in Figure 7.20; all kinetic parameters are set to 1, and $LB = 1, UB = 10$.

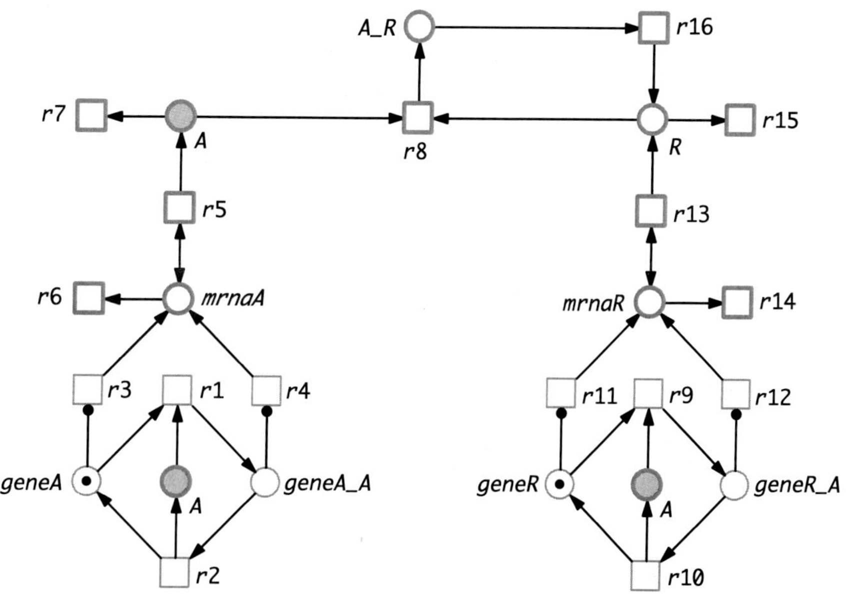

FIGURE 7.22 Reading the running example as $\mathcal{HPN}$. All gene-related places and their pre- and post-transitions are treated as discrete nodes. Please note that read arcs connecting continuous nodes have been replaced by two opposite arcs.

Running Example

For illustration, we now interpret our running example as $\mathcal{HPN}$ model. The 1-bounded places as determined by P-invariant analysis and all related pre- and post-transitions are treated as discrete nodes. The unbounded places and all other transitions are approximated by continuous places and transitions, respectively. In other words, the places *geneA*, *geneA_A*, and *geneR*, *geneR_A*, and the transitions *r*1-*r*4 and *r*9-*r*12 are treated as discrete nodes, and all other as continuous nodes; see Figure 7.22.

7.6.2 Analysis

Basics

Hybrid models have the benefit of encompassing a large class of models, but suffer (so far?) from deficiencies in exact analysis techniques ensuring exhaustive state-space exploration in the general case. However, what usually works is hybrid simulation and evaluating hybrid simulation traces either visually or by simulative model checking.

Hybrid Simulation

The semantics (behavior) of $\mathcal{HPN}$ can be produced via simulation. Snoopy's hybrid simulation engine builds on Gillespie's SSA to simulate the stochastic net parts, and on continuous simulation to integrate the ODEs induced by the continuous net parts. The interplay is summarized in Algorithm 5.

The partitioning of the net elements has an important influence on both the accuracy and the efficiency of $\mathcal{HPN}$ simulation. The more continuous transitions we use, the faster the simulation speed we get. However, this will generally have an impact on the accuracy of the results.

$\mathcal{HPN}$ simulation can be performed using either static or dynamic partitioning. In the former case, transition types are determined by the user before the simulation starts, while in the latter case, the running simulation decides on the fly which transitions are considered stochastic or continuous based on their current rates.

When there is no continuous transition, the $\mathcal{HPN}$ simulation boils down to $\mathcal{SPN}$ simulation. Likewise, when there is no stochastic transition, the $\mathcal{HPN}$ simulation equals $\mathcal{CPN}$ simulation.

ALGORITHM 5 **Basic $\mathcal{HPN}$ simulation algorithm**

Require: $\mathcal{HPN}$ with initial state s_0;
simulation interval $[\tau_0, \tau_{end}]$;
step size h *with* $h < (\tau_{end} - \tau_0)$;

define function f by constructing the ODEs induced by the continuous part of $\mathcal{HPN}$
time $\tau \leftarrow \tau_0$
state $s \leftarrow s_0$
write(τ, s) ▷ add s_0 to trace

while $\tau < \tau_{end}$ **do**
    ensure ODE solver is initialized with s
    determine duration $\Delta\tau$ *until next stochastic event*
    $\tau' \leftarrow \tau + \Delta\tau$
    while $\tau < \tau'$ **do** ▷ continuous simulation
        $\tau \leftarrow \tau + h$
        $s \leftarrow s + h \cdot f(s)$
        write(τ, s) ▷ add to trace
    end while
    ensure $\tau = \tau'$
    determine transition t firing at time τ
    $s \leftarrow$ *fire*(s, t) ▷ compute new state by firing of t
    write(τ, s) ▷ add to trace
end while

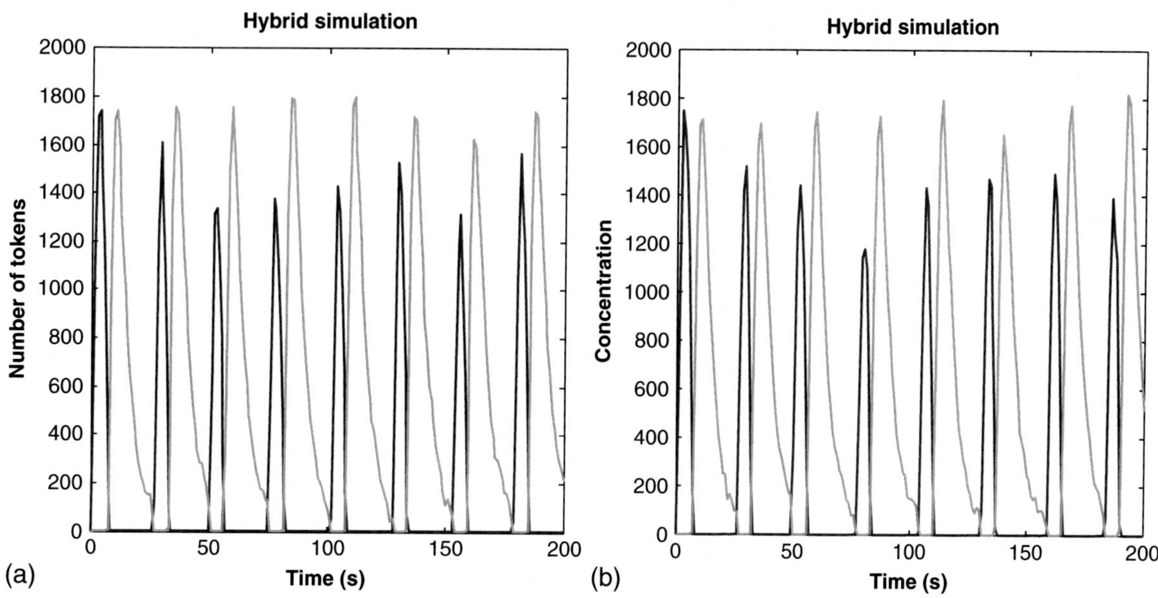

FIGURE 7.23 Two plots of hybrid simulation runs (showing A and R) when reading the running example as $\mathcal{HPN}$.

Simulative Model Checking

With the help of MC2, hybrid simulation traces can equally be subjected to simulative model checking of PLTLc properties, as discussed in the two previous sections.

Running Example

Hybrid simulation of the $\mathcal{HPN}$ in Figure 7.22 yields plots as illustrated in Figure 7.23. These plots suggest that $\mathcal{HPN}$ are able to capture stochasticity; repeated runs look different.

7.6.3 Further Reading

- *A bit of history*: There exists a wide variety of net classes subsumed under the umbrella term *hybrid Petri nets*. Often they just combine discrete and continuous net elements, without involving stochasticity. For example, all $\mathcal{HPN}$ classes covered in the textbook [59] simply enrich $\mathcal{CPN}$ with discrete places and discrete transitions having deterministic firing delay. They exist, as their closely related $\mathcal{CPN}$ classes, as autonomous (time-free) or nonautonomous (timed) versions with constant or variable rates.

  In contrast, the $\mathcal{HPN}$ class introduced in this chapter and supported by Snoopy combines all features of $\mathcal{SPN}$ (to be precise, $\mathcal{XSPN}$) and $\mathcal{CPN}$, and thus has been previously called $\mathcal{GHPN}$ to underline this distinguished feature. To simplify life, we have chosen to speak of $\mathcal{HPN}$, while actually always meaning $\mathcal{GHPN}$.
- *$\mathcal{HPN}$ in Systems Biology*: A net class very popular in Systems Biology is *Hybrid Functional Petri Nets (HFPN)*, which combines $\mathcal{HPN}$ according to [59] with state-dependent arcs known from $\mathcal{XPN}$. This net class is supported by the licensed tool *Cell Illustrator*, which has been deployed in many case studies; see [72] and the references therein.
- *More background*: Please consult [63, 73] for formal definitions, to learn more about how to simulate $\mathcal{HPN}$, or to learn how the dynamic partitioning works.
- *Tool support*: The simulation engines for Snoopy's quantitative Petri nets are directly included in Snoopy, but do also exist as a stand-alone extension called Snoopy Steering and Simulation Server (*S4* for short). The server permits users to share and interactively steer $\mathcal{SPN}$, $\mathcal{CPN}$, and $\mathcal{HPN}$ models during a running

simulation, and to collaborate by controlling model simulation remotely from different clients. The core features of *S4* are outlined in [74]; for more details consult [75] and the *S4* manual [76].

- *Further examples*: Three $\mathcal{HPN}$ case studies are elaborated in [63]: the intracellular growth of bacteriophage T7, the eukaryotic cell cycle, and the circadian rhythm. More details on the $\mathcal{HPN}$ model of the eukaryotic cell cycle can be found in [77].

7.6.4 Exercises

Exercise 7.24. Changing the Net Class. As before, an $\mathcal{HPN}$ does not have to be built from scratch if a corresponding $\mathcal{SPN}$ or $\mathcal{CPN}$ already exists; compare Exercises 7.11 and 7.17. However, static partitioning additionally requires us to change the types of nodes and/or arcs appropriately. This option is found in *Edit → Convert To*.

Take the $\mathcal{SPN}$ and $\mathcal{CPN}$ of the running example and find for each a suitable order of changes to be made to produce the $\mathcal{HPN}$ shown in Figure 7.22. □

Exercise 7.25. Hybrid Simulation. Familiarize yourself with Snoopy's hybrid simulator. Like the stochastic simulation of $\mathcal{SPN}$ (see Exercise 7.12) and the deterministic simulation of $\mathcal{CPN}$ (see Exercise 7.18), the hybrid simulation of $\mathcal{HPN}$ can be parameterized in several ways. For more details, see also Snoopy's FAQ webpage.

(a) Additionally, one can choose among the following synchronization principles.
- *Static*: Each transition type is kept as it has been determined by the user.
- *Dynamic*: The types of stochastic and continuous transitions are adjusted on the fly by evaluating the current transition rates.
- *Continuous*: The entire net is considered as $\mathcal{CPN}$ and thus simulated continuously. Any stochastic transition is automatically converted to a continuous one.
- *Stochastic*: The entire net is considered as $\mathcal{SPN}$ and thus simulated stochastically. Any continuous transition is automatically converted to a stochastic one.

Hint. Immediate transitions are not affected. Explore the differences between these synchronization principles by playing with $\mathcal{HPN}_1$ and $\mathcal{HPN}_2$. For this purpose, introduce more constants to increase the flexibility of the models, as demonstrated in Figure 7.18.

(b) The $\mathcal{HPN}$ of the running example uses read arcs for stochastic nodes, and two opposite arcs for continuous nodes; see Figure 7.22. Compare its behavior with the two extreme cases:
- All side conditions are modeled by read arcs. Play also with the weights of the read arcs, which can now be nonnegative real numbers.
- All side conditions are modeled by two opposite arcs.

(c) Take the $\mathcal{HPN}$ of the running example and change δ_R to 0.08. How does the dynamic behavior change? Can oscillations still be obtained after this perturbation? □

Exercise 7.26. Transition Types. The control cycles in $\mathcal{HPN}_1$ and $\mathcal{HPN}_2$ (see Figure 7.20) deploy stochastic or immediate transitions. There are two further transition types.

- *Deterministic transitions*: The firing occurs after a deterministic firing delay, which is specified by an integer constant. The delay is always relative to the time point where the transition gets enabled. The transition may lose its enabledness while waiting for the delay to expire.
- *Scheduled transitions*: The firing occurs according to a schedule specifying absolute points of the simulation time. A schedule can specify just a single time point, or equidistant time points within a given interval, triggering the firing once or periodically. However, transitions only fire at their scheduled time points if they are enabled at this time.

Create $\mathcal{HPN}_3$ and $\mathcal{HPN}_4$ by replacing the transitions in the control cycle with deterministic or scheduled transitions, and explore their behavior. □

7.7 COLORED PETRI NETS

Basics

Colored Petri nets combine Petri nets with the powerful concept of data types as known from programming languages. Tokens can be distinguished via their colors. This allows for the discrimination of species (e.g., genes, mRNA, proteins, metabolites, second messenger, other molecules), or to distinguish between subpopulations of a species in different locations (e.g., cytosol, nucleus). Colors can also be used to describe similar network structures in a concise way. A group of similar model components (subnets) is represented by one component colored with an appropriate color set, and the individual components become distinguishable by a specific color in this color set.

Colored Petri nets consist, as standard Petri nets, of places, transitions, and arcs. Additionally, a colored Petri net is characterized by a set of discrete data types, called color sets, and related net inscriptions.

- *Places*: get assigned a color set and may contain a multiset of distinguishable tokens colored with a color of this color set. Snoopy supports rich data types for color set definitions, including
  - *Simple types*: dot, integer, string, boolean, enumeration, and index;
  - *Compound types*: product and union.

  As there can be several tokens of the same color on a given place, the tokens on a place define a multiset over the place's color set.
- *Transitions*: get assigned a guard, which is a Boolean expression over variables, constants and so on. The guard must be evaluated to true for the enabling of the transition. The trivial guard *true* is usually not explicitly given.
- *Arcs*: get assigned an expression; the result type of this expression is a multiset over the color set of the connected place.
- *Rate functions*: Quantitative nets may require predicates, which are again Boolean expressions. They permit us to assign different rate functions for different colors. Otherwise, the trivial predicate *true* is used.

Snoopy provides various flexible ways to define states for larger color sets in a concise way. Syntax checking ensures the syntactical correctness of constructed models.

Talking about the behavior of a colored Petri net involves the following notions.

- *Transition instance*: The variables associated with a transition consist of the variables in the guard of the transition and in the expressions of adjacent arcs. For the evaluations of guards and expressions, values of suitable data types have to be assigned to all transition variables, which is called *binding*. A specific binding of the transition variables corresponds to a transition instance. Transition instances may have individual rate functions.
- *Enabling*: Enabling and firing of a transition instance are based on the evaluation of its guard and arc expressions. If the guard is evaluated to true and the preplaces have sufficient and appropriately colored tokens, the transition instance is enabled and may fire. In the case of quantitative nets, the rate function belonging to the transition instance is determined by evaluating the predicates in the rate function definition.
- *Firing*: When a transition instance fires, it removes colored tokens from its preplaces and adds colored tokens to its post-places, that is, it changes the current marking to a new one. The colors of the tokens that are removed from preplaces and added to post-places are decided by the arc expressions.

Applying these coloring mechanisms to the Petri net classes discussed so far yields the colored counterparts, that is, the Petri net classes $\mathcal{PN}^{\mathcal{C}}$, $\mathcal{SPN}^{\mathcal{C}}$, $\mathcal{CPN}^{\mathcal{C}}$, and $\mathcal{HPN}^{\mathcal{C}}$, which inherit all features of their corresponding uncolored Petri net class.

Running Example

Figure 7.24 gives two colored Petri net models for our running example, which have been constructed by the following steps. First, we define a color set *Gene* with two colors, *A* and *R*, to distinguish the two genes, and a variable *x* of this color set.

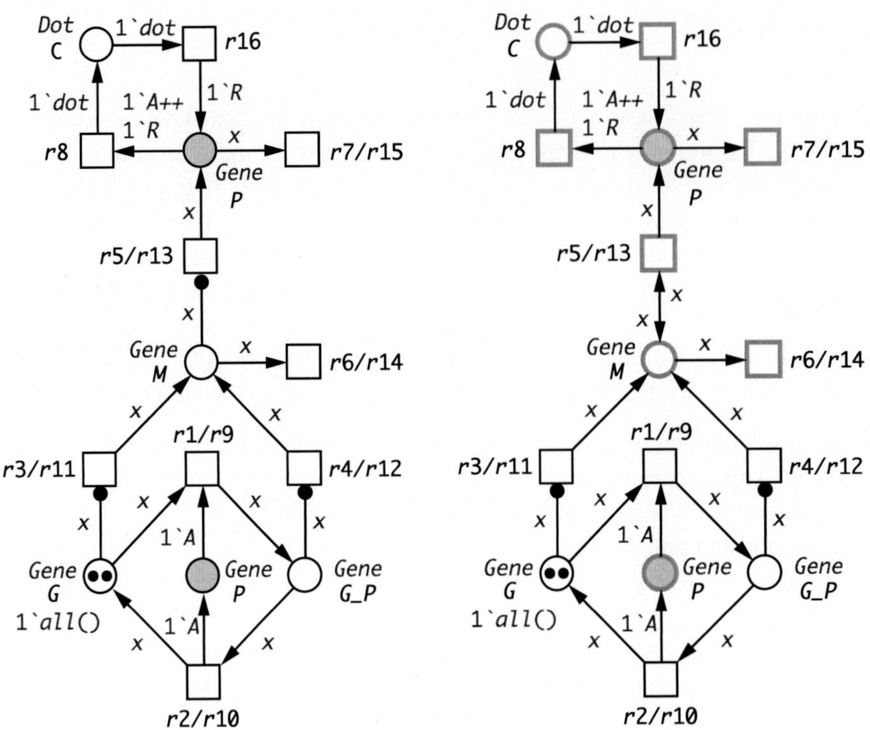

FIGURE 7.24 Colored Petri net models for the running example—$\mathcal{SPN}^{\mathcal{C}}$ (left), $\mathcal{HPN}^{\mathcal{C}}$ (right). These models correspond exactly to the Petri nets in Figures 7.8 and 7.22, respectively. The tokens shown give the total number of tokens of any color. This is replaced by the corresponding multiset upon animation.

```
colorset Gene = enum with A, R;      // enumeration type var x :
var x: Gene;
```

Next, we assign this color set *Gene* to all places. In this way, we use one place to represent two similar objects, that is, to represent two protein objects as one colored place *P*. An exception is *C*, for which we keep the predefined standard color set *Dot*. We specify the initial marking for the place *G* by 1‘*all*(), meaning one token of each color of *Gene*; this corresponds in multiset notation to 1‘*A* ++ 1‘*R*.

Then, we keep for all transitions the trivial guard “true” (not shown).

Afterward, we specify the arc expressions. In the simplest case, an expression consists of a specific color (a constant), such as *dot*, *A*, or *R*, or multiples of them; 1‘*A* just means 1 token of color *A*. If the color of the token does not matter, we use a variable of an appropriate type; in our case we use the variable *x* of the color set *Gene*.

Finally, we take care of the rate functions (not shown). The reactions $r8, r16$ exist in the uncolored model only once, so their colored counterparts have one transition instance each. Thus, we use for them the predicate *true* and the same rate functions as in the uncolored model. All other reactions have two transition instances, one for each possible binding for *x*, that is, *A* or *R*. We use the predicates $x = A$ and $x = R$ to assign different firing rates to these transition instances.

An attractive advantage of colored Petri nets is their potential to easily change the size of a model consisting of many similar subnets by just changing the color set definitions. For example, we could set the color set *Gene* to, let’s say, four colors standing for one activator and three repressors; for details on how to accomplish this, see Exercise 7.28. Then, corresponding stochastic simulation plots can be easily produced and explored, one of which is illustrated in Figure 7.25.

A Petri net can be folded into a colored Petri net if the partitions of the place and transition sets are given. Vice versa, colored Petri nets with finite, discrete color sets can be automatically unfolded into uncolored Petri nets.

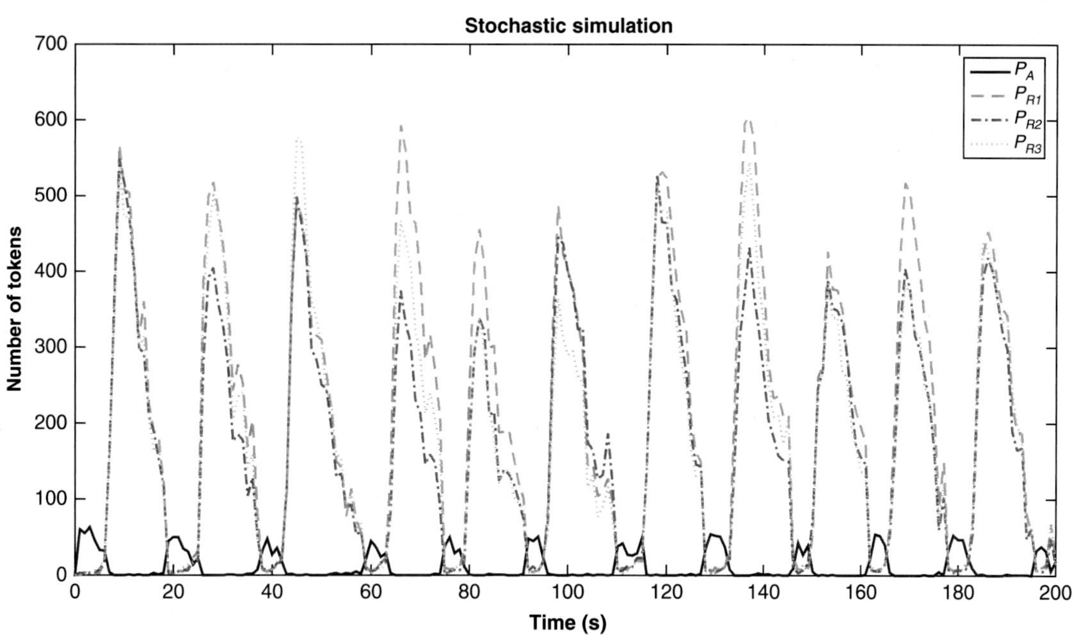

FIGURE 7.25 Plot of a stochastic simulation run for the running example with one activator and three repressors; compare Exercise 7.28.

For example, unfolding the colored Petri nets in Figure 7.24 with the color definitions given above generates the Petri nets in Figures 7.8 and 7.22.

Modeling Procedure

Usually, modeling with colored Petri nets involves the following steps.

1. If you do not start from scratch, convert an uncolored Petri net into a colored Petri net by using the predefined color set *Dot*, which contains only one color called *dot*. This can be easily done with Snoopy's export feature.
2. Identify classes of similar subnets in the uncolored Petri net model. For example, there are two similar subnets in Figure 7.8, corresponding to the two genes.
3. Define suitable color sets to distinguish the identified similar subnets. Use constants, if appropriate.
4. Keep one copy of each class of similar subnets, and delete all others. Assign color sets to places, and define the initial state.
5. Write and assign arc expressions. This may require the definition of appropriate variables.
6. Define guards for transitions.
7. In the case of quantitative nets, define predicates to specify color-dependent rate functions.
8. Check the syntax of inscriptions of the net, that is, arc expressions, guards, and predicates.
9. Check the behavior of your net—first by playing the token game, then by more systematic methods.

If there are any doubts whether you got it right, look at the unfolding, which you get by exporting the colored Petri net to its plain counterpart net class; see Exercise 7.27.

Analysis

The conversion between colored and uncolored net classes is done by means of user-guided folding or automatic unfolding (compare arrows labeled with folding and unfolding in Figure 7.1). Moving between the colored and uncolored levels changes the style of representation, but does not change the actual net structure of the underlying biochemical reaction network. Therefore, all analysis techniques outlined above for each uncolored Petri net class can be equally applied to the colored counterparts.

Snoopy's automatic unfolding of a colored Petri net into its corresponding uncolored Petri net can be triggered either by doing export or starting animation/simulation.

7.7.1 Further Reading

- *A bit of history*: Colored Petri nets ($\mathcal{PN}^\mathcal{C}$), also known as predicate/transition nets, were introduced in [78, 79]. They have been popular since the early 1990s thanks to professional tool support, nowadays distributed as *CPN Tools* [80]. They have been used in various application areas, including communication protocols, distributed systems, automated production systems, work flows, and Very-large-scale integration (VLSI) chips design.
- *Colored Petri nets in Systems Biology*: Genrich et al. [81] were the first to deploy colored Petri net models of metabolic pathways; they encoded the concentration of species as colored tokens in order to implement continuous simulation in the given net annotation language ML. Benefits of colored Petri nets in a stochastic setting were first demonstrated in [82] using a very simple epidemic model.
- *More background*: Please consult the manual [83] for a detailed discussion on how to deal with colored Petri nets in Snoopy, and [84] for formal definitions and an overview of the use of colored Petri nets for Systems Biology.
- *Coloring space*: The use of color to encode Cartesian space is introduced in [85] by means of continuous and stochastic diffusion, which is then applied to construct an $\mathcal{SPN}^\mathcal{C}$ modeling phase variation in bacterial colonies. An extension of this work by a notion of space deploying polar coordinates is given in [86].
- *Reaction diffusion systems*: A tutorial on how to write partial differential equations (PDE) using $\mathcal{CPN}^\mathcal{C}$ can be found in [87], along with a $\mathcal{CPN}^\mathcal{C}$ Brusselator model, which is deployed to explore Turing patterns.
- *Multiscaleness*: In [88], $\mathcal{CPN}^\mathcal{C}$ are applied to model a tissue comprising multiple cells hexagonally packed in a honeycomb formation in order to describe the phenomenon of planar cell polarity (PCP) signaling in Drosophila wing.
- *Unfolding*: The key challenge when unfolding colored Petri nets is the computation of all transition instances, which in fact is a combinatorial problem, suffering from combinatorial explosion. To alleviate the problem, we apply a constraint satisfaction approach. Specifically, we use the efficient search strategies of Gecode [89] to substantially improve the unfolding efficiency; see [84] for details.
- *Further examples*: To learn from case studies, here are some suggestions:
  - $\mathcal{PN}^\mathcal{C}$ are used in [90–92] to discriminate metabolites that follow different T-invariants. These colored models, which have been constructed by hand, could be automatically generated from uncolored nets and their T-invariants.
  - $\mathcal{SPN}^\mathcal{C}$ are used in [93] to describe membrane systems composed of compartments. Each compartment is encoded as a color. Thus, a token changing its compartment changes its color. In [94], coupled Ca^{2+} channels are arranged in two-dimensional space, and each channel is encoded as a color.
  - $\mathcal{SPN}^\mathcal{C}$ and $\mathcal{CPN}^\mathcal{C}$ are applied in [95] to model and analyze *C. elegans* vulval development; each cell is encoded as a color.
  - $\mathcal{HPN}^\mathcal{C}$ are formally introduced in [96], using the repressilator for illustration. Each gene is encoded as a color.

7.7.2 Exercises

Exercise 7.27. Folding and Unfolding. Let's assume we have a network consisting of five components being reversible reactions of the type given in Figure 7.12. We could construct it by copy, paste five times, and appropriate renaming. Instead, we want to use color.

(a) Turn the $\mathcal{SPN}_2$ in Figure 7.12 into a scalable $\mathcal{SPN}^\mathcal{C}$, allowing for a flexible number of reversible reactions.
   1. Start by exporting the $\mathcal{SPN}$ to $\mathcal{SPN}^\mathcal{C}$, which applies a trivial coloring scheme: all places get assigned the color set *Dot*, and all arcs the color value *dot*.
   2. Open the generated $\mathcal{SPN}^\mathcal{C}$ and add color-related definitions as required. Use a constant to specify the number of components (reversible reactions) to keep your model flexible.
   3. Simulate your $\mathcal{SPN}^\mathcal{C}$ for a varying number of components.
   4. Explore different simple data types for this task. Which data type fits your purpose best?

(b) Verify the correctness of your solution by reversing the coloring.
  1. Export your $\mathcal{SPN}^{\mathcal{C}}$ to $\mathcal{SPN}$, which involves unfolding.
  2. Open the generated unfolded $\mathcal{SPN}$, and do *Edit* → *Layout* to automatically generate a more readable net layout.
  3. Do Steps 1 and 2 for different numbers of components. □

Exercise 7.28. Running Example with Multiple Repressors. Take the $\mathcal{HPN}^{\mathcal{C}}$ of the running example in Figure 7.24.

(a) To scale the number of repressors, follow the steps below.
  1. Change the colors of the color set *Gene* to $A, R1, R2, R3$.
  2. Assign the color set *Gene* to the place C.
  3. Change the arc expression of $r8 \rightarrow C$, $C \rightarrow r16$, and $r16 \rightarrow P$ to x.
  4. Change the arc expression of $P \rightarrow r8$ to $1`A ++ \ x$.
  5. Change all predicates of the firing rates for $x = R$ to $x <> A$.
  6. Multiply the constants in the firing rates for the predicate $x = A$ with the number of repressors (e.g., $n = 3$, $MassAction(n * k)$).

(b) Explain the necessity of Steps 5 and 6.

(c) Simulate the scalable $\mathcal{HPN}^{\mathcal{C}}$ constructed in (a) to create traces as shown in Figure 7.25. Try different simulation options.

(d) Explore different simple data types for this task. Which data type fits your purpose best? □

Exercise 7.29. Running Example with Activators/Repressors in a Row or Circle. Instead of having several repressors that act on one activator, let us assume that each protein activates its upstream neighbor and inhibits its downstream neighbor. Two interaction patterns are possible.

(a) *Row*: The first protein has no downstream neighbor, and the last protein has no upstream neighbor.

(b) *Circle*: The last protein is the downstream neighbor of the first protein, and thus the first protein is the upstream neighbor of the last protein.

Start with one of the colored models in Figure 7.24 and modify it to represent the cases (a) and (b). *Hint.* For a variable x of integer data type, $val(x)$ yields the value of the currently bound color. □

7.8 CONCLUSIONS

BioModel Engineering

Modeling and programming have many things in common. Both require abstract reasoning and produce condensed descriptions of behavior that may not always coincide in all aspects with the expected one. Thus, as with programming, modeling should be done with great care. Deviations of expected and observed behavior may be caused by bugs in the software tools used; there is no such thing as error-free software, or in the model itself. So please do not forget to debug your models and to validate their behavior before relying on model-based predictions.

How to Develop a Model—Some Rules

1. Modeling means making abstractions. State explicitly all assumptions underlying your model. Document the literature used to develop the model.
2. If possible, avoid numerical values in the specification of initial states, color sets, and so on. The use of constants makes your model more flexible.
3. Choose appropriate names for constants, variables, color sets, and so on, that which tell something about their meaning. This improves readability of your model.
4. Adopt some naming conventions that allow one to easily distinguish between constants, variables, color sets, functions, and so on. You might want to check popular conventions in coding standards for programming.

Likewise, adopt some drawing conventions, for example, use the same graphical layout for similar components.

5. Try to get your nets nice and tidy. A clear layout usually speaks of a clear mind and helps to avoid silly mistakes.
6. Develop the model step-wise, with slowly increasing model size and complexity of net inscriptions. Validate each modeling step very carefully. Does the model indeed behave as expected?
7. Never blindly trust numerical simulation results.

What We Did

In this chapter, we have outlined a unifying Petri net framework comprising the qualitative, stochastic, continuous, and hybrid paradigms for modeling and analyzing biochemical reaction networks. Each perspective adds specific contributions to the exploration of the network under study. Petri net animation supports intuitive understanding, while the analysis techniques in the various paradigms promote complementary insights and thorough exploration of a given reaction network. The techniques have been presented by means of a specific running example. The mathematical concepts can be equally applied to any kind of network. Thus, Petri nets may serve as umbrella language for different abstraction levels as well as for a pool of interpretation-independent analysis techniques.

Finally, we briefly introduced the orthogonal concept of colored Petri nets, which can be applied to all four modeling paradigms. Colored Petri nets may serve as shorthand notation for uncolored Petri nets; they allow for very concise and flexible Petri net models.

What We Did Not

There are further promising Petri net concepts that have been proven useful for investigating biochemical networks, but that have not been discussed here due to space limitations. Among them are the following.

- *Modular modeling*: The conventional modeling approach, which we also applied in this chapter, yields monolithic models, which are generally hard to curate and maintain. In addition, it is generally impossible to compose these models without manual adjustment.

  In [97], we introduce a modular protein-centered modeling approach, which is illustrated in [98] for the JAK/STAT signaling pathway.
- *Interleaving versus partial-order semantics*: Petri nets represent an inherently concurrent modeling paradigm; they distinguish precisely between alternative and concurrent, that is, independent behavior. However, for analysis purposes, the partial-order (true concurrency) semantics is often reduced to the interleaving semantics, where concurrency of reactions is described by all interleaving sequences.

  Sometimes it is advisable to preserve all concurrency information that a biochemical network model enjoys, for example, to distinguish the subtle difference between read arcs and two opposite arcs. The partial-order semantics is able to give further insights into the dynamic behavior of a network, which may not be apparent from the standard net representation or its interleaving semantics; for examples, see [22, 38].
- *Reaction diffusion systems*: Colored Petri nets permit the convenient encoding of spatial attributes [85, 86], and thus the modeling of processes evolving in time and space, usually considered as stochastic or deterministic reaction-diffusion systems by help of stochastic or deterministic PDE; see [87] for a related tutorial.
- *Multiscaleness*: Further challenges arising from the need to model and analyze complex biological systems at multiple spatial and temporal scales, such as hierarchical organization, mobility, motility, differentiation, or pattern formation of components, are discussed in [99].

Further Reading

Other case studies discussed in various modeling paradigms are the engineering of a novel self-powering electrochemical biosensor [36, 100], the MAPK signaling cascade [38], the influence of the RKIP on the ERK signaling pathway [22], and two versions of a gene regulatory cycle called repressilator [39, 40].

The state of the art in 2011 of how Petri nets might enhance a Systems Biology toolkit is summarized in [67].

Outlook

At the time of writing this chapter, we were about to incorporate a new feature into Snoopy—the definition of functions, usable as rate functions or everywhere where constants are allowed. The use of functions will make models even more flexible and better maintainable than already possible by help of constants. This feature is likely to be supported by Snoopy when this book goes to print.

If you have any comments or suggestions on how to improve our framework and related toolkit, please do not hesitate to contact one of the authors of this chapter.

ACKNOWLEDGMENTS

We would like to thank Mostafa Herajy, Fei Liu, Christian Rohr, and Martin Schwarick for their contributions in developing and supporting the use of Snoopy, Marcie, and Charlie; and Robin Donaldson and David Gilbert for developing and supporting the use of MC2. We appreciate countless productive discussions with all of them.

This work has been partly supported by the Germany Federal Ministry of Education and Research (FKZ0316177E).

7.9 SUPPLEMENTARY MATERIALS

All Petri nets used in this chapter are available at http://www-dssz.informatik.tu-cottbus.de/examples. They can also be found as Supplementary Material on the volume's website.

REFERENCES

[1] Gilbert D, Breitling R, Heiner M, Donaldson R. An introduction to biomodel engineering, illustrated for signal transduction pathways. In: WMC 2008; 5391 of LNCS. Berlin: Springer; 2009. p. 13-28.
[2] Breitling R, Donaldson RA, Gilbert DR, Heiner M. Biomodel engineering—from structure to behavior (position paper). In: Transactions on computational systems biology XII: special issue on modeling methodologies, vol. 5945. Berlin: Springer; 2010. p. 1-12.
[3] Petri CA. Kommunikation mit Automaten (in German) [PhD thesis]. Technische Hochschule Darmstadt; 1962.
[4] Rohr C, Marwan W, Heiner M. Snoopy—a unifying Petri net framework to investigate biomolecular networks. Bioinformatics 2010;26:974-5.
[5] Heiner M, Herajy M, Liu F, Rohr C, Schwarick M. Snoopy—a unifying Petri net tool. In: Proc. Petri nets 2012; 7347 of LNCS. Berlin: Springer; 2012. p. 398-407.
[6] Sandve GK, Nekrutenko A, Taylor J, Hovig E. Ten simple rules for reproducible computational research. PLoS Comput Biol 2013;9:e1003285.
[7] Franzke A. Charlie 2.0—a multithreaded Petri net analyser [Diploma thesis]. BTU Cottbus, Dep. of CS; 2009.
[8] Schwarick M, Heiner M, Rohr C. MARCIE—Model checking And Reachability analysis done effiCIEntly. In: Proc. QEST, Aachen, Germany. IEEE CS Press; 2011. p. 91-100.
[9] Heiner M, Rohr C, Schwarick M. MARCIE—Model checking And Reachability analysis done effiCIEntly. In: Proc. PETRI NETS 2013; 7927 of LNCS. Berlin: Springer; 2013. p. 389-99.
[10] Schwarick M, Rohr C, Heiner M. MARCIE manual—an analysis tool for extended stochastic Petri nets. Tech. Rep. 03-14. BTU Cottbus-Senftenberg, Dep. of CS; 2014.
[11] Donaldson R, Gilbert D. A model checking approach to the parameter estimation of biochemical pathways. In: Proc. CMSB 2008; 5307 of LNCS/LNBI. Berlin: Springer; 2008. p. 269-87.
[12] Edmunds LN. Cellular and molecular bases of biological clocks: models and mechanisms for circadian timekeeping. New York: Springer-Verlag; 1988.
[13] Dunlap JC. Molecular bases for circadian clocks. Cell 1999;96:271-90.

[14] Vilar JMG, Kueh HY, Barkai N, Leibler S. Mechanisms of noise resistance in genetic oscillators. PNAS 2002;99:5988-92.
[15] Murata T. Petri nets: properties, analysis and applications. Proc IEEE 1989;77:541-80.
[16] Silva M. Half a century after Carl Adam Petri's PhD thesis: a perspective on the field. Ann Rev Control 2013;37:191-219.
[17] Petri CA, Reisig W. Petri net. Scholarpedia 2008;3(4):64-77.
[18] Reddy VN, Mavrovouniotis ML, Liebman MN. Petri net representations in metabolic pathways. In: Proc. ISMB-93; 1993. p. 328-36.
[19] Hofestädt R. A Petri net application of metabolic processes. J Syst Anal Model Simul 1994;16:113-22.
[20] Reddy VN. Modeling biological pathways: a discrete event systems approach [Master thesis]. University of Maryland; 1994.
[21] Baldan P, Cocco N, Marin A, Simeoni M. Petri nets for modelling metabolic pathways: a survey. Nat Comput 2010;9:955-89.
[22] Heiner M, Donaldson R, Gilbert D. Petri nets for systems biology. Jones and Bartlett Learning Sudbury, Massachusetts; 2010. Chapter 3. p. 61-97.
[23] Blätke MA, Heiner M, Marwan W. Tutorial—Petri nets in systems biology tech. rep. Otto von Guericke University and Magdeburg, Centre for Systems Biology; 2011.
[24] Heiner M, Mahulea C, Silva M. On the importance of the deadlock trap property for monotonic liveness. In: Int. workshop on biological processes and Petri nets (BioPPN), satellite event of Petri nets 2010; 2010. p. 39-54.
[25] Lautenbach K. Exact liveness conditions of a Petri net class (in German) [PhD thesis]. GMD Report 82, Bonn; 1973.
[26] Heiner M. Understanding network behaviour by structured representations of transition invariants—a Petri net perspective on systems and synthetic biology. Natural computing series. Berlin: Springer; 2009. p. 367-89.
[27] Schuster S, Hilgetag C, Schuster R. Determining elementary modes of functioning in biochemical reaction networks at steady state. In: Proc. 2nd gauss symposium; 1993. p. 101-14.
[28] Palsson BO. Systems biology. Cambridge: Cambridge University Press; 2006.
[29] Larhlimi A, Bockmayr A. On inner and outer descriptions of the steady-state flux cone of a metabolic network. In: Proc. CMSB 2008; 5307 of LNCS/LNBI. Berlin: Springer. 2008. p. 308-27.
[30] Hodge TL. Metabolic pathways analysis: a linear algebraic approach. In: Robeva R, Hodge TL, editors. Mathematical concepts and methods in modern biology: using modern discrete models. Ch. 8. London: Academic Press; 2013. p. 239-66.
[31] Heiner M, Koch I. Petri net based model validation in systems biology. In: Proc. ICATPN 2004; 3099 of LNCS. Berlin: Springer; 2004. p. 216-37.
[32] Heiner M, Sriram K. Structural analysis to determine the core of hypoxia response network. PLoS ONE 2010;5:e8600.
[33] Clarke EM, Grumberg O, Peled DA. Model checking. Cambridge, MA: MIT Press; 2000.
[34] Pnueli A. The temporal semantics of concurrent programs. Theor Comput Sci 1981;13:45-60.
[35] Monteiro PT, Ropers D, Mateescu R, Freitas AT, De Jong H. Temporal logic patterns for querying dynamic models of cellular interaction networks. Bioinformatics 2008;24:i227-33.
[36] Gilbert D, Heiner M, Rosser S, Fulton R, Gu X, Trybiło M. A case study in model-driven synthetic biology. In: IFIP WCC 2008, 2nd IFIP conference on biologically inspired collaborative computing (BICC 2008); 268 of IFIP (Boston). Berlin: Springer; 2008. p. 163-75.
[37] Koch I, Heiner M. Petri nets. In: Junker BH, Schreiber F, editors. Biological network analysis. Ch. 7. Wiley book series on bioinformatics; 2008. p. 139-79.
[38] Heiner M, Gilbert D, Donaldson R. Petri nets for systems and synthetic biology; 5016 of LNCS. Berlin: Springer; 2008. p. 215-64.
[39] Liu F, Heiner M. Petri nets for modeling and analyzing biochemical reaction networks. Ch. 9. Berlin: Springer; 2014.
[40] Blätke MA, Rohr C, Heiner M, Marwan W. A Petri net based framework for biomodel engineering. In: Modeling and simulation in science, engineering and technology. Birkhäuser mathematics. Berlin: Springer; 2014.
[41] Marwan W, Wagler A, Weismantel R. Petri nets as a framework for the reconstruction and analysis of signal transduction pathways and regulatory networks. Nat Comput 2011;10:639–54.
[42] Rao CV, Wolf DM, Arkin AP. Control, exploitation and tolerance of intracellular noise. Nature 2002;420:231–7.
[43] Molloy MK. Performance analysis using stochastic Petri nets. IEEE Trans Comput 1982;100:913-7.
[44] Ajmone Marsan M, Balbo G, Conte G, Donatelli S, Franceschinis G. Modelling with generalized stochastic Petri nets. Wiley series in parallel computing. 2nd ed. London: John Wiley and Sons; 1995.
[45] Bause F, Kritzinger PS. Stochastic Petri nets. Wiesbaden: Vieweg; 2002.
[46] Goss PJE, Peccoud J. Quantitative modeling of stochastic systems in molecular biology by using stochastic Petri nets. Proc Natl Acad Sci 1998;95:6750-5.
[47] Srivastava R, Peterson MS, Bentley WE. Stochastic kinetic analysis of the *Escherichia coli* stress circuit using σ32-targeted antisense. Biotechnol Bioeng 2001;75:120-9.
[48] Marwan W, Sujatha A, Starostzik C. Reconstructing the regulatory network controlling commitment and sporulation in *Physarum polycephalum* based on hierarchical Petri net modeling and simulation. J Theor Biol 2005;236:349-65.

[49] Wilkinson DJ. Stochastic modelling for system biology. 1st ed. New York: CRC Press; 2006.
[50] Heiner M, Lehrack S, Gilbert D, Marwan W. Extended stochastic Petri nets for model-based design of wetlab experiments. In: Transactions on computational systems biology XI; 5750 of LNCS/LNBI. Berlin: Springer; 2009. p. 138-63.
[51] Gillespie DT. Exact stochastic simulation of coupled chemical reactions. J Phys Chem 1977;81:2340-61.
[52] Mauch S, Stalzer M. Efficient formulations for exact stochastic simulation of chemical systems. IEEE/ACM Trans Comput Biol Bioinform 2011;8:27-35.
[53] Stewart WJ. Introduction to the numerical solution of Markov chains. Princeton, NJ: Princeton University Press; 1994.
[54] Schwarick M, Tovchigrechko A. IDD-based model validation of biochemical networks. Theor Comput Sci 2011;412:2884-908.
[55] Heiner M, Rohr C, Schwarick M, Streif S. A comparative study of stochastic analysis techniques. In: Proc. CMSB 2010. ACM Digital Library; 2010. p. 96-106.
[56] Schwarick M. Symbolic on-the-fly analysis of stochastic Petri nets [PhD thesis]. BTU Cottbus, Dep. of CS; 2014.
[57] Rohr C. Simulative model checking of steady-state and time-unbounded temporal operators. In: ToPNoC VIII, LNCS 8100; 2013. p. 142-58.
[58] Marwan W, Rohr C, Heiner M. Petri nets in Snoopy: a unifying framework for the graphical display, computational modelling, and simulation of bacterial regulatory networks; Volume 804 of methods in molecular biology. Ch. 21. New York: Springer; 2012. p. 409-37.
[59] David R, Alla H. Autonomous and timed continuous Petri nets. In: Advances in Petri nets 1993; 674 of LNCS. Berlin: Springer; 1993. p. 71-90.
[60] David R, Alla H. Discrete, continuous, and hybrid Petri nets. Berlin: Springer-Verlag; 2005.
[61] Gilbert D, Heiner M. From Petri nets to differential equations—an integrative approach for biochemical network analysis. In: Proc. ICATPN 2006; 4024 of LNCS. Berlin: Springer; 2006. p. 181-200.
[62] Breitling R, Gilbert D, Heiner M, Orton R. A structured approach for the engineering of biochemical network models, illustrated for signalling pathways. Brief Bioinform 2008;9:404-21.
[63] Herajy M. Computational steering of multi-scale biochemical networks [PhD thesis]. BTU Cottbus, Dep. of CS; 2013.
[64] Hindmarsh AC, Brown PN, Grant KE, Lee SL, Serban R, Shumaker DE, et al. SUNDIALS: suite of nonlinear and differential/algebraic equation solvers. ACM Trans Math Software 2005;31:363-96.
[65] Calzone L, Chabrier-Rivier N, Fages F, Soliman S. Machine learning biochemical networks from temporal logic properties. Trans Comput Syst Biol VI 2006;4220:68-94.
[66] Soliman S, Heiner M. A unique transformation from ordinary differential equations to reaction networks. PLoS ONE 2010;5:e14284.
[67] Heiner M, Gilbert D. How might Petri nets enhance your systems biology toolkit; 6709 of LNCS. Berlin: Springer; 2011. p. 17-37.
[68] Silva M, Recalde L. Petri nets and integrality relaxations: a view of continuous Petri net models. IEEE Trans Syst Man Cybern Part C Appl Rev 2002;32:314-27.
[69] Angeli D, De Leenheer P, Sontag ED. A Petri net approach to persistence analysis in chemical reaction networks. In: Biology and control theory: current challenges; 357 of LNCI. Berlin: Springer; 2007. p. 181-216.
[70] Angeli D. Boundedness analysis for open chemical reaction networks with mass-action kinetics. Nat Comput 2011;10:751-74.
[71] Shinar G, Feinberg M. Structural sources of robustness in biochemical reaction networks. Science 2010;327:1389-91.
[72] Nagasaki M, Saito A, Doi A, Matsuno H, Miyano S. Foundations of systems biology using cell illustrator and pathway databases, vol. 13. Series: computational biology. London: Springer; 2009.
[73] Herajy M, Heiner M. Hybrid representation and simulation of stiff biochemical networks. Nonlinear Anal Hybrid Syst 2012;6:942-59.
[74] Herajy M, Heiner M. A steering server for collaborative simulation of quantitative Petri nets. In: PETRI NETS 2014; 8489 of LNCS. Berlin: Springer; 2014. p. 374-84.
[75] Herajy M, Heiner M. Petri net-based collaborative simulation and steering of biochemical reaction networks. Fund Inform 2014;129:49-67.
[76] Herajy M, Heiner M. Snoopy computational steering framework User manual version 1.0. Tech. Rep. 02-13 BTU Cottbus, Dep. of CS; 2013.
[77] Herajy M, Schwarick M, Heiner M. Hybrid Petri nets for modelling the eukaryotic cell cycle. ToPNoC VIII, LNCS 8100; 2013. p. 123-41.
[78] Genrich HJ, Lautenbach K. The analysis of distributed systems by means of predicate/transition nets. In: Proc. international symposium on semantics of concurrent computation; 70 of LNCS. Berlin: Springer; 1979. p. 123-46.
[79] Genrich HJ, Lautenbach K. System modelling with high-level Petri nets. Theor Comput Sci 1981;13:109-35.
[80] Jensen K, Kristensen LM. Coloured Petri nets: modelling and validation of concurrent systems. Berlin Heidelberg: Springer; 2009.

[81] Genrich H, Küffner R, Voss K. Executable Petri net models for the analysis of metabolic pathways. Int J Software Tools Technol Tran 2001;3(4):394-404.

[82] Bahi-Jaber N, Pontier D. Modeling transmission of directly transmitted infectious diseases using colored stochastic Petri nets. Math Biosci 2003;185:1-13.

[83] Liu F, Heiner M, Rohr C. The manual for colored Petri nets in snoopy—$QPN^C/SPN^C/CPN^C/GHPN^C$. Tech. Rep. 02-12 BTU Cottbus, Dep. of CS; 2012.

[84] Liu F. Colored Petri nets for systems biology [PhD thesis]. BTU Cottbus, Dep. of CS; 2012.

[85] Gilbert D, Heiner M, Liu F, Saunders N. Colouring space—a coloured framework for spatial modelling in systems biology. In: PETRI NETS 2013; 7927 of LNCS. Berlin: Springer; 2013. p. 230-49.

[86] Parvu O, Gilbert D, Heiner M, Liu F, Saunders N. Modelling and analysis of phase variation in bacterial colony growth. In: Proc. CMSB 2013; 8130 of LNCS/LNBI. Berlin: Springer; 2013. p. 78-91.

[87] Liu F, Blätke MA, Heiner M, Yang M. Modelling and simulating reaction-diffusion systems using colored Petri nets. Computers in Biology and Medicine, 53:297–308, October 2014.

[88] Gao Q, Gilbert D, Heiner M, Liu F, Maccagnola D, Tree D. Multiscale modelling and analysis of planar cell polarity in the *Drosophila* wing. IEEE ACM Trans Comput Biol Bioinform 2013;10:337-51.

[89] Gecode. Gecode: an open constraint solving library; 2013. Available from: http://www.gecode.org.

[90] Heiner M, Koch I, Voss K. Analysis and simulation of steady states in metabolic pathways with Petri nets. In: Proc. CPN workshop. University of Aarhus; 2001. p. 15-34.

[91] Voss K, Heiner M, Koch I. Steady state analysis of metabolic pathways using Petri nets. In Silico Biol 2003;3:367-87.

[92] Runge T. Application of coloured Petri nets in systems biology. In: Proc. CPN workshop. University of Aarhus; 2004. p. 77-95.

[93] Liu F, Heiner M. Modeling membrane systems using colored stochastic Petri nets. Nat Computing 2013;12:617-29.

[94] Liu F, Heiner M. Multiscale modelling of coupled Ca^{2+} channels using coloured stochastic Petri nets. IET Syst Biol 2013;7:106-13.

[95] Liu F, Heiner M, Yang M. Modeling and analyzing biological systems using colored hierarchical Petri nets, illustrated by *C. elegans* vulval development. WSPC J Biol Syst 2014;22:463-93.

[96] Herajy M, Liu F, Rohr C. Coloured hybrid Petri nets for systems biology. In: Proc. BioPPN 2014; 1159 of CEUR workshop proceedings. CEUR-WS.org; 2014. p. 60-76.

[97] Blätke MA, Heiner M, Marwan W. Predicting phenotype from genotype through automatically composed Petri nets. In: Proc. CMSB 2012; 7605 of LNCS/LNBI. Berlin: Springer; 2012. p. 87-106.

[98] Blätke MA, Dittrich A, Rohr C, Heiner M, Schaper F, Marwan W. JAK/STAT signalling—an executable model assembled from molecule-centred modules demonstrating a module-oriented database concept for systems and synthetic biology. Mol Biosyst 2013;9:1290-307.

[99] Heiner M, Gilbert D. Biomodel engineering for multiscale systems biology progress. Biophys Mol Biol 2013;111:119-28.

[100] Gu X, Trybilo M, Ramsay S, Jensen M, Fulton R, Rosser S, et al. Engineering a novel self-powering electrochemical biosensor. Syst Synth Biol 2010;4(3):203-14.

Chapter 8

Transmission of Infectious Diseases: Data, Models, and Simulations

Winfried Just[1], Hannah Callender[2], M. Drew LaMar[3] and Natalia Toporikova[4]
[1]Department of Mathematics, Ohio University, Athens, OH, USA, [2]University of Portland, Portland, OR, USA,
[3]The College of William and Mary, Williamsburg, VA, USA, [4]Washington and Lee University, Lexington, VA, USA

8.1 INTRODUCTION: WHY DO WE WANT TO MODEL INFECTIOUS DISEASES?

From the Black Death of the Middle Ages, through the 1918 "Spanish Flu" pandemic, to the current HIV/AIDS pandemic, infectious diseases have been a constant threat to humanity. They are the second leading cause of death worldwide and the third leading cause of death in the United States [1–3]. In low-income countries, lower respiratory infections, HIV/AIDS, diarrheal diseases, malaria, and tuberculosis collectively account for almost one-third of all deaths, and nearly 4 in every 10 deaths are among children under 15 years old [4].

In addition to directly affecting human health, infectious diseases can cause severe economic damage by affecting crops or livestock [5, 6]. The spread through Europe in 1845-1852 of potato late blight, caused by the fungus-like microorganism *Phytophthora infestans*, decimated potato harvests, caused the Irish potato famine, and forced more than 4 million people to emigrate. More recently, an outbreak of bovine spongiform encephalopathy, a disorder affecting the central nervous system in cattle, caused devastating economic losses and widespread panic in the late 1980s and early 1990s [7, 8].

Why do we want to model infectious diseases? A *model* in our sense is a greatly simplified representation of a real system as a mathematical construct or a computer program. Instead of passively observing the outbreak of a disease with all the associated suffering and loss, we can study with pencil and paper or with the help of a computer what happens in the model, and sometimes predict what is likely to happen in the real world. This information can help us prepare for the case of a possible outbreak, for example, by allocating adequate resources to the medical system. More importantly, the model can inform us about the most effective *control measures* for alleviating the severity of an outbreak or, even better, preventing it in the first place if possible.

Let us take a closer look at some examples of infectious diseases and control measures. Cocoa swollen-shoot virus (CSSV) is a plant pathogenic virus that primarily infects cocoa trees. More than 200 million cocoa trees have already been claimed by this disease in Togo, Ghana, the Ivory Coast, and Nigeria. Infected trees usually die within 3-4 years [9]. There is no known method for saving an infected tree. Farmers are being advised to cut down infected trees and trees in their immediate neighborhood. This type of control measure is called *culling.* It can be used to control some plant and animal diseases, but it involves an immediate economic loss. Farmers would want to know how likely it is that this immediate cost will save the plantation and prevent greater losses in the future.

As a second example, consider the flu. The Centers for Disease Control and Prevention (CDC) estimate that the number of deaths in the United States that are related to the *seasonal flu* ranges from 3,000 to 49,000 in a given year [10]. The simplest and least expensive control measures involve *behavior modification*: people can avoid shaking hands and use hand sanitizer more often during flu season. This will not confer absolute protection, but

Algebraic and Discrete Mathematical Methods for Modern Biology. http://dx.doi.org/10.1016/B978-0-12-801213-0.00008-3
Copyright © 2015 Elsevier Inc. All rights reserved.

will diminish the likelihood of infection. Timely *vaccination* may give near complete protection to those who are vaccinated. But each year, a slightly different strain of the seasonal flu evolves. It takes time and considerable financial resources to develop and produce vaccines. Therefore, not enough vaccine may be available so that everybody who would like to receive a flu shot can be vaccinated. In such situations, health care authorities may need to develop vaccination strategies that would give the greatest expected reduction in the number of cases and deaths given the available amount of vaccine.

In addition to the seasonal flu, sometimes a particularly nasty strain of influenza arises and spreads rapidly, such as the 2009 H1N1 "swine flu" pandemic. Vaccines are not available immediately after the first cases are reported. In such cases, health authorities may consider control measures such as imposing *quarantine* on actual or suspected carriers of the disease or tightening border controls so as to prevent individuals who show symptoms or are known to have had contact with infected persons from entering the country. During the 2009 H1N1 pandemic, several countries in fact implemented such measures [11, 12]. Both quarantine and tighter border control are costly to implement and impose a large burden on the affected individuals. It would be important to understand how effective these measures would be in terms of preventing or reducing the spread of a disease.

All three examples above led to questions of great importance for people's health, well-being, and livelihood. These questions exemplify the subject matter of the science of *epidemiology.* A branch of this science, *mathematical epidemiology*, tries to find answers by studying mathematical models of disease transmission.

Before we can tell you more about epidemiology, we need to introduce some terminology. First we need a name for the kind of organism that may catch the disease, which could be a human, an animal, or a plant. The established term in the literature is *host*. Infectious diseases of hosts are caused by *pathogens*, that is, other organisms, such as viruses, bacteria, protozoans, or fungi that grow and reproduce inside their hosts (at the expense of the host's resources) and can spread from host to host.

Epidemiology studies how pathogens spread in *populations* of hosts, where the term "population" refers to the group of hosts for which we want to study or predict the spread of a given disease. In our first example, the population might comprise all cocoa trees of a given plantation; in the second example, it might comprise all humans; and when considering whether or not to impose border controls, as in our third example, the population would comprise all humans in a given country. The choice of the population depends both on the particular disease and the particular questions in which we are interested.

The letter N will always denote the total number of hosts in the population. When building and analyzing models, we will enumerate the whole population with integers from the set $\{1, \ldots, N\}$. We will simply write "host i" instead of "host number i," where i ranges from 1 to N.

The questions studied by epidemiology can often be phrased in terms of an outbreak of a given disease. An outbreak may start by infection of at least one host from *outside the population*. A cocoa tree may become infected by transmission of the virus from a nearby plantation, and new strains of the flu often evolve in domestic animals before they infect the first host in a human population. Epidemiology then studies how the disease subsequently spreads *within the given population* during an outbreak. While the term "outbreak" is also used in the literature in a broader sense, throughout this chapter and the follow-up Chapter 9, we will restrict our attention to outbreaks that are started by hosts who got infected from outside the population. For the most part, we will focus on outbreaks with exactly one initially infected host. This host will be called *the index case*.

Mathematical models can then be used for predicting the future course of the observed outbreak or the pattern of spread in another population that might be affected at a future time. However, models can make realistic predictions only if they are based on reliable data. Much of epidemiology is concerned with collecting and analyzing data about observed outbreaks. There are many kinds of data that epidemiologists rely on and many methods for obtaining them. While we cannot review all or most of them here, let us discuss two examples.

A very important quantity in epidemiology is the *rate* nI at which infections occur in the population. You can think of nI standing for "new infections per time unit." nI is also a good mnemonic for the term *incidence* that professional epidemiologists use for this rate. A related but different number is the *prevalence* $|\mathbf{I}|$. It stands for the total number of infectious hosts at a given time. Both the incidence and the prevalence change over time during an outbreak; we will write $nI(t)$ and $|\mathbf{I}(t)|$ to indicate this dependence on time.

For human diseases, the most obvious data sources are reports on the number of confirmed or suspected cases of new infections. Such reports are filed by health care providers to government agencies (for example, the CDC in the United States).

Exercise 8.1. It might appear that the number of reported cases of new infections that are filed for a given day or week would give a good estimate of the incidence during that day or week. However, this will not always be the case. Why not? □

The graph of the function $nI(t)$, or sometimes the function $nI(t)$ itself, is called the *epidemic curve* of a given outbreak in a given population. Sophisticated mathematical and statistical techniques may be needed to infer reliable estimates of $nI(t)$ from reports on the number of observed cases. The top panel of Figure 8.1 shows the numbers of confirmed cases that were reported on a day-to-day basis for the 2009 H1N1 "swine flu" outbreak in Hong Kong, while the bottom panel shows the reconstruction of the epidemic curve. The top panel has a lot of sharp little peaks. It seems extremely unlikely that these reflect actual properties of nI. According to [13],

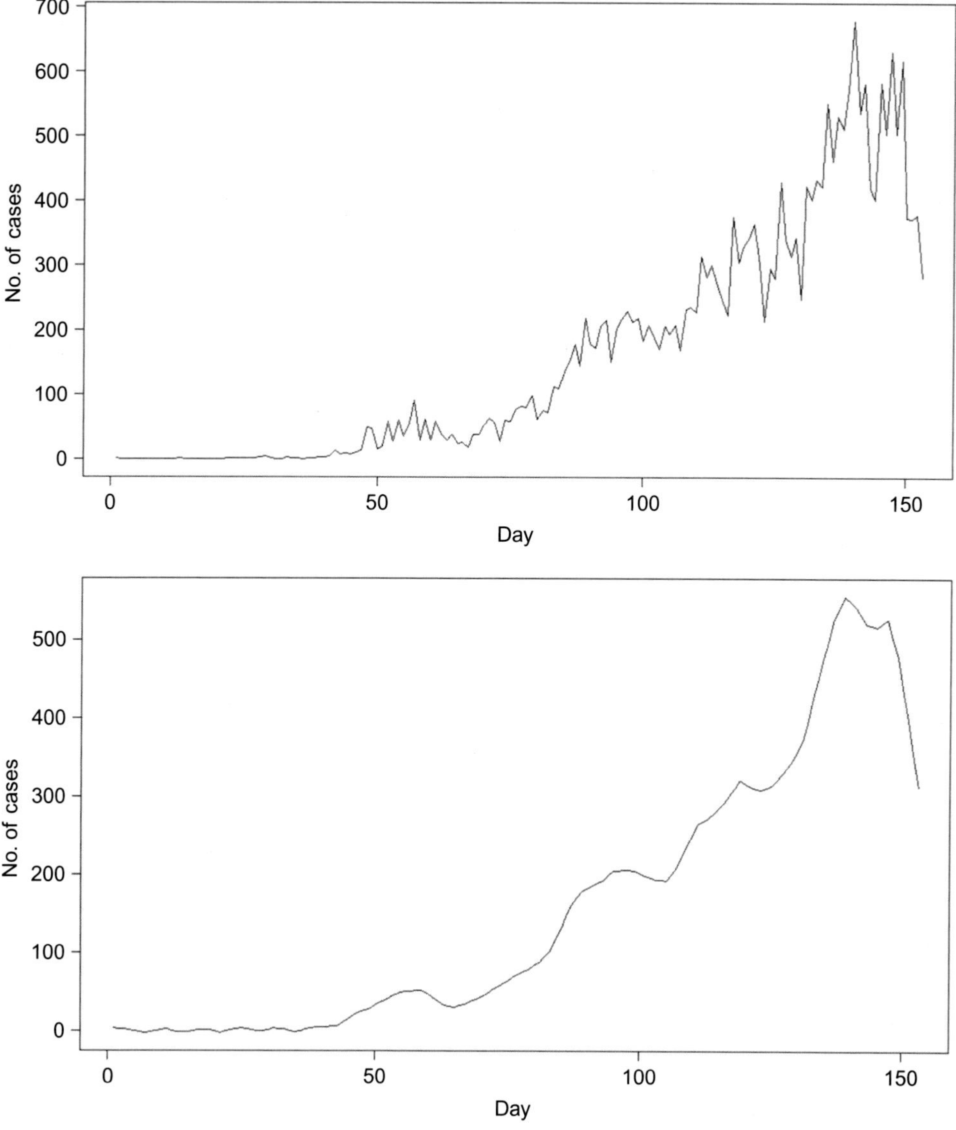

FIGURE 8.1 Top panel: Number of confirmed new cases of H1N1 infections during the 2009 outbreak in Hong Kong. Bottom panel: Epidemic curve for this outbreak deduced from the data shown in the top panel. Reproduced from [13].

they may have been caused by such factors as reporting delay, discrepant access to diagnostics, and varied public perception and its influence on health-seeking behaviors. The bottom panel of Figure 8.1 shows a likely epidemic curve that was reconstructed from the raw data by using mathematical smoothing techniques. It certainly looks like a much more plausible picture of an epidemic curve than the top panel does.

Let us take a closer look at the hypothetical epidemic curve in the bottom panel of Figure 8.1. By and large, it first increases slowly, then more rapidly, then reaches a peak on day 135, and subsequently decreases. This general shape of the epidemic curve is what mathematical models predict for certain types of infectious diseases. However, the curve that was constructed for the real outbreak does not fit this description exactly. It shows a number of additional smaller peaks and valleys, that is, local minima and local maxima. These may simply be due to random fluctuations that we always should expect to see in real data sets. But they may also signify more interesting effects: note the small peak around days 55-66. This is quite pronounced and suggests that the infection spread in two distinct waves. There are various possible explanations for such a pattern [11]. In this particular outbreak, it may be attributable to school closure in response to the outbreak [14]. If this interpretation is correct, then the epidemic curve directly shows the effect of school closure (a control measure that causes behavior modification) on the time course of the outbreak.

Of particular interest to epidemiologists is the total number of hosts who will experience infection during the outbreak. In order to compare the severity of outbreaks in populations of different sizes, it is more informative to express this quantity in terms of the proportion of hosts in the population who will eventually experience infection rather than in terms of absolute numbers. This proportion is a number between 0 and 1 and is called the *final size* of the outbreak. If the final size is close to 0, then only a very small fraction of the population will become infected, and we will have a *minor outbreak*. On the other hand, if the final size is, for example, 0.3, then 30% of the population will experience infection and we will have a *major outbreak* or *epidemic*.[1] An epidemic on the scale of the entire human population is often called a *pandemic*. The final size is the number that we want to reduce by control measures. In particular, we would want to design and implement control measures that confine outbreaks to minor ones.

If each host can be infected only once, as will be the case for influenza and other infections that confer permanent immunity to the given strain upon recovery, the final size corresponds to the area under the epidemic curve. Because nI is the rate at which hosts become infected, this follows from the Fundamental Theorem of Calculus.

The notion of *effectiveness* of a control measure can be understood in terms of how it would alter the epidemic curve. Figure 8.2 shows three epidemic curves[2] for simulated outbreaks in the same small population. As we

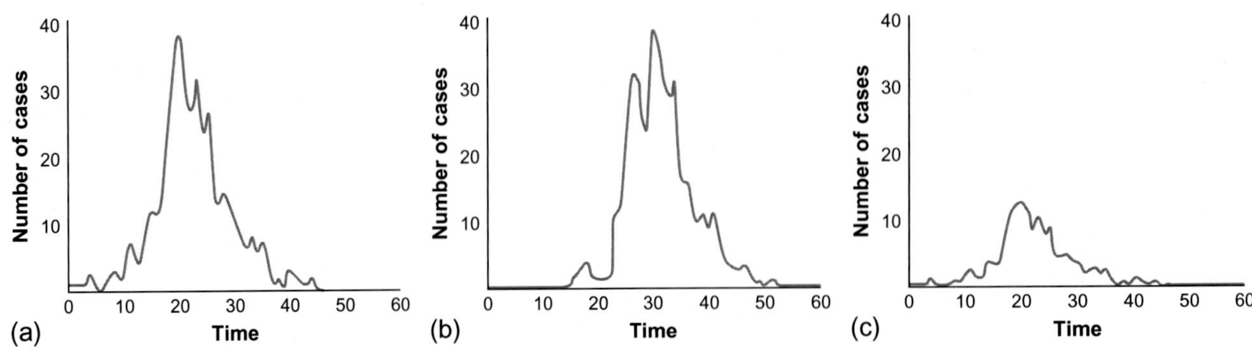

FIGURE 8.2 New infections per day in three simulated outbreaks. Panel a: No control measures in place. Panel b: Control measure B implemented. Panel c: Control measure C implemented.

1. There is no universally accepted definition of the term *epidemic*. In the literature, different authors define it in subtly different ways. To avoid possible confusion, we prefer using the phrase "major outbreak."

2. More precisely: simulated numbers of daily new infections as in the top panel of Figure 8.1. But for the purpose of this discussion, we can treat them as epidemic curves.

mentioned, simulations can give us an idea about the likely effects of control measures. The simulation for panel a assumed that no control measures were taken, the simulation for panel b assumed a certain control measure B was in place, and the simulation for panel c assumed a different control measure C was implemented.

First, look at the curves in panels a and b of Figure 8.2. The curve in panel b is roughly a shifted version of the curve in panel a. The shapes of the two curves are otherwise similar; the differences we see can be attributed to random fluctuations rather than to the effects of control measure B. In particular, it appears that the areas under the two curves are roughly the same. Thus, control measure B had the effect of *delaying* the outbreak, but once the outbreak got started, a similar number of hosts experienced infection as in an outbreak where no control measures at all were taken. In particular, control measure B did not reduce the final size.

In contrast, the area under the graph in panel c is much smaller than for the other two panels. We can see from the panels that control measure C did not delay the onset of the outbreak, but did reduce the all-important final size. Figure 8.2 shows only one simulation for each control measure. But if these are representative for the predictions of a realistic mathematical model, epidemiologists can make inferences about the expected effectiveness of control measures before health authorities decide on their implementation.

Now consider a situation like in our example of the 2009 H1N1 outbreak. We will see in Project 9.1 (see Online Appendix of Chapter 9 [15]) that tighter border control measures, all by themselves, have an effect similar to the one shown in panel b. In contrast, behavior modification, quarantine, or vaccination often work like control measure C: they are likely to reduce the final size. Delaying the onset of an outbreak all by itself does not have this effect; at best the delay can buy some valuable time for the health care system to better prepare for the outbreak. This may be crucial if development and timely administration of vaccines is an option within a relatively short time frame, so that a combination of control measures b and c becomes feasible. But in almost all other cases, delay of the outbreak has no effect on the final size.

We have argued that the ultimate goal of control measures is a reduction of the expected final size of an outbreak and, ideally, preventing the occurrence of major outbreaks. In general, there are several different types of control measures one might consider. Most of them are costly, both in monetary terms and inconvenience to (human) hosts. What epidemiologists often can do is simulate or mathematically analyze models based on hypothetical scenarios: **If** such and such control measures were implemented, **then** certain mathematical quantities, such as $nI(t)$ or the final size, would be predicted to be .... Models allow us to compare the predictions for several possible control measures. The costs of modeling are very small compared with the costs and suffering associated with an actual outbreak. Thus, mathematical epidemiology can give guidance toward optimal choices of control measures.

It is not immediately clear, though, how small the final size needs to be for an outbreak to qualify as minor. For example, the HIV/AIDS pandemic has taken a terrible toll in deaths and human suffering over the last 30 years. It has been estimated that at the end of 2000, there were 36 million people worldwide living with HIV infection [16]. While the fraction of all humans living in 2000 that had experienced HIV infection was "only" about 0.005, on a worldwide scale this translated to 36 million affected people. Thus, there is no universal cutoff proportion below which an outbreak would qualify as minor. Nevertheless, it is possible to define the distinction between minor and major outbreaks in a mathematically precise way. In Chapter 9, we will show how.

The incidence function $nI(t)$ and the prevalence function $|\mathbf{I}(t)|$ tell us everything about the overall numbers of affected hosts, but they do not give us information about the exact pattern of spread from host to host. What makes a given host more prone to infection? What makes a given infected host prone to cause *secondary infections*, that is, to transmit the pathogen to other hosts in the population? One cannot answer these questions based on overall numbers of cases alone. Thus, epidemiologists need to collect other types of data in addition to the reported numbers of new infections.

In Project 8.1 of the Online Appendix [17], we will examine in detail an example of a minor outbreak of measles in Australia that was confined to nine hosts who experienced infection. The index case (host 1) arrived on an intercontinental flight and transmitted the virus directly to hosts 2, 3, 4, and 6. Host 6 transmitted the virus to hosts 5 and 7, and host 7 in turn infected hosts 8 and 9. Thus, the index case caused four secondary infections, each of hosts 6 and 7 caused two secondary infections, and the other hosts caused none. These facts were determined by *contact tracing*, which is a powerful tool of epidemiology. It was found that hosts 2, 3, 4, and 6 arrived with the same flight as the index case, hosts 5 and 7 were staff members at the hospital that treated host 6, while both hosts 8 and 9 were at the same medical facilities at the same time as host 7. Contact tracing uses methods such as analyzing passenger lists and seating charts in airplanes, hospital logs that show which patients and staff members were likely to interact, lists of attendees of certain social functions, and interviews with infected hosts. If this can be done sufficiently rapidly, it may be possible to notify persons who have been at risk of exposure so that they can seek preventive treatment and take other steps to reduce the risk of further spreading the pathogen.

In this chapter and the follow-up Chapter 9, we will illustrate how one can build mathematical models of disease dynamics and use them to derive predictions about outbreaks. Here we will show in detail how to construct such models, while in Chapter 9, we will focus mainly on exploring their predictions. The models we will be working with in both chapters are not entirely abstract mathematical constructs. Instead, they are embodied in computer code that allows us to run simulations. The software that we developed for this purpose is posted in the Online Appendix [17] for this chapter. There you also can find modules that demonstrate how to use the software for deriving valid predictions, some additional material that is related to but goes beyond the confines of the text, and suggestions for further reading. Additional material that complements this chapter and the follow-up Chapter 9 can be found at our website on exploring transmission of infectious diseases on networks with NETLOGO [18].

Take-Home Message:

- *Epidemiology* studies the spread of diseases caused by *pathogens* such as viruses or bacteria, in *populations* of *hosts*, which can be humans, animals, or plants.
- The goal is to *predict* the time course of an *outbreak* of a given disease in a population and the effect of conceivable *control measures*, such as vaccination, quarantine, culling, or behavior modification, on the severity of the outbreak. Of particular importance is the question of how such control measures would affect the *final size* of the outbreak, that is, the proportion of hosts in the population who will eventually experience infection. Ideally, an effective control measure would prevent *major outbreaks*, also known as *epidemics*, from occurring.
- Mathematical models are greatly simplified representations of reality. Based on such models, epidemiologists can *simulate* hypothetical outbreaks under various assumptions about control measures that might be implemented and derive predictions about their likely effectiveness.
- Realistic models need to be based on *epidemiological data.*
- Reports on the overall number of cases can be used to reconstruct the *epidemic curve* that plots the *incidence* as a function of time.
- Many quantities of interest to epidemiologists, such as the final size of an outbreak, can be expressed either in terms of the incidence function $nI(t)$ or the related *prevalence function* $|\mathbf{I}(t)|$.
- *Contact tracing* allows epidemiologists to get more detailed insight into the pattern of transmission than the epidemic curve alone.

8.2 MATHEMATICAL MODELS OF DISEASE TRANSMISSION

Modeling is a process of selective ignorance [19]. Mathematical models are useful tools for understanding important mechanisms and deriving predictions precisely *because* they are greatly simplified versions of the real world. We will need to make very careful decisions about which aspects of the real world we want to ignore, and which ones we want to incorporate into our models. These choices need to be guided by our goal of understanding how pathogens spread between hosts of a given population.

In order for a disease to spread in a population, some pathogens need to leave one host, and they, or their offspring, need to subsequently enter another host. The biological mechanisms involved are often highly complex. For example, the pathogens of malaria, protozoans of the genus *Plasmodium*, cannot be transmitted directly from human to human, but first need to enter the organism of a mosquito that bites an infected human host and then transmits the pathogen to another human host during a subsequent bite. These details matter a lot in epidemiology and add complications to mathematical models. To keep matters reasonably simple, throughout this chapter and its companion Chapter 9 we will assume that the disease we are studying can be transmitted only during a *direct contact* between hosts. Thus, the models we will construct here will be relevant to many, but by far not all, infectious diseases.

8.2.1 Transmission Probabilities

The precise interpretation of the phrase "direct contact" varies greatly from disease to disease. In the case of the flu a direct contact can be defined as temporary close physical proximity, while in sexually transmitted diseases a contact requires intimate physical interaction. In plant diseases, the definition of "direct contact" is less intuitive, as plants do not move, but pathogens may be transmitted by various mechanisms over some distance. We might consider two plants in direct contact if their distance is below a threshold that depends on the particular disease.

A contact in this sense will not always lead to a transmission of the disease, though. You will not automatically catch the flu if you stand in a bus next to an infected person. If that person sneezes into your face though, you most likely will.

The phrase "most likely" brings us to a crucial point: transmission of the disease between any two hosts is a random event. There is only a certain probability that you will end up standing in the bus next to somebody with the flu, and even if you do, you may or may not get infected. This observation has an important consequence: you would really like to know whether and when you will catch the flu. But as you can see from the preceding paragraph, epidemiology cannot give a definite answer to this question. Similarly, epidemiology cannot tell a cocoa planter in Ghana with certainty whether or not his plantation will be destroyed by CSSV within the next 5 years unless he cuts down some of its trees. At best, epidemiology can inform you and the cocoa planter about the probabilities.

There are many different ways in which hosts can make contact and pathogens can be transmitted during a contact. What matters for our mathematical models are the probabilities of these events. Let us introduce some notation. Think about host i as being yourself and host j as another person who at the current time t_{curr} sheds flu viruses. In order for host i to catch the flu *from host* j during the next Δt time units, that is, at some time t with $t_{\text{curr}} < t \leq t_{\text{curr}} + \Delta t$, three conditions must be satisfied at time t: (a) A direct contact between i and j must occur, (b) host i can neither be already infected nor immune at time t, and (c) a sufficient number of flu viruses must be transmitted from j to i during the contact. In particular, host j cannot already have stopped shedding viruses at time t.

Let $c_{i,j}$ denote the probability of a direct contact between hosts i and j during the time interval $(t_{\text{curr}}, t_{\text{curr}} + \Delta t]$, and let $v_{i,j}$ denote the probability that a sufficient number of pathogens, such as viruses, will be transmitted from j to i during a contact *given* that j sheds pathogens at the time of the contact. Technically, $v_{i,j}$ is a *conditional probability*. You can find a brief review of this and all other concepts of probability theory that we will use here at our website [18].

Note that $c_{i,j}$ and $v_{i,j}$ reflect entirely different aspects of how hosts make contact: $c_{i,j}$ depends on the *pattern of mixing* between hosts, that is, on who makes contact, how often, and with whom. For example, if i and j are neighbors who meet each other about six times per month by chance and Δt represents 1 day, then $c_{i,j} \approx 0.2$. The value of $c_{i,j}$ depends on the particular disease only in terms of the nature of the relevant contacts; it will be different for the flu and a sexually transmitted infection (STI aka STD), but the same for all STIs. In contrast, $v_{i,j}$, the probability of transmission per contact with an infectious host, depends strongly on the particular disease. An extreme case is measles, where up to 90% of people who have contact with an infectious person may become infected [20].

Now let $b_{i,j}$ denote the probability that a given host i who is neither infected nor immune at time t_{curr} will be infected at some time t with $t_{\text{curr}} < t \leq t_{\text{curr}} + \Delta t$ by a given host j who currently sheds pathogens. This definition of $b_{i,j}$ is preliminary, as are the definitions of $c_{i,j}$ and $v_{i,j}$ that we gave above. We will slightly modify these definitions later, but for now they will do to build up some intuitions. Note that $b_{i,j}$ is the probability of the event that a contact occurs *and* a sufficient number of pathogens are transmitted during the contact.

Thus, the multiplication law for conditional probabilities would suggest that $b_{i,j}$ should be the product of the probabilities $c_{i,j}$ of a contact and the conditional probability $v_{i,j}$ of transmission. Under our preliminary definitions, this will be approximately true if the length Δt of the time interval is sufficiently short:

$$b_{i,j} \approx c_{i,j} v_{i,j}. \tag{8.1}$$

Exercise 8.2. Why did we need to assume that Δt was "sufficiently short"? More generally, why can we have only approximate equality in (8.1)? □

Clearly, $b_{i,j}$ will depend on the length Δt of the time interval and on the particular disease. Later in this chapter we will give a slightly revised and mathematically more convenient interpretation of the probabilities $b_{i,j}$. These probabilities will play a crucial role in the models that we are going to build and explore here. It is essential that we understand exactly what they mean.

It is not hard to see from (8.1) how the probabilities $b_{i,j}$ may be different for different pairs of hosts. In general, they depend both on the frequencies of contact between a given pair of hosts (which influences $c_{i,j}$) and on the intensity of their contacts (which influences $v_{i,j}$). Both the frequency and the intensity of contacts depend on the particular details of how hosts in a population interact. The power of mathematical modeling is based on our ability to distill all these often confusing aspects into a single number for each pair of hosts: the probability $b_{i,j}$.

In order to build realistic models, epidemiologists need to estimate probabilities such as $b_{i,j}$ from data. This is not easy. Let us take a few minutes to think about the kind of data one would want to collect about individual human hosts so that one could estimate $b_{i,j}$ for a given pair of hosts. Resources are limited, and one needs to make a judicious choice. Clearly, place of residence and employer could matter, as neighbors and coworkers tend to meet more often. Similarly, marital status could matter. Host i may talk to his boss more often than to his wife, but will kiss and hug her more often than his boss, or so one might assume.

Exercise 8.3. There are a lot of other things that we may find interesting about real people: whether they are shy or assertive, if they wear fashionable clothes, whether they have a standoffish or emotionally expressive demeanor, how much they pay for their car insurance, whether they ever received an organ transplant, or whether both of their grandfathers died at the age of 97. Could any of these personal characteristics actually *matter* in terms of $b_{i,j}$? □

You can see from Exercise 8.3 how $b_{i,j}$ distills a lot of information into a single number. In practice, one will need to restrict data collection to the most important aspects of host behavior, so there will always remain some uncertainty about how closely the parameters that are used in models match reality. Online Section 8.3 describes some methods for dealing with these uncertainties.

The probabilities $b_{i,j}$ depend on the *length* Δt of the time interval $(t_{\text{curr}}, t_{\text{curr}} + \Delta t]$ for which they are defined. Throughout these chapters we will assume that Δt is fixed and given by the context, and that $b_{i,j}$ does not depend on the left endpoint t_{curr}. This greatly simplifies our modeling.

Exercise 8.4. Can you think of situations when even the probabilities $b_{i,j}$ would vary over time, that is, depend on t_{curr} for a fixed Δt? When would one want to incorporate this dependence on time into our models? □

Take-Home Message:

- In this chapter and the follow-up Chapter 9, we restrict ourselves to modeling infectious diseases that are transmitted during a *direct contact* between two hosts.
- Of crucial importance for building our models are probabilities $b_{i,j}$. One can roughly think of $b_{i,j}$ as the probability that host j will infect host i during a time interval of a given length Δt, *given* that j sheds pathogens and i is neither infected nor immune at the beginning of the time interval.
- If Δt is sufficiently small, $b_{i,j}$ is approximately equal to the product of the probability $c_{i,j}$ that a contact between i and j occurs during the time interval and the conditional probability $v_{i,j}$ that a sufficient number of pathogens will be transferred from j to i during the contact *given* that j still sheds pathogens at the time of the contact.

- Slightly revised and more precise definitions of $b_{i,j}, c_{i,j}, v_{i,j}$ will be given later in this chapter.
- The probability $b_{i,j}$ distills the pattern of contacts between hosts i and j into a single parameter. For realistic epidemiological models, it needs to be estimated from data, which is not easy.

8.2.2 The Time Line of Within-Host Dynamics

The growth and reproduction of pathogens inside their hosts and their effect on the host are called the *within-host dynamics*. The biological details can be complex and are very important for designing effective ways for treating infected hosts. For human hosts, this is the realm of medicine. In contrast, epidemiology studies patterns of transmission, not treatment. For our purposes, the within-host dynamics will be of interest only inasmuch as they affect probabilities of transmission.

Let us describe, in general terms, the time line of typical within-host dynamics. Suppose a contact occurs between host i and host j at time t. Let us call this contact *successful* (from the point of view of the pathogens) if it leads to infection of i. This terminology may sound perverse, but recall that the ultimate goal of epidemiology is to design effective strategies for fighting the spread of the disease. As every general will tell you, for that you need to be able to put yourself in the shoes of the enemy.

What does it take for a contact at time t between i and j to be successful? First of all, a certain number of pathogens need to be transferred from j to i. Thus, j needs to be infectious at time t. Now the transferred pathogens need to multiply inside their new host i. The immune system of i will start fighting back, and this may cause symptoms, such as fever. But symptoms are not what the pathogens are after: success, in terms of the biology of the pathogens, requires host i to actually start *shedding pathogens*, that is, to become *infectious*. In this case, the pathogens win. Otherwise, the host wins. The odds in this fight may depend on a lot of factors. For example, host i may be immune to infection as a result of prior recovery from the same disease or vaccination. In this case, the contact will always be unsuccessful.

Similarly, a contact that occurs at a time when host i has already been exposed during a prior contact and is headed toward infectiousness, or even is already infectious, will in general not benefit the offspring of the few pathogens that are newly transferred during the current contact. We will consider such a contact unsuccessful. This is a useful simplifying assumption. It seems quite realistic for most diseases, but readers should be aware that sometimes multiple exposures may significantly alter the within-host dynamics. For example, in diseases such as HIV the pathogens quickly mutate and multiple strains eventually overwhelm the immune system of the host. Multiple exposures are likely to accelerate this process.

In summary, a successful contact requires that host i is *susceptible* at time t, that is, neither immune nor already infected at the time of the contact.

If the contact at time t was successful, then we will write $T_i^E = t$, where the superscript E stands for ***Exposure*** of host i to the disease. In more colloquial terms, T_i^E is the time when host i catches the disease. Once a host becomes exposed, this host *will* subsequently develop the disease.

What will happen next to host i? The pathogens that got transferred from j to i will start multiplying inside the organism of i, but most likely, nothing much that we could observe from the outside will happen for a while. At some time $T_i^Y > T_i^E$, host i will develop observable *sYmptoms*. The time interval $[T_i^E, T_i^Y)$ is called the *incubation period*. These symptoms will persist until time T_i^C when host i either ***Ceases*** to show symptoms or dies from the disease. In chronic incurable diseases, there is no time T_i^C; we can mathematically model this possibility by setting $T_i^C = \infty$. The time interval $[T_i^Y, T_i^C)$ is the period during which host i is actually sick and may require treatment.

The top panel of Figure 8.3 below depicts the times and time intervals that we have introduced so far. This time line is of paramount interest to medicine.

But in epidemiology we are most interested in when host i will be able to infect other hosts. Let T_i^I denote the time at which host i becomes ***Infectious***, that is, able to infect other hosts by shedding pathogens. We will call this time the *onset of infectiousness*.[3]

3. Many sources in the literature use *infectivity* instead. But this word usually refers to the ability of pathogens to cause infection; we use "infectiousness" to emphasize that we are writing about a property of hosts.

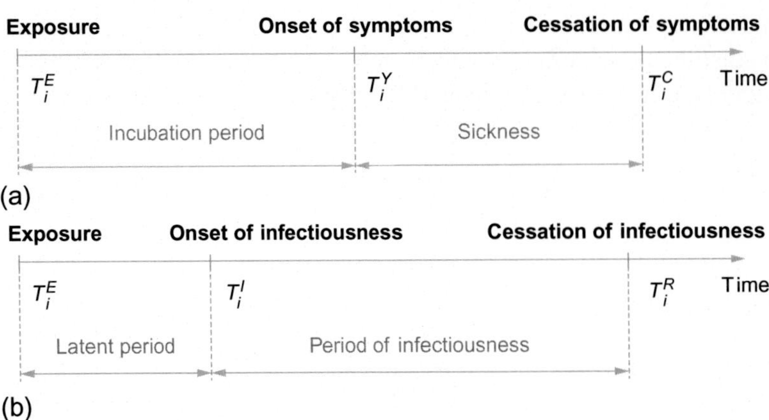

FIGURE 8.3 Schematic representation of transition times.

By definition, $T_i^I \geq T_i^E$. In most cases, pathogens first have to multiply within the organism of the host before being shed, so usually the strict inequality $T_i^I > T_i^E$ will hold. The time interval $[T_i^E, T_i^I)$ is called the *latent* or *preinfectious* period.

Is not this the same as the incubation period? Not necessarily. For example, the onset of infectiousness in measles is believed to occur 2-4 days before the first symptoms (fever, sore eyes, runny nose) appear [20]. Some hosts infected with human papillomavirus (HPV), so-called *carriers*, may never show any symptoms and yet be infectious [23].

Exercise 8.5. To which of the times $T_i^E, T_i^Y, T_i^I, T_i^C$ that we introduced above are the data summarized in the top panel of Figure 8.1 most directly related? □

Eventually, at time $T_i^R > T_i^I$, host i may cease to be infectious. The bottom panel of Figure 8.3 depicts the relation between the times T_i^E, T_i^I, and T_i^R.

The time T_i^R of cessation of infectiousness may or may not coincide with the time T_i^C of cessation of symptoms. For example, in the case of measles it has not yet been definitely established whether infectiousness ends a few days prior to the disappearance of late symptoms (a characteristic rash) or coincides with it [21]. Take a look at both panels of Figure 8.3, which could represent the time line for measles. We see that $T_i^I < T_i^Y$, which means that hosts become infectious some time before showing symptoms. This much has been established for measles. One has to look really closely to see that infectiousness ceases slightly before all symptoms disappear; epidemiologists who study measles are still in the process of looking closely and are not quite sure yet. The real world is messier than our diagrams show.

In many diseases, though, the onset of infectiousness roughly coincides with the onset of symptoms. For example, in the case of the flu, the sneezing and coughing are the most effective mechanisms by which viruses spread from host to host. Throughout the remainder of this chapter and Chapter 9 we will disregard the possible difference between T_i^Y and T_i^I as well as between T_i^C and T_i^R. In epidemiology, only T_i^E, T_i^I, and T_i^R matter, but as we have illustrated above, these times are difficult to estimate directly from data. Sometimes T_i^Y and T_i^C can play the roles of useful stand-ins for T_i^I and T_i^R, respectively, but one needs to exercise some care when relying on such substitutions.

In reality, the onset and cessation of infectiousness may occur gradually. So it is not always clear how to define T_i^I and T_i^R for an actual observed case. These numbers are only mathematical constructs that will greatly help us in building our models. Because successful contacts are random events, the times T_i^E, T_i^I, and T_i^R are *random variables* (abbreviated *r.v.s*). A brief review of r.v.s can be found at our website [18].

Our models will not be able to predict the precise values of T_i^E, T_i^I, and T_i^R. Instead, the models will only give us probabilities that these r.v.s assume values in a given interval. This is technically called the *distribution* of the r.v.s. For example, assume that the current time is early morning on Friday the 13th. If time is measured in days of the month, and you have been designated as host number 7, then the probability that you will catch the flu within the next 10 days can be expressed as the probability $P(13 \le T_7^E < 23)$. The probability that you will *eventually* catch the flu is the probability $P(13 \le T_7^E < \infty)$.

Here is another reason why models cannot possibly predict the exact time T_7^E. If there were a model that would predict that T_7^E will fall for sure into the class period of your biological modeling course, you would want to miss that class and spend that hour outdoors, far away from any people. We might consider your absence excused (your instructor may see it differently, though). But if you do, the model will no longer apply! In effect, you would have adopted a control measure for yourself (behavior modification). However, models can sometimes tell you that T_7^E will fall into that class period with a certain probability p, assuming typical behavior patterns. Depending on p you may decide on a less drastic behavior modification, like sitting in a remote corner all by yourself instead of next to your friends.

But wait a minute. We are talking here about "catching" the flu in terms of being *exposed* to the pathogen. You might have wondered instead whether and when you will start coughing and sneezing.

Exercise 8.6. Suppose the current time is early morning on Friday the 13th and time is measured in days. How would you translate "within the next 10 days, I will start coughing and sneezing" and "eventually I will start coughing and sneezing" into the language of probabilities and r.v.s T_i^E, T_i^I, T_i^R? □

These examples illustrate an important point. The commonly used term "infected" is too vague for our purposes. It may refer to hosts that are preinfectious, sick, or infectious. From now on, we will mostly avoid this term and only talk about *infectious* hosts who shed pathogens and *exposed* hosts to whom pathogens have already been transmitted and who will start shedding them at a future time.

Of particular importance for our modeling will be the time interval $[T_i^I, T_i^R)$. We will refer to it as the *period of infectiousness* of host *i*. Its length $T_i^R - T_i^I$ will be called the *duration of infectiousness of host i* and denoted by τ_i^I. Similarly, the length $T_i^I - T_i^E$ of the latent period will be denoted by τ_i^E.

It is easy to see that τ_i^I is also a r.v. Its value may depend on such factors as the strength of the immune system, lifestyle, or access to medical treatment of host *i*. The *mean value* $\langle \tau_i^I \rangle$ represents the expectation of the value τ_i^I prior to an outbreak. The actual values of τ_i^I may differ from $\langle \tau_i^I \rangle$ depending on circumstances, and host *i* may not even experience infection during a given outbreak. Thus, $\langle \tau_i^I \rangle$ represents a hypothetical estimate of how long host *i* would remain infectious. The mean $\langle \tau^I \rangle$ of all these individual means $\langle \tau_i^I \rangle$, taken over all hosts *i*, represents the expected duration of infectiousness for the given population. The *mean duration of latency* $\langle \tau^E \rangle$ can be defined analogously.

We will use $\langle \tau^I \rangle$ as a parameter in some of our models. The value $\langle \tau^I \rangle$ is disease-specific and can often be fairly reliably estimated from data. For example, for chicken pox, $\langle \tau^I \rangle$ has been estimated as 11 days; for hepatitis A, $\langle \tau^I \rangle$ has been estimated as 3 weeks; and for herpes, $\langle \tau^I \rangle$ has been estimated as 5 days [22].

Take-Home Message:

- Host *i* becomes *exposed* at time T_i^E when a *successful* contact with another host *j* occurs.
- A successful contact requires that host *j* is *infectious* and host *i* is *susceptible* at time T_i^E. It also requires a transmission of a certain number of pathogens from *j* to *i*.
- An exposed host *will* become infectious at some time $T_i^I \ge T_i^E$.
- There *may* be a time $T_i^R > T_i^I$ at which host *i* ceases to be infectious.
- The times T_i^E of exposure, T_i^I of onset of infectiousness, and T_i^R of cessation of infectiousness, as well as the duration τ_i^E of latency and the duration of infectiousness τ_i^I of host *i*, are r.v.s.
- The mean duration of infectiousness $\langle \tau_i^I \rangle$ of host *i* and the overall mean duration of infectiousness $\langle \tau^I \rangle$ may enter our models as parameters.
- The parameter $\langle \tau^I \rangle$ is disease-specific and can be estimated from data.

8.2.3 Movement Between Compartments

Let us make a sweeping claim: the times T_i^E, T_i^I, T_i^R that were defined in the previous subsection contain almost all the information that epidemiologists are trying to predict. Let us illustrate why this claim is at least plausible.

At each such time, a *transition event* happens. We can think about the expressions $T_i^E = t$, $T_i^I = t$, $T_i^R = t$ as specifying transition events; they give us the information *what* happened (a host is exposed to the disease, becomes infectious, or ceases to be infectious), to *whom* the event happened (host i), and *when* the event occurred (at time t). We cannot predict the transition events with certainty for any future outbreak, and obtaining reliable numbers even for a small outbreak that has run its course is difficult in practice (see the second part of Project 8.1 of [17]). But for the purpose of describing our models, let us optimistically assume that we are given a list of these events, ordered from the smallest to the largest times when they occurred. We will refer to such lists as *state transition sequences*. The beginning of the list may look like this:

$$T_1^E = 0, T_1^I = 0.1, T_4^E = 0.9, T_4^I = 1.05, T_7^E = 1.3, T_7^I = 1.4,$$

$$T_1^R = 2.2, T_2^E = 2.4, T_2^I = 2.5, T_7^R = 3, T_4^R = 3.1, T_2^R = 3.9, \ldots. \tag{8.2}$$

Let us assume the population size is actually $N = 7$ in this example.

Now we want you to do a hands-on exercise. Write numbers 1 through 7 on little slips of paper, and prepare four boxes. Label one of the boxes **E**, another one **I**. Let us try to represent the time course of the outbreak by moving the slips between boxes.

Because $T_1^E = 0$, at time 0 you need to put slip 1 into box **E**. Where should you put the other slips? This is not entirely obvious, but a standard assumption is that all other hosts will be susceptible at time 0. Thus you want to label a third box **S** and put the slips numbered 2-7 into it. This gives you a starting point, or, as mathematicians prefer to write, an *initial state*. Note that host 1 must have been exposed through contact with a host who is *not* a member of our population. We will refer to this situation as *introduction of an index case* (host 1 in our example) *into an otherwise susceptible population*. Outbreaks often start like this. A standard assumption is that this will happen only at time $t = 0$, and at future times successful contacts can occur only between hosts of the given population.

Now nothing happens until time $t = 0.1 = T_1^I$. The time interval $[0, 0.1)$ will be the latency period of host 1. At time $t = 0.1$ you want to move slip 1 into box **I**, at time $t = 0.9$ you want to move slip 4 into box **E**, ..., you get the picture. This will safely take you to the end of the first line of (8.2).

At time $t = 2.2 = T_1^R$, host 1 ceases to be infectious. What do you do now? This is not entirely clear. Host 1 may have become susceptible again, may have acquired immunity, or may have died at this time. The data did not tell us which of the above actually happened. As we indicated in the first paragraph of this subsection, these data only give us *almost* all the relevant information.

Let us correct our oversight and give you the missing information right now: a bout of the disease for which the data in (8.2) were collected confers permanent immunity upon each host who is lucky enough to survive it. This will not be true for all diseases, only for *immunizing* diseases such as measles or chicken pox.

So now we know that at time $t = 2.2 = T_1^R$ host 1 will either be dead or permanently immune. The difference may matter a lot to host i, but not to the future course of the outbreak: in either case, host 1 cannot become exposed or infect other hosts at any time $t > 2.2 = T_1^R$. We may as well *remove* host i from further consideration. Thus, let us label the fourth box **R** and bury slip 1 right there at time $t = 2.2$. Now continue moving slips until you run out of data, and do the following exercise before reading on. We mean it. Really take the time to move the paper slips between boxes and do the exercise. It is the most important exercise of this chapter.

Exercise 8.7. The ellipsis at the end of (8.2) suggests that the list goes on. Can you predict what will happen at future times? □

Congratulations! You have built your first epidemiological model, and you have even derived a prediction about the future course of the outbreak, based on partial data. You may not even have been fully aware that what you were doing was building a model, but this is exactly what your paper slips and boxes are: a greatly simplified representation of reality. The movement of paper slips between the boxes represents the change of the *states* of individual hosts from *susceptible* to *exposed* to *infectious* to *removed*. Your model allowed you to *simulate* the time course of the outbreak. The contents of all boxes at any given time t of the simulation represent the *state of the population* at time t in the real outbreak.

A lot can be learned from lists like (8.2) and your simple model.

Exercise 8.8. Based on (8.2) and your simulation, derive each of the following:
(a) The final size of the outbreak. Try to figure this out based only on the state of the population at time $t = 3.9$, as represented by the contents of your boxes at the end of the simulation.
(b) A graph that depicts the prevalence function $|\mathbf{I}(t)|$ by showing how the proportion of infectious hosts in the population changes over time.
(c) Your best guess at the epidemic curve, assuming that time is measured in weeks.
(d) A rough estimate of the mean duration of infectiousness $\langle \tau^I \rangle$.
(e) A rough estimate of the mean duration of the latent period $\langle \tau^E \rangle$. □

In the model that we have described so far, the essential information about the outbreak is embodied in paper slips and boxes. Mathematical models of disease transmission are actually quite similar; the only difference is that the same information is represented by mathematical notation instead of physical objects. The mathematical notation and terminology become quite intuitive once you have done Exercise 8.7. We use the term *compartments* for the boxes in your model and think of hosts moving from compartment to compartment during an outbreak. This is very similar to what is going on in your model, except that there is no physical movement involved, only changes of the *state* of a host. The states will be abbreviated by letters. At a given time t, a susceptible host i will be in state S. A host i who has been exposed to the pathogens but has not yet become infectious will be in state E.[4] An infectious host will be in state I, and a host who plays no role in the propagation of the outbreak at time t can be removed from our consideration and is in state R. In this notation, the beginning of the sequence of states that you simulated with paper slips and boxes would be expressed as follows:

$$(ESSSSSS), (ISSSSSS), (ISSESSS), (ISSISSS), \ldots . \tag{8.3}$$

Exercise 8.9. Complete the list of (8.3) for the whole simulation. □

If this is getting tedious, think about what Exercise 8.9 would have been like for a population of $N = 70$ instead of $N = 7$ hosts. And in practice we want to consider populations of thousands or even millions of hosts. We need to enlist the help of a computer.

IONTW (Infections On NeTWorks), a program that does this job for you, is included in the Online Appendix. Detailed instructions on how to install and work with it can be found in Section 8.1 of [17]. The code was designed in a framework that is known in the literature as *agent-based modeling*. The "agents" in these models are representations of individual hosts in the computer.[5] When executing the code, the computer does the same thing you did in Exercise 8.9. It also displays representations of the hosts as little discs, whose colors code their current states.

At this point you have learned everything that is needed to simulate outbreaks for which we have complete data on the state transition sequence. Such data are difficult to obtain though; see Project 8.1 of [17]. And what we *really* want to do is simulate *future* outbreaks, for which we simply do not have data. But here is what one can do: consider a representative sample of possible sequences of transition times for a hypothetical future outbreak, simulate each of the resulting scenarios, and then make predictions based on what we observe on average. This is, in a nutshell, what we will do throughout the remainder of this chapter and Chapter 9.

Did you notice something odd? In Section 8.2.1, we told you that probabilities $b_{i,j}$ of transmission between pairs of hosts will be of crucial importance in our models. Now we claimed that the current subsection gives you all the know-how you need for simulating outbreaks. But in the previous paragraphs of the current Section 8.2.3, the word "probability" did not appear even once! Before reading on, take a few minutes to do the following exercise.

---

4. Using P (for *preinfectious*) might be more accurate than using E, but E and "exposed" are being used throughout the literature.
5. It is somewhat unfortunate that this established use of the word "agent" clashes with the equally well-established phrase "infectious agent" that is often used as a synonym for "pathogen" in mathematical epidemiology. Here we are using only the term "pathogen" to avoid possible confusion, and reserve the term "agent" for a representation of a real host (human, animal, or plant) in a computer program.

Exercise 8.10. Probabilities must be hiding behind some mask. Can you spot their disguise? □

Probabilities are hidden in the phrase *representative sample*. Of course one can easily dream up some *possible* sequences of transition times, but how likely are those to resemble the ones in an actual future outbreak? A representative sample is one that contains a given sequence with the same probability as it would be observed if the outbreak were to proceed according to the assumptions of our model.

The previous sentence is a mouthful, and it is not at all easy to construct such representative samples. Our computer simulations get around this problem by simultaneously performing the simulations and generating the sequence of transition events. In Section 8.3, we will explain in more detail how this works. But before we do so, let us introduce some important types of disease transmission models, run some actual simulations, and analyze their predictions.

Take-Home Message:

- A *transition event* occurs whenever the state of a host changes.
- Transition events will be denoted by expressions like $T_7^I = 11$, which means that host $i = 7$ became infectious at time $t = 11$.
- At least for certain types of diseases, the sequence of transition events gives complete information about the time course of an outbreak.
- Computer simulations of outbreaks are based on drawing representative samples of such sequences from an appropriate probability distribution.

8.2.4 Basic Model Types: *SEIR*, *SIR*, *SI*, and *SIS*

Mathematical and agent-based models of disease transmission dynamics are based on the notion of *compartments*. We can think of compartments as the formal analogs of our cardboard boxes. Technically, they are the sets of hosts that have the corresponding state. They will be denoted by **S**, **E**, **I**, and **R**, respectively.

There are some restrictions on the movement of hosts from compartment to compartment. Hosts can move into compartment **E** at time $t = T_i^E > 0$ *only from* compartment **S**, and *only if* pathogens are transferred from some host who resides in compartment **I** at the time. Thus, at a time t when compartment **I** is empty, no movement into compartment **E** can occur. After entering it, host i will reside in compartment **E** until time $t = T_i^I$, when the host becomes infectious and moves into compartment **I**. By our definition, there always *will* be such a finite time $T_i^I > T_i^E$. Host i will then reside in compartment **I** during the time interval $[T_i^I, T_i^R)$. Recall that the time interval $[T_i^I, T_i^R)$ is *the period of infectiousness* of host i and its length $\tau_i^I = T_i^R - T_i^I$ is the *duration of infectiousness* of host i.

It is less clear what can happen at time T_i^R. The nature of the particular disease that we want to study may impose some restrictions. In our paper-slip-and-boxes simulation, we made the assumption that hosts will eventually die or cease to be infectious with permanent immunity. Diseases that allow only for these two possibilities are called *immunizing infections*. Prominent examples are childhood infections like measles, scarlet fever, or chicken pox. Under the assumption of an immunizing infection, host i will move to compartment **R** at time T_i^R, and will never leave this compartment at any later time. These conditions describe what it is known in the literature as *SEIR-models*. Panel a of Figure 8.4 shows a schematic representation of *SEIR*-models.

Note that we used the plural "*SEIR*-models." All by itself, panel a of Figure 8.4 does not depict a single model, but a general *type* of models. An actual *SEIR*-model is then constructed by incorporating suitable biological details in the form of *parameters* such as transmission probabilities.

You will by now appreciate how much work is needed to distill the very complex process of disease transmission into a simple model as in Figure 8.4a. You may even wonder whether such simplifications could give us approximately correct ideas of what is going on in real outbreaks. This is a very important concern, as some models make realistic predictions, while other models are far off the mark. We need always be alert to the possibility that we might have made too many simplifying assumptions.

It is easiest though to derive predictions from the most basic models. Mathematicians try to simplify as much as possible. It turns out that sometimes we can pare down the number of features even further!

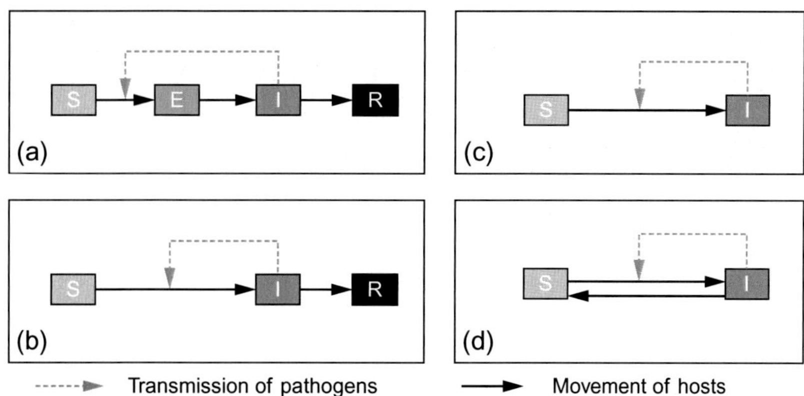

FIGURE 8.4 Schematic representation of model types. Panel a: *SEIR*-models. Panel b: *SIR*-models. Panel c: *SI*-models. Panel d: *SIS*-models.

Look at your solutions to Exercise 8.8(d) and (e). The mean duration of infectiousness could be estimated as approximately 1.8, while the mean duration of latency might be approximately 0.1, an order of magnitude less. In diseases where the mean duration of latency $\langle\tau^E\rangle$ is much shorter than the mean duration of infectiousness $\langle\tau^I\rangle$, we might be able to disregard the **E**-compartment and assume that a host becomes infectious immediately at the time of exposure, that is, $T_i^E = T_i^I$. Strictly speaking, this is false, but on the time scale of $\langle\tau^I\rangle$ we will have $T_i^E \approx T_i^I$, so the counterfactual assumption will perhaps not distort the predictions of the model too much. Under this assumption, we obtain *SIR-models*, as schematically depicted in Figure 8.4b. Surprisingly, *SIR*-models can be useful also for modeling immunizing diseases for which the mean lengths of the periods of latency and infectiousness are of the same order of magnitude. Project 9.2 (see Online Appendix of Chapter 9 [15]) will give you an idea why this is possible.

What about infections such as HIV from which infectious hosts never recover? In this case, we may not need an **R**-compartment. Here we would end up with *SI-models*, as depicted in Figure 8.4c. We do not need times T_i^R for *SI*-models. Or, we might consider setting $T_i^R = \infty$, which makes the duration τ_i^I of infectiousness infinite. This is a favorite trick of mathematicians, but here it should raise eyebrows.

Exercise 8.11. Why? What strikes you as fishy about $\tau_i^I = \infty$? □

In diseases such as gonorrhea, hosts do not acquire immunity when they cease to be infectious. In this case, at time T_i^R, the host becomes susceptible again and moves back to the **S**-compartment. This leads to *SIS-models*, as depicted in Figure 8.4d.

Now let us return to your solution for Exercise 8.11. Consider an *SI*-model (or an *SIS*-model, for that matter). Where do we put a host i who has died? Nowhere, actually. If there is no **R**-compartment, we will make the assumption that hosts never die! More generally, we will entirely disregard *demographics*, that is deaths or emigration of our hosts, and births and immigration of new hosts. We will always assume that the population is fixed at all times and consists of hosts numbered $1, 2, \ldots, N$. This assumption makes modeling much easier. It may not be too unrealistic if the duration of an outbreak is short relative to the average lifespan of hosts (think of the flu for human populations), but will be problematic if the outbreak progresses at a similar time scale as the average life span (think of the HIV pandemic).

Take-Home Message:

- The *type* of a model of disease dynamics depends on which compartments we consider and how hosts can move from one compartment to another.
- The most basic model types are *SEIR*, *SIR*, *SI*, and *SIS*.
- It depends on the nature of the particular disease which model type is appropriate.
- Throughout this chapter and the follow-up Chapter 9 we will ignore demographics and assume that the population is fixed at all times and consists of hosts numbered $1, 2, \ldots, N$.

8.2.5 How to Model Time and Run Simulations

Running simulations is somewhat similar to observing many outbreaks in a tightly controlled artificial environment. Making sense of simulation output is a little bit like drawing conclusions from real-world data. In particular, the observed results will share some common features in different runs of the simulation, but there also may be substantial variability from run to run due to the inherent stochasticity of disease transmission. In Section 8.2 of [17], you will learn how to use our software and how to deal with this inherent messiness so that the common features can be clearly discerned.

However, simulation outputs are also very much *unlike* real data. The environment in which they run is tightly controlled both by the underlying model that is embodied in the code and by the input parameters. It is important to investigate how much the simplifying assumptions that go into the process of building the model might influence its predictions, as reflected in the simulation outputs. In Chapter 9, we will investigate in detail how assumptions about the mixing pattern influence the predictions of the resulting models.

When investigating a given model, one still needs to choose its parameters, such as the mean duration $\langle \tau^I \rangle$ of infectiousness. Epidemiologists need to estimate these parameters from real data, but may not be able to do so with great precision. Moreover, control measures often translate into a change of some of these parameters. It is therefore important to study how the outcomes of simulations change in response to changes in the parameters. Section 8.3 of [17] gives you an introduction to this type of investigation, which goes by the name of *sensitivity analysis*. It contains several relevant simulation exercises. You can work through these exercises at any time after completing Section 8.2 of [17].

A related issue is the fact that the population size in simulations is usually rather limited (to a few hundred hosts in our code), but real populations can have millions of hosts. It is therefore important to study how the predictions of simulations *scale up*, that is, how they change when the population size N becomes large. This is a special case of sensitivity analysis (sensitivity to the parameter N), but the goal here is different: epidemiologists usually know the size of the target population with good precision and are not trying to design control measures that will modify it,[6] but are interested in how likely simulations performed on a small population would give reliable information about outbreaks in large populations. We will return to this issue in Chapter 9.

It is time to run some actual simulations. We strongly recommend that before reading the remainder of this chapter you work through the exercises of Section 8.2 of [17].

When working through the simulation exercises of Section 8.2 of [17], you may have noticed the mysterious drop-down menu **model-time** of IONTW. We perceive time as advancing continuously, and the setting **model-time → Continuous** makes the computer treat time in this way. In terms of our models, this means that state transitions could be happening at any time, as they will in reality. For example, the time T^I_{13} at which host 13 becomes infectious could take any one of the values $3, 4, 3.1, 3.2, 3.14, 3.15, \ldots$. We could even have $T^I_{13} = \pi$.

But suppose the onset of infectiousness of host 13 "really" occurs at time $T^I_{13} = \pi$ and time is measured in days. We could presumably observe that the onset of infectiousness occurred during the fourth day, that is, $3 < T^I_{13} = \pi \leq 4$, or perhaps even that it happened between 2:24 a.m. and 4:48 a.m. On the fourth day, which translates into $3.1 < T^I_{13} \leq 3.2$. But could we possibly tell whether $3.14 < T^I_{13} \leq 3.15$? Note that the difference between 3.14 and 3.15 days is 14.4 min. It is becoming clear that we will never be able to obtain data that pinpoint the transition times with arbitrary precision. An equation like $T^I_{13} = \pi$ is only a mathematical abstraction, albeit a very useful one.

Exercise 8.12. Our definition of transition events allows for the possibility that, for example, $T^I_6 = T^I_{77} = T^R_8 = t$ for some t. In other words, each of these state transitions would happen exactly at the same time. Why can we ignore this possibility in continuous-time models? □

In continuous-time models, the disease transmission parameters need to be specified in terms of two *rates* β and α. If an **E**-compartment is included in the model, we also need a third rate constant γ, but to keep matters

6. It might appear that culling boils down to reducing N, but there are ways to model culling without changing N. In any case, no sane epidemiologist would propose culling an animal population down to a size for which a given simulation tool happens to work.

as simple as possible we will disregard this option in the discussion of this section. The parameter β represents the rate at which two hosts make effective contact, and it allows us, for any given $\Delta t \geq 0$, to calculate the probabilities $b_{i,j}$ that we introduced in Section 8.2.1. The parameter α represents the rate at which hosts cease to be infectious. It allows us, for any given $\Delta t \geq 0$, to calculate the probability a that an infectious host ceases to be infectious during an interval of length Δt as well as the mean duration of infectiousness $\langle \tau^I \rangle$. Details of these calculations will be explained in Section 8.6 of [17].

Often it is easier to pretend that time progresses in discrete, integer-valued steps, and to simply consider the state transitions that are going to happen until the next time step. Note that the epidemic curves of Figure 8.1 of Section 8.1 implicitly treat time in this way: each time step in these graphs corresponds to 1 day, and the curves supposedly tell us about the number of new infections that occurred during a given day. For day $t + 1$ this would be the total number of hosts i for which $t < T_i^I \leq t + 1$. For example, for day 3 one would set $t = 2$ and count that total number of hosts i for which $2 < T_i^I \leq 3$.

We can go one step further and base our simulations on *discrete-time models*. In these models, time is supposed to take only integer values $t = 0, 1, 2, \ldots$, and instead of considering exact transition times T_i^I and T_i^R, we work with the rounded-up versions $\lceil T_i^I \rceil$ and $\lceil T_i^R \rceil$. Recall that the ceiling function $\lceil T \rceil$ is defined so that if $t < T \leq t + 1$, then $\lceil T \rceil = t + 1$. Time $t = 0$ will be used only for designating the initial state. At subsequent times $t + 1$ we proceed as follows. If $\lceil T_i^I \rceil = t + 1$, then we move host i into the **I**-compartment. If $\lceil T_i^R \rceil = t + 1$, then we move host i into the **R**-compartment (in the case of an *SIR*-model) or back into the **S**-compartment (in the case of an *SIS*-model).

Exercise 8.13. Consider the state transition sequence (8.2). Find the state transition events for times $t = 1, 2, 3$ in the corresponding discrete-time model. □

In these models, we lose some information, though. For example, $\lceil T_i^R \rceil = t + 1$ tells us only that $t < T_i^R \leq t + 1$; we no longer know whether removal of infectious host i occurred near the beginning or near the end of the time interval $(t, t + 1]$. This might matter though, as in the latter case host i will still be able to infect other hosts during the time interval $(t, t + 1]$, while in the former case this is very unlikely. We will explore shortly what this loss of information does to the predictions of our model.

When constructing discrete-time models, we need to carefully decide how to choose the length of the interval of physical time that corresponds to one time step of the discrete model. If we consider an interval of 1 week, then t would signify the number of weeks that have elapsed since the start of the outbreak. This may be a good choice if, for example, the mean duration of infectiousness $\langle \tau^I \rangle$ is 1 month, so that hosts will stay infectious for about 4 consecutive time steps. But if, on average, hosts stay infectious only for 2 days, this would be a bad choice, as most hosts would become infectious and recover during the *same* time interval, that is, $t < T_i^I < T_i^R \leq t + 1$ for some t. In an *SIS*-model where hosts move back into the **S**-compartment upon recovery, we would not even *see* most infections.

If the option **model-time → Discrete** is chosen, the input field **time-step** of IONTW allows you to explore various choices of the time scale for discrete-time modeling. We will explore the effects of these choices in Exercise 8.14 below.

What if the mean duration of infectiousness $\langle \tau^I \rangle$ is *exactly* 1 week? There is always *some* variability, so some hosts will still both become infectious and recover during the same time interval, and for some hosts there will be a time t with $T_i^I < t < t + 1 < T_i^R$. But if the variability is small, these two scenarios may happen so infrequently that we might be able to ignore them. At time $t + 1$ we could then move all hosts who reside in the **I**-compartment at time t into the **R**-compartment (in the case of an *SIR*-model) or back into the **S**-compartment (in the case of an *SIS*-model). We will refer to such greatly simplified discrete-time models as *next-generation models*. In Section 8.5 of [17], you can find a detailed explanation where this terminology comes from. For simulating such models with our software IONTW, the parameter **end-infection-prob** that represents the probability a of removal by the next time step needs to be set to 1.

It might appear that these models are useful for predicting the time course of an outbreak only if the mean duration of infectiousness happens to be a natural time unit like 1 week or 1 month. But they could work just fine if the mean duration of infectiousness is, for example, 10.4 days. All we would need to do is rescale our interpretation of physical time to units of 10.4 days; unfamiliar to be sure, but not essentially different from rescaling days into weeks. Our next exercise will illustrate to what extent we might be able to trust the predictions of next-generation and other discrete-time models.

Exercise 8.14. Compare the predictions of four *SIR*-models of the same disease. In one model, choose the continuous-time option; in the second and third models, choose the discrete-time approximation with time steps $\Delta t = 0.02$ and $\Delta t = 1$, respectively. In the fourth model, set the probability of removal a to 1 so as to obtain a next-generation model. Detailed instructions for the exercise are given in Section 8.4 of [17]. □

Take-Home Message:

- *Continuous-time* models keep track of the exact moments at which state transition events occur. Time t is represented by nonnegative real numbers. In simulations of continuous-time models without an **E**-compartment, the parameters are specified in terms of two rates β and α. In our software IONTW, the input fields **infection-rate** and **end-of-infection-rate** control the values of these parameters.
- In *discrete-time models*, time t is represented by nonnegative integers that reflect scaling of time in suitable units Δt. All transition events that occur in the interval $(t, t+1]$ of real time are treated as if they occurred simultaneously at time $t+1$. In simulations of discrete-time models, the parameters are specified in terms of transition probabilities and the length of the time step Δt. In our software, the input fields **infection-prob** and **end-of-infection-prob** control the values of these parameters, while **time-step** controls the value of the parameter Δt.
- *Next-generation models* are discrete-time models based on time units $\Delta t = \langle \tau^I \rangle$. In these models, it is assumed that infectious hosts cease to be infectious after exactly one time unit. In order to simulate such models with our software, we need to set the value of **end-of-infection-prob** to 1 and the value of **time-step** to $\langle \tau^I \rangle$.

8.3 HOW DOES THE COMPUTER RUN SIMULATIONS?

8.3.1 Meet the Simulator

All of us are working with computers every day. More precisely, we interact with the *user interface* that is displayed on the screen. In the simulation exercises, you have already seen the user-interface of our software IONTW. It displays colorful graphics and even pictures of buttons that can be pressed. But what is going on behind the screen?

Let us consider an analogy. Most of you probably have visited a biology lab and noticed that there are two kinds of people in the lab: the students who are performing the same experiment over and over and over again, following established protocol, and the lab director who looks over the students' shoulders and keeps track of what is going on. Actually, chances are you will not even meet the director, because she is away at a conference in Hawaii giving a PowerPoint presentation. If you were to watch the presentation, you would be interacting with the lab's public interface. But with the director out of the picture, back home in the lab we have a chance to meet one of the students. Let's say "Hi."

Analogously, we can think of our simulations as being performed by two parts of the computer program. One part, let's call it the *simulator*, will perform the actual simulation. Another part, let's call it the *director*, will record what happened during the simulation, handle the user interface, and perform some other tasks. You already met the director; now let's meet the simulator. Here is what it will be doing when simulating an *SIR*-model with continuous time:

0. Set t_{curr} to 0. Initialize the state of the population.
   Until instructed to stop **repeat**
   1. Generate the next transition time t_{next} and transition event.
   2. **If** $t_{\text{next}} < T_{\max}$ **then** advance t_{curr} to t_{next} **else** stop.

3. **If** $t_{\text{next}} = T_i^I$ **then** change the state of host i to I.
4. **If** $t_{\text{next}} = T_i^R$ **then** change the state of host i to R.

At each step of our simulation, the computer will contain an internal representation of the state of the population that we are trying to model, as well as a representation of the *physical time* (in days, weeks, or months perhaps) that has elapsed since the start of the outbreak. The latter is what t_{curr} stands for. While all digital computers do their simulations in a sequence of discrete time steps, t_{curr} is in general *not* the same as the number of the current step of the simulation. Think about your paper-slips-and-boxes simulation: a simulation step would correspond to moving a slip of paper from one box to another, and would supposedly represent an event that occurred at a physical time $t = T_i^I$ or $t = T_i^R$ as specified in the list (8.2).

Line 0 is called *initialization* and specifies the *initial state*, that is, the state at time 0. For example, if the initial state is supposed to represent introduction of one infectious host into an otherwise susceptible population, then the simulator will set the state of exactly one host (the presumed index case) to I and the state of all other hosts to S.

Lines 1-4 show how the simulator updates the state of the population. These lines really embody the *SIR*-model in the form of computer code, or rather *pseudocode* that is often used for improved human readability. The simulator will carry out these instructions over and over again, until it is instructed to stop—just like our student back in the lab. These instructions are the analog of experimental protocol that needs to be followed to the letter.

Line 1 instructs it to generate the time $t_{\text{next}} > t_{\text{curr}}$ of the next state transition, and gives information about what is supposed to occur to whom at that time. Recall that we may think of transition events as coded by a symbolic expression like $t_{\text{next}} = T_6^I$, which would mean that host 6 becomes infectious at time t_{next}. In this line our simulator will use a *random number generator*. You can think of this as rolling a kind of virtual die, not necessarily with six sides. But in order to make sure that the state transition sequences that we will get from running many simulations form a "representative sample," the die has to be very cleverly loaded! We will explain in the next subsection how this can be done.

Lines 3 and 4 show how the simulator updates the states of hosts. These lines are written in the form of **If-then** statements. Such instructions are commonly used in agent-based modeling: **If** (some condition is satisfied) **then** (an agent is supposed to perform a certain action or something is going to happen to the agent). This is very intuitive and makes agent-based modeling easy to learn and to use. **If** a condition applies to multiple agents simultaneously, **then** all of them will undergo the specified change. In continuous-time models, this feature is not needed, as by Exercise 8.12 we can assume that only one transition event can happen at any given time. However, we will make use of this feature shortly when we will discuss the case of next-generation models. As Exercise 8.13 shows, these models do allow for possibilities such as $T_6^I = T_{77}^I = T_8^R = t_{\text{next}}$.

Exercise 8.15. How would you modify the above pseudocode for simulating an *SIS*-model instead of an *SIR*-model? □

Line 2 shows how the simulator updates the representation of physical time. This is given in the form of an **If-then-else** statement, where the **else**-clause shows when the simulation will terminate. The number $T_{\max}$ in this line is a constant parameter that represents the maximum time horizon of the simulation.

In our actual code, the simulation can also be terminated by the user; this option would be part of the director's job. But here is a problem that is of more immediate concern to the simulator: there may be no more infectious hosts left, even though $t_{\text{curr}} < T_{\max}$. The model tells us that there cannot be any future transition times. Could we simply instruct the simulator to not generate anything in line 1 in this case?

Exercise 8.16. What would happen in the simulation if we did not generate a new t_{next} and went straight to line 2 again? How can we fix this problem? □

Now let us consider what the simulator needs to do for a next-generation *SIR*-model. Here is the pseudocode:

0n. Set t_{curr} to 0. Initialize the state of the population.

Until instructed to stop **repeat**
1n. Generate the set *New* of hosts that will be infectious at time $t_{\text{curr}} + 1$.
2n. **If** $t_{\text{curr}} + 1 < T_{\max}$ **then** advance t_{curr} to $t + 1$ **else** stop.
3n. **If** the current state of host i is I **then** change the state of host i to R.
4n. **If** $i \in New$ **then** change the state of host i to I.

Notice a few important differences from the previous instructions. The model now specifies that $t_{\text{next}} = t_{\text{curr}} + 1$, and line 2n is the exact analog of line 2 above for this choice of t_{next}. In line 1n, the simulator will no longer need to generate the next time. Instead, the simulator will need to generate a *set* of hosts, represented by integers, that are supposed to *become* infectious during the real-time interval $(t_{\text{curr}}, t_{\text{curr}} + 1]$ and will *be* infectious at time $t_{\text{curr}} + 1$, according to the assumptions of the model. For example, $New = \{3, 71, 208\}$ means that hosts 3, 71, and 208 will be infectious at time $t_{\text{curr}} + 1$. Line 4n resets the states of all these hosts to I. According to the assumptions of the next-generation model, hosts who are infectious at time t_{curr} will no longer be infectious at time $t_{\text{curr}} + 1$. Because we are simulating a next-generation *SIR*-model, the state of all these will need to be changed to R. This is what line 3n does.

Note that we have switched the order of lines, so that 3n corresponds to line 4 of the previous pseudocode and line 4n corresponds to line 3.

Exercise 8.17. What would happen if we left the order of lines as in the previous pseudocode? Why did the previous arrangement of the code work just fine for the model with continuous time? □

Take-Home Message:

- Computer simulations of agent-based models proceed by successively changing states of agents according to a sequence of transition events.
- As the simulation progresses, the sequence of transition events is randomly generated by rolling a kind of virtual die. In order to assure that this sequence is representative for a typical outbreak, the die needs to be loaded in a certain way.

8.3.2 How to Load the Die

Now let us take another look at line 1n of the pseudocode for simulating next-generation models. Line 1n instructs the simulator to *randomly generate* the set *New* of hosts that will be infectious at time $t_{\text{curr}} + 1$. Computers generate random objects by *calling the random number generator*, which is a kind of built-in virtual die. Random number generators are unbiased, but we need a loaded virtual die. The difficult part is figuring out how to load the die in the right way.

Actually, there is another difficulty: in line 1n, our simulator will need to choose the set *New*. There are as many as 2^N possibilities for this set, and already for a population with $N = 20$ hosts we would need a die with more than a million faces! We can sidestep this problem by tossing N biased virtual coins instead and including host i in the set *New* if, and only if, coin number i comes up heads. Note that for $N = 20$, this trick allows us to construct a "virtual die" with more than one million faces out of 20 virtual coins. The random number generator of your computer is very helpful here: it can toss biased coins that come up heads with probability p_i and tails with probability $1 - p_i$. The computer will happily toss the coins for you, but first you need to tell it what probability p_i of heads you want for coin number i. In mathematical terms, "loading the die" boils down to specifying the probabilities p_i. Before reading on, please take a few minutes to do the following exercise.

Exercise 8.18. Note that p_i represents the probability that host i will be included in the set *New*, that is, that host i will be infectious at the next time step. What do these probabilities depend on? Will they be fixed throughout the simulation? Will p_i be the same for each i? □

How should we calculate these probabilities? Note that p_i will depend in general on how many other hosts are infectious at time t_{curr}: the more infectious hosts j there will be, the higher the probability that host i will be the recipient of a successful transmission from at least one of the hosts j who are infectious at time t_{curr}. Thus the probabilities p_i may change at each step of the simulation.

In general it also matters *who* is infectious at time t_{curr}. If hosts i and j live far apart and never make contact, then it does not matter at all whether host j is infectious or not. But if i and j make frequent contact, the status of j will matter. It also matters what kind of contact hosts i and j make. In the terminology of Section 8.2.1, p_i will depend both on the probabilities $c_{i,j}$ of hosts i and j making contact during the time interval $(t_{\text{curr}}, t_{\text{curr}} + 1]$ and $v_{i,j}$ of a successful contact *given* that a contact occurred during the interval and i was still susceptible and j still infectious at the time of the contact. In other words, p_i will depend on the probabilities $b_{i,j}$, which according to (8.1) are approximately equal to the product $c_{i,j}v_{i,j}$. Because for a given j the probabilities $b_{i,j}$ do not need to be the same for each i, neither do the probabilities p_i need to be all equal.

Now let us formulate slightly revised definitions of the probabilities $b_{i,j}, c_{i,j}, v_{i,j}$ that are less intuitive than the ones given in Section 8.2.1 but are easier to work with. Note that the probability of a *successful* contact between hosts i and j is always 0 unless one of these hosts is susceptible and the other infectious. Thus, the probability of a successful contact may change over time. But we want to use $b_{i,j}$ in our models as a fixed parameter that does not change over time!

Let us define an *effective* contact between hosts i and j as one that *would* be successful *if i were* susceptible and j *were* infectious at the time of the contact. The distinction between successful and effective contacts is subtle, but necessary for us. Note that in view of this definition, all successful contacts are effective, but not all effective contacts are successful.

Now we define $b_{i,j}$ as the probability of *at least one effective contact* between hosts i and j during a time interval $(t, t + \Delta t]$. Similarly, we define $c_{i,j}$ as the probability that hosts i and j have *at least one* contact during the time interval and $v_{i,j}$ as the conditional probability that *at least one* contact between i and j during the time interval $(t, t + \Delta t]$ will be effective, *given* that *at least one* contact occurred. As before, under these new definitions all three probabilities $b_{i,j}$, $v_{i,j}$, and $c_{i,j}$ can be assumed fixed over time t. They will depend on the length of the time step Δt, though. However, no matter how Δt is chosen, under the new definitions (8.1) turns into an exact equality:

$$b_{i,j} = c_{i,j}v_{i,j}. \tag{8.4}$$

Now let us see how we can use the probabilities $b_{i,j}$ to calculate the probability p_i that host i will be infected at time $t_{\text{curr}} + 1$. In our construction of next-generation models, we assumed that time was scaled in such a way that $\Delta t = 1$. The models then tell us that a host i who is not susceptible at time t cannot be infectious at time $t + 1$. Thus if the state of host i at time t is I or R, then we must have $p_i = 0$. If host i is susceptible at time t, then host i will be infectious if, and only if, an effective contact between host i and *at least one* host j who is infectious at time t occurs in the time interval $(t, t + 1]$. The previous sentence may take a couple of minutes to digest, but it is very important that you fully understand it.

Think about it: host i can become infectious only during a successful contact with another host j. Thus the contact must be effective *and* host j must be infectious at the time of the contact. In view of the assumptions of the next-generation model, this requires that host j be infectious at time t. In view of our definition of successful contact in Section 8.2.2, if more than one effective contact of host i with infectious hosts j_1, j_2 occurs during the time interval, only the first of these will count as successful.

It follows that p_i will in general depend on the current state of the population, that is, which hosts are susceptible, infectious, or removed. Now we could obtain the values of p_i for the current state by using the same trick of tossing multiple biased coins that we described earlier. Our simulator might execute the following lines of code for each i separately:

i1. **If** host i is not susceptible **then** set $p_i = 0$ **else**.
  i2. List all hosts $Curr = \{j_1, \ldots, j_K\}$ that are currently infectious.
  i3. For each host $j_k \in Curr$, toss a virtual coin that comes up heads with probability $b_{i,j}$.
  i4. **If** at least one of the coins comes up heads **then** update the state of i to I **else** retain the state S for i.

This is not how the computer would actually compute p_i; one can often use simple formulas instead. It is nevertheless worth contemplating the pseudocode: this is pretty much what is going on in real outbreaks!

There is, however, an important aspect in which real outbreaks may differ from the procedure described in the pseudocode. Suppose we toss several coins, let us say one coin for each host j_k in the set *Curr*. Then, for example, the outcomes of coins $j_1, j_2, \ldots, j_5$ will not give us any idea of how coins $j_6, \ldots, j_K$ fall. Knowing that none of the first five coins came up heads will not influence our estimate of the likelihood that at least one of the coins $j_6, \ldots, j_K$ will come up heads. Mathematicians refer to this phenomenon by saying that the events "coin j_k comes up heads" are all *independent*. In the literature on disease modeling, this assumption is usually made. It greatly simplifies the modeling and calculations, and we are using it in the models that are embodied in our code IONTW. The assumption was already used implicitly at the beginning of this subsection in our method for generating the set *New* of hosts who will be infectious at time t. This was also based on tosses of biased virtual coins, one for each host. While in the literature on disease modeling the assumption of independence is rarely spelled out explicitly, modelers need to be aware of it and always need to carefully consider whether the assumption is warranted for the particular population that they wish to study. Project 8.2 of [17] provides examples that illustrate when the assumption of independence is roughly realistic and when it might not be.

Finally, let us briefly discuss continuous-time models. In line 1 of the pseudocode for these models, the simulator needs to do two things: "generate" a time t_{next} and a symbolic expression like $t_{\text{next}} = T_6^I$ that signifies the next transition event and indicates what is going to happen and to whom at time t_{next}. The latter is somewhat similar to what it needs to do in line 1n for next-generation models. However, because the time t_{next} could be any real number $> t_{\text{curr}}$, generating t_{next} is akin to rolling a carefully loaded die with infinitely many sides. Thus the precise implementation of the continuous model requires more sophisticated mathematical tools than implementation of discrete-time models. We will explain details in Section 8.6 of [17]. Here we only want to mention that the relevant calculations need to be performed in terms of rates β and α, the parameters of continuous-time *SIR*-models, instead of probabilities $b_{i,j}$ and a.

Take-Home Message:

- For simulations of next-generation models, one can think about generating the set *New* of hosts who will become infectious at the next time step in terms of independent coin tosses. The probability p_i that host i will be included in the set *New* depends both on the current state of the population and on the probabilities $b_{i,j}$.
- We redefined $b_{i,j}$ as the probability that at least one *effective* contact between hosts i and j will occur over a time interval of length Δt. These probabilities remain fixed during a simulation and enter discrete-time models as parameters.
- Similarly, we redefined $c_{i,j}$ as the probability that hosts i and j have at least one contact during the time interval and $v_{i,j}$ as the conditional probability that at least one contact between i and j during a time interval of length Δt will be effective, *given* that at least one contact occurred. Under these new definitions, $b_{i,j} = c_{i,j}v_{i,j}$.

REFERENCES

[1] Reinhardt UE, Cheng T-M. The World Health Report 2000: health systems: improving performance. Geneva: World Health Organization; 2000.

[2] Bryce J, Boschi-Pinto C, Shibuya K, Black RE. WHO estimates of the causes of death in children. Geneva: World Health Organization; 2005.

[3] Pinner RW, Teutsch SM, Simonsen L, Klug LA, Graber JM, Clarke MJ, et al. Trends in infectious diseases mortality in the United States. JAMA 1996;275(3):189-93.

[4] WHO Fact Sheet. The top 10 causes of death. Available from: http://www.who.int/mediacentre/factsheets/fs310/en/index1.html [accessed 2014.09.18].

[5] Anderson PK, Cunningham AA, Patel NG, Morales FJ, Epstein PR, Daszak P. Emerging infectious diseases of plants: pathogen pollution, climate change and agrotechnology drivers. Trends Ecol Evol 2004;19(10):535-44.

[6] Daszak P, Cunningham AA, Hyatt AD. Emerging infectious diseases of wildlife—threats to biodiversity and human health. Science 2000;287(5452):443-9.

[7] Henson S, Mazzocchi M. Impact of bovine spongiform encephalopathy on agribusiness in the United Kingdom: results of an event study of equity prices. Am J Agr Econ 2002;84(2):370-86.
[8] Schlenker W, Villas-Boas SB. Consumer and market responses to mad cow disease. Am J Agr Econ 2009;91:1140-52.
[9] Crowdy SH, Posnette AF. Virus diseases of cacao in West Africa; cross-immunity experiments with viruses 1A, 1B, and 1C. Ann Appl Biol 1947;3(3):403-11.
[10] Centers for Disease Control and Prevention. Seasonal influenza (flu). Available from: http://www.cdc.gov/flu/about/disease/us_flu-related_deaths.htm [accessed 2014.09.18].
[11] Mummert A, Weiss H, Long L-P, Amigó José M, Wan X-F. A perspective on multiple waves of influenza pandemics. PLoS ONE 2013;8(4):e60343.
[12] Fujita M, Sato H, Kaku K, Tokuno S, Kanatani Y, Suzuki S, et al. Airport quarantine inspection, follow-up observation, and the prevention of pandemic influenza. Aviat Space Environ Med 2011;82(8):782-9.
[13] Lee SS, Wong NS. Reconstruction of epidemic curves for pandemic influenza A (H1N1) 2009 at city and sub-city levels. Virol J 2010;7:321.
[14] Wu JT, Cowling BJ, Lau EHY, Ip DK, Ho LM, Tsang T, et al. School closure and mitigation of pandemic (H1N1) 2009, Hong Kong. Emerg Infect Dis 2010;16(3):538-41.
[15] Just W, Callender H, LaMar MD. Online Appendix: Disease transmission dynamics on networks: network structure vs. disease dynamics. In: Robeva, R, editor. Algebraic and discrete mathematical methods for modern biology. New York: Academic Press; 2015.
[16] Piot P, Bartos M, Ghys PD, Walker N, Schwartländer B. The global impact of HIV/AIDS. Nature 2001;410(6831):968-73.
[17] Just W, Callender H, LaMar MD, Toporikova N. Online Appendix: Transmission of infectious diseases: data, models, and simulations. In: Robeva, R, editor. Algebraic and discrete mathematical methods for modern biology. New York: Academic Press; 2015.
[18] Just W, Callender H, LaMar MD. Exploring transmission of infectious diseases on networks with NetLogo. Available from: https://qubeshub.org/iontw and also http://www.ohio.edu/people/just/IONTW/.
[19] Gross LJ. Selective ignorance and multiple scales in biology: deciding on criteria for model utility. Biological Theory; 2013 8(1): 74-79.
[20] Centers for Disease Control and Prevention. Measles. Available from: http://www.cdc.gov/measles/ [accessed 2014.08.14].
[21] Wikipedia Entry . Measles. Available from: http://en.wikipedia.org/wiki/Measles [accessed 2014.09.21].
[22] Umatilla-Morrow Head Start. Common communicable diseases. Available from: http://www.umchs.org/umchsresources/administration/workplan/Health/ [accessed 2014.09.18].
[23] Dunne EF, Nielson CM, Stone KM, Markowitz LE, Giuliano AR. Prevalence of HPV infection among men: a systematic review of the literature. J Infect Dis 2006;194(8):1044-57.

Chapter 9

Disease Transmission Dynamics on Networks: Network Structure Versus Disease Dynamics

Winfried Just[1], Hannah Callender[2] and M. Drew LaMar[3]
[1]Department of Mathematics, Ohio University, Athens, OH, USA, [2]University of Portland, Portland, OR, USA,
[3]The College of William and Mary, Williamsburg, VA, USA

9.1 INTRODUCTION

This chapter is a continuation of Chapter 8. The level of the exposition is similar in the beginning of each section, but gradually we develop mathematically more advanced material. The format is slightly different from Chapter 8 in that most parts are designed for reading in parallel with online modules that illustrate the concepts with NETLOGO exercises. *It is crucial that readers do work through these modules and exercises within the text when instructed to do so. Subsequent text often depends on conceptual understanding that is developed in the modules. Within each part of a module the exercises should be done in the order in which they appear, as subsequent exercises build on previous ones and often use the same parameter settings.*

In Chapter 8 we discussed in detail how to build models of disease transmission and use them to simulate outbreaks of diseases. These simulations can give us predictions about outbreaks of actual diseases in real populations, but we can rely on these predictions only if we have some assurance that the model is sufficiently realistic. In particular, we would need to convince ourselves that the simplifying assumptions of the model do not significantly distort its predictions.

In building our models we made many assumptions, and we cannot examine all of them here in detail. In this chapter we focus on assumptions about the mixing patterns of hosts. In Section 9.2 we discuss predictions that can be derived under the *uniform mixing assumption.* In Section 9.3 we describe the construction of *network-based models* that allow us to consider more realistic mixing patterns by taking into account the underlying *contact network.* Section 9.4 contains some additional suggestions for further reading.

In the companion modules, we investigate for some simple cases how the structure of the contact network influences the spread of disease on the network. The modules themselves are provided in the online appendix [1]. Whenever we instruct the reader to work through to a particular module, the implicit understanding is that it can be found in Ref. [1]. These modules form a logical sequence that gradually progresses from simpler to more sophisticated concepts. Our website [2] contains material that continues these explorations for more sophisticated types of networks. The modules in Ref. [1] and the material on the website [2] could serve as a basis for undergraduate research projects and lead into the exploration of graduate-level open problems.

Algebraic and Discrete Mathematical Methods for Modern Biology. http://dx.doi.org/10.1016/B978-0-12-801213-0.00009-5
Copyright © 2015 Elsevier Inc. All rights reserved.

9.2 MODELS BASED ON THE UNIFORM MIXING ASSUMPTION

9.2.1 Compartment-Based Models

In Section 8.2, we described how the computer simulates an outbreak in a given agent-based model. Before we can run a simulation though, we need to specify the model's parameters. The description that we gave for next-generation models requires that we input parameters $b_{i,j}$ that represent the probabilities of an effective contact between hosts i and j over a time interval of unit length.[1] For simulating more general discrete-time models and continuous-time *SIR*-models, we would also need parameters that specify the mean duration of infectiousness $\langle\tau_i^I\rangle$ for host i (see Section 8.6 of the online appendix for Chapter 8 [3]).

Sources of variability of these parameters were discussed in Chapter 8. We could incorporate a lot of potentially relevant biological details into our models by choosing separate values of $b_{i,j}$ and $\langle\tau_i^I\rangle$ for each pair of hosts (i,j) and host i, respectively.

There is a big problem though: We would need to specify $N^2 - N$ parameters $b_{i,j}$ (one for each pair (i,j) with $i \neq j$) plus N parameters $\langle\tau_i^I\rangle$. Even for a modest population size of $N = 100$, we would need to input $10,000$ parameters! Moreover, for biologically realistic predictions, all parameters would need to be estimated from data. It is simply not feasible to estimate that many parameters from any real data set.

In order to build and explore tractable models, we need to reduce the number of parameters by making additional simplifying assumptions. It may be feasible to estimate two or three parameters from data, but not 10,000 different ones. For example, we might be able to estimate the mean value b of $b_{i,j}$ for all pairs (i,j) and the mean value $\langle\tau^I\rangle$ of $\langle\tau_i^I\rangle$ for all hosts.

It is possible to build models based on the mean values b and $\langle\tau^I\rangle$. Most mathematical models of disease transmission are in fact based on such averages. Agent-based models of this kind are obtained by setting $\langle\tau_i^I\rangle = \langle\tau^I\rangle$ for every host i and $b_{i,j} = b$ for each pair of hosts (i,j). This is exactly what IONTW does when **network-type** is set to **Complete graph.**

Models that are entirely based on mean values make no distinctions between individual hosts. In a sense, each host is treated as a mathematical version of "Joe Average."

Some readers may object to our use of the phrase "Joe Average" and point out that we should have written "Joe or Jane Average." They are correct; we will call our hosts *Joene Average.* But don't think of Joene as "Joe *or* Jane." S&He is actually Joe *and* Jane. Think about him&er as flipping a suitably biased coin once in a while and deciding whether to be Joe the curmudgeon or Jane the social butterfly, another coin to decide whether to be Jenny the daredevil or Jack the careful, yet another coin ... you name it. Joene is neither boring nor plain; we bet you really would enjoy meeting him&er. Joene is all of us. Unfortunately, s&he is only a mathematical construct. But take heart: When you come to think of it, each one of us is really like Joene, only that our coins are more strongly biased than his&ers.[2]

There are actually two distinct versions of the assumption that "every host is exactly like Joene Average." The first one is called the *assumption of homogeneity of hosts*; the second one is called the *uniform mixing assumption.* Before giving precise mathematical definitions, let us first describe them informally.

The first assumption refers to individual characteristics that are relevant to the within-host dynamics, such as the strength of the immune system (which relates to $\langle\tau_i^I\rangle$) or proclivity to risky behavior, which relates to the probabilities of transmission per contact $v_{i,j}$.

In contrast, the uniform mixing assumption refers to the pattern of making contacts in the first place, which relates to the probabilities $c_{i,j}$.

To illustrate the difference, consider the question, what would it be like to be married to Joene? If Joene is Average in the first sense, then Joene will sometimes recover quickly from a bout of the flu and sometimes slowly, will sometimes be warm and fuzzy and sometimes be emotionally withdrawn, will sometimes act in a stereotypically male and sometimes in a stereotypically female way. There's nothing unusual about that, really, is there? But what about sex and commitment? If Joene is Average *only* in the sense of homogeneity, then Joene

1. Strictly speaking, we may also need a parameter for the maximal time horizon T_{max}, but we will ignore it here for greater transparency.
2. This nonstandard convention is used to highlight the concept that Joene embodies.

may still be quite choosy about with whom and how frequently s&he makes contact. In particular, s&he may be choosy in exactly the way you would hope for in a spouse.

But if Joene is Average in the sense of the uniform mixing assumption, then for any given time interval Joene will be equally likely to make contact with every other host of the population. Forget about getting a commitment from Joene. You can have some exciting moments with Joene at a party where people constantly mingle, but you shouldn't even invite him&er to dinner. Joene just will not sit in one place for two hours.

We will refer to Joene Average who embodies the assumption of homogeneity of hosts but not the uniform mixing assumption as *Joene Choosy* and to Joene Average who embodies both assumptions as *Joene Butterfly*.

Exercise 9.1. For what kind of populations would the uniform mixing assumption be nearly satisfied? When would it not be realistic? □

In mathematical terms, the assumption of homogeneity of hosts boils down to setting $\langle \tau_i^I \rangle = \langle \tau^I \rangle$ for each host i and $v_{i,j} = v$ for each pair of hosts (i,j), where v denotes the mean of the probabilities $v_{i,j}$, taken over all pairs (i,j). In actual populations, the duration of infectiousness for a given real host will in general not be fixed but will show some variability depending on when infection occurs. There will also be some variability between hosts, due to both genetic and environmental factors. If we assume homogeneity of hosts, we ignore this second source of variability, but not the first one. Technically speaking, the probability distribution for the duration of infectiousness τ_i^I will be the same for each copy of Joene; in particular, $\langle \tau_i^I \rangle = \langle \tau^I \rangle$. But in general this will only be a probability distribution, so that the actual value of the random variable τ_i^I may differ from the mean value $\langle \tau_i^I \rangle = \langle \tau^I \rangle$. Thus, homogeneity of hosts does *not* imply that each time a copy of Joene experiences infection, s&he will remain infectious for exactly the same length of time. Notice that the assumption about the duration of infectiousness that we made in Chapter 8 for next-generation models is actually *stronger* than what homogeneity of hosts implies.

Under the assumption of homogeneity of hosts, Eq. (8.4) of Chapter 8 simplifies to:

$$b_{i,j} = v c_{i,j}. \tag{9.1}$$

Homogeneity of hosts all by itself does not imply anything about $c_{i,j}$. In a population of copies of Joene Choosy, hosts i' and j' may still be in a strictly monogamous relationship, and for a sexually transmitted infection (STI) we would then have $c_{i',j'} > 0$, while $c_{i',j} = c_{i,j'} = 0$ for all $i \neq i'$ and $j \neq j'$. In contrast, the assumption of *uniform mixing* that the probabilities $c_{i,j}$ will satisfy: In a population of copies of Joene Butterfly, we will have $c_{i,j} = c$ whenever $i \neq j$. Here c denotes the mean of $c_{i,j}$ taken over all pairs (i,j) with $i \neq j$.

We will always assume that $c_{i,i} = 0$. Perhaps hosts frequently interact with themselves, but they cannot catch an infectious disease this way. Thus if we make *both* the assumptions of homogeneity of hosts and uniform mixing, then (9.1) simplifies to

$$b_{i,j} = \begin{cases} b = vc & \text{if } i \neq j, \\ 0 & \text{if } i = j. \end{cases} \tag{9.2}$$

In the remainder of this section we will make both assumptions: Homogeneity of hosts and uniform mixing. This greatly simplifies our modeling. For example, consider the pseudocode for simulating a next-generation model that is given in Section 8.3.2. In line i3 of this pseudocode, we can substitute the single parameter b for all parameters $b_{i,j}$. By analyzing the pseudocodes, you can see that the predicted future of the outbreak no longer depends on *which* particular hosts are infectious at the current time step. The predicted future only depends on the *numbers* of hosts that reside in each compartment at the current time step. Because we are using the symbol $|A|$ for the size of a set A, these numbers will be denoted by $|\mathbf{S}(t)|, |E(t)|, |\mathbf{I}(t)|, |\mathbf{R}(t)|$. They are random variables (r.v.s) for sure, but their current values completely determine our *expectations* about the future of the outbreak.

As you already saw in Chapter 8, many properties that epidemiologists are trying to predict, such as the epidemic curve or the final size, can be expressed in terms of how the numbers $|\mathbf{S}(t)|, |E(t)|, |\mathbf{I}(t)|, |\mathbf{R}(t)|$ will change over time t. The combined assumptions of homogeneity of hosts and uniform mixing allow us to construct *compartment-based* or *compartment-level* models. While the phrase "compartment-based model" is more entrenched in the literature, we will mostly use the phrase "compartment-level models," which is more accurate. Compartments are implicit in all epidemiological models, but in compartment-level models we predict the future *only* on the basis of the current numbers $|\mathbf{S}(t)|, |E(t)|, |\mathbf{I}(t)|, |\mathbf{R}(t)|$. In contrast, *host-level* models take into account *which particular hosts* reside in which compartment at time t. Notice that the agent-based models we described in Chapter 8 are technically host-level models and potentially allow for more accurate predictions. We will tap into their full potential in Section 9.3.

Let us also mention that many compartment-level models that are studied in the literature assume homogeneity of hosts and uniform mixing only *within* the same compartment. But here we will always make this assumption for the entire population.

You saw in Chapter 8 that mathematical models cannot *precisely* predict the future course of an outbreak. But we can try to predict how the *expected values* of numbers of hosts in the compartments will change. You already explored such predictions numerically in the NETLOGO simulations of Chapter 8, where homogeneity of hosts and uniform mixing were implicitly assumed. In the next subsection we will discuss some mathematical predictions about these expected numbers for compartment-level models.

If Joene Average is all of us and if we are really only trying to predict average numbers of infections, would a model populated with identical copies of Joene give us the same, or very similar predictions as a model that is based on one copy of each of us? This seems at least plausible and is the motivation for studying compartment-level models. Often this does work, but not always, as we will see in Section 9.3.

Take-home message:

- The *assumption of homogeneity of hosts* postulates that all hosts in the population have the same individual characteristics. The *uniform mixing assumption* postulates that each host is equally likely to have contact with each other host.
- The cartoon character Joene Average embodies the assumption of homogeneity of hosts. More specifically, Joene Choosy embodies homogeneity of hosts in the absence of uniform mixing and Joene Butterfly embodies both assumptions.
- In mathematical terms, homogeneity of hosts translates into $\langle \tau_i^I \rangle = \langle \tau^I \rangle$ for all hosts i and $v_{i,j} = v$ for all pairs (i,j) of different hosts. The uniform mixing assumption translates into $c_{i,j} = c$ for all pairs (i,j) of different hosts. If both homogeneity of hosts and uniform mixing are assumed, then $b_{i,j} = b$ for all pairs (i,j) of different hosts.
- *Compartment-level* models (aka *compartment-based* models) make both assumptions of homogeneity of hosts and uniform mixing and allow us to study the change of *expected values* of the numbers of hosts in each compartment over time.

9.2.2 The Basic Reproductive Ratio R_0

Unless explicitly stated otherwise, here we will explore compartment-level *SIR*-models. Thus, there are no **E**-compartments, and the state of the system at time t will be entirely determined by the numbers $|\mathbf{S}(t)|, |\mathbf{I}(t)|, |\mathbf{R}(t)|$ of hosts who reside in the respective compartments at time t. Due to the stochastic nature of disease transmission, these numbers are r.v.s.

Let us consider an initial state $(|\mathbf{S}(0)|, |\mathbf{I}(0)|, |\mathbf{R}(0)|) = (N-1, 1, 0)$ that corresponds to the introduction of one index case, usually denoted by j^*, into an otherwise susceptible population. At some time $t^* = T_{j^*}^R > 0$ the index case j^* gets moved to the **R**-compartment.

What are the possibilities for the state $(|\mathbf{S}(t^*)|, |\mathbf{I}(t^*)|, |\mathbf{R}(t^*)|)$? If we assume a next-generation model, the answer is simple: By definition, $t^* = 1$ and $|\mathbf{R}(t^*)| = |\mathbf{R}(1)| = 1$. The number $|\mathbf{I}(t^*)| = |\mathbf{I}(1)|$ is equal to the *number of secondary infections caused by the index case.* Because we assume a constant population size, we must have $|\mathbf{S}(1)| = N - |\mathbf{I}(1)| - 1$.

Exercise 9.2. What could we say about the state $(|\mathbf{S}(t^*)|, |\mathbf{I}(t^*)|, |\mathbf{R}(t^*)|)$ in a continuous-time model? Would it still always be true that $|\mathbf{I}(t^*)|$ is equal to the number of secondary infections caused by the index case? □

The number of secondary infections caused by the index case j^* is a random variable. Its mean value is the most important parameter in disease modeling. It has a special name and a special symbol R_0 is reserved for it.

Definition 9.1. The *basic reproductive ratio* or *basic reproductive number* R_0 is the mean number of secondary infections caused by an average index case in a large entirely susceptible population. □

The value of R_0 depends on the particular disease. The literature gives estimates of R_0 for pertussis between 12 and 17, for chicken pox between 8 and 9, for rubella between 6 and 7, and for mumps between 4 and 7 [4]. On our website [2], you will find a module that shows you a method for estimating R_0 from data.

Note that this definition uses the phrase "average index case," which in general can be a bit tricky to make sense of. But here we are talking about compartment-level models where *every* host is assumed to be average—more precisely, a copy of Joene Butterfly. Under these assumptions, we can estimate R_0 from the model parameters.

In the context of next-generation models, we can argue as follows: For any *given* host i, the probability that i will be infected by j^* is equal to b. There are a total of $N - 1$ hosts susceptible to secondary infection by j^*. We can think about secondary infections as in the pseudocodes that we discussed in Chapter 8: For each susceptible host i, toss a biased coin that comes up heads with probability b. Then $|\mathbf{I}(1)|$ becomes the number of "successes" in $N - 1$ tosses of such coins. If we assume independence, then the r.v. $|\mathbf{I}(t)|$ has a binomial distribution with parameters $N - 1$ and b (see Ref. [2]). It follows that

$$R_0 = \langle |\mathbf{I}(1)| \rangle = b(N-1) \approx bN. \tag{9.3}$$

The approximation in the second part of (9.3) is used to make the formula a little simpler. It is valid if b is small and the population size N is very large, which is usually satisfied in situations where we want to use compartment-level models. Formula (9.3) doesn't tell us the whole story though, as it is valid *only* in next-generation models. In continuous-time models we need to work with rates β and α instead of probabilities b. We can then calculate $\langle \tau^I \rangle = \frac{1}{\alpha}$, and as long as $\beta \Delta t$ is very small, the probability b of an effective contact during a time interval of length Δt is approximately equal to $\beta \Delta t$. In these types of models, R_0 will be approximately equal to

$$R_0 \approx \frac{\beta N}{\alpha} = \beta \langle \tau^I \rangle N \approx \frac{b \langle \tau^I \rangle N}{\Delta t}. \tag{9.4}$$

The first approximation in Eq. (9.4) is valid if N is large and β is small relative to N and α, and the second approximation will be valid if $\beta \langle \tau^I \rangle$ is sufficiently small. In Module 9.5 you can find more details about these approximations. Note that in next-generation models, we implicitly assume that $\Delta t = \langle \tau^I \rangle$ so that the expression on the right of (9.4) coincides with the right-hand side of (9.3).

The alert reader will have noticed something puzzling: The examples of empirically estimated values of R_0 for the selected diseases that we gave above do not mention population size, but formulas (9.3) and (9.4) contain a factor N. So does R_0 depend on N or does it not? Well, it may or it may not. The constant b in (9.3) may itself depend on population size. If b is inversely proportional to $N - 1$, then the term $N - 1$ cancels out in (9.3) and (9.4) so that R_0 becomes independent of N. This will happen when the mode of transmission of the particular disease can be assumed *frequency-dependent.* For the remainder of this chapter, we will assume for simplicity that R_0 does not depend on N.

As we already mentioned, the initial state $(|\mathbf{S}(0)|, |\mathbf{I}(0)|, |\mathbf{R}(0)|)$ in general does not uniquely determine the states $(|\mathbf{S}(t)|, |\mathbf{I}(t)|, |\mathbf{R}(t)|)$ at future times $t > 0$. However, the initial state determines the expected values of the numbers $|\mathbf{S}(t)|, |\mathbf{I}(t)|, |\mathbf{R}(t)|$ at all future times. In order to keep the notation reasonably uncluttered, these expected numbers will be denoted by $S(t), I(t), R(t)$.

Exercise 9.3.

(a) Assume a next-generation *SIR*-model and an initial state that corresponds to the introduction of one index case into an otherwise susceptible population. Find $(S(1), I(1), R(1))$.

(b) How would the result change if we were considering an *SIS*-model instead? □

The basic reproductive number R_0 does not need to be an integer. If, for example, $R_0 = 0.85$, then our *SIR*-model predicts that $(S(1), I(1), R(1)) = (N - 1.85, 0.85, 1)$. Of course we cannot have 0.85 infectious hosts at any given time. But recall that $I(1)$ is a *mean value*, and may take on fractional values even if the actual numbers that will be observed in each given outbreak are integers.

Let us take this opportunity to observe something very important: If $R_0 = 0.85$, the expected number infectious hosts at time $t = 1$ is *smaller* than the number $I(0) = 1$ of infectious hosts in the initial state. We made this observation in just one special case, but it has a far-reaching generalization.

Theorem 9.1. *Assume homogeneity of hosts and uniform mixing in an SIR- or SIS-model. Consider two times t, t^+ with $0 \leq t < t^+$. If $R_0 < 1$ and $0 < I(t)$, then $I(t^+) < I(t)$.*

You will be able to find further information on Theorem 9.1 and a proof of it on our website [2]; see also the discussion at the end of this subsection. This theorem is a very general result. It applies to all times t, t^+ with $0 \leq t < t^+$ and all values of $R_0 < 1$, not just to the particular choices $t = 0, t^+ = 1$, and $R_0 = 0.85$ in our example. This is how mathematical theorems work: They tell us what we should expect in a wide variety of situations, not just in some specific cases that we might have observed in nature or in simulations. But we can rely on the predictions of the theorems only in situations where their assumptions are satisfied.

Exercise 9.4. Explain why the conclusion of the theorem will not in general hold for *SI*- or *SEIR*-models. □

As long as the assumptions of Theorem 9.1 are satisfied, we should expect that the number of infectious hosts will *decrease* over time. When drawing conclusions from theorems, it is important to pay careful attention to what the theorem is actually saying.

Exercise 9.5.

(a) Use IONTW to simulate a compartment-level next-generation *SIR*-model with **infection-prob** = 0.09, population size $N = 10$, and one index case. Find R_0. Run 8 repetitions in slow motion. For detailed instructions, see Module 9.1 of [1].

(b) Describe your results in terms of $|\mathbf{I}(t)|$ for $t = 0, 1, 2, \ldots$. Do you always observe that $|\mathbf{I}(t+1)| < |\mathbf{I}(t)|$? Do your results confirm Theorem 9.1, or do they appear to conflict with it? How would you explain the relation of your observations to the theorem? □

Exercise 9.6. Use IONTW to simulate a compartment-level next-generation *SIR*-model with **infection-prob** = 0.005, population size $N = 100$, and one index case. Run 100 repetitions. Find the largest and the mean number of hosts who experienced infection in the 100 simulated outbreaks. For detailed instructions, see Module 9.1 of [1]. □

In the settings that were suggested for Exercise 9.6, we have $R_0 = 0.495 < 1$. We do not know the exact results that you obtained, as they depend on the random number generator of your computer. We would be ready to bet, though, that in all of your simulated outbreaks, only a relatively small proportion of the population was affected. Mathematicians can derive this prediction as another theorem.

Theorem 9.2. *Assume homogeneity of hosts and uniform mixing in an SEIR-, SIR-, or SIS-model. For any given probability $p < 1$ there exists a constant $B(p)$ such that whenever $R_0 \leq 1$, then with probability at least p the number of hosts who will experience infection at some time during the outbreak will not exceed $B(p)$, regardless of the population size N. Thus if the population size is large, then with probability very close to 1, introduction of a single index case into an otherwise susceptible population will result only in a minor outbreak.*

You will be able to find further information on Theorem 9.2 and a sketch of its proof on our website [2]; see also the discussion at the end of this subsection.

The theorem gives a mathematically precise meaning to the phrase *minor outbreak* that was defined somewhat informally in Chapter 8. Let us slowly dissect its meaning: With probability p that can be chosen arbitrarily close to 1, the total number of affected hosts will be bounded from above by some constant $B(p)$, regardless of total population size N. For small N, the fraction $\frac{B(p)}{N}$ that gives an upper bound on the expected final size could still be substantial, but for very large N this fraction will be very close to zero. This property really distinguishes minor outbreaks from major ones, where the final size comprises a substantial fraction of the overall population, regardless of the actual population size N. Due to stochastic effects, the actual number of hosts that will be affected may substantially exceed $B(p)$ in some outbreaks. The theorem tells us that this will happen only with probability $1 - p$, which is very small if p is chosen close to 1.

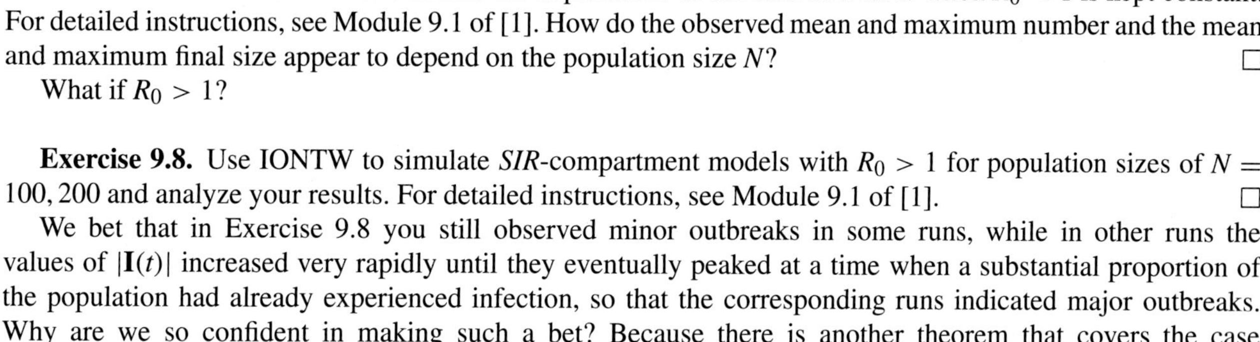

Exercise 9.7. Use IONTW to examine the dependence of the final size on N when $R_0 < 1$ is kept constant. For detailed instructions, see Module 9.1 of [1]. How do the observed mean and maximum number and the mean and maximum final size appear to depend on the population size N? □

What if $R_0 > 1$?

Exercise 9.8. Use IONTW to simulate *SIR*-compartment models with $R_0 > 1$ for population sizes of $N = 100, 200$ and analyze your results. For detailed instructions, see Module 9.1 of [1]. □

We bet that in Exercise 9.8 you still observed minor outbreaks in some runs, while in other runs the values of $|\mathbf{I}(t)|$ increased very rapidly until they eventually peaked at a time when a substantial proportion of the population had already experienced infection, so that the corresponding runs indicated major outbreaks. Why are we so confident in making such a bet? Because there is another theorem that covers the case of $R_0 > 1$.

Theorem 9.3. *Assume homogeneity of hosts and uniform mixing in an SIR- or SEIR-model. If $R_0 > 1$, then there are numbers $r(\infty)$ and z_∞ that satisfy the inequalities $0 < r(\infty), z_\infty < 1$ such that as long as the population size is large, then with probability very close to $1 - z_\infty$, introduction of a single index case into an otherwise susceptible population will result in a major outbreak with final size close to $r(\infty)$.*

The number $r(\infty)$ will be larger for larger values of R_0 and the number z_∞ will be smaller for larger values of R_0.

You will be able to find further information on Theorem 9.3 and its proof on our website [2]; see also the discussion at the end of this subsection.

This theorem is certainly a mouthful. Let us slowly dissect its text and try to understand what it says. The number z_∞ represents the probability of a minor outbreak for very large population sizes. The notation can be understood by thinking of z_∞ as the probability that a near zero proportion of hosts in a near infinite population will experience infection during the outbreak. One important piece of information here is that $z_\infty < 1$, so that the probability $1 - z_\infty$ of a *major* outbreak is positive. Moreover, these probabilities are supposed to depend only on R_0, but not (much) on the actual population size N, so that if we run many simulations for a variety of very large population sizes N, we should observe major outbreaks with approximately the same probability $1 - z_\infty$. The probabilities z_∞ may be significantly different though for different values of R_0; the larger R_0, the less likely it is that the outbreak will only be a minor one.

The number $r(\infty)$ represents the predicted final size for large populations. As you already saw in the solution to Exercise 8.8(a) of Chapter 8, for an *SIR*-model this is the proportion $\frac{|\mathbf{R}(t)|}{N}$ for a very large t, after the outbreak has run its course. Mathematicians are fond of thinking of very large times as $t = \infty$, which explains the notation. Note that $r(\infty)$ gives the approximate proportion only if the observed outbreak was in fact a major one. Again, this proportion is not supposed to depend significantly on N, as long as N is large, but it may be significantly different for different values R_0. In general, the final size will be larger for larger values of R_0. It is very interesting to notice that $r(\infty)$ is always predicted to be less than 1. This means that even if R_0 is very large, a proportion $1 - r(\infty) > 0$ of hosts is predicted to escape infection.

Exercise 9.9. Based on the above reading of Theorem 9.3, try to estimate z_∞ and $r(\infty)$ for $R_0 = 1.5, 2$ based on the data you collected in Exercise 9.8. Do your findings appear to confirm the predictions of Theorem 9.3? □

Theorems 9.1–9.3 are classical results. In the literature on transmission of infectious diseases, they are usually stated in somewhat different forms than in this section. For example, the theorems can be derived from the results that are presented in Chapter 1 of [5], but they are not explicitly stated there. The wordings we chose here omits some mathematical details that are beyond the scope of this chapter; readers will be able to find more information on this on our website [2]. These theorems show why the value of R_0 is so important in disease modeling: Under the assumptions of our theorems, it determines whether any major outbreaks should be expected at all (not if $R_0 \leq 1$), with what probability, and what proportion of hosts will likely experience infection during a major outbreak (if $R_0 > 1$).

Project 9.1 of [1] will show how R_0 relates to the effectiveness of certain possible control measures. As you will see, the abstract Theorems 9.1–9.3 are enormously useful for designing public policy!

Before moving on to the next section, work through the projects in part 9.1.2 of Module 9.1.

Take-home message:

- The *basic reproductive ratio* or *basic reproductive number* R_0 is the mean number of secondary infections caused by an average index case in an entirely susceptible population. The value of R_0 is disease-specific.
- Under the assumptions of uniform mixing and homogeneity of hosts, the value of R_0 predicts the probability of major outbreaks and their expected final size. If $R_0 \leq 1$, then introduction of a single infectious host into an otherwise susceptible population will cause only minor outbreaks in large populations. If $R_0 > 1$, major outbreaks will occur with probability $1 - z_\infty$, which is strictly between 0 and 1. The final size of a major outbreak depends on R_0. It will be a fraction strictly between 0 and 1 and will be larger for larger values of R_0.
- These results can be proved as mathematical theorems. Such theorems are applicable to real populations if, and only if, their assumptions approximate the relevant biological facts with reasonable accuracy.

9.3 NETWORK-BASED MODELS

You saw in Section 9.2 that the assumptions of homogeneity of hosts and uniform mixing are very convenient. They allowed us to construct compartment-level models and derive important predictions about the probability of outbreaks, their final size, and effectiveness of certain types of control measures. These predictions usually depend on only one parameter: The basic reproductive ratio R_0.

But how reliable are these predictions? This is an important concern. Recall the discussion in Project 9.1 of [1] about whether to vaccinate as many hosts as possible or just a sufficient fraction of the population to achieve herd immunity. The tradeoffs involved in this decision pose serious ethical problems. It is of crucial importance to correctly predict the herd immunity threshold before deciding on a policy. The threshold that we derived was based on the uniform mixing assumption. In real populations, the threshold may be lower or higher. We can rely on the predicted value only if we have some assurance that this simplifying assumption does not significantly distort the predictions of the model.

You saw in Exercise 9.1 that mixing may be nearly uniform if hosts move around a lot relative to the size of the habitat, if they encounter each other rarely, and if there is no social structure. In populations with a well-defined social or territorial structure though, some pairs of individuals will have contact relatively frequently (think of coworkers or neighbors in human populations), while other pairs of individuals will almost certainly never encounter each other (think of your likelihood to ever meet the Supreme Leader of North Korea). We can approximate the latter situation by assuming the existence of a *contact network* that determines whether it is even *possible* for the pathogens to be transmitted between two given hosts. The nature of the required contact, and thus the relevant contact network, may depend on the particular disease. Think of the flu versus a computer virus versus a sexually transmitted infection.

In this section we will show you how to build models for disease transmission on contact networks. These models are often more realistic than the corresponding compartment-level models. In the online modules, we will explore with simulations how the predictions depend on the structure of the underlying contact network and how they compare with the predictions of compartment-level models.

9.3.1 Networks and Graphs

First, we need to translate the notion of a *network* into mathematical language. Networks are all around us. People form social networks, for transportation we use road networks, and human brains are giant networks of neurons. What do all of these networks have in common? Each of them consists of a set of entities that mathematicians call *nodes* or *vertices,* together with *links* that represent connections between some (but usually not all) nodes. In social networks, the nodes are individual people, and they may be connected by friendships. We may think of the road network in terms of cities (nodes) that are connected by roads. In the brain, the nodes are neurons, and some of them are connected by axons and dendrites.

Human friendships are usually *symmetric,* that is, if i is a friend of j, then j is also a friend of i. On roads between cities one can usually drive in both directions. But some roads within a city may be one-way, and connections in the brain are definitely asymmetric: The signal can only travel from the axon of a presynaptic neuron via a synapse to the dendrite of a postsynaptic neuron. You can see from these examples that there are at least two distinct types of networks. Mathematicians use two distinct structures for modeling them: *Graphs* for networks with symmetric links that are called *edges*, and *directed graphs* for networks with asymmetric links that are called *arcs*. In the study of disease dynamics, we are interested in *contact networks* that indicate whether two given hosts (nodes of the network) will be likely to make effective contact. For many types of contacts, such as a handshake, it may be just as likely that pathogens will be transferred from host i to host j (if i happens to be infectious) as it is that they will be transferred from host j to host i (if j happens to be infectious).

Exercise 9.10. Can you think of some examples where the assumption of symmetric transmission probabilities of pathogens would not be justified? □

Here we will always assume that the contact network has symmetric links and model it as a graph. Technically, a graph is a pair $G = (V(G), E(G))$, where $V(G)$ is a nonempty set of nodes or vertices, and $E(G)$ is a set of unordered pairs of nodes that represent the edges. In our models of contact networks, $V(G)$ will always be the set $\{1, 2, \ldots, N\}$ that represents the hosts in our population, and a pair $e = \{i, j\}$ will belong to the set $E(G)$ if, and only if, hosts i and j are sufficiently likely to make effective contact. In Section 9.3.3, we will illustrate how one may go about constructing contact networks that are relevant for the spread of infections in certain populations. But first let us get a little more familiar with the terminology of graphs.

Consider a triangle as in Figure 9.1. This is a very simple graph; let us name it K_3. The set $V(K_3)$ is equal to $\{1, 2, 3\}$ and each pair of vertices is connected by an edge. You can see from this example where the names "vertices" and "edges" come from. This terminology is used by extension for any graph G, even when there is no apparent edginess to the relationships (think of friendships) that G models.

In formal graph-theoretic terminology, we would write $E(K_3) = \{\{1, 2\}, \{1, 3\}, \{2, 3\}\}$. Note that we don't need to list the pairs $\{2, 1\}, \{3, 1\}, \{3, 2\}$ separately, because we are assuming symmetry and considering *unordered* pairs of nodes, which means that $\{i, j\} = \{j, i\}$. If you prefer, you can write $E(K_3) = \{\{2, 1\}, \{3, 1\}, \{3, 2\}\}$ instead; this is the same set of pairs. Pictures like Figure 9.1 are often the most convenient representations of graphs, at least if the set $V(G)$ is small. Figure 9.1 is certainly more intuitively appealing than the expression $K_3 = (\{1, 2, 3\}, \{\{1, 2\}, \{1, 3\}, \{2, 3\}\})$ that gives the formal mathematical definition of the same graph.

Figure 9.2 shows a different graph that we will refer to as G_1.

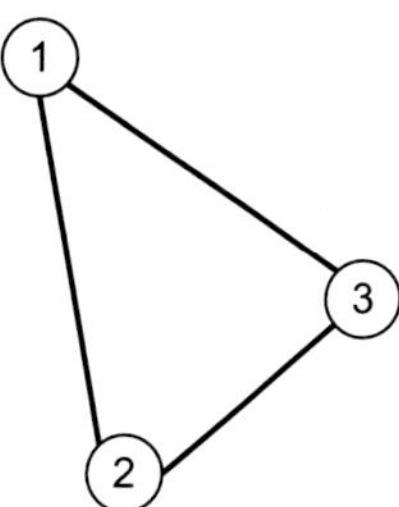

FIGURE 9.1 The triangle is a graph K_3 with three vertices.

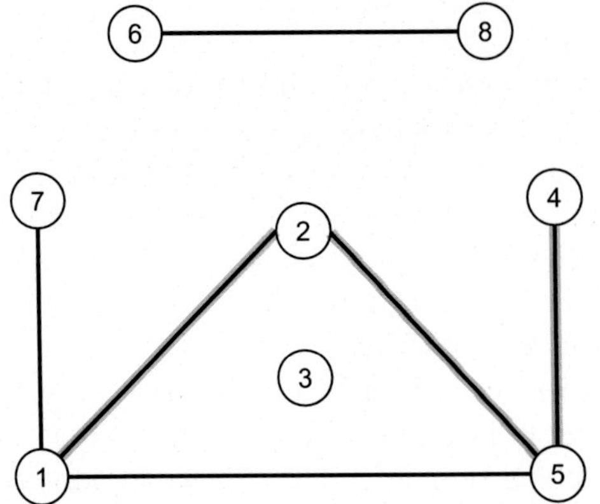

FIGURE 9.2 The graph G_1. The edges traversed by the path $P = (1, 2, 5, 4)$ are highlighted.

Exercise 9.11. Find the sets $V(G_1)$ and $E(G_1)$. □

There are some important differences between the graphs K_3 of Figure 9.1 and G_1 of Figure 9.2. Because in the graph K_3 every pair of distinct vertices is linked by an edge, this graph contains all *possible* edges of any graph with the given vertex set. Such graphs are called *complete.* The graphs that you saw in the **World** window in all exercises of Chapter 8 and Module 9.1 were complete. Traditionally, the notation K_N is used in graph theory for the complete graph with vertex set $\{1, 2, \ldots, N\}$. Notice that we don't allow edges of the form $\{i, i\}$. In contact networks, such "edges" (sometimes called *loops* in graph theory) would represent contact of a given host with itself. As such contacts play no role in disease transmission, we don't count loops as edges. Our graphs are *loop-free* or *simple* graphs.

The graph G_1 is not complete. For example, $\{1, 3\}$ would be a possible edge, but it is missing from the picture. In fact, node 3 is rather an oddball, as it is not connected to any other node. If G_1 represents a social network, then node 3 must feel rather lonely; it is what we call an *isolated node.*

When looking at Figure 9.2, you may think that it shows three graphs: One with vertices $1, 2, 4, 5, 7$; one with vertices $6, 8$ and a single edge $\{6, 8\}$; and one that has only vertex 3 and no edges. Technically, G_1 would be considered a single graph, though. The three distinct substructures, or *subgraphs,* that you can make out in the picture are called *connected components* of G.

Now work through part 9.2.1 of Module 9.2.

A *path* in a graph G is a sequence of nodes $P = (i_1, i_2, \ldots, i_m)$ such that each of the pairs $\{i_1, i_2\}, \{i_2, i_3\}, \ldots,$ $\{i_{m-1}, i_m\}$ is an edge of G. A path $P = (i_1, i_2, \ldots, i_m)$ is *simple* if the vertices $i_1, \ldots, i_m$ are all distinct. For example, $(1, 2, 5, 4)$ is a simple path in G_1 (Figure 9.2), the sequence $(1, 2, 5, 1, 7)$ is a path in G_1 but not a simple path, and $(1, 2, 4, 5)$ is not a path in this graph.

The *length* of a path $P = (i_1, \ldots, i_m)$ is $\ell = m - 1$, that is, the number of edges that we traverse along the path. There may be paths of different lengths between a given pair of nodes. For example, both $P_1 = (1, 5)$ and $P_2 = (1, 2, 5)$ are simple paths from 1 to 2 in G_1. The length of P_1 is equal to 1, while the length of P_2 is equal to 2.

The *costardom network* is an example of a graph whose nodes are movie actors, and two actors are linked by an edge if they ever appeared in the same movie. In the *Kevin Bacon game,* one player names an actor, and the other players try to find a path in this graph from this actor to Kevin Bacon. In fact, players try to find the shortest such path. You can play this game at the website [6].

The length of the shortest path between two distinct nodes i and j in a graph G is called the *distance* between i and j in G and denoted by $d(i,j)$. For example, the distance between node 7 and node 4 in G_1 is equal to 3, as the shortest path between these nodes is $P = (7, 1, 5, 4)$. We will consider sequences of the form (i) as legitimate paths from i to i. Because no edges are traversed, such paths have length 0. It follows that the distance of every node from itself is 0, which makes intuitive sense. It is not hard to see that $d(i,j)$ must be equal to the length of the shortest simple path: As we can cut out redundant cycles, a path that traverses the same node twice cannot be the shortest one between i and j.

Now work through part 9.2.2 of Module 9.2.

Exercise 9.12. Mathematicians have their own version of the Kevin Bacon game, where the network is the graph of scientific collaborations. Two nodes (mathematicians) are connected by an edge if they are coauthors of a joint scientific publication. The *Erdős number* is the distance of a node from Paul Erdős, who (co)authored around 1525 papers.

What is the Erdős number of Paul Erdős himself? What does it mean that the Erdős number of W. Just is 2? What can you deduce from this information about the Erdős numbers of the other authors of this chapter? □

In G_1, there is no path from node 1 to node 8 whatsoever. We will consider the distance between nodes 1 and 8 as infinite. If the distance between node i and node j is finite, we will say that node j is *reachable* from node i. We can now formally define the notion of the *connected component of a node i in a graph G* as the subgraph whose vertex set comprises all nodes at a finite distance from i, that is, all nodes that can be reached by a path in G that starts at i. This includes i itself. The set of edges of the connected component are all those edges of the original graph G that connect vertices of the connected component. A graph G is *connected* if it has exactly one connected component, that is, if every vertex can be reached via a path in G from every other vertex. For example, K_3 is connected (all complete graphs are), while G_1 is not.

The *diameter* of a graph G, denoted by diam(G), is the largest distance between any two of its nodes. The diameter of G_1 is infinite; only connected graphs have finite diameter. It is not hard to see that the diameter of the connected component of node 1 in G_1 is 3, as nodes 7 and 4 are at the largest distance in this subgraph. The diameter of K_3 is 1, and so is the diameter of the connected component of node 8 in G_1. Both of these graphs are complete.

Exercise 9.13. Is it true that a graph G is complete if, and only if, $\text{diam}(G) = 1$? □

Two nodes that are connected by an edge in a given graph are called *adjacent*. The total number of nodes j that are adjacent to node i is called the *degree* of i and will be denoted by k_i. In particular, a vertex is isolated if, and only if, it has degree zero.

The *mean degree* will be denoted by $\langle k \rangle$. For a network of size N it is equal to

$$\langle k \rangle = \frac{1}{N}\sum_{i=1}^{N} k_i = \frac{k_1 + k_2 + \cdots + k_N}{N}. \tag{9.5}$$

Because every edge contains two nodes, it gets counted twice in the numerator of (9.5). This implies the following alternative formula for calculating the mean degree:

$$\langle k \rangle = \frac{2|E(G)|}{N}. \tag{9.6}$$

Exercise 9.14. Find the degrees k_i of all nodes in G_1 and the mean degree $\langle k \rangle$. □

Graphs in which all nodes have the same degree k are called *k-regular graphs* or simply *regular graphs*. Complete graphs are examples of such graphs.

Exercise 9.15. Find a formula for the degree of each node and the total number of edges of a complete graph K_N. □

Now work through part 9.2.3 of Module 9.2.

We have already informally used the notion of a *subgraph*. Connected components are rather intuitive examples. They are examples of subgraphs that are *induced* by a subset of the vertices. In general, for a given graph $G = (V(G), E(G))$ and a subset $V^- \subseteq V(G)$, the *induced subgraph* $G^{\text{ind}}(V^-)$ is the graph with vertex set $V(G^{\text{ind}}(V^-)) = V^-$ whose edge set $E(G^{\text{ind}}(V^-))$ consists of *all* edges in the original graph G that connect vertices in V^-.

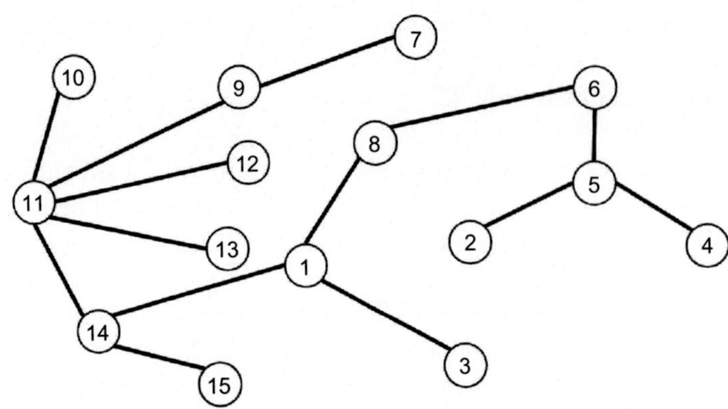

FIGURE 9.3 The graph G_T. It is a tree with set of leaves $\{2, 3, 4, 7, 10, 12, 13, 15\}$.

Not all subgraphs are induced subgraphs. For example, consider the subgraph $G_2 = (\{1, 2, 5\}, \{\{1, 2\}, \{1, 5\}\})$ of the graph of G_1. This is not an induced subgraph, as $\{2, 5\}$ is an edge in G_1 but not an edge in G_2. In general, a subgraph of a graph G is a structure that can be obtained from G by possibly removing some edges, and possibly some nodes, together with the edges to which they belong.

Exercise 9.16. Find all subgraphs and induced subgraphs of the complete graph K_2. □

In our definition of a path $P = (i_1, \ldots, i_m)$, we did not require that the nodes are all distinct, nor did we rule out repeated travel along the same edge. In particular, we can have $i_1 = i_m$. For example, $P_2 = (1, 2, 1)$ and $P_3 = (1, 2, 3, 1)$ are paths from 1 to 1 of length 2 and 3, respectively, in K_3.

A *cycle* is a path $P = (i_1, i_2, \ldots, i_m = i_1)$ with $m > 1$ from i_1 to i_1 such that the nodes $i_1, i_2, \ldots, i_{m-1}$ are all distinct and no edge is traversed twice along the path. Thus P_3 is a cycle in K_3, while P_2 is not, as it traverses the same edge twice. Similarly, $(1, 2, 5, 1)$ is a cycle in G_1, but $(1, 2, 5, 1, 2, 5, 1)$ is not, as $(1, 2, 5, 1, 2, 5)$ contains repeated nodes.

Graphs without cycles are called *acyclic*. Figure 9.3 gives an example.

This picture resembles a fallen tree, with nodes representing branching points or leaves. If you draw graphs of real trees in this way, you will not get any cycles, as distinct branches don't grow together higher up in the tree. Also, you will get a connected graph. For this reason, acyclic connected graphs are called *trees*. Acyclic graphs are often called *forests*, as each of their connected components must be a tree. In a tree or forest, nodes with degree 1 are called *leaves*.

Exercise 9.17. Show that a graph G is a tree if, and only if, for every pair of nodes $i \neq j$ there exists exactly one simple path from i to j in G. □

Now work through part 9.2.4 of Module 9.2.

Take-home message:

- Contact networks can be represented by mathematical structures called *graphs*. A graph is a pair $G = (V(G), E(G))$, where $V(G)$ is a nonempty set of *nodes* or *vertices*, and $E(G)$ is a set of unordered pairs of nodes that represent the *edges*. A graph that contains all possible edges between its N vertices is called *complete* and denoted by K_N.
- A *path* in a graph G is a sequence of nodes so that each pair of subsequent nodes is connected by an edge. The *length* of the path is the number of edges that are traversed by it.
- The *distance* $d(i, j)$ between any two nodes i, j in a graph G is the smallest length of any path that connects these nodes. If no such path exists, the distance is infinite. The *connected component* of a node i in a graph G is the *subgraph* of G that is *induced* by all nodes that are at a finite distance from i. A graph that has exactly one connected component is *connected*.

- The *degree* k_i of a node i in a graph G is the number of nodes j that are *adjacent* to i. Graphs in which all nodes have the same degree k are called *k-regular*.
- Graphs that do not contain *cycles* are called *acyclic* or *forests*. Connected acyclic graphs are called *trees*. Nodes of degree 1 in trees are called *leaves*.

9.3.2 Disease Transmission on Networks

Let us now describe how our models of disease transmission on networks are constructed. We assume a population of hosts that are numbered $1, \ldots, N$. Time may be discrete or continuous. As in compartment-level models, we assume homogeneity of hosts; that is, the length τ_i^I of the interval of infectiousness will have the same distribution for all hosts i, and the probability v that a given contact is effective will be the same for all contacts. In other words, we still consider a population of copies of Joene Average, but in the version exemplified by Joene Choosy. Each of them may be quite selective about with whom s&he tends to make contact.

More precisely, we will assume that we are given a contact network G with $V(G) = \{1, \ldots, N\}$, and contacts occur *only* between hosts i,j for which $\{i,j\} \in E(G)$, that is, between hosts that are represented by *adjacent nodes* in the graph G. If such an edge is present, then we assume that the frequency of contact does not depend on the particular hosts i,j that are connected by the edge. In the context of a discrete-time model, Eq. (9.1) simplifies to

$$b_{i,j} = \begin{cases} b = cv & \text{if } \{i,j\} \in E(G), \\ 0 & \text{if } \{i,j\} \notin E(G). \end{cases} \tag{9.7}$$

Recall that c stands for the probability that there will be at least one contact between i and j during a unit time interval, and v stands for the conditional probability that at least one of the contacts will be effective *given* that there will be at least one contact.

You will have noticed the similarity of (9.7) to (9.2): If G happens to be the complete graph K_N, then $\{i,j\} \in E(G)$ if, and only if, $i \neq j$. Thus, network models on complete graphs are in effect models that make the uniform mixing assumption.

However, all network-based models, even the ones based on complete graphs, are *host-level models,* as they incorporate representations of individual hosts. This makes them conceptually different from compartment-level models that are based only on counts of the number of hosts that reside in each compartment at any given time. In a host-level model, the *state* $\mathrm{St}(t)$ *of the population* at any given time t can be represented by a sequence of symbols as in Eq. (8.3) of Chapter 8. For large populations, these sequences would be too tedious to write out by hand, but the computer will happily keep internal records of such long sequences.

To see how these models work, assume that $N = 8$ and the contact network is described by the graph G_1 of Figure 9.2. Assume that we are considering an *immunizing disease* where hosts gain permanent immunity upon recovery, and let the initial state be $\mathrm{St}(0) = (I, S, S, S, S, S, S, S)$. Consider the following hypothetical recordings of the states at times 1 and 2:

- Scenario 1: $\mathrm{St}(1) = (RISSISSS)$, $\mathrm{St}(2) = (RRSIRSSS)$.
- Scenario 2: $\mathrm{St}(1) = (RIISSSSS)$, $\mathrm{St}(2) = (RRISSSSI)$.
- Scenario 3: $\mathrm{St}(1) = (RISSSSSS)$, $\mathrm{St}(2) = (RRSISSSS)$.
- Scenario 4: $\mathrm{St}(1) = (RISSSSSS)$, $\mathrm{St}(2) = (RRSIRSSS)$.
- Scenario 5: $\mathrm{St}(1) = (RISSSSSS)$, $\mathrm{St}(2) = (RRSISSIS)$.

Exercise 9.18. Which of these scenarios could actually be observed in a network-based model for this disease with contact network G_1? □

Now work through Module 9.3.

Take-home message:

- *Network-based* models of disease transmission are built on the assumption that pathogens can be transmitted only between hosts that are connected by an edge in a given graph G that represents the *contact network.*
- In the network-based models that we consider here, we make the assumption of homogeneity of hosts. Moreover, we assume that the probability of an effective contact over a time interval of specified length Δt between two hosts i and j is equal to a fixed parameter b if $\{i,j\}$ is an edge in G and is 0 otherwise.

9.3.3 Examples of Contact Networks

Network-based models are often more realistic than compartment-level models as they allow us to study, to some extent, how the pattern of mixing influences the expected dynamics of disease transmission. However, the construction of network-based models still involves two simplifying assumptions about the mixing pattern: One is that the probability of a contact between i and j over a given time interval is either 0 (if there is no edge in the network) or fixed and positive. The other assumption is that the network itself does not change over time. In this subsection we will illustrate how one can construct some simple network models that may approximate real-world situations reasonably well within these constraints.

Let us start by considering the monastic order of the Sisters of the Round Table. The sisters spend most of their lives in their individual cells, where they devote themselves to prayer and meditation. The only time they have contact with each other is during meals that they take seated in a fixed order around a giant round table. Within this community, diseases can be transmitted only during mealtime.

The probability of transmission will be largest between sisters who sit next to each other, and then decrease with the distance at the table. It may depend on the particular nature of the disease how far the infectious agents can travel. When constructing a network model, we need to make a decision on how we want to set the cutoff. Making a reasonable choice here is part of the art of modeling; there are no fixed rules. Let us assume that there is a significant probability of transmission from sister i to sister j if at most one sister sits in between. This will be the case if $|i-j| \leq 2$. But because the table is round, this will also be the case if, for example, $i = 2$ and $j = N$. Now we can construct the corresponding contact network $G^1_{\text{NN}}(N,2)$ by defining the set of edges $E(G^1_{\text{NN}}(N,2))$ as follows:

$$\{i,j\} \in E(G^1_{\text{NN}}(N,2)) \text{ if, and only if, } 1 \leq \min\{|i-j|, |i+N-j|, |N+j-i|\} \leq 2. \tag{9.8}$$

More generally, if there is a significant probability of transmission between sisters that have at most $d-1$ seats between them, we would replace the number 2 in (9.8) by d. The resulting graph will be denoted by $G^1_{\text{NN}}(N,d)$ and called *one-dimensional nearest-neighbor graphs* in this text. In IONTW, these graphs are implemented by the option **network-type → Nearest-neighbor 1**.

Now work through part 9.4.1 of Module 9.4 and take a look at these graphs.

For our next example, let us consider a population of trees that are planted in a regular grid, at one unit of length apart. Consider a slow-spreading fungal disease. Trees don't make "contact" in any obvious way, but we can still consider a contact network of sorts by drawing an edge between two trees in this plantation if there is a significant probability that the fungus can spread directly from tree i to tree j. Again, this probability will decrease with increasing distance, and we need to decide on a cutoff. If we choose the cutoff at a distance of ≤ 1, we end up with a ***R**ectangular g**R**id* $G_{\text{RR}}(m,n)$, where the parameters m,n denote the numbers of rows and columns of trees in the plantation. If we choose the cutoff at Euclidean distance of ≤ 1.5 instead, we end up with a rectangular *g**R**id with **D**iagonals* $G_{\text{RD}}(m,n)$.

Similar grids may be relevant to the example of Cocoa swollen-shoot virus (CSSV) that was discussed in Section 8.1 of Chapter 8. In this disease, the virus is spread by mealybugs between close-by cocoa trees. The bugs can travel farther than a distance of 1.5, though. If we were to model this disease, we would need to use more general versions of grids. Such grids will be denoted in the text by $G^2_{\text{NN}}(N,d)$ and called *two-dimensional nearest-neighbor graphs.* Here N is the total number of nodes and the parameter d controls the cutoff distance. The details will be investigated in Exercise 9.57 of Module 9.4. For at least some pairs (m,n) with $mn = N$, the rectangular grid $G_{RR}(m,n)$ is a special case of $G^2_{NN}(N,1)$, and the rectangular grid with diagonals $G_{RD}(m,n)$ is a special case of $G^2_{NN}(N,2)$.

Now work through part 9.4.2 of Module 9.4 and take a look at these graphs.

The networks we have constructed so far are examples of *spatial networks*, where nodes are located in physical space and edges are drawn, at least preferentially, between nearby nodes.

For an example of a different kind, consider the mathematicians at the Ivory Tower Research Institute. Similar to the Sisters of the Round Table, they lead solitary lives. They work mostly in the privacy of their offices, but each of them belongs to at least one research group that sometimes meets for seminars and discussion. The analysis group comprises Drs. 1, 2, and 3; Drs. 4 and 5 work in algebra; the probability and statistics group consists of Drs. 6, 7, 8, and 9; Drs. 2, 10, and 11 work on differential equations; and Drs. 5, 7, 8, and 11 work on applications to biology.

Exercise 9.19. Construct the contact network G_{IV} that, according to the story, seems most appropriate for transmission of diseases in this community. Spell out the assumptions that you made in constructing this network. □

Now work through part 9.4.3 of Module 9.4.

9.3.4 Additional Graph-Theoretic Notions

9.3.4.1 The Degree Distribution of a Graph

Suppose G is a given graph with vertex set $V(G) = \{1,\dots,N\}$. For each nonnegative integer k, let Q_k count the number of nodes i with degree k, and let $q_k = \frac{Q_k}{N}$. Thus Q_k represents the *number* of nodes with degree k, while q_k represents the *proportion* of nodes with degree k.

For example, for the graph G_1 of Section 9.3.1, we get $Q_0 = 1, Q_1 = 4, Q_2 = 1, Q_3 = 2$ and $q_0 = \frac{1}{8}, q_1 = \frac{1}{2}, q_2 = \frac{1}{8}, q_3 = \frac{1}{4}$. For another example, notice that G is k-regular if, and only if, $Q_k = N$ and $q_k = 1$. Because the sequences of counts $\bar{Q} = (Q_0, Q_1, \dots)$ and proportions $\bar{q} = (q_0, q_1, \dots)$ are just two different ways of representing the same information, we will refer to both of them as the *degree distribution of G* and use whichever representation is more convenient for a given purpose.

Because the maximum possible degree in any graph with N nodes is $N-1$, the mean degree $\langle k \rangle$ can be computed as

$$\langle k \rangle = \frac{1}{N}\sum_{k=0}^{N-1} kQ_k = \sum_{k=0}^{N-1} kq_k = \sum_{k=0}^{\infty} kq_k. \tag{9.9}$$

You will recognize the first part of (9.9) as the formula that you discovered in Exercise 9.44 of Module 9.2.

Exercise 9.20.

(a) Find $\langle k \rangle$ for the networks $G_{RR}(3,5)$, $G_{RD}(3,5)$, and G_{IV} that were introduced in Exercise 9.19.

(b) Many species of Gibbons are believed to be sexually entirely monogamous. What would this imply about $\langle k \rangle$ for the sexual contact network in a population of Gibbons? □

For large networks, the parameter $\langle k \rangle$ can often be meaningfully estimated by closely studying a small sample of the network. Performing such estimates in practice poses considerable challenges, though.

Now work through part 9.5.1 of Module 9.5.

9.3.4.2 Neighborhoods

A very useful notion for disease modeling is that of *neighborhoods* of a given node i. Let $i \in V(G)$, and let $\mathcal{N}_1(i) = \{j \in V(G) : \{i,j\} \in E(G)\}$ denote the set of all nodes j that are adjacent to i in G. We can think of $\mathcal{N}_1(i)$

as the set of nearest *neighbors* of i in G. The subscript 1 signifies that the neighbors in this set all have a distance of 1 from i in G. We will also use the more general notions of the sets $\mathcal{N}_\ell(i)$ and $\mathcal{N}_{\leq \ell}(i)$ that consist of all vertices that are at a distance of *exactly* ℓ and *at most* ℓ respectively from i. In particular, $\mathcal{N}_{\leq 1}(i) = \mathcal{N}_1(i) \cup \{i\}$.

We already used this notation in part 9.2.4 of Module 9.2. Recall that if G is a tree with root i, then $\mathcal{N}_\ell(i)$ is the set of nodes at level ℓ of the tree.

Exercise 9.21.

(a) Find the neighborhoods $\mathcal{N}_1(3), \mathcal{N}_{\leq 1}(3), \mathcal{N}_2(3)$, and $\mathcal{N}_{\leq 2}(3)$ in the graph $G^1_{NN}(9, 2)$.

(b) How are the Erdős numbers that were mentioned in Section 9.3.1 related to the neighborhoods in the scientific collaboration graph?

(c) Suppose G is a graph with vertex set $\{1, \ldots, N\}$. Show that for every node i the connected component of i in G is equal to the set $\mathcal{N}_{\leq N-1}(i)$. □

Now work through the remainder of Module 9.5.

9.3.4.3 *Random Contact Networks*

In all the examples of the previous subsection, the populations were small and contacts were rather limited and rigidly structured. For most real-world situations, the picture is much more complicated. Mathematicians at a university may secretly yearn for the quasi-monastic lifestyle of the Ivory Tower Research Institute, but apart from like-minded colleagues, they will meet regularly with students, and, believe it or not, most mathematicians do have a social life.

For transmission of most diseases in large real communities, it seems impossible to construct the actual relevant contact network. The difficulty here goes far beyond deciding on the cutoff transmission probability for including an edge. The flu, for example, can be transmitted during a variety of contacts and in various settings: In the workplace, at home, or during a public-transportation commute. It is not currently feasible to trace the frequencies of all these contacts for, say, the population of an entire city. This may be for the better. As some countries such as ■■■■■[3] are making rapid progress toward universal surveillance, the prospect looks technically feasible in the near future. Make up your own mind whether such a future is desirable.

STIs are an interesting case that falls somewhere between the rigid structure of the toy networks of Section 9.3.3 and the spread of the flu in New York City. The nature of the required contact is rather clear-cut. Most people would be able to name their sex partners, while they may have no clue who was sneezing next to them on subway line *B*. Human societies spend enormous energy on efforts to impose a rigid structure on sexual networks, but individual humans spend similar amounts of energy on evading the societal sanctions. For this reason, many people value their privacy in sexual matters and it seems rather difficult to collect reliable data on the sexual contact network. It is somewhat surprising that even before ■■■■■ started collecting data on all online activities and phone conversations, fairly reliable data on sexual contact networks could be compiled with the help of old-fashioned questionnaires [7, 8].

But how could the predictions of network models of disease dynamics possibly be relevant to real outbreaks if in most cases we don't even know the actual contact network? Even if we trust the data on networks of sexual contacts for the communities studied in [7, 8], how could we use them to predict the spread of an STI in a different population? This seems quite impossible at first blush, but let's not blush so easily. Recall that most epidemiological studies aim to predict the course of a future outbreak in one population based on data collected for *another* population. In the context of compartment-level models, this requires estimating parameters such as R_0 for previously observed outbreaks in other populations. We can then make predictions for our target population by assuming that the parameters will be very similar and using results such as the theorems of Section 9.2.2.

We can choose the graphs G for our modeling in an analogous way: Collect some data on actual populations as best as we can and determine what kind of properties these contact networks have. The "properties" we are

3. The authors apologize for the typesetting error. We explicitly deny baseless rumors about intervention of a government agency.

talking about here usually can be expressed in terms of numerical parameters, such as the mean degree $\langle k \rangle$. Then choose *randomly* a graph that models the contact network. Do this in such a way that with high probability it will have properties similar to the ones that have been empirically observed for contact networks of interest. Finally, simulate the dynamics of disease transmission on one or more such *random graphs.*

Suppose we have estimated some network parameters, such as the mean degree $\langle k \rangle$, from data on real contact networks. Moreover, suppose we have a method for sampling random graphs with roughly these parameters. If we choose several such graphs, we might reasonably expect that they form a *representative sample* of actual contact networks. Essentially this is the same trick that we are using throughout when we run simulations. During the run of multiple simulations, the computer produces in effect a representative sample of state transition sequences that could occur in a given model, and we can use the outcomes of simulations to form reasonably reliable hypotheses about what should happen in real outbreaks. Considering "random" objects (graphs or state transition sequences) is a mathematician's way of dealing with uncertainty. The only novelty is that we are now considering two distinct sources of uncertainty: Uncertainty about the actual contact network, and uncertainty about the actual sequence of state transitions that will happen in a model based on a given network. In order to draw valid conclusions, we will need to take into account both the variability that we see between multiple runs of simulations of a model that is based on a fixed random graph (variability *within* a fixed model) as well as the variability that we observe when we average the results of multiple runs performed for fixed models, and compare these averages for multiple draws of random graphs (variability *between* models).

Constructing representative samples of random graphs whose expected properties closely mirror data from real networks is in general not easy. There are various ways of constructing random graphs. These constructions give us networks that usually share some, but not all properties of real networks. In the next section we will describe the most basic of these constructions.

Take-home message:

- The *degree distribution* of a given graph G can be specified either in terms of the counts Q_k or proportions q_k of nodes with degree k.
- The *neighborhood* $\mathcal{N}_{\leq \ell}(i)$ of a node i consists of all nodes j that are at a distance of at most ℓ from i in a given graph G. Similarly, the neighborhood $\mathcal{N}_{\ell}(i)$ of a node i consists of all nodes j that are at a distance of exactly ℓ from i in G.
- *Random graphs* can be used as models of contact networks when the actual network is not known, but some estimates of its structural properties, such as the mean degree $\langle k \rangle$, are available.

9.3.5 Erdős-Rényi Random Graphs

The simplest type of random graphs are *Erdős-Rényi* random graphs, named after the two Hungarian mathematicians who first studied them systematically in [9]. To construct such a graph, we first decide on the set of nodes $\{1, 2, \ldots, N\}$. Then we list all *possible* edges $e_1, \ldots, e_{(N(N-1))/2}$ and repeatedly toss a biased coin that comes up heads with probability p. We include edge e_m as an actual edge of the random graph if, and only if, the coin comes up heads in toss number m. The material of Chapter 8 will have convinced you that this procedure is very easy to implement in computer code.

We will show now that by choosing a suitable value of the *connection probability* p, one can ensure that the mean degrees $\langle k \rangle$ of the resulting random graphs will be close to the values that one might have estimated from data on real networks. Assume N and p are fixed and consider a node i. There are $N - 1$ other nodes j that could be adjacent to i. For each of these nodes j, membership of the pair $\{i, j\}$ in the edge set will be decided by tossing a biased coin that comes up heads with probability p. It follows that the degree of node i will be equal to the number of "successes" in these independent coin tosses. This implies that the degree of i will be an r.v. with a binomial distribution with parameters $N - 1$ and p (see the review on probability at our website [2]). It follows that the mean degree $\langle k \rangle$ of the resulting graph will satisfy

$$\langle k \rangle = \lambda = p(N - 1). \tag{9.10}$$

It will be more convenient for us to think of Erdős-Rényi random graphs in terms of the parameter λ instead of the parameter p. The connection probability p can then be easily calculated as $p = \frac{\lambda}{N-1}$. The symbol $G_{\text{ER}}(N, \lambda)$ will denote an Erdős-Rényi random graph that is constructed with parameters N and λ.

Note that we used the indefinite article *an* in the previous sentence. Until now, all our symbols for graphs referred to one specific graph. But $G_{\text{ER}}(N, \lambda)$ is no longer uniquely determined; in fact, it could be *any* graph G with vertex set $V(G) = \{1, \ldots, N\}$. The symbol $G_{\text{ER}}(N, \lambda)$ only signifies that the graph is randomly drawn from a specific probability distribution. We will call a particular graph that has been constructed by the method described above *an instance of* $G_{\text{ER}}(N, \lambda)$. For example, if λ is much smaller than $\frac{N-1}{2}$, the probability distribution makes it a lot more likely that we draw instances that are *sparse* graphs (with relatively few edges) than *dense* graphs (with a lot of edges), but it does not even entirely rule out the possibility that $G_{\text{ER}}(N, \lambda)$ will be the complete graph K_N. It only makes the latter possibility very, very unlikely. Thus, all the properties of random graphs that we will be considering are *average* properties.

In particular, the mean degree of a given $G_{\text{ER}}(N, \lambda)$ will usually *not exactly* satisfy (9.10). But if N is large, the mean degree will satisfy (9.10) with *very good approximation.* The formula becomes exact if we take the mean over all graphs $G_{\text{ER}}(N, \lambda)$ according to the probability distribution for parameters N, λ.

Binomial degree distributions are a bit cumbersome to work with. For large N and relatively small λ (as a rule of thumb, if $\lambda < 10$), a binomial distribution can be approximated reasonably well with a *Poisson distribution* with parameter λ (see the review on probability at our website [2]). In this case we should expect a degree distribution in $G_{\text{ER}}(N, \lambda)$ with

$$q_k \approx \frac{\lambda^k \mathrm{e}^{-\lambda}}{k!}. \tag{9.11}$$

These probabilities q_k decrease rapidly to 0 as $k \to \infty$. In particular, as $N \to \infty$, the expected maximum degree in $G_{\text{ER}}(N, \lambda)$ will grow a little slower than $\ln(N)$.

For larger values of λ, the degree distribution in $G_{\text{ER}}(N, \lambda)$ will be approximately normal, with a distinct peak near $\langle k \rangle = \lambda$ and the vast majority of nodes having degrees that are close to the mean degree. In particular, if we keep p fixed and increase N, the mean degree $\lambda = p(N-1)$ will scale roughly linearly.

The construction of Erdős-Rényi random contact networks implies that the nodes j that are adjacent to a given node i are chosen randomly. This looks almost identical to the uniform mixing assumption, which postulates that the hosts j with whom host i makes contact during i's period of infectiousness are randomly chosen from the population. Thus, one might conjecture that disease transmission on an Erdős-Rényi random contact network should be very similar to what is predicted under the uniform mixing assumption. Online Module 9.6 will lead you to discovering to what extent this conjecture is true.

Now work through Module 9.6.

Take-home message:

- Erdős-Rényi random graphs $G_{\text{ER}}(N, \lambda)$ are the simplest type of random graphs.
- The parameter λ denotes the mean degree for the probability distribution of all *instances* in a given construction of Erdős-Rényi graphs with N nodes.

9.4 SUGGESTIONS FOR FURTHER STUDY

Erdős-Rényi random graphs are important as baseline models for statistical comparisons. But their applicability to modeling actual contact networks is rather limited, as they are in some respects very much unlike real contact networks. In particular, degree distributions in most actual contact networks are usually very different from binomial distributions. Often, these degree distributions approximate *scale-free* distributions. Real contact networks also tend have much higher *clustering coefficients* than do Erdős-Rényi networks.

There are alternative constructions of random networks that have these properties, and their study is a vibrant area of current research. At our website [2], the reader can find additional modules that provide a natural continuation of the material in this chapter and a good entry point into the advanced literature on these types of random networks.

At the very end of the last century, two seminal papers appeared in *Science* and *Nature:* Barabási and Albert [10] described a mechanism by which random networks with approximately scale-free degree distributions might evolve, while Watts and Strogatz [11] described a construction of *small-world networks* that simultaneously have high clustering coefficients and small diameters. Modifications of these and other constructions of random graphs that share important properties with empirically studied contact networks have been extensively studied ever since. A collection of important classical papers on the subject can be found in [12]; the survey [13] and the book [14] give comprehensive introductions. Chapter 9 of [15], Chapter 17 of [14], and Chapter 7 of [16] cover advanced material on disease transmission on networks that may be of interest to students who have worked through the modules in the present chapter and the follow-up modules on our website [2].

ACKNOWLEDGMENTS

We are greatly indebted to Natalia Toporikova for valuable suggestions that contributed to the design and editing of this chapter, as well as for the figures. Special thanks are due to Rebecca Driessen, Todd Graham, Claire Seibold, and Ying Xin for valuable comments on earlier versions of this chapter and Chapter 8.

REFERENCES

[1] Just W, Callender H, LaMar MD. Online Appendix: Disease transmission dynamics on networks: Network structure vs. disease dynamics.. In: Robeva, R, editor. Algebraic and discrete mathematical methods for modern biology. New York: Academic Press; 2015.

[2] Just W, Callender H, LaMar MD. Exploring transmission of infectious diseases on networks with NetLogo. Available from https://qubeshub.org/iontw and also http://www.ohio.edu/people/just/IONTW/.

[3] Just W, Callender H, LaMar MD, Toporikova N. Online Appendix: Transmission of infectious diseases: data, models, and simulations. In: Robeva, R, editor. Algebraic and discrete mathematical methods for modern biology. New York: Academic Press; 2015.

[4] Anderson RM, May RM. Directly transmitted infectious diseases: control by vaccination. Science 1982;215:1053-1060.

[5] Diekmann O, Heesterbeek H, Britton T. Mathematical tools for understanding infectious disease dynamics. Princeton, NJ: Princeton University Press; 2012.

[6] The oracle of Bacon. http://oracleofbacon.org/. Accessed: 2014-09-18.

[7] Jolly AM, Muth SQ, Wylie JL, Potterat JJ. Sexual networks and sexually transmitted infections: a tale of two cities. J Urban Health 2001;78:433-445.

[8] Lewin BF-M, Helmius K, Lalos G, Ann MS-A. Sex in Sweden-on the Swedish sexual life. Stockholm, Sweden: The Swedish National Institute of Public Health, 2000.

[9] Erdős P, Rényi A. On the evolution of random graphs. Rényi, A. Selected Papers of Alfréd Rényi, Budapest, Hungary: Akadémiai Kiadó 1976;2:482-525.

[10] Barabási A-L, Albert R. Emergence of scaling in random networks. Science 1999;286:509-512.

[11] Watts DJ, Strogatz SH. Collective dynamics of 'small-world' networks. Nature 1998;393:440-442.

[12] Newman M, Barabási A-L, Watts DJ. The structure and dynamics of networks. Princeton, NJ: Princeton University Press; 2006.

[13] Newman MEJ. The structure and function of complex networks. SIAM Rev 2003;45:167-256.

[14] Newman MEJ. Networks: an introduction. Oxford, UK: Oxford University Press; 2010.

[15] Barrat A, Barthelemy M, Vespignani A. Dynamical processes on complex networks. Cambridge, UK: Cambridge University Press; 2008.

[16] Pastor-Satorras R, Rubí M, Diaz-Guilera A. Statistical mechanics of complex networks. Berlin, Germany: Springer, p. 625; 2003.

Chapter 10

Predicting Correlated Responses in Quantitative Traits Under Selection: A Linear Algebra Approach

Janet Steven[1] and Bessie Kirkwood[2]

[1] *Christopher Newport University, Newport News, VA, USA,* [2] *Sweet Briar College, Sweet Briar, VA, USA*

10.1 INTRODUCTION

In 1858, Charles Darwin and Alfred Russel Wallace presented an idea to the scientific community that provided the framework for understanding how the diversity of life on earth came to be [1]. That idea, evolutionary change through natural selection, explains how lineages of organisms can change over time. Both Darwin and Wallace were inspired by voyages around the world, and subsequent observations of slight differences in closely related species. They also were influenced by an increasing acceptance in the scientific community that the earth was very old, and that geological processes generated changes in landforms over time. Both scientists independently developed an explanation for how change could occur in lineages of living organisms: they recognized that organisms with a greater ability to survive and reproduce would have more offspring in the next generation, and therefore those traits that increased survival and reproduction would increase in the population over time.

Darwin applied the fundamental principles of natural selection to many aspects of natural history [2]. He recognized that for natural selection to occur in a lineage, the trait or characteristic that is changing must be connected with survival and reproduction, not all individuals in a population can be identical for the trait, and the trait must be passed from parent to offspring. For example, Darwin argued that grouse that were of the same color as their habitat would be less likely to be seen by hawks, and conversely any bird whose color deviated from the optimal camouflage would be destroyed and unable to pass on its genetic information. While Darwin developed many excellent verbal arguments, an examination of natural selection in a quantitative framework did not come until later, and is connected with the development of statistical techniques. In 1898, Hermon Bumpus published one of the first detailed datasets quantifying traits related to survival in a population of sparrows [3]. His study carefully documented body size differences between sparrows surviving a severe winter storm and those that died. He provided quantitative evidence for two of the three critical aspects of Darwin's theory: variation in the traits in a population, and a clear connection between survival and some of the traits measured. However, at the time of Bumpus' study, scientists did not yet understand how traits were passed from parents to offspring.

In this chapter, we examine how natural selection on traits within a population leads to evolutionary change by altering the mean trait value from one generation to the next. We will see that the amount of that change depends on the strength of selection and the proportion of variance in the trait that is controlled by genetics. Both of these factors can be quantified and used to predict the outcome of selection.

Algebraic and Discrete Mathematical Methods for Modern Biology. http://dx.doi.org/10.1016/B978-0-12-801213-0.00010-1
Copyright © 2015 Elsevier Inc. All rights reserved.

This chapter lays out the mathematical background necessary to address these questions and provides multiple examples as illustrations. The chapter also includes exercises and projects, which use the language R for all statistical analyses, although one can also use any other statistical package. An introductory tutorial on R is provided as an online supplement and can be accessed from the volume's companion website. Readers who have not used R before and wish to use it here will need to work through the first two sections of the online R tutorial before beginning the exercises.

10.2 QUANTIFYING SELECTION ON QUANTITATIVE TRAITS

A *quantitative trait* is a characteristic of an organism that can be measured as a continuous variable; that is, it can take on any value within a range. Height and weight are both quantitative traits. The value of a quantitative trait in a particular organism can be determined by many genes, each contributing to the trait value, and may also be influenced by nongenetic factors. Unlike some traits controlled by a single gene, the genes controlling a continuous trait are difficult to determine directly from the measurement of the trait. However, biologists can estimate the genetic control over a trait by comparing parents and offspring.

In this section, we introduce some basic statistical approaches used for describing quantitative traits, use covariance to determine whether a trait is acted on by natural selection, and explore methods for determining the proportion of variance in a trait that is controlled by genetics.

10.2.1 Describing Traits Mathematically

Since the beginnings of agriculture, humans have known that offspring tend to look like their parents, and have been breeding plants and animals based on that understanding. The quantitative description of the genetic basis of inheritance began with Gregor Mendel in 1865. Mendel discovered that inheritance did not involve blending of traits in the offspring, but rather the inheritance of discrete "particles" of information that stayed intact across generations [4]. His work laid the foundation for understanding genes and allelic variants of those genes. Mendel's ideas didn't gain widespread acceptance in the scientific community until 1900, when they were rediscovered and became incorporated into a growing study of discrete variants in traits. Mendel's work, and the work of many turn-of-the-century geneticists, involved examining traits with two distinct forms in the population: wrinkled vs. round peas, for example, or fruit flies with red or white eyes. Inheritance in these traits was easy to observe in breeding experiments. But scientists still did not have an explanation for how continuous traits, like body size in sparrows, were controlled by genetics.

For natural selection to operate, individuals in a population must differ in the particular trait of interest, and therefore a quantitative approach to studying selection must involve measurement of those differences. In *continuous traits*, individuals can take on any value in a given range, and every individual in a population may be slightly different. In 1919, geneticist Ronald Fisher developed a mathematical model that demonstrated how a continuous trait could be controlled genetically by many genes, each contributing in small part to the value of the trait [5]. In the process, he also introduced the statistical concept of variance. This understanding of continuous traits was crucial for scientists in the 1930s and 1940s interested in merging the discoveries made for discrete traits by geneticists and the concept of natural selection proposed by Darwin.

In the last 30 years, evolutionary biologists have developed many quantitative techniques for examining natural selection, and have also amassed considerable empirical data that show patterns of natural selection in populations. Often, studies begin by identifying a trait in a population that is potentially under selection, and quantifying the mean and variance of that trait. Biologists define a population as a group of potentially or actually interbreeding organisms. Individuals in this group of organisms could be related and also may be living in the same area, subject to the same selection pressures. Therefore, the population is a natural unit of study in evolutionary biology, and it is the population itself that evolves as traits change over generations.

When a trait in a population has a symmetric, bell-shaped distribution, it may be well described using the normal distribution as a model. The normal distribution, in turn, is described by specifying its center, or location, and its spread, or variability. The center of the distribution of a variable X is its *mean* (also called expected value); it is usually denoted by the Greek letter μ, by $E(X)$, or μ_X. These symbols represent the true mean of the population; frequently, biologists do not measure every single individual in a population, but rather a sample of them. If the distribution of a sample of data is fairly symmetric and bell-shaped, the *sample mean* describes the center of the distribution of the sample. If the variable is X, its sample mean is denoted by $\overline{x}$ and is calculated as the arithmetic average:

$$\overline{x} = \frac{1}{n}\sum_{i=1}^{n} x_i \tag{10.1}$$

where n is the number of individuals in the sample.

The variability is given by the *standard deviation*, denoted by σ. The standard deviation is an average deviation (distance) of the population away from the mean [6]. Formally,

$$\sigma = \sqrt{E[(X-\mu)^2]} \tag{10.2}$$

A useful rule for interpreting the standard deviation is that in a population having a normal distribution, about 68% of the population values are within one standard deviation of the mean, about 95% are within two standard deviations of the mean, and rarely is a member of the population more than three standard deviations from the mean (Figure 10.1). Again, when the distribution of a sample of data is fairly symmetric and bell-shaped, the sample standard deviation describes the variability of the distribution of the sample. The standard deviation is an average of the deviation of the sample about the sample mean:

$$s = \sqrt{\frac{\sum (x_i - \overline{x})^2}{n-1}} \tag{10.3}$$

The variability of a distribution is also described by the *variance*, which is the square of the standard deviation. Population variance is denoted Var(X) or σ^2, and the latter notation is literal because the variance is the square of the standard deviation. Sample variance is denoted s^2.

While both variance and standard deviation are measures of variability, each has distinct advantages. Variance has mathematical properties that simplify some analyses. For example, if X and Y are independent variables, Var($X + Y$) = Var(X) + Var(Y) but the standard deviation of $X + Y$ does *not* equal (standard deviation of X) + (standard deviation of Y). On the other hand, variance units are square units but standard deviation is in the same units as the variable. So if X is in units of inches, the standard deviation of X is in inches while Var(X) is in square inches. Fortunately, it is easy to switch back and forth between the variance and standard deviation as convenience dictates.

The sample mean is commonly used as an estimator of the mean of the population from which the sample came, and it has some nice statistical properties. One weakness, however, is that it is sensitive to extreme values; that is, an outlier value will have disproportionate influence on the value of the sample mean. The sample standard deviation has the same weakness. For this reason, it is good practice to check

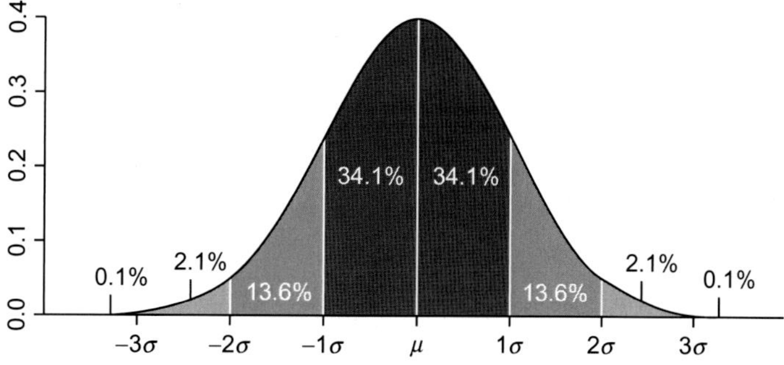

FIGURE 10.1 The normal distribution. By Mwtoews [CC-BY-2.5 (http://creativecommons.org/licenses/by/2.5)], via Wikimedia Commons.

the shape of a data distribution before making statistical inferences based on mean and standard deviation. Sometimes a skewed distribution can be transformed into a normal distribution by applying a well-chosen function, called a normalizing transformation, so that standard, normal-theory-based statistical methods can be applied.

Example 10.1. To explore these concepts further, we will work through an example involving the selective influence of hummingbirds on the flowers they pollinate. Hummingbirds visit flowers to drink nectar, and often pick up pollen from the anthers in the flower on their forehead and chin in the process. When they visit another flower, they can transfer the pollen to the female structure in the flower (the stigma). In this way, the hummingbird facilitates the reproductive success of the plant through both male function (because the sperm in the pollen grain fertilize an ovule in the female structures) and female function (because the ovule needs to be fertilized to develop into a seed). Therefore, floral traits that increase the amount of pollen collected and deposited during a hummingbird visit are likely to be favored by natural selection. In a study on *Polemonium brandegeei* (Brandegee's jacob's-ladder), Kulbaba and Worley measured floral traits that were likely to influence the effectiveness of hummingbird visits [7]. One of the traits they measured was the depth of the petal tube in the flower; deeper tubes mean hummingbirds may be more likely to come into contact with the pollen in the flower as they feed on the nectar at the base of the tube. In one of the five trials they conducted, plants had the following petal tube lengths (in mm) [8]:

$$25.45, 24.68, 19.35, 22.55, 23.25, 29.92, 27.30, 25.19, 26.87, 28.62, 27.60, 18.43$$

Using Eq. (10.1), we find that the sample mean floral tube length was 24.93 mm. This is useful as a general descriptor of the sample, but doesn't tell us much about variability. Some plants had floral tubes that were much shorter than others; the smallest tube was only 18 mm, while the longest was 30 mm. In contrast, hummingbird bills are 15-20 mm long. Therefore, the variation in floral tube length appears to be in the right range for selection to act on it. We can describe the population variation by calculating sample variance using Eq. (10.3); s^2 is 12.48 mm^2. The square root of the sample variance gives us the sample standard deviation, which is 3.53 mm. □

Exercise 10.1. On February 1, 1898, Hermon Bumpus gathered sparrows that had been harmed by a severe winter storm outside of his laboratory at Brown University. Of the 136 birds brought inside, 72 revived and the rest died. In order to examine whether survival was related to characteristics of the birds, Bumpus measured several aspects of the birds' size. Some of the traits Bumpus measured included length from the tip of the beak to the tip of the tail, weight, and the length of the humerus bone in the wing. All of Bumpus' data is presented in the text file *bumpus.txt* in the companion website for this volume. (Data were obtained from the Field Museum website: http://www.fieldmuseum.org/explore/hermon-bumpus-and-house-sparrows.)

The variables in the data set are

line	line number from Bumpus' tables
sex	m = male; f = female
age	for males only, a = adult, y = young; for females, the value is set to NA because this is how missing values are indicated in R
survival	TRUE if survived, FALSE if perished
totallength	"from tip of the beak to the tip of the tail" (mm)
alarextent	"from tip to tip of the extended wings" (mm)
weight	weight of the bird (g)
beakhead	length of beak and head (mm), "from tip of the beak to the occiput"
humerus	length of humerus (in.)
femur	length of femur (in.)
tibiotarsus	length of tibiotarsus (in.)
skullwidth	width of skull (in.), "from the postorbital bone of one side to the postorbital bone of the other"
sternum	length of keel of sternum (in.)

The following R commands will help get you started:

1. Save the data file in a convenient location and read the data into your R workspace using
```
>bumpus=read.table(file=file.choose(), header=TRUE)
```
2. A column of the data array is referenced using both the table name and the column name. For example, the weight variable is `bumpus$weight`. You may prefer to break out the columns into separate vectors with shorter names, for example,
```
>weight=bumpus$weight
```
3. Make histograms for weight and length of the humerus bone for all of the birds, and then for just the surviving birds. To limit data to only the surviving birds, use
```
>hist(bumpus$weight[bumpus$survival ==TRUE])
```
To help with comparisons, you can specify the x-axis scale. For example, set the range from 22 to 32 this way:
```
>hist(bumpus$weight[bumpus$survival ==TRUE],xlim=c(22,32))
```
What differences do you note in shape, center, and spread? Use mean and standard deviation in your comparisons.
4. Calculate the means of weight and length of the humerus bone for males and females separately. How do males and females differ?
R note: For the mean weight of females, use
```
>mean(bumpus$weight[bumpus$sex == "f"])
```
5. Comparing means for weight and length of the humerus bone, are there differences between birds that survived and all birds? Make comparisons separately for males and females.
R note: To specify females that survived, use the logical "and" operator, denoted by `&`. The mean weight of that subgroup, for example, is obtained with
```
> mean(bumpus$weight[bumpus$sex=="f" & bumpus$survival==TRUE])
```
6. Calculate the variance and standard deviation for weight and length of the humerus bone for males and females separately.
7. Are the variances in weight and humerus length the same for surviving and all sparrows? (Again, make separate comparisons for males and females.) If not, speculate on how natural selection might be at work.
8. Use the following line of R code to create a boxplot. Does it support your response to the previous question? What additional information does it provide? By modifying the code, create comparable boxplots for weight and humerus length for both males and females.
```
> boxplot(bumpus$weight[bumpus$sex=="f"],
bumpus$weight[bumpus$sex=="f" &  bumpus$survival == TRUE], names=c("all
females","surviving females"),  main="weight in g")
```
□

10.2.2 Quantifying Reproduction and Survival

Natural selection depends on the relationship between the variation in a trait and reproductive success. Biologists use the term *fitness* to describe the reproductive output of an individual. Darwin used the term "survival of the fittest" as shorthand for the idea of natural selection, but he was using the idea of "fitness" in a different sense than we use the word today. He was considering how well an organism "fit" its environment; organisms with variants of traits that made them better competitors for food, or more camouflaged from predators, "fit" their environment better and were therefore more likely to survive and have more offspring. At its heart, natural selection is about how much each individual contributes to the genetic information in the next generation, and therefore biologists studying evolution often want to know how many offspring an individual has over the course of its life. In theory, measuring fitness in a population of organisms should involve following each individual over its life and recording how many offspring it has before it dies. In practice, these data are difficult to collect in most species, and biologists may estimate fitness in a way that is more appropriate for their study organism.

The fitness term quantifies the contribution of an individual to the next generation of the population. Fitness can be estimated by counting the number of offspring an individual has over its lifetime. In this way, both survival and reproduction can be encapsulated into one term: an individual that doesn't survive to sexual maturity has no offspring and therefore has a fitness of 0. Biologists use the term w to indicate fitness. However, because natural selection changes an entire population over generations rather than changing an individual over its lifetime, the absolute number of offspring isn't as important as the relative number compared to other individuals in the population. A frog that sires 210 tadpoles might appear to have high fitness, but if the average number of tadpoles sired in the population is 300, then he will have a smaller contribution to the next generation than most frogs in the population. Selection favors individuals who have a greater than average contribution to the next generation; therefore, biologists often calculate *relative fitness* [9]. If $\overline{w}$ is the average fitness for a sample from a population, and the fitness of any given individual i in the sample is w_i, then the relative fitness of an individual is $w_i/\overline{w} = \hat{w}$. This term is useful because it includes both survival and reproduction of an individual and the relative contribution of genetic information by that individual in the next generation.

Example 10.2. In the hummingbird pollination study described above, the researchers also determined the number of seeds each plant sired in the experimental population during the study trials by using DNA markers [7]. The number of seeds sired is a measure of male fitness. The data on seeds sired for the first trial is as follows: [8]

$$4, \quad 11, \quad 6, \quad 3, \quad 8, \quad 11, \quad 12, \quad 2, \quad 9, \quad 6, \quad 4, \quad 4$$

Much like petal tube length, there is a considerable amount of variation in the number of seeds sired: one plant only sired two seeds during the study, while another plant sired six times that many. The average male fitness, $\overline{w}$, is 6.7. Therefore, the plant that only sired 2 seeds has a relative fitness $\hat{w}$ of 2/6.7, or 0.30, while the plant that sired 12 seeds has a relative fitness of 1.8. Note that the average relative fitness is 1. □

In order to operate, natural selection requires a connection between fitness and the trait potentially under selection. In other words, the trait must influence fitness in some way. Up to this point, we have used variance to describe the trait and we have quantified fitness. Next, we will examine the way we can describe the relationship between fitness and the trait of interest.

10.2.3 Describing the Relationship Between Fitness and a Trait

If there is a relationship in a population between variance in a trait and variance in fitness, then it indicates that the trait is under selection. We can quantify the relationship between two traits using *covariance*. While variance measures the variability of a single variable, covariance is a measure of the extent to which two variables "vary together." Formally [6], if X and Y are variables with means μ_x and μ_y, the covariance of X and Y, denoted Cov(X,Y), is the average or mean of the product $(X - \mu_x)(Y - \mu_y)$. That is,

$$\mathrm{Cov}(X,Y) = E\big[(X-\mu_x)\left(Y-\mu_y\right)\big] \tag{10.4}$$

If there is a positive association between X and Y, then larger values of X and larger values of Y tend to occur together, and smaller values of X tend to occur with smaller values of Y (Figure 10.2). Then the product $(X - \mu_x)$ $(Y - \mu_y)$ tends to be positive more often than not, and Cov(X,Y) is positive. The stronger the association, the larger the covariance will be. On the other hand, if there is a negative association between X and Y, smaller values of one tend to occur at the same time as larger values of the other, making the products $(X - \mu_x)(Y - \mu_y)$ tend to be negative, and hence Cov(X,Y) will be negative. If X and Y are measured with the same units, then Cov(X,Y) is in square units, as variance is.

One of the ways that covariance arises naturally in statistics is when a variable is the sum of two other variables, say $Z = X + Y$. Then

$$\mathrm{Var}(Z) = \mathrm{Var}(X+Y) = \mathrm{Var}(X) + \mathrm{Var}\,(Y) + 2\,\mathrm{Cov}(X,Y)$$

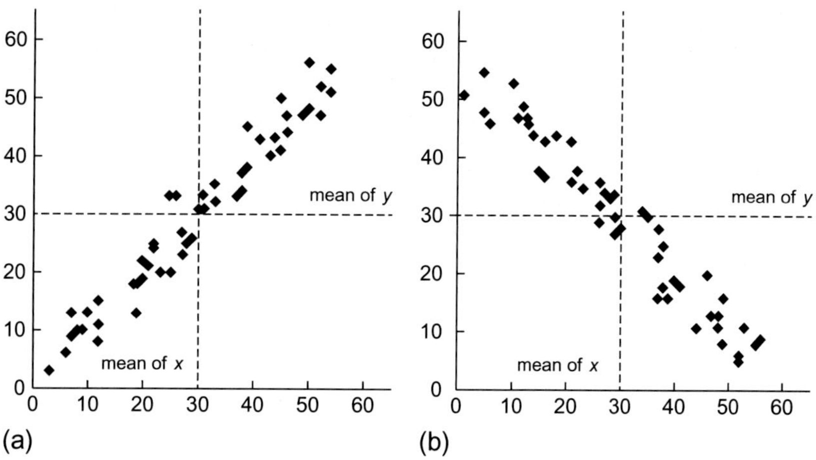

FIGURE 10.2 Scatterplots illustrating a positive (a) and negative (b) covariance. The vertical line $x = \mu_x$ and the horizontal line $y = \mu_y$ divide the plane into four quadrants. When Cov(X,Y) is positive, most of the data points fall into the quadrants where $(x_i - \mu_x)$ and $(y_i - \mu_y)$ are both positive *or* where they are both negative, so that the products $(x_i - \mu_x)(y_i - \mu_y)$ are more often positive. When Cov(X,Y) is negative, most of the points fall into the quadrants where one of $(x_i - \mu_x)$ and $(y_i - \mu_y)$ is positive and the other negative, so that the products $(x_i - \mu_x)(y_i - \mu_y)$ are more often negative.

If X and Y are independent, $\text{Cov}(X, Y) = 0$, from which it follows that

$$\text{Var}(X + Y) = \text{Var}(X) + \text{Var}(Y)$$

Note that the value of Cov(X,Y) will change if the units of measurement are changed. However, if covariance is divided by $\sigma_X\sigma_Y$, the resulting quantity is scale free [6]. This is the *correlation* between X and Y, denoted ρ for the true value of a population or r for a sample:

$$\rho = \frac{\text{Cov}(X, Y)}{\sigma_x\sigma_y} \tag{10.5}$$

$$r = \frac{\text{Cov}(X, Y)}{s_x s_y} \tag{10.6}$$

This adjustment guarantees that $-1 \leq r \leq 1$, but there are no similar limits on the range of values of covariance.

The relationship between two variables can also be described statistically using linear regression. The connections between covariance and simple linear regression are discussed in a later section of this chapter.

We have already determined that the fitness of an individual relative to its population is the most useful way to describe fitness, giving us the term $\hat{w}$. We will use the term z to indicate the value of the trait of interest. To determine whether the trait is changing due to natural selection, we look at the covariance between relative fitness, $\hat{w}$, and the trait value, z. The value we obtain is the average amount of change we expect in the trait from one generation to the next; we will represent this change with the term $\Delta\bar{z}$. Therefore, the equation to describe change due to natural selection on a trait becomes [10]:

$$\Delta\bar{z} = \text{Cov}(\hat{w}, z) \tag{10.7}$$

Example 10.3. In our hummingbird example, we have values for a trait potentially under selection, petal tube length, and values for fitness. Therefore, we can examine how relative fitness covaries with petal tube length to learn whether the trait is under selection. The covariance between relative male fitness and petal tube length is 0.72 mm, suggesting that the trait could experience selection. However, a change of 0.72 mm from one generation to the next is not very large, considering that the sample mean is 25 mm. In addition, we have yet to consider the importance of genetic and environmental effects in determining the trait value. □

Exercise 10.2. Animals vary in the size of the meals they eat: Some animals, like snakes, eat only every few months but consume a prey item that is very large in comparison to their body size. Other animals, like many bird species, eat several small meals a day. Eating a large meal could mean more energy and resources for producing offspring, but it could also mean decreased mobility and therefore greater susceptibility to predation. Therefore, the amount of food consumed at one time could potentially be a trait correlated with fitness because it might affect both reproduction and survival. In 2010, Pruitt and Krauel investigated these ideas in a study on meal size in wolf spiders [11]. The researchers observed wolf spiders in the lab, and determined how much weight they gained after eating as many crickets as they wanted in a single feeding bout. Then they individually marked the same spiders, and released them into enclosures in the forest understory. In some enclosures, they removed or excluded all possible predatory insects and birds. In other enclosures, the spiders were exposed to predation. The researchers returned to the field site every other day for 40 days to look for egg cases, which the female carries on her abdomen. If a female had an egg case, the researchers recorded the day it was observed and its weight. The weight of the egg case is highly correlated with the number of eggs and the quality of offspring, and therefore is a good measure of fitness.

The data collected in this study are in the file *wolfspiders.txt* [12], available on the companion website for this volume. The columns are:

treatment	whether the spiders were exposed to predation or not
population	the enclosed area where the spiders were kept in the field
startmass	how much the spiders weighed before eating (g)
change	how much weight the spiders gained after eating as many crickets as they chose (gluttonous feeding) (g)
threshold	the satiation threshold, measured as the percent weight gain following gluttonous feeding
numattacked	the number of crickets a spider attacked during gluttonous feeding
eggcase	whether or not a spider produced an egg case (1 if yes, 0 if not)
dayeggcase	the first day that an egg case was observed
masseggcase	the mass of the egg case made by the spider (g)
	NA indicates missing data.

In R, to simplify subscripting in the exercises, you may want to create a logical vector to distinguish the safe spiders from spiders under predation, as follows:

```
>safe=(wolfspiders$treatment=="safe")
```

This vector has the value TRUE for each safe spider and FALSE for each spider under predation.

The logical operator NOT is represented in R by `!`. The values of `!safe` are opposites of the corresponding values of `safe`. Now `wolfspiders$threshold[safe]` is a vector of threshold values for the safe spiders and `wolfspiders$threshold[!safe]` is a vector of threshold values for the spiders under predation.

The R tutorial provides more information about and practice with logical vectors.

Using the dataset, do the following data analysis steps:

1. Calculate relative fitness for each individual in the dataset, using egg case mass as a measure of fitness.
2. Make a histogram of relative fitness. What is the shape of the distribution?
3. Make separate histograms of relative fitness for the spiders that were exposed to predation and those that were safe. Does it look like the risk of predation affected fitness?
4. Make a histogram of the satiation threshold. What is the shape of the distribution?
5. What is the covariance between relative fitness and satiation threshold? Do you think this trait is under selection?
6. Now, calculate the covariance between relative fitness and satiation threshold separately for safe spiders and spiders at risk of predation. How does the relationship differ between the two groups? What does this tell you about selection? □

Equation (10.7) tells us how we expect a population to change due to natural selection. However, it is only part of the story; factors beyond the relationship between fitness and a trait may affect what actually happens from one generation to the next. For example, if the trait isn't controlled by genetics, we don't expect change in the next generation. In the next section, we will explore ways to take genetics into account.

10.2.4 Determining the Genetic Component of Quantitative Traits

While Fisher demonstrated that continuous traits can be controlled by genetics, the actual trait value for an individual can also be determined by nongenetic effects, and genetic effects that are not passed from parent to offspring. For example, running speed in horses is controlled both by genetic factors and by training. Any increase in speed due to training contributes to the overall speed of the individual, but won't be passed on to the individual's offspring. Scientists studying patterns of natural selection are interested in estimating the amount of variation in a trait that is due to inherited genetic factors, because the genetic factors are responsible for changing the population mean from one generation to the next. Biologists separate the total variance of a trait in a population (called the phenotypic variance) into two main components: *additive genetic variation* and variation due to other factors. Additive genetic variation involves only the genetic effects that remain intact when transmitted from parent to offspring; they "add" to the increase or decrease of the trait value for the offspring. Additive genetic variance is indicated with the symbol V_A. The remainder of the variance is usually lumped into "environmental variance," V_E, although environmental variance also includes nonadditive genetic effects, such as dominance relationships between alleles and interactions between genes. If the total phenotypic variance is indicated by V_P, then [9]

$$V_P = V_A + V_E \tag{10.8}$$

To determine the proportion of variance in the population that is due to additive genetic effects, and therefore the variance that is likely to contribute to evolutionary change, we can divide V_A by V_P. This term is indicated by h^2, and called *heritability*. (The convention is to indicate that the heritability is related to variance, the square of the standard deviation, by including a superscript 2; the value achieved by dividing V_A by V_P does not actually get squared.)

$$h^2 = V_A/V_P \tag{10.9}$$

Estimating heritability is important for understanding how a population might respond to selection; traits with high heritability have a high proportion of variance explained by genetics, and therefore individuals more likely to survive or reproduce due to a variant in that trait are also more likely to pass at least part of their trait value to their offspring. If heritability is low, offspring are less likely to look like their parents, and an individual with high survival and reproduction has a smaller chance of passing that trait value on to the next generation, resulting in less change in the population over time.

Example 10.4. In the case of petal tube length in *P. brandegeei*, heritability for the trait was estimated to be 0.13 [13]. This value is relatively low, suggesting that while there is some genetic control over the trait, much of the variation in the trait is due to the environment, or to genetic factors that aren't always passed directly to offspring, such as interactions between genes. Therefore, despite the covariance between this trait and fitness, not all the plants with longer tubes will pass that trait on to their offspring reliably. In contrast, the researchers measured the distance between the male and female structures in the flower, another trait relevant to the hummingbird population. The male structures (the anthers, where the pollen is made) and the female structure (the stigma, where the pollen lands) are clustered together in the middle of the petal tube. If the male structures stick out beyond the female structure, the hummingbird may pick up a lot of pollen but not transfer the pollen it is carrying to the female structure, resulting in low seed production. The heritability of stigma-anther separation distance was 0.85, suggesting that additive genetic effects play a large role in determining trait values for individual plants, and contribute greatly to the variation observed in the trait. Therefore, selection by hummingbirds on this trait may result in a significant trait change. □

10.2.5 Estimating Heritability in a Trait

Heritability can be estimated in multiple ways; in this section, we will estimate heritability using the relationship between parents and offspring [9, 14]. Estimating heritability from parent-offspring data takes advantage of the fact that the degree to which offspring look like their parents is an indicator of the additive genetic control over a trait. This relationship is (almost always) linear, and therefore regression analysis with the trait value of a parent as a predictor of the trait value of the offspring is used to determine the genetic control over the trait. Either the average trait value of the two parents (the *midparent* value), or the trait value of one parent if the other is unknown, is used. If the parent has more than one offspring, the y value is the average trait value of all offspring. The coefficient of regression is also the heritability of the trait if the midparent value is used, or two times the regression coefficient if a single parent's trait value is used. More formally, we can define the coefficients in the model as follows:

i indicates the ith family in the dataset.
y_i is the mean trait value for all offspring in family i.
β_0 is the intercept.
β_1 is the regression coefficient.
x_i is the trait value for a single parent in family i or the mean trait value for the two parents in family i.
ε_i is the error term.

Therefore, the model is

$$y_i = \beta_0 + \beta_1 x_i + \varepsilon_i \tag{10.10}$$

With this relationship between x and y, Eq. (10.11) follows:

$$\beta_1 = \frac{\text{Cov}(P, O)}{V_P} \tag{10.11}$$

where P is the trait value of the single parent or midparent in a family, O is the mean trait value of the offspring in a family, and the phenotypic variance in the denominator is for the parents.

Example 10.5. Heritability may be determined for any trait measurable in both parents and offspring. Fruit flies, like most insects, experience stress at high temperatures and in dry environments. Heritability for the ability to tolerate this stress was measured by Bubliy et al. by timing how long it took for a fly to stop moving in conditions of 37 °C and less than 10% humidity [15]. The state of not being able to move is called "knockdown," and the time it takes a fly to get to that state is knockdown time, measured in minutes. To estimate heritability, the authors regressed knockdown time for an average of eight offspring against the midparent (Figure 10.3).

The equation of the regression line for these data is

$$y = 0.23x + 41.02$$

Because both parents were used in the x value, the regression coefficient is directly equal to the heritability. Therefore, $h^2 = 0.23$.

We should also obtain the same heritability by taking the covariance approach. The covariance between the average values for the parents and the offspring is 9.70 min^2 and the variance for the parents is 41.68 min^2. Therefore,

$$\beta_1 = \frac{9.70}{41.68} = 0.23$$

□

Exercise 10.3. Exercise 10.2 demonstrated that there is variability in a wild population of wolf spiders for satiation threshold. It also established a relationship between satiation threshold and gluttony for wolf spiders that aren't threatened by predators. Both of these are required for natural selection to change satiation threshold over time. Natural selection also has a third requirement: a genetic basis for the trait. Pruitt and Krauel also collected data in the lab on satiation thresholds in mothers and their offspring [11]. The mother wolf spiders

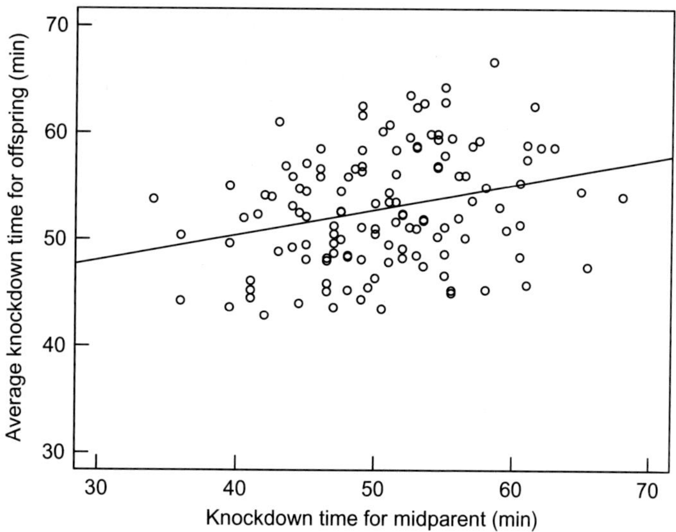

FIGURE 10.3 Heritability of knockdown time at 37 °C and less than 10% humidity in *Drosophila melanogaster*. Knockdown time for offspring was averaged across eight individuals per family. The midparent value is the average knockdown time of the mother and father. Data from Bubliy et al. [15].

carry their egg case until it hatches, and then the spiderlings climb up her legs and ride on her back and abdomen until they are big enough to live independently. Therefore, it was easy for the researchers to identify mother-offspring pairs. They measured satiation threshold in two offspring from each mother. The data are in the file *spiderheritability.txt* [12]. The first column is labeled "dam" and is the satiation threshold value for the mother. The second column is labeled "offspring" and is the mean satiation threshold for two adult female offspring.

1. Make a scatterplot with maternal satiation threshold on the x-axis and average offspring satiation threshold on the y-axis. Do the mothers and offspring appear to have similar variability? What is the relationship between mothers and offspring?
2. Use regression analysis to determine whether the relationship between the mother's satiation threshold and that of her offspring is significant. Report the estimated equation for the line, the p-value for the regression, and the correlation.
3. Use the regression coefficient to determine heritability for satiation threshold.
4. Estimate the heritability by dividing the covariance by the maternal variance. Do you get the same value? □

10.2.6 The Breeder's Equation

At this point, we have now mathematically defined the components of a trait critical to selection: the relationship between the trait and fitness, and the proportion of variation in the trait that can be passed on to offspring. Therefore, if we combine these two terms, we can more accurately predict the change in a trait from one generation to the next. The difference between the average value of the trait in the parental generation and the offspring generation is the response to selection [10]. We defined this earlier as $\Delta\bar{z}$. Therefore,

$$\Delta\bar{z} = \mathrm{Cov}\left(\hat{w}, z\right) h^2 \tag{10.12}$$

Example 10.6. In our *P. brandegeei* example, we found that the covariance between relative fitness and the trait value was 0.72 mm. However, the heritability for the trait was only 0.13. Therefore, the response to selection after one generation for this trait is predicted to be 0.094 mm, a very small amount given that the mean trait value for the sample is 25 mm. □

Equation (10.12) is an alternative way of writing an equation fundamental to modern plant and animal breeding: the *breeder's equation*. In the breeder's equation, the covariance between relative fitness and the trait value is called the *selection differential* (S). The change in the mean trait value in the next generation, $\Delta\bar{z}$, is called the *response to selection* in the breeder's equation and notated with R. If we substitute S for the covariance term in Eq. (10.12), and R for $\Delta\bar{z}$, we obtain the classical form of the equation [9]:

$$R = h^2 S. \tag{10.13}$$

The breeder's equation was first developed as a tool for use in controlled breeding applications. In an experimental population, a subset of individuals with extreme trait values are interbred, and the mean trait value of their offspring is compared with the mean of the original population. In experiments where the breeding individuals each produce the same number of offspring, the difference between the mean of the population as a whole and the mean of the interbreeding individuals is the selection differential (S). The difference between the parental generation and the offspring generation is the response to selection (R).

Evolutionary biologists use the breeder's equation in experimental situations to quantify the consequences of selection on a trait, and to estimate heritability. The selection differential is determined by the choices the experimenter makes, and then the response to selection can be measured in the next generation. As a result, two of the three terms in the equation are known, and we can solve for the third and obtain an estimate of heritability.

Example 10.7. Song and Chen selected to increase oil content in a variety of maize [16]. They started with an experimental population of plants, selected the 100 ears of corn that contained the highest oil content, and grew kernels from them. They bred these plants to each other, and measured oil content in them to determine the response to selection. The mean oil content of the entire experimental population before selection was 71.6%, while the mean oil content of the 100 ears chosen for breeding was 87.9%. Therefore, the selection differential S is 87.9–71.6%, or 16.3%. The offspring had a lower mean oil content than their parents; oil content in the next generation averaged 79.0%. The response to selection, R, is 79.0–71.6%, or 7.4%. We can rearrange Eq. (10.13) to calculate realized heritability:

$$h^2 = \frac{R}{S}$$

Substituting the experimental values, we get

$$h^2 = \frac{7.4\%}{16.3\%}$$

$$h^2 = 0.45$$

□

Exercise 10.4. Theodore Garland and colleagues selected on voluntary wheel-running behavior for several generations in a laboratory strain of house mice [17]. The mice were permitted to run as much as they liked on a wheel in their cage, and a sensor in the wheel counted the number of wheel revolutions per day. Each generation, they selected the ten families that ran the most, and bred between families to obtain the next generation. Four different selection lines were established. Data for the first generation [18] can be found in the wheelrunning.txt file on the companion website for this volume. The file contains the following variables:

animal	a unique ID number for each mouse
sex	whether the mouse was male (1) or female (0)
gen	the generation. Generation 0 was the parental generation before selection, and generation 1 is the product of selection imposed in generation 0.
line	the name for each line
revol	the average number of revolutions per day that the mouse ran in the wheel
pups	the number of pups each individual had
bred	whether the animal was bred (1) or not bred (0)

R note: The function `tapply()` provides an efficient way to calculate means for all four lines of mice with one line of code. The generic form of `tapply` that we will use is `tapply(datavector, groupvector, statistic)`.

To simplify notation, assign to the name `line` the value of `wheelrunning$line`, let `revol = wheelrunning$revol`, and let `gen = wheelrunning$gen`. If we wanted the mean number of revolutions for each line, regardless of generation, we would use

```
>tapply(revol, line, mean).
```

In order to specify a particular generation, say generation 0, we would use

```
>tapply(revol[gen == 0], line[gen  == 0], mean).
```

1. Calculate the average number of revolutions for generation 0 for each line.
2. Calculate the average number of revolutions for generation 1 for each line.
3. For each line, calculate *R*, the response to selection. Did wheel running increase from one generation to the next?
4. Calculate the relative fitness for each mouse in line 6 and generation 0, then calculate the covariance between relative fitness and number of revolutions for line 6, generation 0. This value is the selection differential, *S*, for line 6.
5. Using the breeder's equation, calculate the realized heritability for wheel running behavior in selection line 6.

□

10.2.7 The Price Equation

The breeder's equation and Eq. (10.12) contain a term for explaining how selection changes a trait, and a term for the influence of heredity on selection. In 1972, George Price developed a more general equation that predicts the outcome of selection and includes the same term for selection, but a more general term instead of heritability. The Price equation is [19, 20]

$$\Delta\bar{z} = \mathrm{Cov}(\hat{w}, z) + E\left(\hat{w}\Delta z\right) \tag{10.14}$$

where the second term on the right-hand side is the expected value for the product of relative fitness and Δz, the change in the trait during transmission from one generation to the next. The second term allows the user to model a variety of factors that might affect inheritance, not just genetic variance. If offspring are identical to their parents, the second term becomes 0. This is analogous to a heritability of 1; in both Eqs. (10.12) and (10.14), only the covariance between relative fitness and the trait would determine the outcome of selection.

The breeder's equation makes some assumptions that the Price equation does not. The breeder's equation assumes that the trait mean does not change from one generation to the next due to factors other than selection, such as the resource environment. However, a change in the mean of this nature could be included in the second term of the Price equation. Additionally, the breeder's equation assumes that the relationship between parents and offspring is linear due to genetic effects, while a nonlinear relationship could be modeled with the Price equation [10]. The Price equation can also be modified to include indirect fitness in relatives [21]. In practice, the breeder's equation is much more widely used to predict the outcome of selection and quantify realized heritability, but increasingly authors in the field are advocating for and using the Price equation [22]. For more on the Price equation, see Frank [21].

10.3 COVARIANCE AMONG TRAITS UNDER SELECTION

Both Eqs. (10.13) and (10.14) examine the change in a single trait in a population over time; however, individuals express many traits that may be shaped by similar genetic factors and affected by the same selection pressures. Biologists frequently examine groups of traits simultaneously, and use matrix algebra to keep track of the relationships among them. The relationships between all traits are described using a variance-covariance matrix. Direct selection on each trait is quantified using partial regression coefficients, and is used along with the genetic variance-covariance matrix to predict the outcome of multivariate selection.

10.3.1 Nonindependence of Multiple Traits

When a trait shares a genetic basis with another trait in the same individual, the evolutionary fates of the two traits are linked. A single gene that contributes to variation in both traits will result in a covariance between the traits. For example, in the three-spined stickleback, fish that are more aggressive toward other fish of the same species are also more likely to pursue food in the possible presence of a predator [23]. Genes that affect how fearful or bold a fish is generally could potentially be determining trait values for both of these traits, therefore making them likely to covary in a population.

We can use a matrix to keep track of multiple traits and the pairwise linear relationships between them. Suppose we are interested in linear relationships among three variables, call them X_1, X_2, X_3. We could examine $\mathrm{Cov}(X_1,X_2)$, $\mathrm{Cov}(X_1,X_3)$, and $\mathrm{Cov}(X_2,X_3)$. These, along with the three variances, can be organized into a 3×3 matrix, called the covariance matrix, or the *variance-covariance matrix*:

$$\begin{bmatrix} \mathrm{Var}(X_1) & \mathrm{Cov}(X_1,X_2) & \mathrm{Cov}(X_1,X_3) \\ \mathrm{Cov}(X_1,X_2) & \mathrm{Var}(X_2) & \mathrm{Cov}(X_2,X_3) \\ \mathrm{Cov}(X_1,X_3) & \mathrm{Cov}(X_2,X_3) & \mathrm{Var}(X_3) \end{bmatrix}$$

In this way, the matrix includes all possible pairwise linear relationships between the variables. In the first row, the variable X_1 is related to itself in the first column, to variable X_2 in the second column, and variable X_3 in the third column. The second row shows how X_2 is related to all three variables, and the third row does the same for X_3. The matrix is symmetrical about the diagonal; for example, the upper-right and bottom-left corners are the same value. For this reason, often only half of the matrix is represented in a table of values, because it still provides a complete reporting of the matrix.

Example 10.8. As an example, recall Hermon Bumpus' data on sparrows in the first exercise set of this chapter, and consider the three variables X_1 = weight, X_2 = total length, and X_3 = length of humerus. Calculations with software give us

Cov(weight, total length) = 3.067 g mm
Cov(weight, humerus length) = 0.0177 g in.
Cov(total length, humerus length) = 0.0398 mm in.
Var(weight) = 2.176 g^2
Var(total length) = 12.68 mm^2
Var(humerus length) = 0.0005 in.
Therefore, the phenotypic variance-covariance matrix is

$$\begin{bmatrix} 2.176 & 3.067 & 0.0177 \\ 3.067 & 12.68 & 0.0398 \\ 0.0177 & 0.0398 & 0.0005 \end{bmatrix}$$

The variance-covariance matrix itself is informative, but can be hard to interpret because different units of measurement are used for different traits and traits have differing magnitudes of variance. If we calculate correlations between traits using Eq. (10.6), the values will be easier to compare. The correlation matrix for the relationships between total weight, total length, and humerus length in sparrows is

$$\begin{bmatrix} 1 & 0.584 & 0.519 \\ 0.584 & 1 & 0.485 \\ 0.519 & 0.485 & 1 \end{bmatrix}$$

All three traits are moderately correlated with one another: birds that are larger in weight are also larger in length and have longer humerus bones. The values on the diagonal are 1, because a trait is always perfectly correlated with itself. In this case, positive correlations between size traits is not surprising; what is interesting is the fact that correlations are not particularly close to 1. Birds of the same length may vary considerably in weight, or in other words, size in terms of length is only one of the factors determining the overall weight of the bird. □

The variance-covariance matrix easily generalizes to larger sets of variables. The variance-covariance matrix of the variables X_1, X_2,..., X_k is a $k \times k$ matrix in which the row i, column j entry is $\mathrm{Cov}(X_i, X_j)$. It is also worthwhile to observe that if X_1, X_2, and X_3 are independent, then their covariances are 0 and the covariance matrix is simply a diagonal matrix with variances of the variables in order down the diagonal:

$$\begin{bmatrix} \mathrm{Var}(X_1) & 0 & 0 \\ 0 & \mathrm{Var}(X_2) & 0 \\ 0 & 0 & \mathrm{Var}(X_3) \end{bmatrix}$$

Matrix notation and matrix multiplication provide a very efficient way to express systems of linear equations, and therefore work with multiple traits in equations describing selection outcomes. In the case of three traits, suppose $\mathbf{B}$ is a column vector

$$\mathbf{B} = \begin{bmatrix} b_1 \\ b_2 \\ b_3 \end{bmatrix}$$

and suppose $\mathbf{G}$ is the 3×3 variance-covariance matrix of X_1, X_2, X_3.

Then the product $\mathbf{GB}$ is a 3×1 vector

$$\begin{bmatrix} \mathrm{Var}(X_1)\, b_1 + \mathrm{Cov}(X_1, X_2)\, b_2 + \mathrm{Cov}(X_1, X_3)\, b_3 \\ \mathrm{Cov}(X_2, X_1)\, b_1 + \mathrm{Var}(X_2)\, b_2 + \mathrm{Cov}(X_2, X_3)\, b_3 \\ \mathrm{Cov}(X_3, X_1)\, b_1 + \mathrm{Cov}(X_3, X_2)\, b_2 + \mathrm{Var}(X_3)\, b_3 \end{bmatrix}$$

Furthermore, if $\mathbf{Y} = \begin{bmatrix} Y_1 \\ Y_2 \\ Y_3 \end{bmatrix}$, then the equation $\mathbf{Y} = \mathbf{GB}$ is mathematically equivalent to the system of equations

$$\begin{aligned} Y_1 &= \mathrm{Var}(X_1)\, b_1 + \mathrm{Cov}(X_1, X_2)\, b_2 + \mathrm{Cov}(X_1, X_3)\, b_3 \\ Y_2 &= \mathrm{Cov}(X_2, X_1)\, b_1 + \mathrm{Var}(X_2)\, b_2 + \mathrm{Cov}(X_2, X_3)\, b_3 \\ Y_3 &= \mathrm{Cov}(X_3, X_1)\, b_1 + \mathrm{Cov}(X_3, X_2)\, b_2 + \mathrm{Var}(X_3)\, b_3. \end{aligned}$$

In the special case that X_1, X_2, and X_3 are independent, the equation $\mathbf{Y} = \mathbf{GB}$ simplifies to

$$\begin{aligned} Y_1 &= \mathrm{Var}(X_1)\, b_1 \\ Y_2 &= \mathrm{Var}(X_2)\, b_2 \\ Y_3 &= \mathrm{Var}(X_3)\, b_3 \end{aligned}$$

Exercise 10.5. White settlers have introduced many new plant species to North America over the last 400 years. Some of those plants grow very well in their new habitat, and outcompete native plants and disrupt ecosystems by growing aggressively. Reed canary grass was introduced from Europe around 1850, and has become a problematic invasive species in many parts of the United States. Once the plant establishes from seed, it sends out tillers that branch and root, increasing the size and number of stems on the plant. Calsbeek and colleagues measured several traits in plants from four populations of reed canary grass in an attempt to identify how the relationships between traits might change during an invasion [24]. They measured traits that help a plant grow more aggressively, including how long it takes for the plant to emerge from an underground stem, how tall a plant can grow, and the number of tillers it has. They collected seeds from four source populations: one in the Czech Republic and one at a similar latitude in Vermont, one in France, and another comparison population in North Carolina. By using these four populations, they could compare differences between the native and invasive ranges of the species, as well as differences between populations at different latitudes. They grew and measured all plants in the same greenhouse. Data are in the file *reedcanary.txt* on the companion website for this volume [25]. The columns in the data file are

ID	the identification number for each plant
range	whether the population was in the native or invasive range for the species
location	the source population for the seeds (Czech, France, NCarol, or Vermont)
emertime	the number of days it took for the plant to emerge above the soil from an underground stem
maxht	the maximum height of a stem at 78 days (cm)
tillernum	the number of tillers on the plant at 78 days
abovewt	the aboveground dry-weight biomass of the plant (g)
rootwt	the belowground dry-weight biomass of the plant (g)
totalwt	the total dry-weight biomass of the plant (g)

R note: As in some previous exercises, references to variables will be greatly simplified by the use of logical vectors. For example, create a logical string that is TRUE for plants located in France and FALSE for other locations:

```
>France=reedcanary$location=="France"
```

Now, create a matrix of data for plants from France with columns for emergence time, maximum height, and tiller number with the assignment

```
>France.data=cbind(reedcanary$emertime[France],reedcanary$maxht[France],reedcanary
$tillernum[France])
```

Refer to the R tutorial for help with creating matrices with `cbind()`, and help with the `cov()` and `cor()` functions.

1. Calculate the phenotypic variance-covariance matrix and the correlation matrix for emergence time, maximum height, and tiller number in plants from the population in France.
2. Calculate the phenotypic variance-covariance matrix and the correlation matrix for the same traits in plants from the population in North Carolina.
3. Within each matrix, which traits covary with each other? Can you give a biological explanation for the correlations?
4. Compare the variance-covariance matrices for the two populations. In what ways are they similar and different? □

10.3.2 The Genetic Variance-Covariance Matrix

Traits that share genetic factors are expected to demonstrate phenotypic covariance between trait values, but sometimes environmental factors can generate phenotypic correlations between traits, and occasionally traits under some shared genetic control are not correlated in the phenotype due to opposing environmental factors [26]. As a result, biologists studying the evolution of traits estimate the *additive genetic covariance* between traits. The additive genetic covariance, like additive genetic variance, describes the covariance between traits that can be passed on to the next generation. Additive genetic covariance can be estimated in several ways, and requires some knowledge of how individuals are related; see Lynch and Walsh for a thorough treatment of techniques [26]. Biologists call the phenotypic variance-covariance matrix the **P** *matrix* and variance-covariance matrix the **G** *matrix*.

Example 10.9. In the study described in Exercise 10.5, the researchers also estimated the genetic variance-covariance matrix for emergence time, maximum height, and tiller number in each population of reed canary grass. The researchers were able to estimate the **G** matrix because they grew groups of plants that were genetic clones of one another. The **G** matrix for plants in the France population was [24]

$$\begin{bmatrix} 0 & 0 & 0 \\ 0 & 149.5 & -7.69 \\ 0 & -7.69 & 10.27 \end{bmatrix}$$

The genetic variation for emergence time was 0; as a result, there was no genetic covariance with the other two traits. However, maximum height and tiller number have a slight negative genetic covariance. This value suggests that plants that achieve a greater height are genetically constrained to produce fewer tillers, and vice versa. □

10.3.3 Simultaneous Selection on Multiple Traits

The genetic variance-covariance matrix essentially allows us to describe the genetics of multiple traits; similarly, we can use a vector to explain the selection differential for multiple traits simultaneously. If S_j represents the selection differential on a single trait j, a vector **S** can include selection differentials for multiple traits:

$$\boldsymbol{S} = \begin{bmatrix} S_1 \\ S_2 \\ S_3 \\ \vdots \\ S_i \end{bmatrix}$$

Each selection differential can be determined empirically by the same methods as for a single trait; for instance, $\mathbf{S}_1$ is the difference between the mean trait value for the entire population and the interbreeding individuals, or it is the covariance between relative fitness and the trait value, $\mathrm{Cov}\left(\hat{w}, z\right)$. If we use the covariance term to describe fitness **S**, we get a vector of covariances between relative fitness and each trait value z_i. However, the values in the vector **S** include correlated changes between traits due to phenotypic correlations. If trait values z_1 and z_2 are positively correlated, selection on one will generate a covariance between fitness and the trait in the other. To eliminate this indirect effect, biologists use multiple regression analysis to calculate partial regression coefficients for the relationship between each trait and relative fitness [10, 27]. This approach gives values for each trait that reflect the effect of only direct selection on the trait. From the following multiple regression model using least-squares estimates, the resulting vector of partial regression coefficients is called the *selection gradient*, and designated as $\boldsymbol{\beta}$:

$$\boldsymbol{\beta} = [b_1, b_2, \ldots, b_i]$$

where

$$\hat{w} = b_0 + b_1 z_1 + b_2 z_2 + b_3 z_3 + \cdots + b_i z_i + \varepsilon \tag{10.15}$$

Additionally, $\boldsymbol{\beta}$ can be determined by multiplying the vector of selection differentials by the inverse of the phenotypic variance-covariance matrix:

$$\boldsymbol{\beta} = \mathbf{P}^{-1}\mathbf{S}. \tag{10.16}$$

Using linear regression to calculate $\boldsymbol{\beta}$ makes the assumption that the relationship between fitness and the trait is linear, and implies that selection favors one set of extreme trait values. However, as we saw with the sparrow data Bumpus collected, selection sometimes favors the individuals with intermediate values. Quadratic regression coefficients can be used to detect selection of this nature [28].

Example 10.10. In earlier examples, we explored the variance, heritability, and relative fitness of floral traits in *P. brandegeei*. Now we have the tools to examine selection on two floral traits in a multivariate context. We have data on two floral traits that may affect fitness: the length of the petal tube and the separation distance between the stigma and the anther in the flower. In our data subset of 12 individuals, petal tube length averaged 24.9 mm, with a variance of 12.48 mm^2, and the heritability was estimated to be 0.13. For stigma-anther separation, the sample had a mean of 0.0125 mm, and a variance of 5.48 mm^2. On average, the stigma and anther were about the same length, but the variance shows that the stigma stuck out beyond the anthers in some flowers, and was shorter than the anthers in others. The heritability of stigma-anther separation was 0.85. The phenotypic covariance between petal tube length and stigma-anther separation is −6.01 mm^2, and the correlation between the two traits is −0.73. Plants with long petal tubes also tend to have negative values for stigma-anther separation, and therefore have stigmas that are much shorter than the anthers. Conversely, plants with short petal tubes tend to have stigmas that stick out beyond the anthers. The phenotypic variance-covariance matrix for these two traits is

$$\mathbf{P} = \begin{bmatrix} 12.48 & -6.01 \\ -6.01 & 5.48 \end{bmatrix}$$

Earlier, we calculated a covariance between relative male fitness and petal tube length of 0.724 mm. Because petal tube length is correlated with stigma-anther separation, selection that favors long petal tubes will also generate a covariance between relative fitness and petal tube length. Additionally, stigma-anther separation may be under selection directly, and could potentially favor stigmas that are longer than the anthers, which would oppose the change generated by the correlation. The covariance between stigma-anther separation and relative male fitness is − 0.077 mm. Therefore, the vector **S** of selection differentials for these two traits is

$$\begin{bmatrix} 0.724 \\ -0.077 \end{bmatrix}$$

Using Eq. (10.16), we can determine $\boldsymbol{\beta}$:

$$\boldsymbol{\beta} = \mathbf{P}^{-1}\mathbf{S} = \begin{bmatrix} 0.109 \\ 0.105 \end{bmatrix}$$

Linear regression with the following model also generates values for $\boldsymbol{\beta}$:

$$\hat{w} = b_0 + b_1\,(\text{tube length}) + b_2\,(\text{separation}) + \varepsilon.$$

The analysis gives us the following partial regression coefficients, which are equivalent to those obtained above:

$$\hat{Y} = -1.701 + 0.109\,(\text{tube length}) + 0.105\,(\text{separation})$$ □

Exercise 10.6. Traits that are associated with achieving successful matings are a special case of natural selection called sexual selection. Often, traits of this nature are under relatively strong selection. In a study on introduced populations of guppies in Australia, Lindholm and colleagues measured relative male fitness and a variety of traits that female guppies find attractive [29]. Females are generally attracted to male guppies that have larger areas of orange coloration and a greater number of black spots. The researchers collected guppies from the wild, and then measured the traits of interest and male fitness in controlled laboratory conditions. Data from one of the study populations [30], Mena Creek, are in the file *guppydata.txt*.

Columns in the file are

ID	male ID number
offspring	number of offspring sired by a given male
relfit	relative fitness = number of offspring sired by a given male/mean fitness (average number of offspring per male within a trial)
body	body area (mm^2)
tail	tail area (mm^2)
black	area of black color (mm^2)
fuzzbl	area of fuzzy black coloration (mm^2)
orange	area of carotenoid coloration (mm^2)
totalirid	area of combined iridescent coloration (mm^2)

R note: Refer to the R tutorial for help with inverting matrices and solving simple matrix equations.

1. Calculate the phenotypic variance-covariance matrix for the area of black color, the area of orange color, and the area of combined iridescent coloration.
2. Calculate the phenotypic correlation matrix for the same traits. Which traits are correlated?
3. Using the covariance between relative fitness and each trait, calculate **S** for the area of black color, the area of orange color, and the area of combined iridescent coloration.
4. Using Eq. (10.16), calculate $\boldsymbol{\beta}$ for the area of black color, the area of orange color, and the area of combined iridescent coloration.
5. How do the values in the **S** and $\boldsymbol{\beta}$ vectors compare? Was accounting for covariance between traits in our selection gradient important? □

10.3.4 Predicting the Outcome of Selection on Covarying Traits

When two traits are affected by the same genetic factors, selection on one trait can change the other trait, even when it is not under selection or when it is under selection in opposing directions. For two traits linked by shared genetics, the change in each trait depends on selection on both traits. For example, in reed canary grass, selection that favors increased tiller production may result in shorter plants, even if being shorter has a neutral or somewhat negative effect on the plant. The magnitude of selection on each trait, the heritability of a trait, and the genetic covariance between traits all interact to generate the outcome of selection, often in unpredictable ways.

By keeping track of multiple traits simultaneously in a matrix context, we can predict how selection on each trait, and the covariance among all pairs of traits, affect the outcome of selection. In Eq. (10.13), we multiplied the selection differential by heritability to determine the change in the trait value in the next generation. We can modify Eq. (10.13) in a matrix context to accommodate multiple traits. We replace the selection differential with $\boldsymbol{\beta}$. Similarly, the change in the mean of each trait as a result of selection becomes a vector indicated with $\Delta\bar{\mathbf{z}}$. The heritability term for a single trait is replaced with the genetic variance-covariance matrix, **G**. These terms together give us the *multivariate breeder's equation* [10]:

$$\boldsymbol{\Delta\bar{z}} = \mathbf{G}\boldsymbol{\beta} \tag{10.17}$$

Example 10.11. McGlothlin and colleagues used the multivariate breeder's equation to explore selection on attractiveness traits in male juncos [31]. Males have patches of white on their tail feathers, and males with more tail white are considered to be more attractive. Males with a larger body size and more tail white also tend to win dominance contests with other males. The researchers measured three traits in males: wing length, tail length, and the amount of white on the tail. The percentage of white on each tail feather on the right side of the tail was estimated, and then the percentages were added together to give an index for the relative amount of tail white. Wing length in males averaged 82.2 mm, tail length averaged 71.3 mm, and tail white averaged 2.18. The researchers also collected data on mating success in the same birds by counting the number of matings that produced an offspring; mating success ranged from 0 to 3. They also collected data on relatedness among the study individuals that enabled them to estimate the **G** matrix for wing length, tail length, and tail white:

$$\mathbf{G} = \begin{bmatrix} 0.67 & 0.94 & 0.06 \\ 0.94 & 2.29 & 0.01 \\ 0.06 & 0.01 & 0.03 \end{bmatrix}$$

The associated matrix of heritabilities (on the diagonal) and genetic correlations makes it easier to interpret relationships between traits:

$$\begin{bmatrix} 0.33 & 0.76 & 0.41 \\ 0.76 & 0.53 & 0.04 \\ 0.41 & 0.04 & 0.50 \end{bmatrix}$$

All three traits have moderate heritabilities, indicating the presence of additive genetic variation for each trait. The genetic correlation between wing length and tail length is large, suggesting shared genetic control over the two traits. Tail white was correlated with wing length but not with tail length; therefore, we expect that selection on tail white might affect the size of the wings but not the tail itself.

We can use Eq. (10.17) to explore different scenarios of selection. For example, let us assume that wing length covaries with mating success, and the covariance value is 10 mm. Wing length is therefore under direct selection. If no other traits are experiencing direct selection, then $\boldsymbol{\beta}$ becomes

$$\begin{bmatrix} 10 \\ 0 \\ 0 \end{bmatrix}$$

If we multiply this vector by **G**, we get

$$\begin{bmatrix} 6.7 \\ 9.4 \\ 0.6 \end{bmatrix}$$

Even though tail length is not under direct selection, we predict that the change in tail length from one generation to the next will be greater than the change in wing length. This is because tail length has a higher heritability, and is highly correlated with wing length.

The researchers also used regression analysis to calculate partial selection differentials for the three traits in males, with mating success as the fitness measure. They standardized the values to a mean of 0 and a variance of 1 prior to analysis to make them easier to compare to each other. The estimate was

$$\boldsymbol{\beta} = \begin{bmatrix} 0.173 \\ -0.117 \\ 0.014 \end{bmatrix}$$

Using Eq. (10.17) and the standardized **G** matrix, we can calculate the response to selection for each trait:

$$\Delta\bar{\mathbf{z}} = \begin{bmatrix} -0.026 \\ 0.070 \\ 0.073 \end{bmatrix}$$

Although selection on body length was strong and positive in relation to the other traits, the predicted response to selection is small and slightly negative. In addition, correlations between wing length and tail white generate a response in tail white that is much greater than would be expected by direct selection on the trait. □

Exercise 10.7. In the study examined in Exercise 10.5, the researchers also estimated **G** for the study populations from France and North Carolina (Table 10.1).

1. Using these estimates and Eq. (10.17), calculate the outcome of selection on emergence time, maximum height, and tiller number in these two populations for the following selection scenarios. Because comparisons

TABLE 10.1 G Matrix for Emergence Time, Height, and Tiller Number in Reed Canary Grass Plants from France and North Carolina

Location	Trait	Emergence Time	Height	Tiller Number
France	Emergence time	0	0	0
	Height	0	149.487902	−7.690694
	Tiller number	0	−7.690694	10.269224
North Carolina	Emergence time	0	0	0
	Height	0	171.80604	−35.41625
	Tiller number	0	−35.41625	24.71871

From Calsbeek et al. [24]

are relative, the magnitude of values in β is not important; values between 1 and 10 will give answers that are intuitive to interpret.

Equal selection on each trait (values of β are equal)
Selection on height only
Selection on tiller number only

2. How do the outcomes of selection differ for each scenario within each population?
3. Do the two populations differ in the way they respond to the same selection scenarios? □

10.3.5 Evolution of the G Matrix Itself

The structure of the **G** matrix and its ability to generate a response to selection that is not well matched to the phenotype associated with the highest fitness potentially represents a constraint to adaptive evolution [32]. The extent to which genetic covariances function as a constraint to adaptive change, however, depends on how easily the **G** matrix itself can change. Changes in **G** matrices may be important for releasing populations from the constraints they impose, and allowing populations to evolve in new directions. For example, Dany Garant and colleagues examined a 40-year dataset in great tits for possible changes in the **G** matrix over time [33]. The timing of nesting in the spring is highly dependent on the weather, and birds are now laying eggs an average of 12 days earlier in the spring, due to an increase in mean annual temperatures. The researchers estimated the **G** matrix for laying date, number of eggs in the clutch, and egg weight for the first, cooler, 20 years of data and the more recent, warmer 20 years. They found a negative correlation between clutch size and laying date for the cooler years, and no significant change in the **G** matrix over time, despite the fact that selection on laying date has likely changed as temperatures have increased. The persistence of the correlation may represent a constraint on adaptive evolution, if novel selection differentials in new temperature conditions are offset by the persistent negative correlation with clutch size.

The rate and extent of change that can occur in a **G** matrix is not entirely known. Some of the genetic relationships that generate covariance, such as genes that are near each other on a chromosome, can change within a relatively small number of generations [26]. Covariance can also be shaped by genes that directly control variation in both traits, which is predicted to change much less quickly [34]. Fixation of alleles and consequent loss of genetic variation decrease genetic covariances, and could change relatively rapidly under certain selection regimes [35, 36]. Studies in both experimental and wild populations provide evidence for unchanging **G** matrix structure in some cases and rapid changes in others. Three species of crickets were found to have similar **G** matrices for five size characteristics despite overall differences in size, suggesting a relatively static **G** matrix [37]. Another study on the vinegar fly *Drosophila serrata* found that genetic constraints restrict the adaptive response of populations to sexual selection on male pheromones [38]. In contrast, the **G** matrix for four size traits in a population of collared flycatchers varied significantly over a 25-year time period [39]. In *Arabidopsis thaliana*, alleles of a single gene influence the covariance between branch number and flowering time, and the fixation or loss of an allele will significantly affect the structure of the **G** matrix in a few generations [40]. As more studies are conducted and we refine our understanding of the genetic factors shaping the **G** matrix, we will gain a better understanding of whether the structure of **G** significantly influences the long-term evolution of suites of traits, or whether it evolves alongside those traits.

Project 10.1.

1. Many published studies on multivariate selection include estimates of the **G** matrix. Search for a paper that includes an estimate of the **G** matrix for a suite of traits.
2. Research possible selection pressures on the system, and then estimate selection differentials that are plausible.
3. Using the multivariate breeder's equation, explore the consequences of your selection scenarios. Identify traits where the outcome of selection is very different than expected given its selection differential, and identify the parts of the variance-covariance matrix responsible for the outcome.
4. Try altering values in the **G** matrix, and exploring the effect on the outcome of selection. Identify places where altering **G** results in the relaxation of a constraint. □

REFERENCES

[1] Darwin C, Wallace A. On the tendency of species to form varieties; and on the perpetuation of varieties and species by natural means of selection. J Proc Linnean Soc Lond Zoo 1858;3:45-62.

[2] Darwin C. On the origin of species by means of natural selection. London: Murray; 1859.

[3] Bumpus HC. The elimination of the unfit as illustrated by the introduced sparrow, *Passer domesticus*. Biological Lectures, Woods Hole Marine Biological Station 6: pp. 209-225; 1898.

[4] Bateson W, Mendel G. Mendel's principles of heredity. Cambridge: Cambridge University Press; 1909.

[5] Fisher RA. XV—the correlation between relatives on the supposition of Mendelian inheritance. Trans R Soc Edinburgh 1919;52:399-433.

[6] Devore JL, Berk KN. Modern mathematical statistics with applications. Belmont, CA: Brooks/Cole; 2007.

[7] Kulbaba MW, Worley AC. Selection on *Polemonium brandegeei* (Polemoniaceae) flowers under hummingbird pollination: in opposition, parallel, or independent of selection by hawkmoths? Evolution 2013;67:2194-206.

[8] Kulbaba MW, Worley AC. Data from: selection on Polemonium brandegeei (Polemoniaceae) flowers under hummingbird pollination: in opposition, parallel, or independent of selection by hawkmoths? Dryad Dig Repository 2013. doi:10.5061/dryad.5f1t0.

[9] Falconer DS, Mackay TFC. Introduction to quantitative genetics. 4th ed. New York: Prentice Hall; 1996.

[10] Walsh B, Lynch M. Evolution and selection of quantitative traits: I. Foundations. http://nitro.biosci.arizona.edu/zbook/NewVolume_2/newvol2.html#2A;2011.

[11] Pruitt JN, Krauel JJ. The adaptive value of gluttony: predators mediate the life history trade offs of satiation threshold. J Evol Biol 2010;23:2104-11.

[12] Pruitt JN, Krauel JJ. Data from: the adaptive value of gluttony: predators mediate the life history trade-offs of satiation threshold. Dryad Dig Repository 2012. doi:10.5061/dryad.1763.

[13] Kulbaba MW, Worley AC. Floral design in *Polemonium brandegeei* (Polemoniaceae): genetic and phenotypic variation under hawkmoth and hummingbird pollination. Int J Plant Sci 2008;169:509-22.

[14] Fernandez GCJ, Miller Jr JC. Estimation of heritability by parent-offspring regression. Theor Appl Genet 1985;70:650-4.

[15] Bubliy OA, Kristensen TN, Kellermann V, Loeschcke V. Humidity affects genetic architecture of heat resistance in *Drosophila melanogaster*. J Evol Biol 2012;25:1180-8.

[16] Song TM, Chen SJ. Long term selection for oil concentration in five maize populations. Maydica 2004;49:9-14.

[17] Careau V, Wolak ME, Carter PA, Garland T. Limits to behavioral evolution: the quantitative genetics of a complex trait under directional selection. Evolution 2013;67:3102-19.

[18] Careau V, Wolak ME, Carter PA, Garland Jr T. Data from: limits to behavioral evolution: the quantitative genetics of a complex trait under directional selection. Dryad Dig Repository 2013. doi:10.5061/dryad.37b28.

[19] Price GR. Selection and covariance. Nature 1970;227:520-1.

[20] Gardner A. The Price equation. Curr Biol 2008;18:R198-202.

[21] Frank SA. Natural selection. IV The Price equation. J Evol Biol 2012;25:1002-19.

[22] Morrissey MB, Kruuk LEB, Wilson AJ. The danger of applying the breeder's equation in observational studies of natural populations. J Evol Biol 2010;23:2277-88.

[23] Huntingford FA. The relationship between anti-predator behaviour and aggression among conspecifics in the three-spined stickleback. *Gasterosteus aculeatus*. Anim Behav 1976;24:245-60.

[24] Calsbeek B, Patel M, Lavergne S, Molofsky J. Comparing the genetic architecture and potential response to selection of native and invasive populations of reed canary grass. Evol Appl 2011;4:726-35.

[25] Calsbeek B, Patel M, Lavergne S, Molofsky J. Data from: Comparing the genetic architecture and potential response to selection of native and invasive populations of reed canary grass. Dryad Dig Repository 2011. doi:10.5061/dryad.38d8k.

[26] Lynch M, Walsh B. Genetics and analysis of quantitative traits. Sunderland, MA: Sinauer; 1998.

[27] Lande R, Arnold SJ. The measurement of selection on correlated characters. Evolution 1983;37:1210-26.

[28] Brodie III ED, Moore AJ, Janzen FJ. Visualizing and quantifying natural selection. Trends Ecol Evol 1995;10:313-8.

[29] Lindholm AK, Head ML, Brooks RC, Rollins LA, Ingleby FC, Zajitschek SRK. Causes of male sexual trait divergence in introduced populations of guppies. J Evol Biol 2014;27:437-48.

[30] Lindholm AK, Head ML, Brooks RC, Rollins LA, Ingleby FC, Zajitschek SRK. Data from: causes of male sexual trait divergence in introduced populations of guppies. Dryad Dig Repository 2014. doi:10.5061/dryad.dg261.

[31] McGlothlin JW, Parker PG, Nolan Jr V, Ketterson ED. Correlational selection leads to genetic integration of body size and an attractive plumage trait in dark eyed juncos. Evolution 2005;59:658-71.

[32] Steppan SJ, Phillips PC, Houle D. Comparative quantitative genetics: evolution of the G matrix. Trends Ecol Evol 2002;17:320-7.

[33] Garant D, Hadfield JD, Kruuk LEB, Sheldon BC. Stability of genetic variance and covariance for reproductive characters in the face of climate change in a wild bird population. Mol Ecol 2008;17:179-88.
[34] Walsh B, Blows MW. Abundant genetic variation + strong selection = multivariate genetic constraints: a geometric view of adaptation. Annu Rev Ecol Evol Syst 2009;40:41-59.
[35] Gromko MH. Unpredictability of correlated response to selection: pleiotropy and sampling interact. Evolution 1995;49:685-93.
[36] Houle D. Genetic covariance of fitness correlates: what genetic correlations are made of and why it matters. Evolution 1991;45: 630-48.
[37] Bégin M, Roff DA. The constancy of the G matrix through species divergence and the effects of quantitative genetic constraints on phenotypic evolution: a case study in crickets. Evolution 2003;57:1107-20.
[38] Chenoweth SF, Rundle HD, Blows MW. The contribution of selection and genetic constraints to phenotypic divergence. Am Nat 2010;175:186-96.
[39] Björklund M, Husby A, Gustafsson L. Rapid and unpredictable changes of the G- matrix in a natural bird population over 25 years. J Evol Biol 2012;26:1-13.
[40] Stinchcombe JR, Weinig C, Heath KD, Brock MT, Schmitt J. Polymorphic genes of major effect: consequences for variation, selection and evolution in *Arabidopsis thaliana*. Genetics 2009;182:911-22.

Chapter 11

Metabolic Analysis: Algebraic and Geometric Methods

Terrell L. Hodge[1], Blair R. Szymczyna[2] and Todd J. Barkman[3]

[1]Department of Mathematics, Western Michigan University, Kalamazoo, MI, USA, [2]Department of Chemistry, Western Michigan University, Kalamazoo, MI, USA, [3]Department of Biological Sciences, Western Michigan University, Kalamazoo, MI, USA

11.1 INTRODUCTION

At the molecular level, life is maintained through a series of biochemical reactions, enacted via enzymes, in which the components are the *metabolites* or *species*, with the biochemical reactions chained together into so-called *metabolic pathways*. The systematic study of metabolites and metabolic pathways has wide-ranging applications, including agriculture, pharmacology, metabolic engineering, disease diagnosis, transmission, and prevention, and foundational research in multiple areas of biology and biochemistry. Among such contexts and topics, treated in this module are caffeine biosynthesis, an application to green chemical engineering in the form of analysis of the biodegradation in wastewater of certain green chemicals (ionic liquids), and use of metabolite profiles in urine and other bodily fluids to identify biomarkers for health factors, such as glue sniffing. Such applications benefit from mathematical treatments of biochemical reaction networks using linear algebra (the mathematics of systems of linear equations), along with more general algebraic methods, as are introduced in the first half of this chapter. Providing evidence for metabolites and changes in the products of metabolism requires analytical techniques, among the most predominant of which is nuclear magnetic resonance (NMR) spectroscopy, the focus of the second half of this chapter. Mathematical methods intrinsic to the implementation of NMR and the further determination and characterization of NMR data, including metabolite profiles from samples, build on and then extend the same linear algebraic framework introduced in the first part of this chapter, including principal component analysis (PCA) via singular value decomposition (SVD).

In this chapter, we take a hands-on approach to develop an understanding of some mathematical frameworks applied in this biological context, and to use the databases and tools introduced at appropriate levels of formality. Except where otherwise indicated, reading and working through the exercises as they appear will be essential to build the foundation for subsequent topics. This allows us to start with a very elementary approach, while at times delving into topics with a greater level of mathematical sophistication. This is especially the case in the projects given, where a greater level of independence and involvement to complete readings and generally a greater familiarity with concepts from linear algebra will be required. While this chapter could accompany a first course in linear algebra, linear algebra is not a prerequisite for this chapter. It is hoped that this treatment will provide a small window into applications of this field, and higher algebraic ones building on it, of importance to biologists, biochemists, and anyone interested in metabolic processes as biochemical reaction networks.

Algebraic and Discrete Mathematical Methods for Modern Biology. http://dx.doi.org/10.1016/B978-0-12-801213-0.00011-3
Copyright © 2015 Elsevier Inc. All rights reserved.

11.2 ENCODING THE REACTIONS: LINEAR ALGEBRAIC MODELING

In this section, we use caffeine biosynthesis as one context for introducing and describing one mathematical framework for capturing systems of biochemical reaction equations that constitute metabolic pathways, and metabolic networks more broadly. In general, mathematical modeling can complement biological research and modeling by providing a consistent means to capture information, to position it for *in silico* studies, to identify potential new targets for study or verify experimental hypotheses, and to further reveal hidden patterns underlying the data.

Every day, millions of people start, support, and end their days with an intake of caffeine. Whether it is a mug of coffee in the morning, a cup of tea in the afternoon, a chocolate dessert in the evening, or via a near-infinite combination of other forms varying in popularity and by culture over time, caffeine is the most widely used psychoactive drug, with uses in some forms possibly dating to antiquity. However, still long before human consumption, the plant families upon which we draw for our caffeine, such as coffee, tea, cacao, maté, and guaraná, evolved the ability to synthesize and degrade it. Understanding the mechanisms by which they do so could, from a metabolic engineering standpoint, improve the production of caffeine and aid in the development of "naturally decaffeinated" drinks [1]. Caffeine is a purine alkaloid, derived from the purine nucleotides adenine and guanine. Despite their commercial importance, the original physiological role of plants of caffeine and related purine alkaloids, like theobromine, is not yet fully understood. Possibly an evolutionary waste product, hypothesized roles include an allelopathic role, for example, in which caffeine in leaves or seeds falling to earth around a parent plant inhibit the growth of other plants nearby, and a chemical defense role, for example, in which caffeine and other purine alkaloids inhibit infestation by insects or limit consumption by herbivores due to toxic effects [2]. Recent research on transgenic tobacco and chrysanthemum plants supports the latter, suggesting the introduction of genes associated with caffeine and other purine alkaloid production into plants which do not currently produce these could be a valuable biological mechanism for improving agricultural production and reducing pesticide use [2–4].

The first reaction step in the most well-known pathway for the biosynthesis of caffeine in *Coffea arabica* is the methylation of xanthosine by the enzyme xanthosine methyltransferase (XMT). Methylation steps are common in metabolism and seem to be a way to limit the reactivity of molecules and give them different properties, as will be discussed further below when the full set of reactions are introduced. To get started with helpful representations, notation, and terminology, the first reaction step alone is presented via the reaction equation (and its slight variation) in Eq. (11.1).

$$\text{Xanthosine} + \text{SAM} \longrightarrow \text{7-methylxanthosine} + \text{SAH} + \text{H}^+, \tag{11.1}$$

or perhaps

$$\text{Xanthosine} + \text{SAM} \overset{XMT}{\longrightarrow} \text{7-methylxanthosine} + \text{SAH} + \text{H}^+.$$

The meaning of the specific remaining components of Eq. (11.1) (e.g., SAM, SAH) is explored as part of Exercise 11.1. In the reaction representation given by Eq. (11.1), the *compounds* (also called *species*, or *metabolites*[1]) are xanthosine, SAM, 7-methlyxanthosine, SAH, and H^+. Each reaction equation defines two *complexes*, each a subcollection[2] of compounds, where the first consists of the compounds forming the inputs to the reaction (commonly called the complex of *substrates* or *reactants*) and the second consists of the compounds forming the outputs of the reaction (commonly called the *products*). In the reaction representation given by Eq. (11.1), the two complexes are given (loosely) by the two sets of species $\{xanthosine, SAM\}$, and $\{7\text{-}methlyxanthosine, SAH, H^+\}$. In Eq. (11.1), the catalyzing enzyme XMT is not included as an element of the reaction. Enzymes remain unchanged during a metabolic reaction, and hence, to avoid redundancy, are generally

1. Generally, context will prevent confusion with actual species in a taxonomical sense, for example, *C. arabica* or *Camellia sinensis*, though we may favor the term "compounds" or "metabolites" in this chapter to avoid such potential ambiguity.

2. This is speaking rather loosely for the moment; presently we will refine this notion of complexes by putting an ordering on all of the compounds and then representing each complex by an ordered m-tuple, where m is the total number of relevant metabolites/compounds/species.

not included among reactants and products, even though the reaction may not effectively proceed without the enzyme, and in this sense is a necessary "input."

Building on the single reaction in Eq. (11.1), using the abbreviations to be explored in Exercise 11.1 below, and ignoring the enzymes catalyzing each step, the putative full caffeine biosynthesis pathway in *C. arabica*, as per [1], is given by the following system:

$$\text{Xanthosine} + \text{SAM} \longrightarrow \text{7-methylxanthosine} + \text{SAH} + \text{H}^+ \quad (11.2)$$

$$\text{7-Methylxanthosine} + \text{H}_2\text{O} \longrightarrow \text{7-methylxanthine} + \text{D-ribofuranose} \quad (11.3)$$

$$\text{7-Methylxanthine} + \text{SAM} \longrightarrow \text{theobromine} + \text{SAH} + \text{H}^+ \quad (11.4)$$

$$\text{Theobromine} + \text{SAM} \longrightarrow \text{caffeine} + \text{SAH} + \text{H}^+ \quad (11.5)$$

Of these four reactions, three are methylation steps. Methylation can change the solubility of molecules in cells and caffeine probably has an easier time passing through a cell membrane than xanthosine. Thus, one overall description for the role of methylation here, and the cumulative effect of this set of reactions, is that xanthosine is methylated in order to change its chemical properties so that it can be a less toxic molecule to the cell that can be transported. More generally, building on earlier observations, nucleotides are degraded (the purine bases, in this case) as part of normal metabolism of all organisms. (For example, specifically, guanosine and adenosine are degraded to uric acid and excreted in the urine of humans.) Xanthosine is part of that breakdown or salvage pathway, and appears to be an intermediate product that some plants (coffee in this case) have diverted into a different biochemical pathway because it can be converted into ecologically useful chemicals like caffeine. Currently, xanthosine (or its derivative, xanthine) is the only known chemical link or bridge from normal nitrogenous waste metabolism to the production of the ecologically useful molecule, caffeine.

With this example in mind, there are a number of databases that collect, assemble, and display information about metabolites, metabolic processes, and metabolic pathways and their analyses more specifically. Among these are general-purpose sites like KEGG [5] and MetaCyc [6], as well as sites affiliated with particular mathematical analysis methods, such as BIGG [7], or particular scientific emphases, such as the Human Metabolome Database (HMDB), and others to be noted later in this chapter. In Exercise 11.1, in the context of Eqs. (11.2–11.5), we utilize just two of these databases to illustrate some of the types of information, including pictorial representations of pathways, commonly used in biochemistry, and to provide access to further context for more mathematical treatments of metabolism and metabolic pathways as well as a tool that can be used for additional independent exploration.

Exercise 11.1 (MetaCyc). For a pictorial view of this pathway in standard biochemical reaction diagram form (in mathematical format, a *hypergraph*), go to http://metacyc.org [6]. In the search box in the site's upper right-hand corner, type "caffeine biosynthesis." From the search results, choose "caffeine biosynthesis 1." At the time of writing, this will produce the interactive picture shown in Figure 11.1.

In the style of the biochemical diagram shown above, each single reaction, represented notationally by a single arrow "$\longrightarrow$" in equations like Eq. (11.1), is now graphically represented by a multi-pronged arrow that, in mathematical language, is an example of a *hyperedge*.

1. Mouse over each hyperedge to see the standard-form equation for the corresponding reaction. What do the abbreviations SAH and SAM in Eq. (11.1) represent? Are they abbreviations that appear in MetaCyc?
2. Does the set of reactions appearing on this MetaCyc page correspond to the system in (11.2–11.5) labeled above?
3. To see a representation in terms of the molecular structure of each principle compound, select the pull-down box "More Detail" next to "Enzyme View" at the top middle of this MetaCyc page. What is the impact of the reaction on the molecule of xanthosine in the first step (i.e., first reaction)? That is, what does "methylation" seem to correspond to, here, at a molecular level? *Hints*: It may help to recall that three of the four reactions in (11.2–11.5) are methylation steps, so one may try comparing across all the reaction steps. It may also help to look up the relevant molecules in a different database. For example, for the first reaction step, go to http://http://www.genome.jp/kegg/ [5]. Enter "xanthosine" in the search box, scroll down to "KEGG

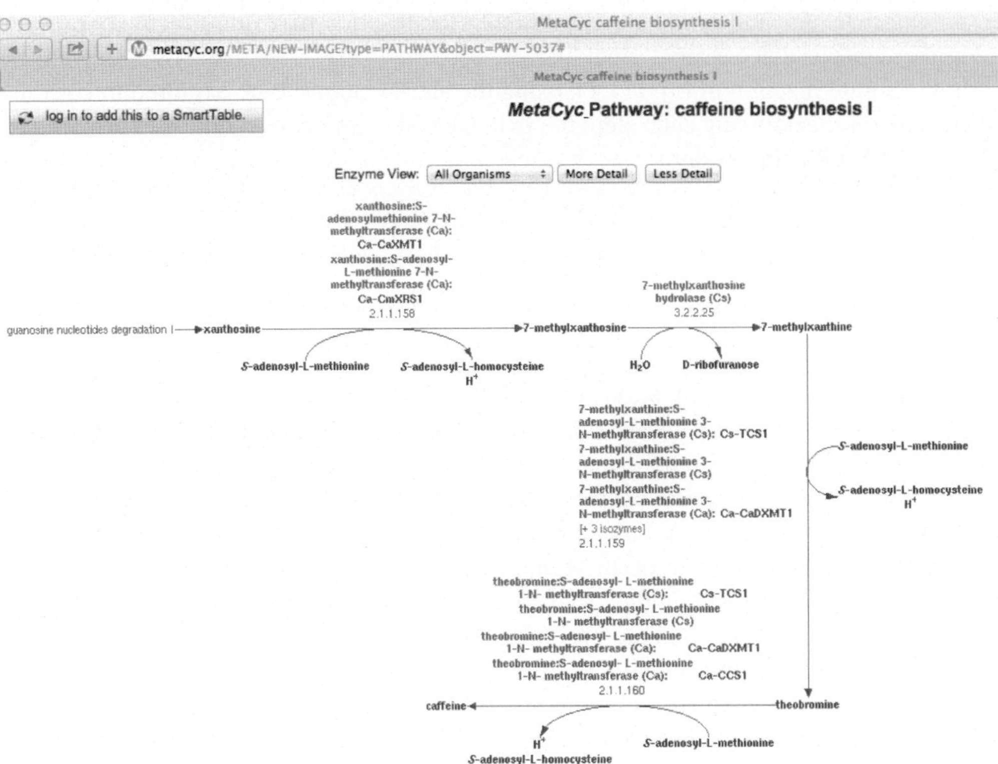

FIGURE 11.1 A graphical representation of a putative major pathway for caffeine biosyntheses as per MetaCyc [6].

COMPOUND" and click on the link for xanthosine ("C01762"); among other information, a picture of the molecule is displayed. Likewise, use KEGG to find a picture of the molecular structure of 7-methylxanthosine (also available from the compound list for xanthosine). Compare these diagrams with those in MetaCyc to inform your deductions.

4. Continuing with the previous MetaCyc page for caffeine biosynthesis 1, select "Less Detail" to return to the previous view. Above each hyperedge, a list of enzymes is provided. Mouse over an enzyme to see a description of the related standard taxonomic species in which this enzyme has been demonstrated to catalyze the reaction (including the nature of evidence for the entry) and one or more genes associated with these enzymes. For which of the reactions listed in (11.2–11.5) is *C. arabica* experimentally indicated?
5. Compare the results of your previous answer with the outcome of selecting *C. arabica* from the pull-down menu for the button currently labeled "All Organisms" next to "Enzyme View." Then, change to *C. sinensis*. Does the set of reactions (i.e., the *pathway*) change? And what is the experimental evidence for this pathway?
6. Instead of going straight to the "caffeine biosynthesis 1" link from the search page, one can select the "temporary Smart Table" option.[3] When this opens, try using the pull-down menu on the resulting table page to add various columns to the table that consolidate the information listed above, including lists of the relevant compounds, enzymes, reactants, and products associated with each system of reactions, taken in their totality. Note that the reactions are labeled by their enzymes; this is way to record and include the enzyme data, although they are omitted from the complexes. □

Having explored some common biochemical representations of reaction systems, we turn now to some more mathematical treatments. If there are a total of m compounds in a system of reaction equations under review, fix an

3. *Warning*: In current practice, this is a bit buggy after the initial request.

ordering (listing) of them. Then each complex in a given reaction can be represented by an m-tuple $(y_1, \ldots, y_m)$ wherein, for $1 \leq i \leq m$, the entry y_i is the number of moles of the ith compound in our fixed list. For example, suppose we fix the following ordering for the $m = 10$ compounds in the caffeine biosynthesis pathway given by (11.2–11.5), with the accompanying abbreviations shown:

1. Xanthosine (XR)
2. *S*-adenosyl-L-methionine (SAM)
3. 7-methylxanthosine (SevenMXR)
4. *S*-adenosyl-L-homocysteine (SAH)
5. H^+ (Hplus)
6. H_2O (H2O)
7. 7-methylxanthine (SevenMX)
8. D-ribofuranose (Drib)
9. Theobromine (Tb)
10. Caffeine (Cf)

Then relative to this fixed ordering, for the first reaction (11.2), the complex of reactants is given by the 10-tuple $\mathbf{y}_1 := (1, 1, 0, 0, 0, 0, 0, 0, 0, 0)$, while the complex of products can be represented by $\mathbf{y}'_1 := (0, 0, 1, 1, 1, 0, 0, 0, 0, 0)$. With the idea that As substrates are consumed in a reaction, and hence subtracted, while the products are produced, and hence added, we may represent the entire reaction (11.2) as the 10-tuple given by the formal difference $\mathbf{y}'_1 - \mathbf{y}_1 = (-1, -1, 1, 1, 1, 0, 0, 0, 0, 0)$. Similarly, reaction (11.3) can be given by the 10-tuple $\mathbf{y}'_2 - \mathbf{y}_2 = (0, 0, -1, 0, 0, -1, 1, 1, 0, 0)$. The outcome of combining the two reactions, reaction (11.2) followed by reaction (11.3), can be represented by the sum

$$\begin{aligned}(\mathbf{y}'_1 - \mathbf{y}_1) + (\mathbf{y}'_2 - \mathbf{y}_2) &= (-1, -1, 1, 1, 1, 0, 0, 0, 0, 0) + (0, 0, -1, 0, 0, -1, 1, 1, 0, 0)\\ &= (-1 + 0, -1 + 0, 1 + (-1), 1 + 0, 1 + 0, 0\\ &\quad +(-1), 0 + 1, 0 + 1, 0 + 0, 0 + 0)\\ &= (-1, -1, 0, 1, 1, -1, 1, 1, 0, 0),\end{aligned}$$

representing the fact that the total output of the two reaction steps consumed one molecule each of XR, SAM, and H2O, while producing one molecule each of SAH, SevenMX, and Drib.

Note that one mole of SevenMXR was produced and then consumed, yielding a corresponding entry of 0 in the sum $(\mathbf{y}'_1 - \mathbf{y}_1) + (\mathbf{y}'_2 - \mathbf{y}_2) = (-1, -1, 0, 1, -1, 1, 1, 0, 0)$. There is, in this kind of arithmetic of tuples, no way to determine from the output $(-1, -1, 0, 1, -1, 1, 1, 0, 0)$ alone whether SevenMXR (or indeed any of the compounds for which the corresponding entries are 0) was involved in intermediate steps of a reaction chain with more than one reaction. In a general reaction system, compounds that are produced and then consumed in equal quantities and those which were never involved will both have entries of 0 in the m-tuple obtained from adding those m-tuples that correspond to each reaction step. For a reaction $\mathbf{y} \to \mathbf{y}'$, we call $\mathbf{y}' - \mathbf{y}$ its *reaction vector.*

Exercise 11.2.

1. Write out the two reaction vectors (10-tuples) $\mathbf{y}'_3 - \mathbf{y}_3$, $\mathbf{y}'_4 - \mathbf{y}_4$ that describe the remaining two reactions in the caffeine biosynthesis pathway from the equation system (11.2–11.5).
2. Take the sum of the four reaction vectors $(\mathbf{y}'_1 - \mathbf{y}_1) + (\mathbf{y}'_2 - \mathbf{y}_2) + (\mathbf{y}'_3 - \mathbf{y}_3) + (\mathbf{y}'_4 - \mathbf{y}_4)$ and interpret the outcome in English. □

Exercise 11.3 (Hess's Law and Spanning Sets). The arithmetic of tuples, as in Exercise 11.2 and the material preceding it, appears in many useful contexts. For example, Hess's Law (in physical chemistry) ensures that enthalpy of a given chemical reaction is constant, regardless of the number of steps taken in the reaction. In practice, that lets one solve for the enthalpy of a reaction for which one may not have the experimental data by relating it to other reactions in ways that we can now state in terms of linear combinations of complexes and/or reaction vectors.[4]

1. As an illustration, consider the following typical sample problem, and an accompanying solution, for calculating enthalpy using Hess's Law. These are posed as Problem 11.1 at http://www.chemteam.info/Thermochem/HessLawIntro1a.html. They are also available with the Supplementary Materials for this chapter. Note that "???" denotes the unknown enthalpy value ΔH^o sought for the given reaction.
Problem 11.1. Calculate the enthalpy for this reaction:

$$2\text{C}(s) + \text{H}_2(g) \rightarrow \text{C}_2\text{H}_2(g) \quad \Delta H^o = ??? \text{ kJ}$$

given the following thermochemical equations:

$$\text{C}_2\text{H}_2(g) + (5/2)\text{O}_2(g) \rightarrow 2\text{CO}_2(g) + \text{H}_2\text{O}(l) \quad \Delta H^o = -1299.5 \text{ kJ}$$

$$\text{C}(s) + \text{O}_2(g) \rightarrow \text{CO}_2(g) \quad \Delta H^o = -393.5 \text{ kJ}$$

$$\text{H}_2(g) + (1/2)\text{O}_2(g) \rightarrow \text{H}_2\text{O}(l) \quad \Delta H^o = -285.8 \text{ kJ}$$

Solution.
(1) Determine what we must do to the three given equations to get our target equation:
  (a) first equation: flip it so as to put C_2H_2 on the product side;
  (b) second equation: multiply it by 2 to get 2C; and
  (c) third equation: do nothing. We need one H_2 on the reactant side and that is what we have.
(2) Rewrite all three equations with changes applied:

$$2\text{CO}_2(g) + \text{H}_2\text{O}(l) \rightarrow \text{C}_2\text{H}_2(g) + (5/2)\text{O}_2(g) \quad \Delta H^o = +1299.5 \text{ kJ}$$

$$2\text{C}(s) + 2\text{O}_2(g) \rightarrow 2\text{CO}_2(g) \quad \Delta H^o = -787 \text{ kJ}$$

$$\text{H}_2(g) + (1/2)\text{O}_2(g) \rightarrow \text{H}_2\text{O}(l) \quad \Delta H^o = -285.8 \text{ kJ}$$

Notice that the ΔH values changed as well.
(3) Examine what cancels: $2CO_2$ from first and second equations H_2O from first and third equations $(5/2)O_2$ from first equation and sum of second and third equations
(4) Add up ΔH values for our answer:

$$+1299.5 \text{ kJ} + (-787 \text{ kJ}) + (-285.8 \text{ kJ}) = +226.7 \text{ kJ}$$

In the language of this chapter, the metabolites are C, H_2, C_2H_2, O_2, CO_2, H_2O; with this order, the target reaction $2C(s) + H_2(g) \rightarrow C_2H_2(g)$ is represented by $\mathbf{d} = (-2, -1, 1, 0, 0, 0)$. The three given equations can be represented as $\mathbf{a} = (0, 0, -1, -\frac{5}{2}, 2, 1)$, $\mathbf{b} = (-1, 0, 0, -1, 1, 0)$, and $\mathbf{c} = (0, -1, 0, -\frac{1}{2}, 0, 1)$. The effect of the operations on the equations suggested here is to produce $-\mathbf{a} = (0, 0, 1, \frac{5}{2}, -2, -1), 2\mathbf{b} = (-2, 0, 0, -2, 2, 0)$, and $1\mathbf{c} = (0, -1, 0, -\frac{1}{2}, 0, 1)$ as before. Then $\mathbf{d} = -\mathbf{a} + 2\mathbf{b} + 1$, so according to Hess's Law, the corresponding enthalpies share the same relationship: $\Delta H_{\mathbf{d}} = -\Delta H_{\mathbf{a}} + 2\Delta H_{\mathbf{b}} + 1\Delta H_{\mathbf{c}}$. This enables the calculation of $\Delta H_{\mathbf{d}}$ from the other reactions.

Using a similar format as in Problem 11.1, see Problem 11.2 below and try to rework the solution. For additional practice, consider likewise (re)working one or more other problems at http://www.chemteam.info/Thermochem/HessLawIntro1a.html; see also the Supplementary Materials for this chapter. □

4. *Note*: Mathematically inclined students can work out the exercises below without even knowing what enthalpy is, only that it is preserved as shown therein. However, you may wish to start this exercise by doing a web search on Hess's Law, for additional background, if it is unfamiliar.

Problem 11.2. The standard molar enthalpy of formation, $\Delta_f H^o$, of diborane cannot be determined directly because the compound cannot be prepared by reaction of boron and hydrogen. However, the value can be calculated. Calculate the standard enthalpy of formation of gaseous diborane (B_2H_6) using the following thermochemical information:

(a) $4B(s) + 3O_2(g) \rightarrow 2B_2O_3(s) \quad \Delta H^o = -2509.1$ kJ
(b) $2H_2(g) + O_2(g) \rightarrow 2H_2O(l) \quad \Delta H^o = -571.7$ kJ
(c) $B_2H_6(g) + 3O_2(g) \rightarrow B_2O_3(s) + 3H_2O(l) \quad \Delta H^o = -2147.5$ kJ □

Solution.

(1) An important key is to know what equation we are aiming for. The answer is in the word *formation*:

$$2B + 3H_2 \rightarrow B_2H_6$$

Remember that formation means forming 1 mole of our target substance. This means that a one MUST be in front of the B_2H_6.

(2) In order to get to our formation reaction, the following must happen to equations (a), (b), and (c):
equation (a): divide through by 2
equation (b): multiply through by 3/2
equations (c): flip
Why?
equation (a): this gives us 2B (from 4B) for our final equation
equation (b): this gives us $3H_2$ for our final equation
equations (c): this puts B_2H_6 on the right-hand side of the final equation

(3) The above manipulations have consequences for the coefficients AND the ΔH^o values. Rewrite equations (a), (b), and (c):
(a) $2B(s) + (3/2)O_2(g) \rightarrow B_2O_3(s) \quad \Delta H^o = -1254.55$ kJ
(b) $3H_2(g) + (3/2)O_2(g) \rightarrow 3H_2O(l) \quad \Delta H^o = -857.55$ kJ
(c) $B_2O_3(s) + 3H_2O(l) \rightarrow B_2H_6(g) + 3O_2(g) \Delta H^o = +2147.5$ kJ
Add the three equations and the ΔH^o values: $2B + 3H_2 \rightarrow B_2H_6 \quad \Delta H^o = +35.4$ kJ

2. Return again to part (1) (Problem 11.1). In the formal language of linear algebra, the *span* $Span(\mathbf{a}, \mathbf{b}, \mathbf{c})$ consists of all $\mathbb{R}$-linear combinations $\alpha_1\mathbf{a}_1 + \alpha_2\mathbf{b} + \alpha_3\mathbf{c}$ of the reaction vectors $\mathbf{a}, \mathbf{b}, \mathbf{c}$. We showed the reaction vector $\mathbf{d} \in Span(\mathbf{a}, \mathbf{b}, \mathbf{c})$. Reframe Hess's Law in this light.
3. Suppose we take instead $\mathbf{u} = -\mathbf{a} + 3\mathbf{b}$, and $\mathbf{v} = 1\mathbf{b} + (-1)\mathbf{c} = \mathbf{b} - \mathbf{c}$. Then $\mathbf{d} \in Span(\mathbf{u}, \mathbf{v})$, with $\mathbf{d} = 1\mathbf{u} - 1\mathbf{v}$. In terms of the original metabolites, to what biochemical reactions do $\mathbf{u}, \mathbf{v}$ correspond? □

The formalization in Exercise 11.3, in terms of vectors and spanning sets, confers certain advantages, especially as the number of reactions grows and ad hoc procedures become less transparent. For example: (1) Can one even tell if a given set of reactions can be used to produce a desired reaction, and if so, how? (2) Can one determine whether a proposed collection of reactions is redundant (i.e., some are already linear combinations of the others, and hence could be eliminated without loss of generality)?

Both questions have affirmative answers using well-known linear algebraic/matrix formulations. Question (1)'s goal can be rephrased as: Determine whether a reaction vector $\mathbf{d} \in Span(\mathbf{a}_1, \ldots, \mathbf{a}_t)$ for some set of reaction vectors $\mathbf{a}_1, \ldots, \mathbf{a}_t$. This can be accomplished algorithmically (see [8, Sections 2.1-2.3] or any introductory matrix analysis/linear algebra text) by (i) taking the $m \times t$ matrix A with *columns* given by $\mathbf{a}_i$, $1 \le i \le t$, and creating the augmented matrix $[A|\mathbf{d}]$ with $\mathbf{d}$ as an additional *column* vector; (ii) computing[5] the *row-reduced echelon form (RREF)* of $[A|\mathbf{d}]$; (iii) observing the number/location of the *pivots*, where a *pivot* is the first nonzero entry in each row of $RREF([A|\mathbf{d}])$. The number of pivots in $RREF(B)$, for any matrix B, is the *rank*, $rank(B)$, of B. If no pivot of $RREF([A|\mathbf{d}])$ appears in the last (i.e., augmented) column, that is, if $rank(A) = rank([A|\mathbf{d}])$, then

5. For example via *Gaussian elimination*, a technique now often taught even in high school, and implementable on many calculators, but also illustrated and carried out using technology in Exercise 11.4 below.

$\mathbf{d} \in Span(\mathbf{a}_1, \ldots, \mathbf{a}_t)$. Moreover, in this case, the linear combination producing **d** satisfies $\mathbf{d} = \alpha_1\mathbf{a}_1 + \cdots + \alpha_t\mathbf{a}_t$ where $\alpha_i = 0$ if and only if the ith column of $RREF(A)$ does not have a pivot and $\alpha_i = b_i$, the ith entry in the final (i.e., augmented) column of $RREF([A|\mathbf{d}])$.

Question (2) can be answered by the same mechanisms: the *nonbasic columns* of $RREF(A)$ (i.e., those without pivots), are linear combinations of the *basic columns* (i.e., those with pivots) of $RREF(A)$. The same exact linear relations present in the columns of $RREF(A)$ hold for the original columns of A, and these linear relationships are determined in the very same way as they were for **d** above, by referring to the entries of the corresponding nonbasic column of $RREF(A)$. Thus, the *basic columns* of A, that is, those corresponding to basic columns of $RREF(A)$, form a set of vectors whose linear combinations still give $Span(A) = Span(\mathbf{a}_1, \ldots, \mathbf{a}_t)$, the span of the columns of A. Formally, we say the basic columns of A *generate* $Span(A)$. Also, no fewer will do so; otherwise, the additional linear dependencies would result in more nonbasic columns in $RREF(A)$.) Thus, $rank(A)$ is the minimal number of linearly independent vectors which generate $Span(A)$.

We illustrate these ideas below and provide further opportunities to test them in Exercise 11.4. Although the examples are relatively small, it is important to remember that, in real life, biochemical reaction systems can be enormous, with hundreds or thousands of metabolites and/or reactions.[6] The linear algebraic and algebraic reformulations in this chapter provide the foundations for practical and effective algorithms that can be implemented on a computer to handle these extremely large systems, and to take the grunt work out of small ones.

Exercise 11.4 (Determining Spanning and Redundancy (i.e., Linear Dependence) Algorithmically)**.** For Problem 11.1 as in Exercise 11.3 part (1),

$$[A|\mathbf{d}] = \left[\begin{array}{ccc|c} 0 & -1 & 0 & -2 \\ 0 & 0 & -1 & -1 \\ -1 & 0 & 0 & 1 \\ \frac{-5}{2} & -1 & \frac{-1}{2} & 0 \\ 2 & 1 & 0 & 0 \\ 1 & 0 & 1 & 0 \end{array}\right]; RREF([A|\mathbf{d}]) = \left[\begin{array}{ccc|c} 1 & 0 & 0 & -1 \\ 0 & 1 & 0 & 2 \\ 0 & 0 & 1 & 1 \\ 0 & 0 & 0 & 0 \\ 0 & 0 & 0 & 0 \\ 0 & 0 & 0 & 0 \end{array}\right]; \text{and } RREF(A) = \begin{bmatrix} 1 & 0 & 0 \\ 0 & 1 & 0 \\ 0 & 0 & 1 \\ 0 & 0 & 0 \\ 0 & 0 & 0 \\ 0 & 0 & 0 \end{bmatrix};$$

implying $rank(A) = 3 = rank([A|\mathbf{d}])$. Thus, all columns of A are basic columns, and, from the last column of $RREF([A|\mathbf{d}])$, $\mathbf{d} = -1\mathbf{a} + 2\mathbf{b} + 1\mathbf{c}$, as we saw before. The bars separating elements of A from **d** in $[A|\mathbf{d}]$ help us recall the special role of the last column but do not otherwise affect the matrix calculations of the RREFs, and could be omitted. Calculation of the RREFs may be done using many calculators or MATLAB.[7]

1. Relative to Exercise 11.3 part (2), following the discussion above, use the appropriate corresponding $RREF([A|\mathbf{d}])$ and $RREF(A)$ to compute the desired reaction vector **d** as a linear combination of the reaction vectors for the given equations of those problems you had selected from http://www.chemteam.info/Thermochem/HessLawIntro1a.html. Use this matrix formulation to carry out another one of the problems at that site. Better yet, go to the related link http://www.chemteam.info/Thermochem/HessLawIntro1b.html, or the Supplementary Materials for this chapter, for calculations using four reactions, and see how these matrix calculations algorithmically reduce the *ad hoc* work needed for Problem #5 there.
2. Returning to the setting of Problem 11.1 as in Exercise 11.3 part (1), keep **d** as before, but now consider $Span(\mathbf{a}, \mathbf{u}, \mathbf{b}, \mathbf{c}, \mathbf{v})$, for $\mathbf{u}, \mathbf{v}$ as in part (3) of Exercise 11.3. We have already seen redundancy in this system, since by their definition, already $\mathbf{u}, \mathbf{v} \in Span(\mathbf{a}, \mathbf{b}, \mathbf{c})$. Set up a matrix A with columns in the

6. For example, go to the KEGG database and request the metabolism of *Escherichia coli*.
7. If a matrix B has been entered into MATLAB, the needed command is `rref(B)` for a matrix B. For small matrices, the needed Gaussian elimination can also be carried out by hand. (See, for example, [8, Section 1.2], [9, Ch. 1], or any basic reference on Gaussian elimination for linear systems.) Another technology option is to use Octave (http://www.gnu.org/software/octave/), a freely available software that imitates MATLAB.

order $\mathbf{a}, \mathbf{u}, \mathbf{b}, \mathbf{c}, \mathbf{v}$ to represent $Span(\mathbf{a}, \mathbf{u}, \mathbf{b}, \mathbf{c}, \mathbf{v})$, and compute $RREF(A)$. Which columns of A are basic, and which are nonbasic? For those that are nonbasic, what are their representations in terms of the other columns?

3. Continuing Part (2): Furthermore, in Exercise 11.3, we have seen two distinct linear combinations of $\mathbf{d}$ as an element of $Span(\mathbf{a}, \mathbf{u}, \mathbf{b}, \mathbf{c}, \mathbf{v})$, one in terms of $\mathbf{a}, \mathbf{b}, \mathbf{c}$, and the other in terms of $\mathbf{u}, \mathbf{v}$. Is this situation represented by the calculation of $RREF([A|\mathbf{d}])$, or is a different solution produced? What transpires if A is replaced by B, with columns ordered as $\mathbf{a}, \mathbf{b}, \mathbf{c}, \mathbf{u}, \mathbf{v}$?
4. Continuing Part (3): We know $\mathbf{d} \in Span(A)$ and $\mathbf{d} \in Span(\mathbf{u}, \mathbf{v})$. Is $Span(A) = Span(\mathbf{u}, \mathbf{v})$? □

The representation of complexes and reactions via m-tuples is the first step in a mathematical formulation of the stoichiometry of balanced equations, as we now discuss in more detail. For a system of n reactions with m total compounds, we kick our notation up just a notch and use double subscripts to record the entries $y_{j,k}, 1 \leq j \leq m$ of the m-tuple for the reactant complex $(y_{1,k}, y_{2,k}, \ldots, y_{m,k}) = \mathbf{y}_k$ of the kth reaction, for $1 \leq k \leq n$. We likewise record the entries $y'_{j,k}, 1 \leq j \leq m$ for the product complex $(y'_{1,k}, y'_{2,k}, \ldots, y'_{m,k}) = \mathbf{y}'_k$ of the kth reaction, for $1 \leq k \leq n$. Then the m-tuple for the kth reaction step $(1 \leq k \leq n)$ is $\mathbf{s}_k := \mathbf{y}'_k - \mathbf{y}_k = (y'_{1,k} - y_{1,k}, y'_{2,k} - y_{2,k}, \ldots, y'_{m,k} - y_{m,k})$, and consists of the *stoichiometric coefficients* $s_{j,k} := y'_{j,k} - y_{j,k}$, one for each of the $j = 1, \ldots, m$ compounds in the reaction system.

Putting these n m-tuples $\mathbf{s}_k$, $1 \leq j \leq n$ as columns of a matrix produces an $m \times n$ matrix, the *stoichiometry matrix*:

$$S = [s_{i,j}]_{\{1 \leq i \leq m, 1 \leq j \leq n\}} = \begin{bmatrix} s_{1,1} & s_{1,2} & \cdots & s_{1,n} \\ s_{2,1} & s_{2,2} & \cdots & s_{2,n} \\ \vdots & \vdots & \ddots & \vdots \\ s_{m,1} & s_{m,2} & \cdots & s_{m,n} \end{bmatrix}. \tag{11.6}$$

For clarity, we let $\mathbf{s}_{i*}$ denote the ith row of S, while $\mathbf{s}_{*j}$ denotes the jth column (so formally $\mathbf{s}_{*j} = \mathbf{s}^T_{j*}$, the transpose of the row vector s_{j*}). For example, the stoichiometry matrix of Eq. (11.6) for the reaction system consisting only of the first two steps (11.2) and (11.3) of the caffeine biosynthesis pathway is a 2×10 matrix:

$$S = \begin{bmatrix} -1 & 0 \\ -1 & 0 \\ 1 & -1 \\ 1 & 0 \\ 1 & 0 \\ 0 & -1 \\ 0 & 1 \\ 0 & 1 \\ 0 & 0 \\ 0 & 0 \end{bmatrix} \tag{11.7}$$

Exercise 11.5. Write out the stoichiometry matrix for the caffeine biosynthesis pathway (11.2–11.5). □

So far we have considered only the complexes involved in each reaction and have ignored the rate of each reaction. Let v_k denote the reaction rate of the kth reaction, and let $\mathbf{v}$ be the column (i.e., $m \times 1$ matrix) with entries v_k, $1 \leq k \leq n$. Consider for a moment just the first two steps (11.2) and (11.3) in the caffeine biosynthesis pathway we have been discussing; here, without specifying values further, we have $\mathbf{v} = \left[\begin{smallmatrix} v_1 \\ v_2 \end{smallmatrix}\right]$. Then the change, for this two-step system, in the total concentration $[XR]$ of the first compound XR (in our previously chosen ordering) will be $s_{1,1}v_1 + s_{1,2}v_2 = -1v_1 + 0v_2 = -1v_1$, with $S = [s_{i,j}]_{\{1 \leq i \leq 10, 1 \leq j \leq 2\}}$ as in (11.7). Likewise, the change in the total concentration of any of the 10 compounds will take the form $s_{i,1}v_1 + s_{i,2}v_2$, for the appropriate $1 \leq i \leq 10$. By definition, this is the matrix product

$$\mathbf{s}_{i*}\mathbf{v} = \begin{bmatrix} s_{i,1} & s_{i,2} \end{bmatrix} \begin{bmatrix} v_1 \\ v_2 \end{bmatrix} = [s_{i,1}v_1 + s_{i,2}v_2] \tag{11.8}$$

with the corresponding dimensions equation $(1 \times 2) \times (2 \times 1) = (1 \times 1)$. More generally, for the $m \times n$ matrix S and the $n \times 1$ matrix $\mathbf{v}$, their matrix product is size $(m \times n) \times (n \times 1) = m \times 1$, with the form

$$S\mathbf{v} = \begin{bmatrix} \mathbf{s}_{1*} \\ \mathbf{s}_{2*} \\ \vdots \\ \mathbf{s}_{m*} \end{bmatrix} \begin{bmatrix} v_1 \\ v_2 \\ \vdots \\ v_n \end{bmatrix} = \begin{bmatrix} s_{1,1} & s_{1,2} & \cdots & s_{1,n} \\ s_{2,1} & s_{2,2} & \cdots & s_{2,n} \\ \vdots & \vdots & \ddots & \vdots \\ s_{m,1} & s_{m,2} & \cdots & s_{m,n} \end{bmatrix} \begin{bmatrix} v_1 \\ v_2 \\ \vdots \\ v_n \end{bmatrix} \tag{11.9}$$

$$= \begin{bmatrix} s_{1,1}v_1 + s_{1,2}v_2 + \cdots + s_{1,n}v_n \\ s_{2,1}v_1 + s_{2,2}v_2 + \cdots + s_{2,n}v_n \\ \vdots \\ s_{m,1}v_1 + s_{m,2}v_2 + \cdots + s_{m,n}v_n \end{bmatrix} \tag{11.10}$$

$$= \begin{bmatrix} \mathbf{s}_{1*}\mathbf{v} \\ \mathbf{s}_{2*}\mathbf{v} \\ \vdots \\ \mathbf{s}_{m*}\mathbf{v} \end{bmatrix} = \begin{bmatrix} \mathbf{s}_{*1}v_1 & \mathbf{s}_{*2}v_2 & \cdots & \mathbf{s}_{*n}v_n \end{bmatrix}. \tag{11.11}$$

For an arbitrary system of reactions with m compounds and n reactions steps, the principles of stoichiometry, based on conservation of mass, dictate that the ratios of the compounds in an essential reaction remain constant, that is, $y_{j,1} : y_{j,2} : \cdots : y_{j,m}$ is an invariant of the jth reaction, as is $y'_{j,1} : y'_{j,2} : \cdots : y'_{j,m}$. Using molecules or molar proportions, the values $y_{j,k}$, respectively, $y'_{j,k}$, are integers. Hence, the stoichiometric matrix S is a matrix of integer constants. However, each reaction rate v_k, $1 \le k \le n$, is really a function of many factors, including some that can change over time t, such as temperature, pressure, and concentration, and we may denote $v_k = v_k(t)$ to emphasize this. For a reaction system with m compounds and n reactions, with $X_1, \ldots, X_m$ the ordered list of compounds, we take[8] $[X_1], [X_2], \ldots, [X_m]$ to be their concentrations at time t. Thus, also $[X_j] = [X_j](t)$, for $1 \le j \le m$. We can also express this ordered list of concentrations as an $m \times 1$ matrix (i.e., column) $[\mathbf{X}]$. The dynamic mass-balance equation characterizes the total change in concentrations for all the compounds in the system by the equivalent matrix equations:

$$\frac{\mathrm{d}[\mathbf{X}]}{\mathrm{d}t} = S\mathbf{v} \tag{11.12}$$

$$\begin{bmatrix} \dfrac{\mathrm{d}[X_1]}{\mathrm{d}t} \\ \dfrac{\mathrm{d}[X_2]}{\mathrm{d}t} \\ \vdots \\ \dfrac{\mathrm{d}[X_m]}{\mathrm{d}t} \end{bmatrix} = \begin{bmatrix} s_{1,1} & s_{1,2} & \cdots & s_{1,n} \\ s_{2,1} & s_{2,2} & \cdots & s_{2,n} \\ \vdots & \vdots & \ddots & \vdots \\ s_{m,1} & s_{m,2} & \cdots & s_{m,n} \end{bmatrix} \begin{bmatrix} v_1(t) \\ v_2(t) \\ \vdots \\ v_n(t) \end{bmatrix}$$

8. Although this is common biochemical notation, it should not be confused with the square brackets also being used for matrix notation.

For the system given by Eqs. (11.2) and (11.3), Eq. (11.12) looks like

$$\begin{bmatrix} \frac{d_{[XR]}}{d_t} \\ \frac{d_{[SAM]}}{d_t} \\ \frac{d_{[SevenMXR]}}{d_t} \\ \frac{d_{[SAH]}}{d_t} \\ \frac{d_{[Hplus]}}{d_t} \\ \frac{d_{[H2O]}}{d_t} \\ \frac{d_{[SevenMX]}}{d_t} \\ \frac{d_{[Drib]}}{d_t} \\ \frac{d_{[Tb]}}{d_t} \\ \frac{d_{[Cf]}}{d_t} \end{bmatrix} = \begin{bmatrix} -1 & 0 \\ -1 & 0 \\ 1 & -1 \\ 1 & 0 \\ 1 & 0 \\ 0 & -1 \\ 0 & 1 \\ 0 & 1 \\ 0 & 0 \\ 0 & 0 \end{bmatrix} \begin{bmatrix} v_1(t) \\ v_2(t) \end{bmatrix} = \begin{bmatrix} -v_1 \\ -v_2 \\ v_1 - v_2 \\ v_1 \\ v_1 \\ -v_2 \\ v_2 \\ v_2 \\ 0 \\ 0 \end{bmatrix}.$$

Exercise 11.6. Replicate Eq. (11.12), but now for the full caffeine network (11.2–11.5). □

Exercise 11.7 (Re: Stoichiometric Matrix—Exercise and Project).

1. (Exercise) Research in progress suggests that the picture of caffeine biosynthesis, among the whole collection of caffeine-producing plant families, is more complex than current literature, that is, based on (11.2–11.5) and explored in Exercise 11.1, suggests. Determine the dimensions of and write out the stoichiometric matrix S for the sample extended caffeine reaction network in the accompanying picture (caffeine biosynthesis.pdf) posted in the Supplemental Materials for this chapter.
2. (Project) (Optional; Continuation of Exercise 11.7 Part (1), and Referencing [10]) Extend the stoichiometric matrix to include the external input (xanthosine) and output (caffeine) to the system. One can calculate a spanning (generating) set and basis (generating set of minimal size) for the nullspace $N(S) = \{\mathbf{x} | S\mathbf{x} = \mathbf{0}\}$ from $RREF(S)$. (See [10] or [8, Section 4.2].) Is the outcome a "biologically good" basis, as described in [10]? Can you find a set of extreme paths for this system? What, if anything, do you learn? □

11.3 ADDING REACTION KINETICS: ALGEBRAIC FORMULATION OF MASS-ACTION KINETICS

Reaction rates depend, in part, upon concentration levels. Generally, the higher the concentrations, the more molecules collide, which leads to increased reaction rates. According to the law of mass action, at constant temperature[9] we can express the dependency of the reaction rates upon the concentrations of the reactants as monomials in the reactants. Specifically, if an elementary reaction R has reactant complex $\mathbf{y}_R = (y_1, \ldots, y_m)$, then the reaction rate $v = v(t)$ of R satisfies

$$v = k_R[X_1]^{a_1}[X_2]^{a_2} \cdots [X_m]^{a_m} \tag{11.13}$$

where k_R is a constant independent of the concentration, and $a_i = 0$ if $y_i = 0$, that is, if the ith compound is not a reactant. Under the mass action framework, one can assume $a_i = y_i$, $1 \leq i \leq m$. Then upon assigning a *rate constant* $k_{R:\mathbf{y}\to\mathbf{y}'}$ to each reaction $R : \mathbf{y} \to \mathbf{y}'$, one obtains a system of polynomial equations that captures the dynamics of the concentrations over time for the system of reaction equations by

9. And presuming also that the number of molecules is high, say on the order of 10^{23} (Avogadro's number), and that the container in which the reactions are occurring is well mixed.

$$\frac{\mathrm{d}[\mathbf{X}]}{\mathrm{d}t} = \sum_{R:\mathbf{y}\to\mathbf{y}'} k_{R:\mathbf{y}\to\mathbf{y}'}[X_1]^{y_1}[X_2]^{y_2}\cdots[X_m]^{y_m}(\mathbf{y}-\mathbf{y}'). \tag{11.14}$$

Despite the fact that complete knowledge of a full set of rate constants is relatively rare, there is much information about the system and the network's capacity that can be obtained from the qualitative behavior of this network ([11, Lecture 1]). In particular, when $\frac{\mathrm{d}[\mathbf{X}]}{\mathrm{d}t} = 0$, one obtains the *steady-state* equations, and solutions of these are called *steady states*. Positive steady states are those steady-state solutions for which all concentrations are positive. A system with multiple steady states exhibits *multistationarity*, which "in cellular systems provides a mechanism for switching between different cellular responses and can be crucial for cellular decision making," as per [12].

Exercise 11.8. Go to paper [12][10] and examine the small reaction network (futile cycle) there of Example 2.2. Then try to write out the equations for the mass-action kinetics of this reaction system, as per Eq. (11.14). You can check your answer by turning to [12, Example 2.3] (accounting for notational differences: [12] uses **c** in place of [**X**]). □

When *conservation laws* exist, the space $\mathbb{R}^m$ splits up into pieces (the *stoichiometric classes* of steady-state solutions) that are invariant under the dynamics. These stoichiometric classes are defined via the *stoichiometric subspace* of the reaction network, the subspace of $\mathbb{R}^m$ generated by the stoichiometric vectors $\mathbf{y}'-\mathbf{y}$ for the reactions $R:\mathbf{y}\to\mathbf{y}'$. By definition, the stoichiometric subspace is the set of all vectors in $\mathbb{R}^m$ that arise as an $\mathbb{R}$-linear combination $\sum_R \alpha_R(\mathbf{y}'-\mathbf{y})$, running over all reactions $R:\mathbf{y}\to\mathbf{y}'$. Equivalently, the stoichiometric subspace is $Span(S)$, as in (11.6), for the stoichiometric matrix S of the set of reactions R, where, if needed, any reversible reaction $R:\mathbf{y}'\leftrightarrow\mathbf{y}$ is first replaced by two one-way reactions $R_1:\mathbf{y}'\to\mathbf{y}$ and $R_2:\mathbf{y}\to\mathbf{y}'$. As in the material leading to Exercise 11.4, $Span(S)$ is generated by the basic columns of the stoichiometric matrix S.

Exercise 11.9.

1. Write out the vectors that will generate the stoichiometric subspace for the reaction system in Exercise 11.8.
2. What is the *dimension* of this space, that is, the smallest number of linearly independent vectors that span the stoichiometric subspace? Can you find a minimal set of generators for this stoichiometric subspace? (*Hint*: See Exercise 11.4 and the discussion preceding it.)
3. Check your answers via [12, Example 3.2].
4. For the reaction system in Exercise 11.8, check that $\frac{\mathrm{d}[X_1]}{\mathrm{d}t}+\frac{\mathrm{d}[X_5]}{\mathrm{d}t}=0$, so the change in $[X_1]+[X_2]$ must remain constant over time and therefore equals the constant $[X_1](0)+[X_5](0)$, the sum of initial concentrations for the first and fifth metabolites. The equation $[X_1]+[X_2]=[X_1](0)+[X_5](0)$ is an example of a conservation law.
5. Conservation laws correspond to vectors $\mathbf{x}$ in $\mathbb{R}^m$ satisfying $S^T\mathbf{x}=\mathbf{0}$, also called the left-nullspace $N(S^T)$ of S, for S^T the transpose of S. (Recall that, for any $s\times t$ matrix $B=[b_{ij}]=[\mathbf{b}_{*1}\ \ldots\ \mathbf{b}_{*t}]$, its *transpose* B^T is the $t\times s$ matrix $B^T=[b_{j,i}]=\begin{bmatrix}\mathbf{b}_{*1}\\ \vdots\\ \mathbf{b}_{*t}\end{bmatrix}$ obtained by taking ith row of B^T to be the ith column of B.) The number of linearly independent conservation laws is $m-rank(S)$. For the reaction system in Exercise 11.8, how many such independent conservation laws will there be? Can you find any others than those in part (3) of this exercise above? □

The definition of the stoichiometric class of a concentration vector [**X**], as it appears in [12], can be reformulated as the set of all vectors in $\mathbb{R}^m$ with nonnegative entries lying in the set $\{[\mathbf{X}](0)+Span(S)\}$. The solution $[\mathbf{X}]=[\mathbf{X}](t)$ to Eq. (11.14) that starts at $[\mathbf{X}](0)$ must lie in $\{[\mathbf{X}](0)+Span(S)\}$. In this sense, the stoichiometric class is "invariant under the dynamics." Upon fixing choices of the rate constants $k_{R:\mathbf{y}\to\mathbf{y}'}$ and setting

$$f_k([\mathbf{X}]) = \sum_{R:\mathbf{y}\to\mathbf{y}'} k_{R:\mathbf{y}\to\mathbf{y}'}[X_1]^{y_1}[X_2]^{y_2}\cdots[X_m]^{y_m}(\mathbf{y}-\mathbf{y}'),$$

10. Included in the Supplemental Materials for this chapter; preprint also available from the Math ARXIV.

from the dynamic mass-balance equations, one sees that, under the assumption $\frac{d[X]}{dt} = 0, f_k(\mathbf{X})$ is an element of the stoichiometric subspace of the system, and hence the study of steady states and multistationarity is entwined with the study of the stoichiometric subspace for the system. As noted in [12], properties of steady states (such as the number of them) are really set up to their stoichiometric classes.

Exercise 11.10 (Optional Project; Presumes Familiarity with Linear Algebra)**.** Following a large body of research on the topic of determining the steady states (or positive steady states) of (bio)chemical reaction networks, the authors of [12] use further analysis of this setting and the polynomial map $f_k([\mathbf{X}])$ to provide a condition that prevents the existence of multiple positive steady states in terms of the injectivity of $f_k([\mathbf{X}])$ via a condition on the determinant of the associated linear Jacobian map (matrix) of $f_k([\mathbf{X}])$. This condition can be assessed effectively by using symbolic computation software like Mathematica, Maple, or MATLAB; a version of the steps/pseudocode appears in [12, Section 11]. Use MATLAB to implement the procedure outlined in [12, Section 11] to determine that the potential of a biochemical reaction network of your choice has multiple positive steady-state solutions. □

11.4 DIRECTIONS FOR FURTHER READING AND RESEARCH: METABOLIC PATHWAYS

In the previous section, we began to see extensions of the linear algebraic framework (e.g., modeling of reaction systems as systems of linear equations, and introduction of the stoichiometric subspace), capturing aspects of biochemical reaction network systems in a more algebraic mathematical model that utilized polynomials to also capture further dynamic features of the networks. These algebraic formulations are being utilized to pursue questions of network characterization and inference; we list only a few instances here.

For example, in a mass-action kinetics framework, if one has collected some concentration data over time for a set of species expected to be involved in a (complex) biochemical reaction system, how (well) can one determine the most likely pathways/reaction interactions to take place, without an often costly, inefficient, or computationally prohibitive attempt to measure all possible interactions? A confounding issue is that different networks (formed from the same sets of species) can yield the very same deterministic mass-action models; this is the *identifiability problem*, and an algebraic characterization appears in [13]. The difficulty that such a reaction network may not be uniquely identified by a spanning set of its stoichiometric subspace $Span(S)$ is hinted at in parts (2)-(3) of Exercise 11.4, where we already saw nonunique choices for choices of generating sets of reaction vectors, even for a single known system!

However, by introducing a stochastic mass-action model, the problem of network inference can be recast [14, 15] in terms of maximum likelihood of a biologically motivated *algebraic statistical model* [16, Sections 1.1, 1.2, 1.3, 3.3]. The associated algebraic methods, including polyhedral geometry, algebraic geometry, and computational methods encompass a natural generalization of linear algebraic methods, passing from systems of linear equations to systems of linear inequalities and/or polynomial equations. Work in progress (personal communication) should soon provide a software package implementing the improved algorithms in [15] for this algebraic statistical approach to biochemical network inference.[11]

Other essentially algebraic approaches to incorporating mass-action kinetics and combining multilevel systems data, remaining in the framework of linear algebraic methods applied to the stoichiometric matrix and related transformation matrices, are the "mass action stoichiometric simulation" (MASS) models of [17, 18]. For a good graphic of the overall MASS approach, see [18, p. 176]. Enhancing our understanding of current metabolic systems by applying facets of the algebraic and statistical techniques sketched in this section, combined with concepts and methods from molecular phylogenetics[12], to the problem of the evolution of metabolic pathways and networks and their ancestral reconstruction, is one topic of future research interest of the authors (Hodge and Barkman).

11. This updates a previous "Bioreactor" software suite that is not currently supported.

12. The study of evolutionary relationships among genes and organisms utilizing molecular biology/genomics, particularly molecular sequence data, and mathematical and statistical modeling.

11.5 NMR AND LINEAR ALGEBRAIC METHODS

When one suspects the existence of as-yet-uncharacterized metabolic processes, or if one wants to observe differences in metabolite levels over time, or alter pathways to produce or reduce the output (or *flux rate*) of certain reaction products (as in the case of metabolic engineering) how does one physically determine the metabolites in play, pathways being followed, or, more generally, the concentrations over time of various compounds? From a mathematical modeling perspective, how is data obtained for the *in silico* models of the previous sections, or to verify their predictions and utility? There are multiple methods to approach these questions; currently, the analytical techniques most predominantly used for the quantification of metabolites are mass spectrometry (MS) and NMR spectroscopy, each with its own strengths and weaknesses. The construction and use of these methods, particularly NMR, require very sophisticated methods from physics, biochemistry, mathematics, and statistics. Consequently, treatment of NMR spectroscopy is generally undertaken by advanced students and other specialists. However, in the upcoming sections, we provide a self-contained, accessible summary of this fundamental and far-reaching area of modern research, illustrated through several current research applications and accompanied by just one of the multiple strands of mathematical ideas that inform and support the use of NMR spectroscopy, with multiple opportunities to explore others.

11.6 NMR SPECTROSCOPY AND APPLICATIONS TO THE STUDY OF METABOLISM

In the 1960s, the investigation of small organic molecules by NMR spectroscopy required a patient and skilled scientist with a solid knowledge of physical chemistry. Nowadays, NMR spectroscopy is accessible to relatively inexperienced users in many disciplines and is capable of studying biomolecular complexes larger than 1 MDa (megadalton), such as ribosomes and virus capsids [19–21]. Despite NMR being a rather insensitive technique, many technological and experimental advances over the last 50 years have enabled the analysis of molecules at nanomole quantities [22, 23]. NMR also plays a large role in the pharmaceutical sciences and can be used to characterize interactions between biomolecules and weakly binding ligands, identify and design new therapeutics, and assess the effect of therapeutics on their biomolecular targets. In the clinic, solution NMR spectroscopy can be used for the detection, diagnosis, and characterization of many diseases [24]. NMR metabolomic studies have identified biomarkers for asthma, cancers, neurological disorders, inflammatory diseases, metabolic disorders, and viral infections. The analysis of the metabolome in bodily fluids and tissues is a common use of NMR, which is still being actively developed and has the potential to aid in the design of preventative and personalized therapies for several diseases [25, 26]. Solution NMR spectroscopy is ideal for investigating the metabolomes of bodily fluids because minimal sample preparation is required. For samples that are more solid in composition, such as tissue biopsies, soils, and plant material, solid-state NMR can be employed. The direct investigation of metabolomes within living organisms is also now possible using the related magnetic resonance imaging (MRI) technique, which is an extension of NMR spectroscopy. Studies have already been conducted in the investigation of metabolic differences between the brains of healthy people and patients suffering from schizophrenia and those of epilepsy [27, 28].

NMR spectroscopy is one of the main analytical tools of metabolomic studies because it is robust and can reliably provide reproducible, information-rich data [29]. Also, preparation of the NMR samples and collection of the data is fast, effective, and economical. Bodily fluids, such as urine, serum, and sinovial fluid, can be used directly in the NMR spectrometer or require only minimal preparation. Fluid-handling robots, sample changers, and flow probes have been developed to quickly prepare, transfer, and insert many samples into the spectrometer for automatic acquisition, with little intervention from a spectroscopist. Another advantage is that acquisition of the data is nondestructive to the samples, allowing the collection of multiple experiments and subsequent sample analysis by another technique. Some other popular analytical techniques are biased toward certain types of compounds (charged, neutral, hydrophilic, or hydrophobic), but NMR is unbiased and is only limited by the compound's solubility in the solvent. The main disadvantages of NMR spectroscopy are the sensitivity of the instrument and the spectral resolution. Some molecules may not be observed because they are below the detectable limit of the spectrometer or hidden under the signals of more concentrated components.

11.6.1 Principles of NMR Spectroscopy

The nuclei of many atoms, much like the needle of a compass, have a magnetic moment. When these nuclei are placed into a strong magnetic field, such as that associated with an NMR spectrometer, the majority of the magnetic moments become aligned with the magnetic field, analogous to a compass needle pointing toward the magnetic north pole of the earth. The aligned state is the lowest energy state in which the spin can exist. Following quantum mechanics principles, the magnetic moments of nuclei do not align perfectly with the magnet field, but rather precess around its direction with a frequency that is dependent on the nuclei type and the magnetic field strength. Chemists and biophysicists typically name NMR spectrometers after the precessional frequency of the hydrogen nuclei in the associated magnetic field, as opposed to using Tesla, the standard unit for magnetism. For example, a 14.1-Tesla magnet is associated with a 600-MHz NMR spectrometer.

In nuclei of atoms such as hydrogen, ^{1}H, the magnetic moment can also exist in a second, higher energy state. The magnetic moments of the nuclei in this state precess with the same frequency, but the orientation is opposite to that of the low energy state. Because there are more magnetic moments in the low energy state than the high energy state, the net magnetization vector, **M**, is parallel to the magnetic field direction, $\mathbf{B}_o$, and defined as the positive Z-axis. Irradiating the sample with radiofrequency light that matches the precessional frequency of the nuclei's magnetic moment causes nuclei in the two states to switch orientations and the net magnetization vector, **M**, to precess around the incident radiation. If the radiation is applied along the X-axis, the **M** vector will rotate into the Y-axis. This phenomenon is known as magnetic resonance, and is the basis for NMR spectroscopy. In this next exercise, we explore some visualization demonstrations to help make this set-up more concrete, as well as to set the stage for later mathematical formulations relevant to analyzing NMR outputs.

Exercise 11.11 (Magnetic Resonance and $\mathbb{R}^3$).

1. For an overview of the underlying principles of NMR illustrated in the 2D case of a compass needle oscillating back and forth in the plane, please see the short video[13]: http://www.youtube.com/watch?v=1OrPCNVSA4o.
2. Now, to pass from the compass needle to the 3D situation discussed in this chapter, let's get oriented[14] with 3D space. To view and work with a standard orientation (one that obeys the "right-hand rule") of 3D space $\mathbb{R}^3$ given by X-, Y-, and Z-axes, please access the online interactive demonstration "Sets of Linear Combinations and Their Images Under Linear Transformation"[15] at http://demonstrations.wolfram.com/SetsOfLinearCombinationsAndTheirImagesUnderLinear Transformat/. Set the coordinates of vector **a** to be $a_1 = 10, a_2 = 0 = a_3$; those for vector **b** to be $b_1 = 0 = b_3, b_2 = 10$; and for vector **c**, $c_1 = c_2 = 0, c_3 = 10$. Set the coordinates of point **p** to be $p_1 = 0 = p_2 = p_3$, and remove the box checks for "predetermined vector **u**" and "predetermined vector **v**" (so that these do not display). Set the sliders for r_1, r_2, r_3 at 0. The entries for matrix A may be set to any values at present. Then, in the display window, the blue vector lies along the X-axis, the purple vector lies along the Y-axis, and the green vector lies along the Z-axis. Each vector has length[16] 10 (chosen so the visualization will be clear).
   (a) Alter the lengths of vectors **a**, **b**, and **c** by changing the entries a_1, b_2, and c_3, respectively, to see the impact of multiplication by a scalar on these vectors. Changing a_1 to be 20, for instance, scales the previous blue vector by a factor of 2.
   (b) Return again to the settings given initially in part (a) above (e.g., $a_1 = 10, a_2 = 0 = a_3$, etc.). Repeat similar operations, but by using the sliders for r_1, r_2, and r_3. As per the caption on the demonstration, you should see green dots representing the corresponding integer multiples of each of vectors **a**, **b**, **c**. (Notice that $r_1 = 1$ yields two green dots, one for the blue vector **a**, and one for its negative $-\mathbf{a}$. Notice

13. The corresponding software illustrated there is also freely available.
14. Pun intended.
15. If you have not previously used Wolfram demonstrations, you may need to download a free viewer from Wolfram; this is quick and easy. Follow their instructions as prompted.
16. More generally, we take the *length* $||\mathbf{x}||$ of a vector **x** with components $x = (x_1, \ldots, x_n)$ to be $||\mathbf{x}|| = \sqrt{x_1^2 + \cdots + x_n^2}$, generalizing the three familiar notion of the distance between two points in a plane when one of the points is the origin.

also that the green dots reflect the alternate representation of vectors as arrows emanating from the origin and as points corresponding to the locations of the arrowheads.)

(c) Return again to the settings given in part (a) above ($a_1 = 10, a_2 = 0 = a_3$, etc.). Set $r_1 = 1$. Now, implement the stretching (scalar multiplication) of the blue vector **a** by a factor of two, this time by applying the linear transformation associated with taking the matrix product $A\mathbf{x}$ of matrix A with vector **x**, here for $\mathbf{x} = \mathbf{a}$, by setting $A_{1,1} = 2$, and leaving all other $A_{i,j} = 0$. As per the caption on the demonstration, you should see black dots, representing the images of the green dots, appear. Check that this coincides with the effect in part (a) of changing a_1 to equal 20. What vector does the additional black dot represent?

3. Return again to the settings given initially in part (a) above ($a_1 = 10, a_2 = 0 = a_3$, etc.). In the initial description above of magnetic resonance, the vectors **M** and $\mathbf{B}_o$ will lie along the Z-axis, while radiation will be applied along the X-axis. The vector **M** will be some length that differs from that of $\mathbf{B}_o$. Without regard to what might be actual "reasonable" lengths, represent this situation by setting the blue vector and purple vector both to have length 5, with the blue vector lying on the X-axis and setting the purple vector (representing **M**) to lie on the Z-axis. Set the green vector, representing $\mathbf{B}_0$, to have length 10, and keep it on the Z-axis. Set $r_2 = 1$. The (eventual) "rotation" of **M** into the Y-axis is accomplished by a linear transformation.[17] Implement this by setting $A_{1,1} = 1 = A_{2,2} = A_{3,2}$, and all other $A_{i,j} = 0$. Observe the resulting change from green dots to black dots.

4. The columns of A (as in part (b) above) determine the images, under the transformation given by matrix multiplication of the original unit (i.e., length 1) vectors in each coordinate axis direction (respectively, $(1,0,0)$, $(0,1,0)$, and $(0,0,1)$), and the transformation respects the scaling of each of these unit vectors that produces the vectors **a**, **b**, and **c** (i.e., blue, purple, and green, respectively). In essence, the first two columns of A say to leave the unit vectors in the X- and Y-directions as is, but to move the unit vector in the Z-direction into the Y-direction. Since we have used the setting $r_2 = 1$, we see the effect of transforming the purple vector **M** to the Y-direction.
   (a) Implement settings that will demonstrate the impact of the same linear transformation (matrix A as in part (3)) on the green vector.
   (b) Implement settings that would return the blue, purple, and green vectors to the respective X-, Y-, and Z-axis directions, but then "rotate" the blue vector to the Y-axis direction.
   (c) Repeat the previous part, but rotate the blue vector to lie on the line $Y = X$ in the XY-plane.

5. While the effect in the situations above is apparent rotation of a selected vector into another, the full transformations performed in parts (3) and (4) are in fact examples of *projections*: linear transformations that collapse a higher-dimensional space onto a lower-dimensional one inside it (here collapsing $\mathbb{R}^3$ onto a copy of a plane $\mathbb{R}^2$ in $\mathbb{R}^3$). For instance, in part (3), matrix multiplication by A sends $Span((1,0,0),(0,1,0),(0,0,1))$ to $Span((1,0,0),(0,1,0))$. Linear transformations that are *rotations* preserve the length of any vector, a property not shared by projections.

   For an illustration and the definition of true rotation matrices that perform general rotations around X-, Y-, and Z-axes, go to http://demonstrations.wolfram.com/VectorRotationsIn3D/. Two points (blue and red) on a sphere define a vector that can then be rotated around each of the axis by choosing the axis and moving the rotation slider from 0° to 360°. The remaining buttons ("vector start point," "vector end point") and sliders ("ϕ start," "θ start") provide the means of changing to a new vector.[18] To check your understanding of the matrix mechanisms in play, using the rotation matrices provided with the demo:
   (a) To what component of the demo does the γ correspond?
   (b) Write out the matrix that would perform a rotation of 90° around the Y-axis.
   (c) Write out a matrix that would perform the rotation that takes the unit vector $\mathbf{e}_3 = (0,0,1)$ in the Z-direction onto the positive Y-axis. Under the corresponding linear transformation, where would the unit vector $\mathbf{e}_2 = (0,1,0)$ be carried?

17. As are all true rotations, but see part (5) of Exercise 11.11.
18. These move the blue and red points around using *spherical coordinates* that are defined using ϕ and θ, but one does not have to know the theory of spherical coordinates to play with this demo.

(d) Compare the true rotation matrix you just created in part (5(c)) to that of the matrix employed in the setting of part (3) for magnetic resonance, wherein radiation is applied along the X-axis, the magnetic field $\mathbf{B}_o$ lies along the Z-axis, and the vector $\mathbf{M}$ rotates from the $\mathbf{B}_o$ direction to the Y-axis. How do these differ in their effects on the standard unit vectors $\mathbf{e}_1 = (1, 0, 0)$, $\mathbf{e}_2$, and $\mathbf{e}_3$, and on the X-axis, $\mathbf{M}$, and $\mathbf{B}_o$? Is either a desirable representation of this physical situation?

6. Having played with the 3D setting of $\mathbb{R}^3$, now view a more dynamic simulation of the 3D phenomenon of magnetic resonance at http://www.mathworks.com/matlabcentral/fileexchange/41295-nuclear-magnetic-resonance-simulation. The simulation can be downloaded and accessed through MATLAB, but a video of its implementation is also available at the site. □

NMR experiments involve exciting nuclei and exploiting their behavior in the excited state to learn about the connectivity between nuclei and their unique environments in the sample. Modern NMR spectrometers use pulse sequences, a series of radiofrequency pulses and delays, to excite nuclei, move the net magnetization vectors to different orientations in XYZ-space, and manipulate nuclei to communicate with one another through the transfer of magnetization. At the end of the pulse sequence, any magnetic moments that remain in the excited state will undergo relaxation and return to the low energy state.[19] The return of magnetization along the Z-axis to the original low energy state is accomplished by longitudinal (T_1) relaxation mechanisms, while the loss of magnetization in the XY-plane results from transverse (T_2) relaxation mechanisms.

The signals that are observed directly in NMR experiments result from net magnetization that remains in the XY-plane at the conclusion of the pulse sequence and is detected by receivers. The conclusion of radiofrequency pulses leaves the spectrometer's magnetic field as the only force in the system, and the net magnetization vectors in the XY-plane begin to precess around the Z-axis. With the receivers at fixed positions, the information gathered appears as an oscillating function that decays with time. This time domain function, known as a *free induction decay (FID) curve*, can be impossible to interpret directly since there can be many different nuclei in the sample that have different precessional frequencies. Conveniently, by means of the *Fourier transform*,[20] time domain data can be converted into frequency domain data and yield spectra that have signals at the positions of the precessional frequencies of the nuclei in the sample. The distance of the signal away from a standard value, which is set to zero, is known as the *chemical shift* of the associated nuclei.

11.6.2 The NMR Spectrum

The solution NMR spectrum provides many details about the chemical compounds in the sample. The information is encoded in the properties of the resonances (signals) that are observed, including chemical shift, relative peak intensities, peak multiplicity, J-coupling, and the relaxation properties. We discuss these, in turn, below, but the reader may benefit by taking a look at the multiple figures of NMR spectra represented in this chapter, for example, Figures 11.2–11.4.

The chemical shift of a resonance is the position at which the signal of a nucleus appears on the spectrum, and is a measure of the rate, υ, at which the magnetic moment of the nucleus precesses in the magnetic field. Consequently, chemical shift can measured in hertz (Hz, s^{-1}). Since precession rate depends on the magnetic field strength of the spectrometer, υ_O, and field strength varies between magnets, the more convenient parts per million, or ppm, measurement is generally used. The ppm are denoted by δ. This is relative to an accepted chemical shift reference, υ_{ref}, such as sodium 2,2-dimethyl-2-silapentene-5-sulfonate (DSS). One has $\delta = \frac{\upsilon_i - \upsilon_{\text{ref}}}{\upsilon_O}$. The chemical shift of a nucleus is affected by several factors, such as the atoms to which it is bonded, the functional groups in which it is found, and its overall proximity to other magnetic fields within the molecule. Magnetic fields can be generated by the magnetic moments of other nuclei, quadrupolar nuclei, and ring current effects associated with aromatic systems. Solution properties such as pH, ionic strength, temperature, and solute concentration can also

19. These principles were also elucidated in the simpler 2D compass needle case in the video referenced in part (a) of Exercise 11.11.
20. Although for reasons of space we do not treat this excellent topic here, a discrete version of the ideas encapsulated by the Fourier transform, following linear algebraic principals elucidated elsewhere in this chapter, appears in [8, Section 5.8].

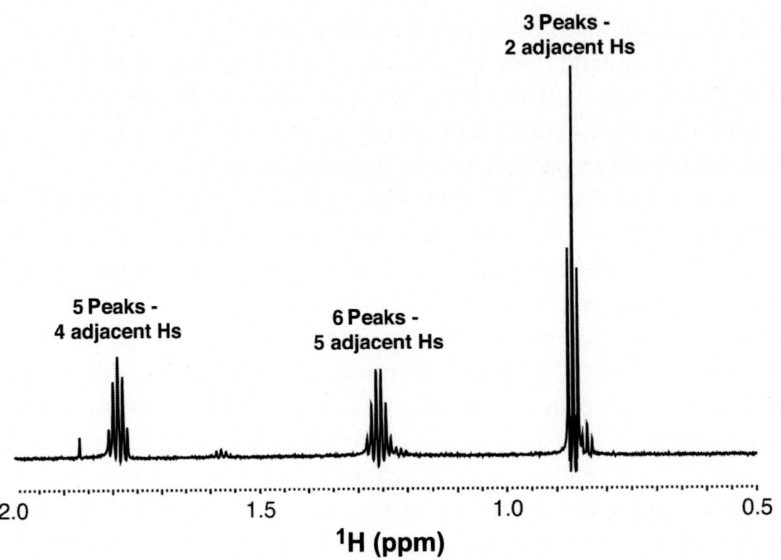

FIGURE 11.2 A typical ^{1}H NMR spectrum. Coupling between hydrogen nuclei that are close together in a molecule leads to peak multiplicity. The number of coupled hydrogen atoms is one less than the number of assocated peaks.

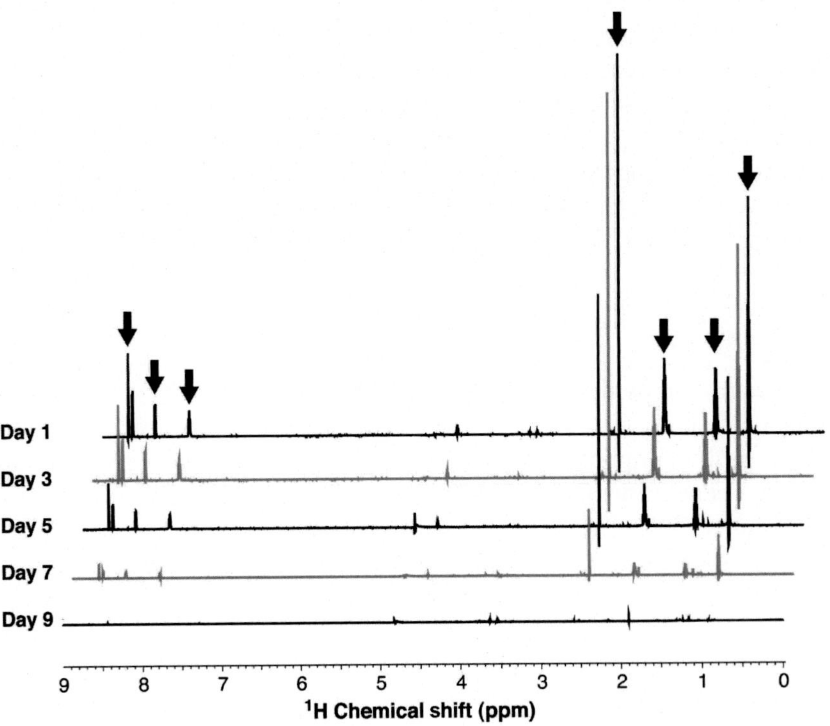

FIGURE 11.3 ^{1}H NMR spectra of an IL being degraded by a bacterial culture. The original compound (black arrows) becomes completely degraded since the corresponding signals disappear by Day 9. No appearance of new signals suggests that the bacteria do not excrete any metabolites of the compound.

effect chemical shift values. Fortunately, for a given set of conditions, chemical shifts are highly reproducible. For a further accessible illustration of chemical shifts for a selection of 11 compounds relative to a different v_{ref} (given by tetramethylsilane, $(CH_3)_4Si$, usually referred to as TMS), go to http://www2.chemistry.msu.edu/faculty/reusch/VirtTxtJml/Spectrpy/nmr/nmr1.htm and scroll down to the interactive chart under "Chemical Shift."

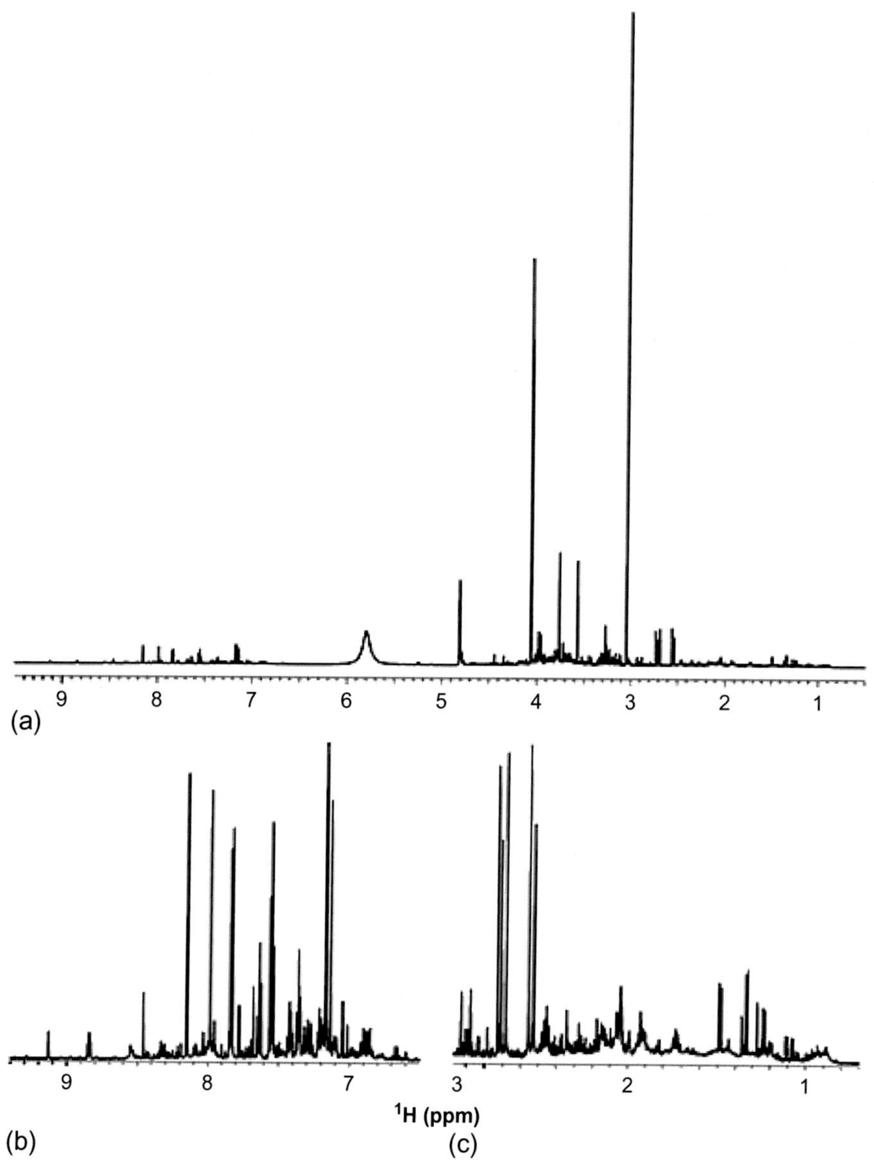

FIGURE 11.4 ^{1}H NMR spectrum of urine. (a) The full spectrum of urine reveals that some components have higher concentrations than others. The intense resonances at 3 and 4.1 ppm are likely associated with creatinine, the most concentrated chemical in urine. (b, c) Exclusion of spectral regions that contain intense resonances reveals the complex chemical makeup of urine. *We thank Claudio Luchinat and Leonardo Tenori at CERM in Florence, Italy for providing this urine NMR spectrum.*

The intensity of a resonance correlates with the number of nuclei that have a specific chemical shift. Since hydrogen atoms attached to a single carbon atom often have identical chemical shifts, the integrated intensities of resonances can help in the assignment of peaks to a specific atom or group of atoms in the chemical structure. In a hydrogen spectrum, methine groups (CH) have a relative intensity of 1, while methylene (CH_2) and methyl groups (CH_3) have intensities of 2 and 3, respectively. Likewise, the intensities of a molecule's resonances are directly proportional to the concentration of the molecule in solution. Quantitative NMR exploits these principles and has the potential to quantify the relative amounts of molecules in a mixture and the degree of contamination. The processing of NMR spectra for quantitative analysis requires great care in order to ensure that all peaks are comparable. Several mathematical functions are required to edit the data. The FID is often subjected to linear prediction, apodization with a sine or cosine function, and zero-filling prior to the Fourier transform. After the frequency domain is calculated, the spectrum is subjected to baseline correction by subtracting a polynomial or another analytical function that is fit to the signal-free regions of the spectrum. Other edits that can be applied include reference deconvolution, which removes spectral artifacts from signals and improves their shape, and normalization, which scales or adjusts a spectrum to enable more meaningful comparisons between samples [30].[21] Creatinine, for example, is generally used to normalize signals observed in urine since its concentration is generally constant and it has a resonance with a chemical shift of 4.06 ppm, which is separate from other signals in the mixture [31]. Only in special cases, such as muscular dystrophy, is the use of creatinine not suitable.

Peak multiplicity in ^{1}H spectra results from neighboring hydrogen atoms and is related to the degree of communication between the nuclei, which is called J-coupling (also known as spin-spin and scalar coupling). In ^{1}H spectra, the resonances only appear as a singlet if no hydrogen atoms are within three covalent bonds. If one hydrogen atom is within three covalent bonds and has a different chemical shift, the resonance is split into two peaks and is referred to as a doublet. The presence of two adjacent hydrogen atoms leads to the formation of a triplet (three peaks), and so on. An example of a typical NMR spectrum is shown in Figure 11.2. The distance in hertz between the peaks in the multiplets is the degree of J-coupling between the nuclei involved. The size of the coupling is largely dependent upon the number of bonds separating the atoms and the average torsion angle of the hydrogen atoms that are separated by three bonds. Since peak multiplicity can lead to complex spectra where signals are overlapped, a decoupling technique can be used to collapse the multiplets into a singlet peak, facilitating data analysis.

Exercise 11.12 (Optional; for Those Not Faint of Heart)**.** To better understand the process and complexity of accounting for J-coupling and to get a taste of some software used to interpret and process NMR spectra more generally, consider downloading a copy of the Chenomx NMR Suite (http://www.chenomx.com). Follow the easy instructions for downloading and selecting the evaluation option.[22] Open the accompanying tutorial PDF to Chapter 4 (Spin Simulator), and read (and, as inclined, follow) the detailed instructions for simulating the spectrum for valine using the sample files that accompany the download. This "simple" example runs over three pages and takes 25 steps. The subsequent tutorial, for a more "complex" example (proline) requires six pages of instructions! □

In addition to the through-bond J-coupling, nuclei can also communicate through space. One method of through-space communication is dipole-dipole cross-relaxation, which results in the Nuclear Overhauser Effect (NOE). Since magnetization is only transferred between spins that are close in space, the NOE is a powerful means of obtaining structural details about molecules. Dipolar coupling is another through-space interaction that averages to zero when molecules are randomly tumbling in solution, but residual dipolar coupling values from molecules that are partially aligned are also valuable in molecular structure elucidation.

The size of molecules also affects the observed signals and is one of the limitations of NMR spectroscopy. Thankfully, the molecules investigated in metabolomic studies have molecular weights between 100 and 1000 Da and are well within the range of molecular sizes easily studied by NMR. As the size of a molecule increases, the rate at which the molecule tumbles in solution decreases, and this affects the rate at which the excited spins undergo relaxation. Chemical or dynamic exchange processes within or between molecules also affect the relaxation rate. Both T_1 and T_2 relaxation rates are influenced by the size of the molecule being investigated, whereas chemical exchange processes have a greater impact on T_2 relaxation. Knowledge of the relaxation rates for the molecules being investigated in metabolomic studies is important so that the resonances are not weakened to varying degrees due to incomplete relaxation of the nuclei. Consequently, the delay between NMR scans must be long enough to ensure that all the nuclei in the sample are fully relaxed, especially in quantitative NMR studies. An analogy is the telling of a story. If given enough time, a listener will hear the story in its entirety (maximum signal), but if the storyteller is constantly cut off, the listener will never get the whole story (less intense signal).

21. Although we do not discuss these here, exploration of these mathematical applications would make excellent additional projects. Commercial software for NMR routinely performs these adjustments, and unpacking the underlying mathematical ideas and principles involves analytical, algebraic, and statistical tools.

22. With this free option, however, work cannot be saved.

The relaxation parameters can be accurately calculated from NMR data, but the addition of these experiments to the already time-consuming collection of metabolomic data sets can be impractical.

11.6.3 NMR Investigations of Metabolism

NMR spectroscopy can investigate metabolism at many different levels, ranging from studies of the synthesis or degradation of a single sample to changes in the metabolome of a biological sample. Metabolomic studies are classified into one of two groups. In metabolic flux experiments, changes in the metabolites are observed over time. Group-based studies involve comparing the metabolomes of people who have been divided into groups based upon the parameters being tested.

11.6.3.1 A Metabolic Flux Experiment: Ionic Liquids

Metabolic flux experiments can investigate temporal changes in an entire metabolome or a single, isolated step of a metabolic process that involves a single molecule. One example is the degradation of ionic liquids (ILs) by bacteria. ILs are low-melting salts that have chemical properties ideal for industrial purposes and the potential to be used in new chemical technologies, since they are temperature-insensitive and stable [32]. Chemical reactions that occur in IL solvents are found to have enhanced reaction rates and selectivity. In addition, ILs are viewed as "green," environmentally friendly chemicals due to their nonvolatile nature and their potential to replace harmful volatile organic solvents in industrial processes [33]. Even though increased IL use would decrease air pollution, their persistence in wastewater could impact the environment if they or their degradation products were toxic to organisms [34]. Several studies have investigated the breakdown of ILs using spectrophotometry or MS techniques, but we recently investigated the rate and mechanism by which ILs are broken down and monitored the presence of intermediate metabolites using NMR spectroscopy (personal communication). One-dimensional spectra were acquired at several time points after an IL was introduced into the bacterial culture that could biodegrade the sample. Alignment of the NMR spectra revealed details of the degradation process (Figure 11.3). The resonances that were present at Day 1 (arrows) no longer remained in the sample on Day 9, suggesting that any metabolic products were not observable by NMR (i.e., CO_2), remained inside the bacteria or were assimilated into biomass. The appearance of the new peaks in the spectrum would suggest that the original molecule was metabolized into different molecules and excreted into the media. The volumes of the peaks in the spectrum can be obtained using an integration function in an NMR processing program. A plot of the intensities as a function of time yields kinetic curves for the degradation and production of molecules.

11.6.3.2 A Group-Based Study: Urine

Urine is the predominant mechanism by which water-soluble wastes are excreted from the body, and has been used to diagnose diseases for several thousand years [35, 36]. Hippocrates played a large role in establishing uroscopy (studying urine to diagnose disease), which, in the early days, involved assessing its color, smell, and taste.[23] Modern analytical methods have greatly helped assess the presence and quantify the components in the metabolome of urine. At present, nearly 220 diseases can be characterized from urinary metabolites, and NMR spectroscopy plays a significant role in urine metabolomics [38]. Urine contains at least 3079 metabolites, and up to 209 of them can be identified by NMR spectroscopy. Consequently, as indicated in Figure 11.4, the 1D ^{1}H NMR spectrum of urine is very complex and information rich.

The challenge that researchers face is to assign each of the resonances in these complicated spectra to a single molecule in the mixture. To complicate the task further, a single resonance in the spectrum can correspond to hydrogen atoms in different molecules that have the same, or similar, chemical shifts.

One method that can be used to tackle such a problem is to start with the 1D ^{1}H NMR spectra of each component in the mixture. The use of databases to identify small molecules in NMR spectra is a very powerful

23. See also [37] for a great survey article on metabonomics that includes an illustration of a medieval urine "color wheel."

approach, since knowing the NMR spectrum of a molecule allows one to identify the corresponding resonances in the spectrum of the sample. Over the years many databases have emerged that provide the NMR spectra for the metabolites found in various biological sources. General data banks that are free to access include the HMDB [39], Biological Magnetic Resonance Data Bank (BMRB) [40, 41], Madison-Qingdao Metabolomics Consortium Database (MMCD) [42], Platform for RIKEN metabolomics (PRIMe) [43], Birmingham Metabolite Library (BML) [44], and MetaboLights [45]. Many biofluid specific databanks also exist, such as the Urine Metabolome Database [38], which can be found at http://www.urinemetabolome.ca. These databases are user friendly and provide access to the ^{1}H NMR spectra for many of the metabolites.

Exercise 11.13.

1. Access the BMRB metabolomics database (http://www.bmrb.wisc.edu/metabolomics/) to easily obtain the NMR spectrum of a molecule, such as glutaconic acid, by entering the name into the "enter search: field" and clicking on the "data" tab.
2. It is important to note the conditions under which the spectrum was acquired, because the conditions of your sample may be different and these variables may affect the spectrum that results. According to the data from part (1) of this exercise, what was the pH at which the entry for glutaconic acid was acquired, and at what temperature? For comparison purposes, an experimenter would collect the urine sample using the same conditions.
3. The use of the same reference molecule is also important, since commonly used reference molecules can have slightly different chemical shifts. According to the entry in the BMRB, what is the reference molecule for the listed acquisition of glutaconic acid? □

Variables one cannot always control, but which can influence the spectrum, include concentration and interactions of the molecule of interest with other molecules in the metabolomic mixtures. In these cases, one can confirm the identity of specific signal assignments by adding a small amount of purified chemical to the metabolomic mixture and repeating the spectrum. Since only the signals from the molecule of interest will increase in intensity, their positions on the spectrum will be obvious. It is also important to note whether the chemical shifts are obtained experimentally or are calculated using empirical or *ab initio* techniques.

Not knowing the identity of molecules in the mixture makes analysis more difficult, and several spectroscopic and mathematical techniques have been developed to identify individual molecules in a mixture. The 1D hydrogen, carbon, and nitrogen spectra can easily be collected, but the severe overlap of signals in the complex mixtures of molecules seen in metabolomic studies often renders ^{1}H 1D spectroscopy difficult, if not impossible, to analyze. (See Appendix [46, Figure A11-1b].) One approach to simplifying the problem is to expand the data into a second dimension (or even a third dimension) through several different methods. The appendix [46] to this chapter discusses these possibilities in more detail, including significant applications, such as the determination of small molecule structure, that go beyond identification of the chemical makeup of a sample and how it changes over time. However, while 2D methods provide powerful approaches, for example, to separating molecules, they are limited to systems where the molecules are few and present in similar concentrations.

In complicated mixtures, such as urine, more sophisticated methods need to be employed to extract the data of interest. In 2011, Bobae Kwon and collaborators presented one example in *Forensic Science International*, regarding the search for a urine biomarker to identify people who sniff glue [47]. The abuse of toluene-based materials, such as glue and other volatile substances, is not highly recognized, but is, in fact, a significant world-wide problem due to ease of access, use by children and adolescents, and potential to cause death [48]. Identification of a biomarker of toluene abuse could help in the identification, treatment, and management of abusers. Exposure-identifying biomarkers would also be helpful for individuals who are potentially exposed to toluene-containing materials in the workplace.

Kwon et al. collected urine samples from people who identified themselves as glue sniffers and a control group of individuals who did not sniff glue, and then collected NMR spectra on the samples. They then used PCA to identify which species in the mixture were most variable among the glue sniffing and control group individuals. The paper used the program Chenomx to identify the molecules prior to using PCA, but PCA also revealed the individual signals that were different between the samples and correlated the signals to individual molecules. PCA, for example, revealed that signals at 7.83, 7.64, 7.55, and 3.96 ppm increased by a similar value while signals at 4.06 and 3.05 ppm decreased in value. We will first use a database to analyze spectra associated with these lists of signals, including formulating and exploring a hypothesis as to the cause(s). We will then return to the notion of spectra and PCA in this context, and provide a linear algebraic/geometric approach to them.

Exercise 11.14 (Biomarkers for Glue Sniffers, I).

1. To identify the molecules that are associated with the set of peaks (at 7.83, 7.64, 7.55, and 3.96 ppm) where the signals increased, log into the metabolomics section of the BMRB http://www.bmrb.wisc.edu/metabolomics/, and "Search 1D Peak Lists." Then enter the ppm values into the search box individually, with only spaces separating the individual numbers: "7.83 7.64 7.55 3.96". After the numbers have been entered, press the "Peak Search" button. How many entries are in the resulting table?
2. Verify that only three of the molecules have an optimal "Peak Match" score of 1.00: guanine, hippuric acid, and glycolate, while all the other hits have a score of 0.67 or lower. To determine which of these three molecules is associated with the peaks, look up the individual spectra of the molecules. Clicking on the name of the molecule links to a page that provides a wealth of information on the molecule.
3. To access the NMR chemical shift data for guanine, for example, click on the "bmse000090" tab near the top of the web page. Guanine is a very common chemical species in biological systems since it is a precursor of guanosine, one of the five nucleotides used to make DNA and RNA molecules. The molecule has the chemical formula $C_5H_5N_5O$, but only one of the hydrogen atoms is attached to a carbon atom and can be observed in the ^1H NMR spectrum. The four hydrogen atoms attached to a nitrogen atom are constantly swapped with the hydrogen atoms in water, and are not observed in NMR experiments where the ^1H water signal is suppressed. Consequently, the ^1H NMR spectrum of guanine consists of a single signal.[24] Explain why the chemical shift of the carbon bound proton is at 7.6 ppm. Since this is similar (<0.05 ppm away) to our signal at 7.64 ppm, we can hypothesize that guanine levels in urine may be affected by a component in glue.
4. Another possibility is that glycolate levels in urine are affected as well. Glycolate is a small molecule with a chemical formula of C_5H_4O, which is an intermediate in the photorespiration pathway of plants but can also be produced by humans, is commonly used in skin care products, and is found in urine. In glycolate, two of the protons are bound to oxygen atoms and not observable by ^1H NMR, but the other two protons are bound to the same carbon and form a methylene group (-CH_2-). Since the two hydrogen atoms are virtually identical, the chemical shift of both is 3.93 ppm, and has a similar chemical shift to our signal (<0.05 ppm away) at 3.96 ppm.
5. Explain why, thus far, neither guanine or glycolate explain all the signals that increase in the urine of the glue-sniffing individuals.
6. Hippuric acid remains another possibility. Hippuric acid is an acyl glycine molecule that is formed when benzoic acid and the amino acid glycine are linked together via an amide bond. A diet that includes a large amount of foods rich in phenolic compounds, such as tea, wine, and fruit juices, leads to increased amounts of hippuric acid in urine. Hippuric acid has a chemical formula of $C_9H_9NO_3$ and has four chemical shifts: 7.808, 7.618, 7.53, and 3.944 ppm. Are these values similar to those observed for the signals that increased in the glue-sniffing individuals?
7. Remarks: Those who have downloaded the evaluation copy of Chenomx can also go to the library manager (see the tutorial PDF in Exercise 11.12 of this chapter) and upload the 500-MHz compound set (the NMR setting used in [47]), and then check the library for spectra for each of guanine, glycolate, and hippuric acid. Currently, one of these is not in the database exactly as is, but you can use BMRB or HMDB to find it.[25] The remaining two are under slightly different names (try clicking on the "alternate names" option).

24. The 1D ^1H, 2D ^1H-^1H TOCSY and 2D ^1H-^{13}C heteronuclear single quantum spectroscopy (HSQC) spectra on the page, also confirm this; see the appendix [46] for additional information about these alternative spectra.

25. A direct link to the latter and other external databases will appear on the Chenomx page for your chosen metabolite.

8. Use MetaCyc, KEGG, or another similar database to investigate links between toluene and hippuric acid via metabolic pathways. Does this provide further evidence for hippuric acid as a biomarker for glue sniffing?
9. Of the three possibilities (guanine, glycolate, hippuric acid), for which one is the hypothesis that it increases among glue-sniffing individuals best supported? Can the hypothesis that the levels of the other two increase be ruled out? □

Investigation into the biology of the three compounds strongly supports the hypothesis that hippuric acid is the compound that changes in response to glue sniffing. First, the concentration of hippuric acid in urine is, on average, 220 mol/mmol creatinine, whereas the concentrations of guanine and glycolate are 0.2 and 42 mol/mmol creatinine, respectively [38]. The 100-fold and 5-fold higher concentrations of hippuric acid relative to guanine and glycolate suggest that hippuric acid signals on the NMR spectra will be that much more intense than the signals of the other compounds. If anything, the signals of guanine and glycolate will be hidden by the signals of hippuric acid, and not observed. Also, since all four hippuric acid signals changed, one can be confident that hippuric acid is a good biomarker of glue sniffing. A literature search also confirms the hypothesis, since it is already known that toluene is converted into hippuric acid via metabolic pathways and benzoic acid [49, 50].

Exercise 11.15 (Biomarkers for Glue-Sniffers, II). Which molecule has decreased levels in glue-sniffing individuals? Recall that signals at 4.06 and 3.05 ppm decreased in value in people identified as glue sniffers. □

In these examples, the analysis is rather straightforward since the signals considered by the researcher were the more prominent signals in the spectra of urine, but the same analysis can also be applied to urine in general. Even though urine from different individuals is similar in chemical composition, the relative concentrations of the components vary based upon several different factors, such as diet, lifestyle, sex, age, and work and living environments. The likelihood that urine from any two individuals has exactly the same composition and relative concentrations of the components is extremely small. Also, the composition of a single person's urine will change. The variations in different urine samples combined with multivariate statistical analysis can tease apart the different components that are observed in urine and other bodily fluids.

11.6.3.3 Linear Algebraic Realizations, SVD, and PCA

Overview: NMR spectra for mixtures of compounds can be modeled as linear combinations of spectra for elementary compounds (e.g., single metabolites). An NMR spectrum is thus an element of an m-dimensional space $\mathbb{R}^m$, where m denotes the total number of metabolites. When comparing spectra from groups (e.g., a control group and glue sniffers) one approach is to seek the metabolites/compounds or combinations thereof that explain the greatest variability in the data (expecting differences between the groups). This variability, at least up to the level of covariance (controlled for the mean) can be encoded by the covariance matrix $C = \frac{1}{\#\text{samples}-1}X^TX$ of the data matrix X. The data of spectra in the high m-dimensional space can be mapped to a lower dimensional space by means of application of SVD to C. This is a method of decomposing any linear transformation by factoring its $m \times n$ matrix A into nice matrices, so that the transformation's behavior is easily described in terms of specialized eigenvalues (singular values) and eigenvectors. Geometrically, in the special case when $n = m$, these show the distortion of a unit sphere in $\mathbb{R}^m$. When SVD is applied to X of arbitrary size $m \times n$, the singular values and corresponding eigenvectors suggest the "directions" (i.e., vectors = spectra or linear combinations of them) explaining the greatest amounts of variation, as given by C. Selecting these directions (utilizing the singular values to create cut-off values as necessary) results in a lower dimensional subspace to which the data is mapped, and which, at some appropriate level, "explains" the data (including separating it into groups). We provide the details below.

Previously, in Exercise 11.11, we saw simple examples in $\mathbb{R}^3$ of how matrix multiplication gave transformations encompassing "stretching" of vectors (scalar multiplication) and rotations around coordinate axes. These two transformations (stretching and rotating) generalize to $\mathbb{R}^m$, for any $m \geq 1$, as encoded by the following definitions, relevant for square matrices.

A square (i.e., $n \times n$) matrix A is *diagonal* if A has the form where

$$A = \begin{bmatrix} d_{1,1} & 0 & 0 & \cdots & 0 \\ 0 & d_{2,2} & 0 & \cdots & 0 \\ 0 & 0 & d_{3,3} & \cdots & 0 \\ \vdots & \vdots & \vdots & \ddots & \vdots \\ 0 & 0 & 0 & \cdots & d_{n,n} \end{bmatrix}.$$

It is common to replace $d_{i,i}$ by d_i, and to write $D = diag(d_1, \ldots, d_n)$.

By the definition of a diagonal matrix D, it is immediate that multiplication $D\mathbf{x} = d_i\mathbf{x}$, for $\mathbf{x} = \mathbf{e}_i$ a standard basis vector having a 1 in the ith coordinate, and 0 elsewhere. In this sense, D gives a linear transformation that stretches (or shrinks) each standard basis vector. That is, each standard basis vector $\mathbf{e}_i$ is an *eigenvector* for D, with *eigenvalue* d_i, that is, the effect of multiplication by D is to simply scale $\mathbf{e}_i$, by the scalar (eigenvalue) d_i.

A square ($n \times n$) matrix A with real number entries is *orthogonal* if $A^T A = I = AA^T$, where I is the $n \times n$ identity matrix and A^T is the transpose of A, that is, the ith column of A^T is exactly the ith row of A. The content of the definition of an orthogonal matrix is less transparent than that for a diagonal matrix, but it has several nice features that are equivalent. First, recall that the standard basis vectors $\mathbf{e}_1, \ldots, \mathbf{e}_n$ of $\mathbb{R}^n$ form an *orthonormal set* of vectors, that is, the inner product $\mathbf{e}_i^T \mathbf{e}_j$ of any two equals 0, unless $i = j$, in which case it equals 1.

Theorem 11.1. *(See Exercise 11.16 for a partial proof, or [8, Section 5.6] for more details.) A square ($n \times n$) matrix A with real number entries is orthogonal if and only if:*

1. *The columns of A form an orthonormal set of vectors.*
2. *The rows of A form an orthonormal set of vectors.*
3. *The length $||A\mathbf{x}||$ equals $||\mathbf{x}||$, for any (column) vector $\mathbf{x}$ with n components.*

Exercise 11.16.

1. Check that the matrices (from part (5) of Exercise 11.11) that implement rotations around the X-, Y-, and Z-axes in $\mathbb{R}^3$ are orthogonal. (*Hint*: Use one or more basic trigonometric identities.)
2. For later use, it is helpful to note[26] that, for any $r \times s$ matrix B and any $s \times t$ matrix C, $(BC)^T = C^T B^T$. Presuming this relationship, if there is a third matrix D of size $t \times u$, find an expression for the $u \times r$ matrix $(BCD)^T$ in terms of B^T, C^T, and D^T.
3. Show that if A is orthogonal, so is A^T.
4. Use $||\mathbf{x}|| = \sqrt{\mathbf{x} \cdot \mathbf{x}^T}$ for any row vector $\mathbf{x}$, or, equivalently, $||\mathbf{x}|| = \sqrt{\mathbf{x}^T \cdot \mathbf{x}}$ for any column vector $\mathbf{x}$, to show Theorem 11.1(3). □

According to Theorem 11.1(3), no stretching or shrinking of $\mathbf{x}$ occurs under multiplication by an orthogonal matrix A. In this sense, multiplication by an orthogonal matrix can perform only the (higher dimensional) equivalents of rotations or reflections. Recall that the images under multiplication by A of the standard basis vectors $\mathbf{e}_1, \ldots, \mathbf{e}_n$ are just the columns of A. Theorem 11.1(2) says these images are simply another set of basis vectors at right angles (think "coordinate axes") for $\mathbb{R}^n$, so, in this sense as well, multiplication by A leads to a rotation or realignment of a coordinate system. For these reasons, transformations by orthogonal matrices are often referred to simply as rotations.

Theorem 11.2 (Singular Value Decomposition). *(For one of several equivalent approaches to a proof, see [8, Section 5.12] or [51] a step-by-step online tutorial with illustrations.) Suppose A is an arbitrary $m \times n$ matrix with real number entries, and $\mathbf{x}$ is a column vector with n entries. Then there is a matrix factorization $A = U\Sigma V^T$ into matrices with real number entries for which*

26. For a proof, one can compare the jth column of each side.

1. *U is $m \times m$, and V is $n \times n$, and both are orthogonal matrices;*
2. *In the factorization of A,* $\Sigma = \begin{bmatrix} D & 0 & \cdots & 0 \\ 0 & 0 & \cdots & 0 \\ \vdots & \vdots & \ddots & \vdots \\ 0 & 0 & \cdots & 0 \end{bmatrix}$ *is an $m \times n$ matrix, $D = diag(\sigma_1, \ldots, \sigma_r)$ is an $r \times r$ diagonal matrix, with $\sigma_1 \geq \sigma_2 \geq \cdots \geq \sigma_r > 0$, and $r \leq min(m, n)$ is the rank of A.*
3. *The matrix V can be partitioned into columns $V = [V_1|V_2]$ so that for $V_1 = [\mathbf{v}_{*1} \cdots \mathbf{v}_{*r}]$ and $V_2 = [\mathbf{v}_{*(r+1)} \cdots \mathbf{v}_{*n}]$:*
   (a) *$||A\mathbf{v}_1||^2$ is the largest value of $||A\mathbf{x}||^2$ for any $\mathbf{x}$ with $||\mathbf{x}|| = 1$;*
   (b) *$||A\mathbf{v}_2||^2$ is the largest value of $||A\mathbf{x}||^2$ for any $\mathbf{x}$ for which both $||\mathbf{x}|| = 1$ and $\mathbf{x}$ is orthogonal to Span($\mathbf{v}_1$) hold true;*
   (c) *$||A\mathbf{v}_3||^2$ is the largest value of $||A\mathbf{x}||^2$ for any $\mathbf{x}$ for which both $||\mathbf{x}|| = 1$ and $\mathbf{x}$ is orthogonal to Span($\mathbf{v}_1, \mathbf{v}_2$) hold true;*
   (d) *and so on, so for all $1 \leq j \leq r$, $||A\mathbf{v}_j||^2$ is the largest value of $||A\mathbf{x}||^2$ for any $\mathbf{x}$ for which both $||\mathbf{x}|| = 1$ and $\mathbf{x}$ is orthogonal to Span($\mathbf{v}_1, \mathbf{v}_2, \ldots, \mathbf{v}_{j-1}$).*
4. *The matrix U can be partitioned into columns $U = [U_1|U_2]$ so that $U_1 = [\mathbf{u}_{*1} \cdots \mathbf{u}_{*r}]$ and $U_2 = [\mathbf{u}_{*(r+1)} \cdots \mathbf{u}_{*m}]$, $A\mathbf{v}_{*j} = \mathbf{u}_{*j}\sigma_j = \sigma_j\mathbf{u}_{*j}$, for $1 \leq j \leq r$, and $A\mathbf{v}_{*j} = \mathbf{0}$ for $1 \leq j \leq n$.*

Remarks.

(1) Assuming $A \neq 0$, then $r > 0$ and the values $\sigma_1, \ldots, \sigma_r$ are called the *nonzero singular values of A*. Setting $\sigma_i = 0$ for $r + 1 \leq i \leq n$, one has the *singular values* $\sigma_1 \geq \cdots \geq \sigma_n$ of A. The vectors $\mathbf{v}_{*1}, \ldots, \mathbf{v}_{*n}$ are the *right singular vectors*, while $\mathbf{u}_{*1}, \ldots, \mathbf{u}_{*m}$ are the *left singular vectors of A*.

(2) A special case of the SVD occurs when A is a square $n \times n$ matrix and $rank(A) = n$. An $n \times n$ matrix B is *diagonalized* by an orthogonal matrix[27] if there is a diagonal matrix $D = diag(d_1, \ldots, d_n)$ and an orthogonal matrix Q so that B factors as $B = QDQ^T$. Thus, upon applying a rotation that moves the coordinate axes into the new directions given by the column vectors of $Q = [\mathbf{q}_{*1} \ldots, \mathbf{q}_{*n}]$, transformation by B can be represented by scaling each $\mathbf{q}_{*j}$ by d_j, $1 \leq j \leq n$.

Exercise 11.17 (SVD in MATLAB). Calculation of the SVD of a matrix A entered in MATLAB is given simply by the command `[U,S,V]=svd(A)`; the command `svd(A)` alone produces the list of singular values. In MATLAB, enter `A = [1.4015 -1.0480; -0.4009 1.0133]; eigshow(A)` to obtain an interactive demonstration of the geometric content of the SVD for the 2×2 matrix A in terms of its singular values and right and left singular vectors. In the resulting pop-up, follow the help button instructions (for svd mode) to complete the illustration of a figure illustrating the SVD for A in terms of distortion of the unit circle in $\mathbb{R}^2$. Then calculate the singular values for A and the actual matrix factorization (SVD) of A. Compare your results with the illustration in the summary article [52]. □

Theorem 11.2 is a fundamental result from linear algebra. It says that, in essence, every matrix can be factored into a rotation, then a matrix that does stretching/shrinking, and then another rotation. In the context of this chapter, Theorem 11.2 is important because it provides a means to carry out PCA on a data matrix that captures NMR spectrum data. Let's consider this NMR spectra data context a bit further first, before we discuss PCA and the SVD further.

Previously, in the context of representing biochemical reaction systems, individual metabolites (compounds) m_i were represented by the standard vectors $\mathbf{e}_1 = (1, 0, \ldots, 0, 0), \ldots, \mathbf{e}_m = (0, 0, \ldots, 0, 1)$, under some fixed ordering of all the m metabolites (compounds). This leads to an m-dimensional space $\mathbb{R}^m$ spanned by these vectors. (For example, every reaction $\mathbf{y} \to \mathbf{y}'$ was representable as an element of this space, obtained from a convention of writing $\mathbf{y}' - \mathbf{y}$.) Also, there was no linear dependence among these metabolites as vectors (i.e., no one distinct metabolite could be obtained as a linear combination of the remainders). These two properties

27. This is a special case of diagonalization, the treatment of which generally comes early in a linear algebra course, but we do not need the full measure here.

of a set of vectors determine a *basis* of a vector space, and the number of elements of the basis is an invariant of the space; by definition, the *dimension* of the vector space is this number. Now, in the context of ^{1}H NMR spectra, under a fixed set of sampling and chemical shift conditions,[28] it is possible to represent each elementary metabolite uniquely by its spectrum. These spectra can be scaled (e.g., by concentrations) and added (which is what occurs when two or more metabolites are present in a sample). The set of all the spectra corresponding to the m metabolites generate an m-dimensional vector space, for which nonnegative linear combinations of metabolites' spectra yield all possible spectra, presuming there is no "noise" in the data. It is convenient to also allow for linear combinations with negative coefficients, as in considering the difference between two spectra[29] when trying to resolve one spectrum as a sum of others (as in Exercise 11.14). Carrying out this resolution, adjusting for noise, and grouping potential test subjects by their sample spectra are precisely tasks with which PCA assists.

To better understand and determine spectra and information encoded by them, it can also be useful to take a vector space with a basis that is not the metabolites/spectra themselves, but peaks of the spectra. For example, suppose that[30] among a set of n spectra a total of p peaks are identified. (Some of the spectra may display a given peak, while some may not.) Ordering these p peaks, one can identify each of the n spectra by the ordered p-tuple whose entries are the intensity values at each peak, or alternatively, the concentration (or a normalized concentration) value over the peak. One can take the standard vectors $\mathbf{e}_1, \ldots, \mathbf{e}_p$ to represent the peaks; their span is the *peak space* $\mathbb{R}^p$, and within it, the original n spectra correspond to n particular vectors (linear combinations in peak space). PCA may be used to isolate vectors in peak space of the greatest importance for identifying features of interest encoded by the spectra. In some applications, the vectors thus isolated may be peaks themselves, for example, those for which the greatest amount of change is observed in measurements over time. Projecting $\mathbb{R}^p$ onto the subspace spanned by a set of identified peaks will help identify regions of interest in the spectra for further study. Alternatively, PCA may help identify metabolites or interesting complexes of metabolites appearing in the spectra; these complexes will necessarily be represented by linear combinations of the peaks (i.e., vectors in $\mathbb{R}^p$). As they arise from PCA, these complexes may or may not be among the original given spectra themselves, nor be simply peaks, but form a set of linearly independent vectors in the peak space. Projecting $\mathbb{R}^p$ onto their span will help reveal information encoded by the spectra. In either case, carrying out this projection is another function PCA performs.

Now, suppose X is a matrix which might represent several different formats of ^{1}H NMR data sets, for example,

- columns of X are spectra; rows of X are metabolites; the *ij*th entry $x_{i,j}$ is the relative concentration of metabolite i appearing in spectrum j;
- rows of X are discretized "bins" or "buckets" (measured in ppm); columns of X are spectra; $x_{i,j}$ is the integral intensity or the frequency (or an average frequency) measurement at the ith bucket for the jth spectrum (appropriately normalized);
- columns of X are spectra; rows of X are peaks; $x_{i,j}$ is the intensity or relative concentration of peak i in spectrum j;
- additional data of interest may be added or recorded by X; for example, additional added rows may record measurement types such as gender, height, weight, or other status factors associated with the individuals from which the trials (e.g., columns, here spectra) were drawn.

For example, in [47], there were nine urinary samples; after being processed, the end resulting spectra were reduced to ppm spectral buckets and normalized to the region of 0.5-8.5 ppm. Using tenths (respectively, hundredths) would split the region into 80 (800) parts, yielding a matrix X of size 80×9 (800×9) if the data are recorded as columns of spectra and the rows are "bucket" regions of the chemical shift axis. Alternatively,

---

28. Such as those noted in Exercise 11.13.
29. For example, in Chenomx, "sum lines" and "subtraction lines" are two graphical features corresponding to these types of linear combinations.
30. Upon setting a threshold value, agglomeration of multipeaks, and other processing steps described previously.

for a total of p peaks in the nine spectra, a matrix X recording peak intensity for spectra as columns and peaks as rows would have size $p \times 9$.

Presume that each column vector of an (otherwise arbitrary) $m \times n$ data matrix X has mean 0.[31] In mathematical terms, the goal of PCA, as applied to X, can be succinctly stated as follows:
PCA GOAL: For $r = rank(X)$, find some orthogonal matrix Q so that:

- for $Y = XQ$, the columns of Y are orthogonal, and $Span(Y) = Span(X)$,
- for $1 \leq j \leq r$, the *variances* $\frac{1}{n-1}\mathbf{y}_{*j}^T \cdot \mathbf{y}_{*j} = \frac{1}{n-1}\sum_{k=1}^{n} y_{k,j}^2 = \frac{1}{n-1}||\mathbf{y}_{*j}||^2$ of Y are successively maximized (with the largest variance associated to $\mathbf{y}_{*1}$, the next largest to $\mathbf{y}_{*2}$, and so on), while for all $i \neq j$ the *covariances* $\frac{1}{n-1}\mathbf{y}_{*i}^T \cdot \mathbf{y}_{*j}$ of Y are minimized.[32]

Under these assumptions, the first r columns of Y are the *principal components* of X.

Let's unpack this "goal" statement. Now, the matrix $C_Y = \frac{1}{n-1}Y^TY$ is the *covariance matrix* of Y. For $\mathbf{y}_{*j}$ the jth column of Y, the diagonal entries of C_Y equal $\frac{1}{n-1}\mathbf{y}_{*j}^T \cdot \mathbf{y}_{*j} = \frac{1}{n-1}\sum_{k=1}^{n} y_{k,j}^2 = \frac{1}{n-1}||\mathbf{y}_{*j}||^2$, the variances $var(\mathbf{y}_{*j})$ of Y. The off-diagonal entries $cov(\mathbf{y}_{*i}, \mathbf{y}_{*j}) = \frac{1}{n-1}\mathbf{y}_{*i}^T \cdot \mathbf{y}_{*j}$ are the covariances of Y. By their definitions, the covariances measure the extent to which there is linear dependence among the columns of Y (e.g., if the columns are orthogonal, the covariances are zero). Under the assumption that there is a reasonably low level of noise in the transformed data Y, the variances measure signal strength.[33] If C_Y were already diagonal, then the covariances would all be zero, and the variances would show up as the diagonal entries, with the largest diagonal entries demonstrating the greatest amount of variance and the most importance in terms of signal strength.

By taking only the vectors (directions) corresponding to the largest diagonal entries and looking at the subspace the corresponding eigenvectors generate, one could account for the regions, metabolites, etc. that are of greatest import to understanding the data. (This is the idea of *dimension reduction*.) One could sacrifice some accuracy (or, depending upon the context, cast off some "noise"; see [8, Example 5.12.3] for a general application PCA in this context; see [53] for an application of SVD to denoising FIDs) by ignoring the others. In this sense, maximizing variance and minimizing covariance[34] are achieved through diagonalization of C_Y. Under the assumption $rank(X) = r$, $Span(X)$ has dimension r, and if $Y = XQ$ for Q orthogonal, with $Span(XQ) = Span(X)$, then one is not changing the information in the data carried in (the columns of) X, except to reveal it through the application of a rotation of a basis of $Span(X)$. In combination, PCA aims to re-represent the data given by X in such a fashion that its signal can be maximized and directions of highest importance identified. This is illustrated in Exercise 11.18 for some particular concrete examples and by revisiting our glue-sniffing example [47] in Exercise 11.19. The final exercise, Exercise 11.20, shows how Theorem 11.2 accomplishes PCA. In it, one shows that for X, Y, Q, as in the "PCA Goal" statement, one can take $Q = V$, for $X = A$ and V, as in Theorem 11.2.

Exercise 11.18 (PCA Example with ^{1}H NMR Data)**.** Refer to [54, Section 2.3].[35] Three types of samples (plasma, urine, and saliva) were drawn once from 150 subjects.

1. For each type of sample, the relevant data matrix was either 150×409 or 150×493. What formed the rows and columns of each data matrix X, what was the size of the data matrix, and how was the data entry $x_{i,j}$ defined?
2. These data were further separated into a testing set of 50 subjects and a remaining training set of 100. After some rescaling, PCA was applied to the training data for the remaining three types of sample data, so to X of size 100×493 or 100×409, as relevant. We do not learn the value of $r = rank(X)$, but it is no larger than $100 = \min(100, 493)$ $(\min(100, 409))$, so there are at most $r = 100$ principal components; likely many

31. This is not a real constraint; if the means are not 0, we could repeat the analysis below by creating a new matrix with column means 0 simply by subtracting the mean from each. This linear operation creates the more general definition of a covariance matrix, and one then simply applies PCA to it.
32. The value 0 is the best (i.e., least) one can get.
33. Under, for example, the signal-to-noise ratio given as the ratio of the variance of signal to the variance of noise.
34. This can also be thought of as measuring redundancy in the data.
35. Available free via Googling or from this chapter's Supplementary Materials.

fewer. The first two principal components, $PC1$ and $PC2$, were examined. We do not know what the vector representations of these as elements of $\mathbb{R}^{100}$ would be, but by SVD $PC1$ and $PC2$ are orthogonal vectors and it makes sense to represent them as two new coordinate axes vectors (e.g., the x-axis for $PC1$ and y-axis for $PC2$). Refer to [54, Figure 1]. As shown there, the *scores plots* relative to $PC1$ and $PC2$ are given as follows: Each point on the scores plot corresponds to a subject/spectrum, an element of $\mathbb{R}^{493}$ or $\mathbb{R}^{409}$. PCA projects each spectrum onto the span $Span(U_1)$ for U_1 as in Theorem 11.2. As per previous remarks and Exercise 11.20 below, in fact $Span(Y) = Span(PC1, PC2, \mathbf{y}_3, \ldots, \mathbf{y}_r) = Span(U_1)$, an r-dimensional space, for $Y = XV$, where V arises from Theorem 11.2 by setting $A = X$. In the new principal coordinates, each spectrum is represented uniquely as a linear combination $\sum_{j=1}^{r} \alpha_j \mathbf{y}_{*j}$ of the columns of Y (or U_1, replacing α_j by α_j/σ_j, using the singular values σ_j, $1 \leq j \leq r$). If we take only the first two multiples α_1 and α_2, that is, the coordinates for the principal component axes $PC1$ and $PC2$, then from a spectrum one gets an ordered pair (α_1, α_2). These pairs form the score plots. What information relative to the participants' gender do you glean from these three plots?

3. As per [54, Figure 1], for each sample type, what percent of the variation was explained by the first principal component $PC1$? By $PC2$? Do these results suggest to you that reducing to $Span(PC1, PC2)$ was sufficient to capture the behavior of the data? □

Exercise 11.19 (Application of PCA to Biomarkers for Glue Sniffers). Among the species identified by prior analysis of the urine spectra in [47], PCA identified three of greatest variability. What do the PCA score plots for two of the principal components (see [47, Figure 4a]) and three of the principal components (see [47, Figure 4b]) imply? If PCA ordered metabolites by the greatest variability, why were PC1 and PC3 chosen for Figure 4a instead of PC1 and PC2? Are the results consistent with Exercises 11.14 and 11.15? □

Exercise 11.20 (PCA by SVD). Let X be an $m \times n$ data matrix X of m measurement types (types of metabolites) with n trials (spectra), and let $X = U\Sigma V^T$ be the SVD of X as in Theorem 11.2. Set $Y = XV$. Recall that by the assumptions of Theorem 11.2, V is orthogonal.

1. Explain why, for each j with $1 \leq j \leq r$, the variances $var(\mathbf{y}_{*j})$ are successively maximized (with the largest variance associated to $\mathbf{y}_{*1}$, the next largest to $\mathbf{y}_{*2}$, and so on. (*Hint*: $\mathbf{y}_{*j} = X\mathbf{v}_{*j}$. Now compare with Theorem 11.2(3).)
2. Explain why the columns of Y are orthogonal. What does this imply about the covariances $cov(\mathbf{y}_{*i}, \mathbf{y}_{*j})$?
3. Explain why $Span(Y) = Span(X)$. (*Hint*: Show $Span(Y) \subseteq Span(X)$. Explain why $dim(Span(Y)) = r$, and also $dim(Span(X)) = r$, so then $Span(Y) = Span(X)$ is forced.)
4. Parts (1)-(4) above verify that $Y = XV$ satisfies the conditions for PCA, hence the columns $\mathbf{y}_{*j}$, $1 \leq j \leq r$, are, by definition, the principal components of X. Note that for the purpose of determining the principal components of X, it suffices to consider the matrix product $Y = XV_1$. Explain why the columns of U_1 are also often called the *normalized principal components of* X, and give an orthonormal set of vectors lying in the directions of greatest variation for the data X, given by $\sigma_1 \geq \cdots \geq \sigma_r > 0$. Thus, the left singular vectors of X and the right singular vectors of X, both associated to the nonzero singular values $\sigma_1, \ldots, \sigma_r$, provide the mechanisms for PCA. □

11.7 NMR FOR METABOLIC ANALYSIS AND MATHEMATICAL METHODS: DIRECTIONS OF FURTHER RESEARCH

In summary, the mathematical methods used to analyze NMR spectra can be classified in two different categories: quantitative NMR spectroscopy and chemometrics [55]. Quantitative NMR spectroscopy is the more complete yet time consuming of the two methods, since it results in all the molecules in the mixture being identified and quantified, as discussed above. Multiple 1D and 2D spectra must be collected, and the data analyzed and compared to standard molecules. Spectral features such as the number of peaks, J-coupling values, peak multiplicity, and relative peak intensities are all used to identify the molecules in the sample. The task is complete when all the peaks in the sample are assigned to a specific molecule.

Chemometrics is a quicker approach to the problem since identification of all the components in the mixture is not necessary and is more flexible since both supervised and unsupervised methods of analysis are possible. Pattern recognition by PCA (originating with [56]) is the unsupervised technique currently used most since it can easily handle large multidimensional data sets, summarize the data, and reveal how features are related. The features are scalar values that are representative of a region of the spectrum, and can be NMR signals, small regions of the spectrum (bins), amplitudes, or integrals. Other unsupervised techniques that have been applied to NMR data include hierarchical cluster analysis (HCA) [57, 58] and K-means [59, 60], and statistical total correlation spectroscopy (STOCSY) [61]. Supervised methods of analysis include partial least squared discriminate analysis (PLS-DA) and orthogonal projections to latent structures (OPLS) [32, 62, 63]. For example, following up on Exercise 11.18, PLS-DA was used in [54] to improve on an initial PCA analysis indicating clustering of spectral data by gender, and the recovery, through ^{1}H NMR urine, saliva, and plasma data, of the metabolites responsible for the clustering effects.

As follows from SVD, PCA depends on assumptions that the data measurements are linearly associated, but such assumptions correspond to a presumption that the data approximately follows a normal distribution. However, there are arguments that actual real metabolite data is not normally distributed [64]. Replacing the assumption of "zero covariances" by one of "statistical independence," leads similarly to independent component analysis (ICA), applicable to non-Gaussian data and utilized for metabolic analysis in [64]. For future research, it would be of interest to apply a simultaneous algebraic and geometric generalization of PCA and ICA, deemed *principal cumulant component analysis* (PCCA). Whereas PCA uses second-order statistics (mean, covariance), PCCA utilizes multivariate cumulants, "statistical objects generalizing covariance matrices to higher-order information," and "analysis of the cumulants via a multilinear model suggested by how cumulants transform" without requiring the statistical independence assumption of ICA [65].

11.8 SUPPLEMENTARY MATERIALS

The following references and materials utilized in chapter exercises are available as Supplementary Materials from the book's companion website: [12, 47, 54], caffeine biosynthesis.pdf, and PDF copies of materials at the following two websites relative to Hess's Law: http://www.chemteam.info/Thermochem/HessLawIntro1a.html and http://www.chemteam.info/Thermochem/HessLawIntro1b.html. An appendix [46] to this chapter is also available from the website.

REFERENCES

[1] Ogita S, Uefuji H, Morimoto M, Sano H. Application of RNAi to confirm theobromine as the major intermediate for caffeine biosynthesis in coffee plants with potential for construction of decaffeinated varieties. Plant Mol Biol 2004;54:931-41.

[2] Ashihara H, Sano H, Crozier A. Caffeine and related purine alkaloids: biosynthesis, catabolism, function and genetic engineering. Phytochemistry 2008;69:841-56. doi:10.1016/j.phytochem.2007.10.029.

[3] Kim YS, Uefuji H, Ogita S, Sano H. Transgenic tobacco plants producing caffeine: a potential new strategy for insect pest control. Transgenic Res 2006;15:667-72. doi:10.1007/s11248-006-9006-6.

[4] Kim YS, Lim S, Yoda H, Choi C-S, Choi Y-E, Sano H. Simultaneous activation of salicylate production and fungal resistance in transgenic chrysanthemum producing caffeine. Plant Signal Behav 2011;6(3):409-12. doi:10.4161/psb.6.3.14353.

[5] Kyoto Encyclopedia of Genes and Genomes (KEGG) pathway database: http://www.genome.jp/kegg/pathway.html, Kanehisa Laboratories.

[6] Caspi R, Foerster H, Fulcher CA, Kaipa P, Krummenacker M, Latendresse M, et al. The MetaCyc database of metabolic pathways and enzymes and the BioCyc collection of Pathway/Genome Databases. Nucleic Acids Res 2014;42:D459-71. Available from: http://metacyc.org.

[7] Schellenberger J, Park JO, Conrad TC, Palsson B. BiGG: a biochemical genetic and genomic knowledge base of large scale metabolic reconstructions. BMC Bioinformatics 2010;11:213. Available from: http://bigg.ucsd.edu.

[8] Meyer C. Matrix analysis and applied linear algebra. Philadelphia, PA: SIAM; 2000.

[9] Leon SJ. Linear algebra with applications. Upper Saddle River, NJ. Pearson/Prentice Hall; 2006.

[10] Hodge TL. Metabolic pathway analysis: a linear algebraic approach. In: Robeva R, Hodge T, editors. Mathematical concepts and methods in modern biology. London, UK: Academic Press/Elsevier; 2013.
[11] Feinberg M. Lectures on chemical reaction networks; 1979. Available from: http://www.crnt.osu.edu/LecturesOnReactionNetworks.
[12] Feliu E, Wulf C. Preclusion of switch behavior in reaction networks. Appl Math Comput 2012;219:1449-67. Available from: http://dx.doi.org/10.1016/j.amc.2012.07.048.
[13] Craciun G, Pantea C. Identifiability of chemical reaction networks. J Math Chem 2008;44:(1):244-59.
[14] Craciun G, Pantea C, Rempala G. Algebraic methods for inferring biochemical networks: a maximum likelihood approach. Comput Biol Chem 2009;33(5):361-7. See also arXiv:0810.0561v2.
[15] Craciun G, Kim J, Pantea C, Rempala G. Statistical model for biochemical network inference. Commun Stat Simul Comput. 2013; 42(1): 121–137. doi:10.1080/03610918.2011.633200.
[16] Pachter L, Sturmfels B. Algebraic statistics for computational biology. Cambridge, UK: Cambridge University Press; 2005.
[17] Jamshidi N, Palsson B. Formulating genome-scale kinetic models in the post-genome era. Mol Syst Biol 4; Article number 171. doi:10.1038/msb.2008.8.
[18] Jamshidi N, Palsson B. Mass action stoichiometric simulation models: incorporating kinetics and regulation into stoichiometric models. Biophys J 2010;98:175-85.
[19] Christodoulou J, Larsson G, Fucini P, Connell SR, Pertinhez TA, Hanson CL, et al. Heteronuclear NMR investigations of dynamic regions of intact *Escherichia coli* ribosomes. Proc Natl Acad Sci USA 2004;101(30):10949-54.
[20] Mulder FA, Bouakaz L, Lundell A, Venkataramana M, Liljas A, Akke M, et al. Conformation and dynamics of ribosomal stalk protein L12 in solution and on the ribosome. Biochemistry 2004;43(20):5930-6.
[21] Szymczyna BR, Gan L, Johnson JE, Williamson JR. Solution NMR studies of the maturation intermediates of a 13 MDa viral capsid. J Am Chem Soc 2007;129(25):7867-76.
[22] Molinski TF. NMR of natural products at the "nanomole-scale". Nat Prod Rep 2010;27(3):321-9.
[23] Breton RC, Reynolds WF. Using NMR to identify and characterize natural products. Nat Prod Rep 2013;30(4):501-24.
[24] Zhang A-H, Sun H, Qiu S, Wang X-J. NMR-based metabolomics coupled with pattern recognition methods in biomarker discovery and disease diagnosis. Magn Reson Chem 2013;51(9):549-56.
[25] Bren L. Metabolomics: working toward personalized medicine. FDA Consum 2005;39(6):28-33.
[26] Clayton TA, Lindon JC, Cloarec O, Antti H, Charuel C, Hanton G, et al. Pharmaco-metabonomic phenotyping and personalized drug treatment. Nature 2013;440(7087):1073-7.
[27] Terpstra M, Vaughan TJ, Ugurbil K, Lim KO, Schulz SC, Gruetter R. Validation of glutathione quantitation from STEAM spectra against edited ^{1}H NMR spectroscopy at 4T: application to schizophrenia. MAGMA 2005;18(5):276-82.
[28] Simister RJ, McLean MA, Barker GJ, Duncan JS. A proton magnetic resonance spectroscopy study of metabolites in the occipital lobes in epilepsy. Epilepsia 2003;44(4):550-8.
[29] Martinez-Granados B, Morales JM, Rodrigo JM, Del Olmo J, Serra MA, Ferrndez A, et al. Metabolic profile of chronic liver disease by NMR spectroscopy of human biopsies. Int J Mol Med 2011;27(1):111-7.
[30] Morris GA. Compensation of instrumental imperfections by deconvolution using an internal reference signal. J Magn Reson 1988;80(3):547-55.
[31] Akira K, Masu S, Imachi M, Mitome H, Hashimoto M, Hashimoto T. ^{1}H NMR-based metabonomic analysis of urine from young spontaneously hypertensive rats. J Pharm Biomed 2008;46(3):550-6.
[32] Gordon CM. New developments in catalysis using ionic liquids. Appl Catal Gen 2001;222(1-2):101-17.
[33] Allen DT, Shonnard DR. Green chemistry. Upper Saddle River, NJ: Prentice Hall; 2002.
[34] Ranke J, Stolte S, Stoermann R, Arning J, Jastorff B. Design of sustainable chemical products—the example of ionic liquids. Chem Rev 2007;107(6):2183-206.
[35] Kouba E, Wallen EM, Pruthi RS. Uroscopy by Hippocrates and Theophilus: prognosis versus diagnosis. J Urol 2007;177(1):50-2.
[36] Echeverry G, Hortin GL, Rai AJ. Introduction to urinalysis: historical perspectives and clinical application. Methods Mol Biol 2010;641:1-12.
[37] Nicholson JK, Lindon JC. Metabonomics. Nature 2008;453:1054-6.
[38] Bouatra S, Aziat F, Mandal R, Guo AC, Wilson MR, Knox C, et al. The human urine metabolome. PLoS ONE 2013;8(9): e73076.
[39] Wishart DS, Tzur D, Knox C, Eisner R, Guo AC, Young N, et al. HMDB: the Human Metabolome Database. Nucleic Acids Res 2007;35(Database issue):D521-6.
[40] Markley JL, Anderson ME, Cui Q, Eghbalnia HR, Lewis IA, Hegeman AD, et al. New bioinformatics resources for metabolomics. Pac Symp Biocomput 2007;157-68.
[41] Ulrich EL, Akutsu H, Doreleijers JF, Harano Y, Ioannidis YE, Lin J, et al. BioMagResBank. Nucleic Acids Res 2008;36(Database issue):D402-8.

[42] Cui Q, Lewis IA, Hegeman AD, Anderson ME, Li J, Schulte CF, et al. Metabolite identification via the Madison Metabolomics Consortium Database. Nat Biotechnol 2008;26(2):162-4.

[43] Akiyama K, Chikayama E, Yuasa H, Shimada Y, Tohge T, Shinozaki K, et al. PRIMe: a web site that assembles tools for metabolomics and transcriptomics. In Silico Biol 2008;8(3–4):339-45.

[44] Ludwig C, Easton JM, Lodi A, Tiziani S, Manzoor SE, Southam AD, et al. Birmingham Metabolite Library: a publicly accessible database of 1-D H-1 and 2-D H-1 J-resolved NMR spectra of authentic metabolite standards (BML-NMR). Metabolomics 2012;8(1):8-18.

[45] Haug K, Salek RM, Conesa P, Hastings J, de Matos P, Rijnbeek M, et al. MetaboLights—an open-access general-purpose repository for metabolomics studies and associated meta-data. Nucleic Acids Res 2013;41(Database issue):D781.

[46] Szymczyna BR. Appendix to this chapter: two-dimensional NMR.

[47] Kwon B, Kim S, Lee DK, Park YJ, Kim MD, Lee JS. ^{1}H NMR spectroscopic identification of a glue sniffing biomarker. Forensic Sci Int 2011;209(1-3):120-5.

[48] Cruz SL, Orta-Salazar G, Gauthereau MY, Millan-Perez Pena L, Salinas-Stefanon EM. Inhibition of cardiac sodium currents by toluene exposure. Br J Pharmacol 2003;140(4):653-60.

[49] Carlisle EJ, Donnelly SM, Vasuvattakul S, Kamel KS, Tobe S, Halperin ML. Glue-sniffing and distal renal tubular acidosis: sticking to the facts. J Am Soc Nephrol 1991;1(8):1019-27.

[50] Hanioka H, Hamamura M, Kakino K, Ogata H, Jinno H, Takahashi A, et al. Dog liver microsomal P450 enzyme-mediated toluene biotransformation. Xenobiotica 1995;25(11):1207-17.

[51] Will T. Introduction to the singular value decomposition; 1999. (online tutorial). Available from: http://websites.uwlax.edu/twill/svd/index.html.

[52] Moler C. Professor SVD. 2006-91425v00. Available from: http://www.mathworks.com/company/newsletters/articles/professor-svd.html.

[53] Man PP, Bonhomme C, Babonneau F. Denoising NMR time-domain signal by singular-value decomposition accelerated by graphics processing units. Solid State Nucl Magn Reson 2014;61-62:28-34. Available from: http://dx.doi.org/10.1016/j.ssnmr.2014.05.001.

[54] Ramadan Z, Jacobs D, Grigorov M, Kochhar S. Metabolic profiling using principal component analysis, discriminant partial least squares, and genetic algorithms. Talanta 2006;68:1683-91. doi:10.1016/j.talanta.2005.08.042.

[55] Wishart DS. Quantitative metabolomics using NMR. TrAC Trends Anal Chem 2008;27(3):228-37.

[56] Pearson K. On lines and planes of closest fit to systems of points in space. Philos Mag 1901;2(7-12):559-72.

[57] Kim H, Saifullah K, Khan S, Wilson EG, Kricun SDP, Meissner A, et al. Metabolic classification of South American Ilex species by NMR-based metabolomics. Phytochemistry 2010;71(7):773-84.

[58] Mahle DA, Anderson PE, DelRaso NJ, Raymer ML, Neuforth AE, Reo NV. A generalized model for metabolomic analyses: application to dose and time dependent toxicity. Metabolomics 2011;7(2):206-16.

[59] Hageman JA, van den Berg RA, Westerhuis JA, Hoefsloot HCJ, Smilde AK. Bagged K-means clustering of metabolome data. Crit Rev Anal Chem 2006;36(3-4):211-20.

[60] Cuperlovic-Culf M, Belacel N, Cuif AS, Chute IC, Ouellette RJ, Burton IW, et al. NMR metabolic analysis of samples using fuzzy K-means clustering. Magn Reson Chem 2009;47:S96-104.

[61] Cloarec O, Dumas ME, Craig A, Barton RH, Trygg J, Hudson J, et al. Statistical total correlation spectroscopy: an exploratory approach for latent biomarker identification from metabolic ^{1}H NMR data sets. Anal Chem 2005;77(5):1282-9.

[62] Wold S, Sjostrom M, Eriksson L. PLS-regression: a basic tool of chemometrics. Chemom Intell Lab Syst 2001;58(2):109-30.

[63] Trygg J, Wold S. Orthogonal projections to latent structures (O-PLS). J Chemom 2002;16(3):119-28.

[64] Bartel J, Krumsiek J, Theis FJ. Statistical methods for the analysis of high-throughput metabolomics data. Comput Struct Biotechnol J 2013;4. Available from: http://dx.doi.org/10.5936/csbj.201301009.

[65] Morton J, Lim L-H. Principal cumulant component analysis. Available from: http://www.stat.uchicago.edu/~lekheng/work/pcca.pdf. Accessed: 2014-03-25.

Chapter 12

Reconstructing the Phylogeny: Computational Methods

Grady Weyenberg and Ruriko Yoshida
Department of Statistics, University of Kentucky, Lexington, KY

12.1 INTRODUCTION

A *phylogenetic tree* (or *phylogeny*) is a graph that summarizes the relations of evolutionary descent between different species, organisms, or genes. Phylogenetic trees are useful tools for organizing many types of biological information, and for reasoning about events that may have occurred in the evolutionary history of an organism.

Generally, the first step in reconstructing a phylogenetic tree is to obtain a sample of representative individuals from each taxa to be included in the phylogeny, and then observe and quantify in them a number of morphological traits. The traits measured could range from simple physiological observations, for example, skull volume, or they may detail the differences between the biological sequences that comprise the chemistry of life.

Contemporary phylogenetic analysis is generally carried out on data obtained from either nucleic acid sequences (DNA/RNA) or amino acid sequences (proteins). In either case, the data consist of a string of characters from a fixed alphabet. In the case of nucleic acids, the alphabet consists of four characters, most commonly denoted A, C, G, and T.

When a sequence is duplicated, for example, when an organism reproduces, it may be preserved without change, or it may be *mutated.* At any given position in the sequence, a number of things may occur during a mutation. A character may be *substituted* for another character, for example an A may change to become a C. A character may be *deleted*, shortening the sequence overall. Finally, one or more characters may be *inserted*, lengthening the sequence.

The possibility of the latter two types of mutation means that any time we wish to compare multiple sequences we may be first required to *align* them. A sequence alignment identifies characters in each observed sequence that correspond to common characters in the ancestor sequence.

Aligning multiple sequences is known to be a very difficult problem. (An NP-hard problem, to be precise [1].) However, despite the difficulty of the alignment problem, it is generally assumed that sequences used for phylogenetic reconstruction are perfectly aligned. This chapter addresses the possible analyses that can be carried out after sequence alignment(s) have been obtained. At the time of this writing, `MUSCLE` [2] and `MAFFT` [3] are the most commonly used tools to obtain sequence alignments. Interested readers may see [4] for an introduction to the alignment of sequences.

Phylogenetic analysis relies heavily on computers, and a large number of programs have been written to perform various types of analysis. In the interest of uniformity, we will attempt to demonstrate techniques using the `R` statistical system whenever possible. Unfortunately, space constraints preclude any significant discussion of the details of the `R` language in this chapter, but the syntax should be largely legible to anyone with previous

Algebraic and Discrete Mathematical Methods for Modern Biology. http://dx.doi.org/10.1016/B978-0-12-801213-0.00012-5
Copyright © 2015 Elsevier Inc. All rights reserved.

experience in curly-bracket languages. There are a large number of quality publications available for those who wish to familiarize themselves with the R language; the document available at http://cran.r-project.org/doc/manuals/R-intro.pdf is a reasonable place to begin.

A large amount of the functionality of R is provided by optional add-on packages, and in the case of phylogenetics many of these packages add an R interface to some external third-party program. We will provide the names of these add-on packages, and any external programs that they require, before any example that depends on their presence. However, the reader must be prepared to refer to the documentation provided by add-on authors for instructions on their installation and configuration.

Exercise 12.1. Upon completing this exercise, readers should have a base R installation that is ready to be used in further examples and exercises.

(a) Obtain and install the R system from http://cran.r-project.org/ or a local mirror. The basic installation includes little more than a basic console interface to the R interpreter, so you may also wish to download the RStudio IDE from https://www.rstudio.com/ for an improved GUI.

(b) Open R and enter the following command into the console.

```
"Hello, World!"
```

The interpreter should respond with the following.

```
[1] "Hello, World!"
```

(c) Install the ape add-on package by entering

```
install.packages("ape")
```

The system might ask for you to select a download location and will then proceed to download the add-on. When it is complete, you must activate the package by doing the following.

```
library(ape)
```

You should also take a moment to install and load the "phangorn" package, which provides more features used in subsequent examples.

(d) The program below reconstructs and plots a tree from a built-in example dataset.

```
data(woodmouse)
dm <- dist.dna(woodmouse)
trw <- nj(dm)
plot(trw)
```

(e) *Bootstrapping* is a nonparametric technique for exploring variance. A new dataset is generated by uniformly sampling (with replacement) from the observed data. The code chunk below generates a bootstrap resample from the woodmouse data and builds a new tree from the result.

```
i <- sample.int(ncol(woodmouse),replace=TRUE)
wm2 <- woodmouse[, i]
plot(nj(dist.dna(wm2)))
```

Execute the code a few times (a new dataset is generated each time) and observe the changes in the reconstructed tree. Are there any features of the tree that seem to be relatively constant or, conversely, particularly variable? □

From a statistical point of view, we can think of a phylogenetic tree as a graphical model. A phylogeny is described as a binary tree in which the leaves of the tree are the observed values of a given site in the different species and internal nodes take the values of the site for putative ancestral species. For example, in the case of DNA sequences, notes of the phylogenetic tree are multivariate random variables with four states $\{A, C, G, T\}$. A phylogenetic tree is a directed graph (from the common ancestral species of the given species to each observed species) and the evolution is described in terms of a discrete-state continuous-time Markov process on a phylogenetic tree. We introduce ways for quantifying evolutionary change in Section 12.3.

A continuous-time Markov chain is characterized by a substitution rate matrix, and the phylogenetic tree summarizes the relationships between the species in terms of edge lengths (times since divergence) and common ancestors.

The continuous-time Markov process gives rise to a distance measure between any sets of sequences. Pairwise distances can be used, together with a distance-based method that is a way to reconstruct a phylogenetic tree from a set of all pairwise distances, reconstructing the phylogenetic tree that relates the sequences. We present such methods for phylogenetic tree reconstruction in Sections 12.2.

Usually a set of all pairwise distances, which is often called a distance matrix, computed from an alignment does not give a *tree metric*, which is a distance matrix realizing a phylogenetic tree. Thus a distance method tries to find a tree metric closest to the given set of pairwise distances computed from the alignment under some criteria. In Section 12.3.3, we will discuss the evolutionary model on a phylogeny and we will discuss how to infer the maximum likelihood estimator (MLE) on the tree.

Using the MLE under an evolution model, we can compute all pairwise distances from the given alignment. From all pairwise distances, we can also reconstruct a phylogenetic tree via *distance-based methods*. Distance-based methods have been shown to be statistically consistent in all settings (such as the long branch attraction) in contrast with parsimony methods [5–8]. Distance-based methods also have a huge speed advantage over parsimony and likelihood methods in terms of computational time, and hence enable the reconstruction of trees with large numbers of taxa. However, a distance-based method is not a perfect method of reconstructing a phylogenetic tree from the inputs sequence dataset: in the process of computing a pairwise distance, we ignore interior nodes of a tree as well as a tree topology, and we lose information from the input sequence datasets. Therefore, it is important to understand how a distanced-based method works and how robust it is with noisy datasets. In Section 12.3.1, we will discuss details of distance-based methods.

The increased use of multi-locus datasets for phylogenetic reconstruction has increased the need to determine whether a set of gene trees significantly deviates from the phylogenetic patterns of other genes. Motivated by this problem, there has been significant work devoted to the development of statistical methods for testing hypotheses of discordance between the trees in a collection. For example, the Bayesian estimation methods (e.g., [9–11]), the Templeton test implemented in `paup*` [12] (e.g., [13]), the partition-homogeneity test (PHT) also implemented with `paup*` (e.g., [14]), the Kishino-Hasegawa test (e.g., [15]), and the likelihood ratio test (LRT; e.g., [16]) are statistical methods used to determine if there is a "significant" level of incongruence between the trees (these methods are also called partition likelihood support (PLS) [17]). On the other hand, the methods that are used for the host-parasite analysis test whether there is a "significant" level of congruence between the trees. Since [18] (see a summary of works in [19] and the references within), there have been many studies analyzing host-parasite cospeciation. For example, the LRT (e.g., [20]), applying the Markov chain Monte Carlo (MCMC) techniques for estimating lateral transfers [21], methods that compare trees' pairwise distance matrices (e.g., by the Mantel test [22], ParaFit [23], and [24]), Brooks parsimony analysis (PSA) [25–29], and PSA [30] implemented in the software `TreeMap` [31, 32] are statistical methods used to test for codivergence between trees. Therefore, the core idea of these methods for the gene tree analysis and for the host-parasite analysis is to study how and how much these given trees differ.

An interesting biological example of applications of statistical methods to tree discordance and multi-locus phylogenetic analysis is the loline alkaloid gene cluster [33] in fungal endophytes (*Epichloë* and *Neotyphodium* spp.) that are common protective mutualists in temperate grasses. The lolines are very unusual structures with potent insecticidal activity. A cluster of genes encodes biosynthetic enzymes, and of these LolC is believed to catalyze the first biosynthetic step. LolC appears to have evolved from a homolog of *Aspergillus nidulans* CysD, used by many fungi to condense H_2S with *O*-acetylhomoserine in an alternative methionine and cysteine biosynthesis pathway. A neighbor-joining (NJ) analysis (Figure 12.1) indicates fungal cysD orthologs with relationships approximately, but not precisely, in keeping with the fungal housekeeping gene phylogenies. With the *Epichloë festucae* genome and the many other available genomes, we see an opportunity for more systematic survey, looking for genes of unusual evolutionary histories or altered evolution rates that could indicate neofunctionalization.

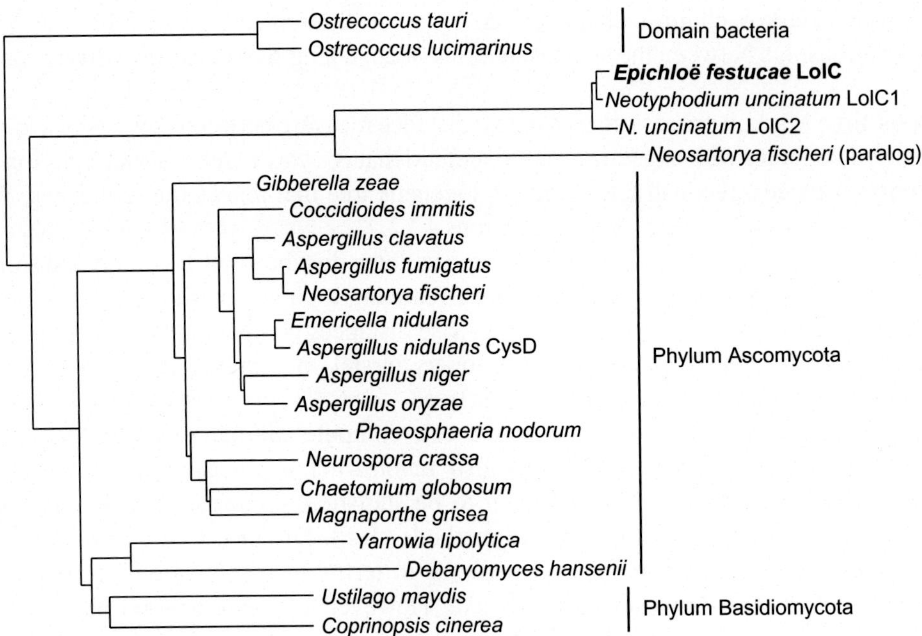

FIGURE 12.1 Example of a phylogenetic tree. This tree is reconstructed by NJ using the data relating LolC from *Epichloë* and *Neotyphodium* spp. to the most closely related sequences identified by blastp of the GenBank database. Taxonomic relationships are indicated for the CysD orthologs.

Exercise 12.2. In this exercise, you are asked to conduct the following procedure.

1. Download file: lolC introns trm.fasta at http://cophylogeny.net/courses/S12/lolC
2. Open the file in a text editor. Select all and copy. In a browser, go to http://www.phylogeny.fr/.
3. Click on Phylogeny analysis and select "One Click."
4. Go to the bottom of the page and click Create Workflow.
5. Paste text (from lolC introns trm.fasta) into the text box.
6. Scroll to the bottom, and click Submit.
7. After several seconds, a page will appear with Tree Rendering results.
8. You will see several options and buttons on this page. Among those options, try those under Tree style. Also click on the Reroot (outgroup) button, then go to the tree and double-click on some branches. Describe what you observe. Feel free to try other options on this page.
9. Click the Alignment tab. Click (under Outputs) Alignment in FASTA format.

How does "alignment in FASTA format" differ from the input FASTA? □

Exercise 12.3. *Newick format* is a way to represent a tree using parentheses and commas. One can find a nice explanation of Newick format at http://evolution.genetics.washington.edu/phylip/newicktree.html. This format was developed by the English mathematician Arthur Cayley.

For example, the following tree in Newick format:

```
((A:0.7,B:0.4):0.3,(C:0.3, D:0.4):0.5);
```

corresponds to the tree depicted in Figure 12.2.

Write the trees in Figure 12.3 in the Newick format.

Verify with Newick tree viewer at http://www.trex.uqam.ca/index.php?action=newick. □

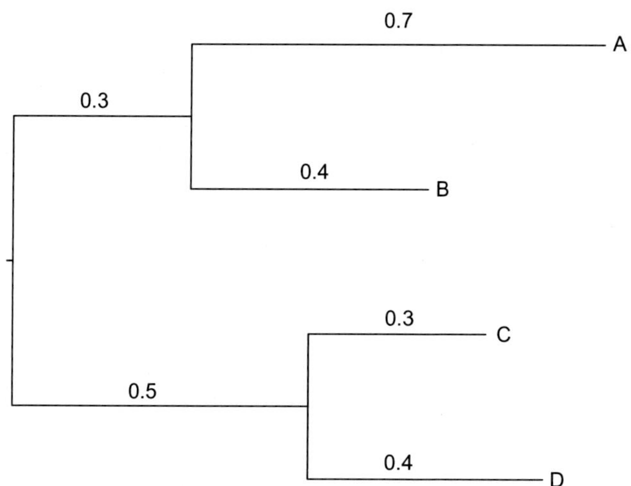

FIGURE 12.2 Tree for the Newick format in the example in Exercise 12.3.

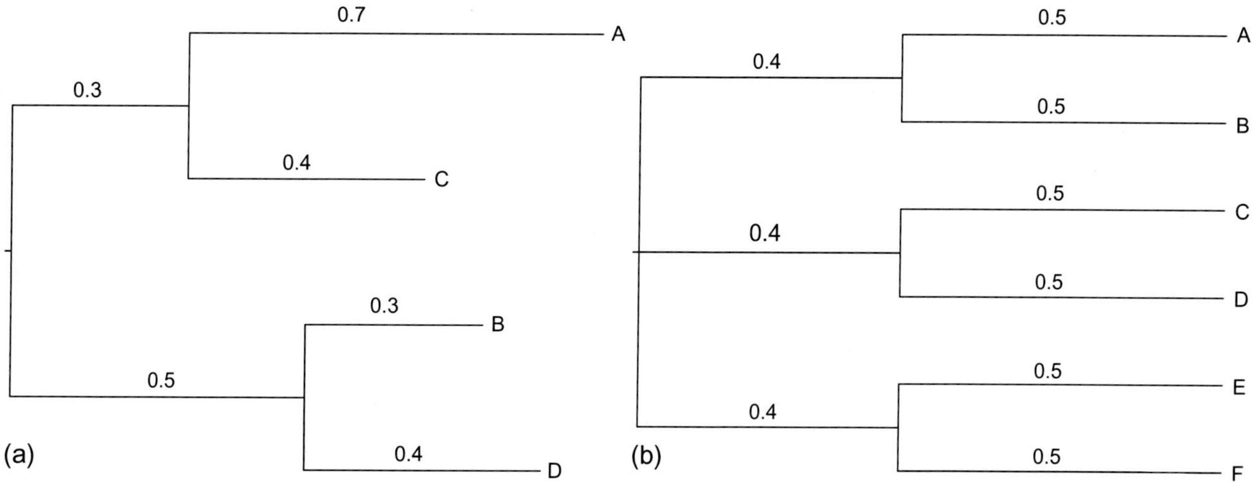

FIGURE 12.3 Trees for Exercise 12.3.

12.1.1 Sequences and Alignments

Contemporary phylogenetic analysis is generally carried out on data obtained from either nucleic acid sequences (DNA/RNA) or amino acid sequences (proteins). In either case, the data consists of a string of characters from a fixed alphabet. In the case of the nucleic acids the alphabet consists of four possible nucleic acid residues, or *character states*, most commonly denoted A, C, G, and T. Proteins are typically understood to be chains consisting of characters from the set of 20 amino acids that are encoded by DNA, although the full picture for proteins is slightly more complicated.

The processes involved in collecting sequence data from organisms is interesting in its own right, but we will not address this problem, except to refer the interested reader to [4]. We will instead assume that the sequences we have are perfectly accurate copies, and concern ourselves with the question of how to analyze the data.

There are many databases of biological sequences currently available, both public and proprietary. Two of the most well-known public databases are GenBank (www.ncbi.nlm.nih.gov/genbank/), curated by the National Institutes of Health in the United States, and UniProt (www.uniprot.org), maintained by the European Bioinformatics Institute (and others). Each database record of a sequence and its associated metadata is identified by a unique *accession number*. These alphanumeric codes are also used to identify datasets in the literature, allowing users to return to the database and retrieve sequences for further analysis.

Hemoglobin is the primary oxygen-binding protein in the blood of vertebrates. Humans have several different hemoglobin genes found in multiple locations in the genome, all believed to be derived from a single ancestral copy [34]. The GenBank database has several hemoglobin gene sequences, which we will use as example data for this chapter. The genes called HBA1 and HBA2 (accession numbers NM_000558 and NM_000517.4) produce so-called α-hemoglobin proteins. The genes HBG1 and HBG2 (NM_000559.2 and NM_000184.2) encode γ-hemoglobin, while HBB (NM_000518.4) produces β-hemoglobin. Finally, HBE1 (NM_005330.3) produces ε-hemoglobin, and HBZ (NM_005332.2) encodes ζ-hemoglobin. These sequences can be obtained by searching the GenBank website, but some programs are also capable of directly accessing the database by accession number (as we will show in a later example).

When a sequence is duplicated, for example, when an organism reproduces, it may be preserved without change, or it may be *mutated*. At any given position in the sequence a number of things may occur during a mutation. A character may be *substituted* for another character, for example, an *A* may change to become a *C*. A character may be *deleted*, shortening the sequence overall. Finally, one or more characters may be *inserted*, lengthening the sequence.

Most methods of analyzing sequences work by comparing the characters found at *homologous sites* in each sequence. A pair of positions in two sequences are homologous if the characters in those positions are both descended from the same character in a shared ancestral sequence. Because sequences can be modified not only by substitution, insertions, and deletions of the character sites themselves, we often cannot be certain of which sites are truly homologous.

The problem of inferring homology in multiple sequences by producing a *sequence alignment* is notoriously difficult, and is unfortunately classified as an NP-hard problem [35]. However, there have been many heuristic methods proposed for inferring homology in a set of sequences. At this time, popular choices include the Clustal family of programs (www.clustal.org), MUSCLE (www.drive5.com/muscle/), and T-Coffee (www.tcoffee.org). Despite the difficulty of the alignment problem, it is generally assumed that sequences used for phylogenetic reconstruction are perfectly aligned, according to the criteria defined by the chosen algorithm, and also that the homologies implied by the alignment are correct.

The R code chunk below retrieves (from GenBank) and aligns (using MUSCLE) the seven hemoglobin sequences discussed previously. After alignment, we find that our sequences all have the same length, due to the introduction of *gap characters* into the sequences. A portion of the hemoglobin sequence alignment is shown in Figure 12.4. Corresponding sites in each sequence now represent putative homologs, with the gap characters indicating that the sequence does not possess a homologous site corresponding to that position in the alignment.

```
##Create vector of Human Hemoglobin GenBank accession numbers
seq.ids <- c(HBA1="NM_000558", HBA2="NM_000517.4", HBG1="NM_000559.2",
             HBG2="NM_000184.2", HBE1="NM_005330.3", HBZ="NM_005332.2",
             HBB="NM_000518.4")
hemoglobin.rna <- read.GenBank(seq.ids) #Fetch sequences by accession
names(hemoglobin.rna) <- names(seq.ids) #Set mnemonic sequence names
## Run MUSCLE sequence alignment program. Needs separate installation from
## (http://www.drive5.com/muscle/downloads.htm)
hemo.aligned <- muscle(hemoglobin.rna)
```

The presence of gap characters poses an interesting problem for many analysis techniques. One common, although not particularly attractive method, is to simply ignore any site without a character recorded for every sequence. Another method instead treats a gap as if it were simply missing data, perhaps treating the gap as if the characters were simply unobserved random variables. After aligning the sequences, the next task is to quantify the similarities or dissimilarities observed in the aligned sequence.

```
HBA1  141  TGTCCTTCCCCACCACCAAGACCTACTTCCCGCACTTCG---ACC
HBA2  170  .......................................---...
HBG1  154  .tgt..a...atgg...c...gg.t...tgacag...t.gca...
HBG2  154  .tgt..a...atgg...c...gg.t...tgacag...t.gca...
HBE1  354  .tgtt.a....tgg...c...ga.tt..tgacag...t.gaa...
HBZ   159  .cag.ca...gcag.........................---...
HBB   151  ..gt..a...ttgg...c...gg.t...tga.tc...t.ggg.t.
```

FIGURE 12.4 A portion of the hemoglobin multiple sequence alignment, with numbers at left indicating the starting position in each sequence. Periods indicate that characters are identical to the character shown in the top sequence. Lowercase characters indicate characters are different than those found in the HBA1 sequence, while the dashes represent gaps in the sequence alignment.

12.2 QUANTIFYING EVOLUTIONARY CHANGE

Once we have a sequence alignment in hand, a natural next question is how one might measure the (dis)similarity between the observed sequences. When observing data that takes the form of real numbers, there is usually an obvious way to quantify the distance between two values, and most often the method is subtraction. However, sequences present a much more complicated problem, and there have been numerous ways proposed to quantify the differences between them.

12.2.1 Probabilistic Models of Molecular Evolution

The most common approach to modeling molecular evolution probabilistically is to treat the evolution of each character as an independent continuous-time Markov process. A Markov process is a stochastic process that has the "memoryless" property: the future evolution of the process is conditional only on the current state of the system. In particular, it is independent of any behavior in the past, given the current state.

This model is motivated by the assumption that when a sequence is duplicated, the character in the new sequence is randomly selected from a distribution that depends only on the current state of the character. The character most likely remains unchanged, but substitutions are a possibility, and the probabilities of various substitutions depend only on the chemical dynamics of sequence replication. The biochemistry is believed to be inherently stochastic in nature, and there is no known mechanism by which the state of the sequences in previous generations could affect the outcome.

Because evolution is assumed to take place slowly over a very large number of generations, and because it simplifies the calculations, it is customary to model time as a continuous variable, rather than by counting discrete generations. There is a good deal of literature on the subject of Markov processes that cannot be properly introduced here, due to space constraints, but the books [36, 37] contain some useful introductory material.

The continuous-time, discrete-space Markov chains that are used to model character evolution can be anthropomorphized in the following way: each character in the sequence alphabet is represented by a company, and each position in the sequence is represented by a phone operator, who is an employee of one of the companies. On each operator's desk are a number of phones, and each phone is connected to one of the companies other than their current employer. The phones all ring independently after an exponentially distributed period of time, with each phone possibly having its own distinct rate constant. Upon answering the first phone that rings, the operator is offered a job by the company that the phone is connected to, and immediately moves to the new call center, where the process repeats. The state of the chain at any point in time is determined by the operator's current employer.

The dynamic behavior of such a system is determined by a *transition rate matrix*, typically denoted Q, which describes the rates at which the different types of substitution occur, and the initial character state. Many classes of transition rate matrices have been proposed, each of which makes slightly different assumptions about the probabilistic nature of molecular evolution. One of the earliest models for nucleotide substitution, the Jukes-Cantor (JC) model [38], is simple to analyze, but makes assumptions about the nature of DNA evolution that are considered to be fairly unreasonable by many biologists. As a result an increasingly complicated series of models have been developed that attempt to better accommodate the properties of empirical sequences. We will discuss a few of the important models in more detail below.

A further important simplifying assumption is that the entire relevant period of evolution under study is understood to be stochastic fluctuation about the Markov chain's equilibrium distribution (a concept we will introduce shortly). The biological implication of this assumption is that we are modeling only what is called *neutral evolution*, those mutations in the genome that become fixed in the population purely by chance, and do not confer any selective [dis]advantages. In particular, we are not modeling the directed processes collectively known as natural selection that are often implied by the term "evolution."

A final drawback worth mentioning, which applies to all of the models we will discuss, is that in practice these models cannot be used to simultaneously estimate both the overall rate of base substitution and the amount of time that the Markov process has been evolving. Much of what we would like do with these models involves attempting to estimate the amount of time that a Markov process has evolved, given only observations of the beginning and ending states. However, the overall rate of substitution and the passage of time are intertwined in such a way that, without imposing additional assumptions, it is only possible to estimate their product, the *evolutionary distance*, which measures the mean number of substitution events expected to occur per site.

12.2.1.1 Introduction to Markov Chains

The behavior of a continuous-time Markov process on a state space with n elements is governed by an $n \times n$ transition rate matrix, Q. The off-diagonal elements of Q represent the rates governing the exponentially distributed variables that are used to describe the amount of time that elapses before a particular type of base substitution occurs. The ijth element of Q represents the rate at which characters in the ith state are replaced with the jth state. The diagonal elements of the rate matrix must be set such that every row in the matrix sums to 0.

The Q matrix can be used to compute a transition probability matrix, $P(d)$. This matrix gives the probability that a character in the ith state at the present time will be in state j at an evolutionary distance d. If we use $X(d)$ to denote the state of a character site at distance d from the present state, then $\mathbb{P}[X(d) = j|X(0) = i] = P_{ij}(d)$.

The probability matrix is related to the rate matrix by the matrix exponential,

$$P(d) = \exp(dQ).$$

An interesting property of these types of stochastic processes is that for certain classes of rate matrices, $P(d)$ converges to a fixed matrix as $d \to \infty$, and furthermore the rows of the limiting matrix may all be identical to a single vector, which we will denote π. When this occurs, it implies that behavior of the process at large distances is independent even of the starting state of the system; for every possible starting state, in the far future the distribution governing the character state has the probability masses specified by the vector π. We call this limiting distribution the stationary-state (or equilibrium) distribution.

Matrix exponentiation can be a somewhat difficult operation to carry out for general matrices, and we will not discuss it in detail here. We will instead use numerical methods to obtain matrix exponentials in our examples. In R, matrix exponentiation is provided by the function `expm`, in the add-on package of the same name. (Note: The standard R exponential function, `exp`, will compute an element-wise exponential when supplied with a matrix argument. This is *not* equivalent to the matrix exponential.)

12.2.1.2 Jukes-Cantor 1969

The simplest model of DNA evolution is the JC (or JC69) model. In addition to the Markov process assumptions, it also assumes that there is only a single transition rate that governs all types of substitution [38]. This assumption implies a transition matrix of the form,

$$Q = \begin{pmatrix} -1 & 1/3 & 1/3 & 1/3 \\ 1/3 & -1 & 1/3 & 1/3 \\ 1/3 & 1/3 & -1 & 1/3 \\ 1/3 & 1/3 & 1/3 & -1 \end{pmatrix}.$$

We index the rows and columns of matrices for nucleotide models in the following order: adenine, guanine, cytosine, and thymine/uracil. In this case it does not matter, but in subsequent models this becomes important.

Below, we implement a function that calculates the transition probability matrix function $P(d)$ and use it to approximate the stationary distribution for the JC model. The characteristic timescale of the system (i.e., the parameter of the time t in the continuous time Markov chain) is 1, and the probability matrix has converged quite well at a distance $d = 100$. (Beware: Attempting to evaluate the function at `Inf`, the R symbol representing infinity, appears to cause R to enter an infinite loop.)

```
library(expm)
DNA.alphabet <- c("a", "g", "c", "t")
Q <- matrix(1/3, nrow = 4, ncol = 4) # create 4x4 matrix and fill with 1/3
diag(Q) <- -1  # set diagonal to -1
colnames(Q) <- rownames(Q) <- DNA.alphabet
P <- function(d) expm(d * Q)  # Implement P(d)
P(100)  # Characteristic timescale  is 1, so 100 is ''close'' to infinity.
```

```
##      a    g    c    t
## a 0.25 0.25 0.25 0.25
## g 0.25 0.25 0.25 0.25
## c 0.25 0.25 0.25 0.25
## t 0.25 0.25 0.25 0.25
```

Thus, we find that the stationary distribution for the JC model is uniform on all possible character states, $\pi = (1/4, 1/4, 1/4, 1/4)$. Under this model, because the Markov chain is time reversible, we can think that the initial probability of this model is $\pi = (1/4, 1/4, 1/4, 1/4)$.

Another interesting and useful thing happens if a Markov process has a stationary distribution π, and the following relationship with the rate matrix Q is true.

$$\pi_i Q_{ij} = \pi_j Q_{ji}, \forall i, j \tag{12.1}$$

This is known as the *detailed balance* equation, and when it holds the process is *reversible*. If a process is reversible, it means that once it has converged to the equilibrium distribution, the "arrow of time" disappears: there is no way to determine if a character's process $X(d)$ is indexed in the proper direction, or if the time index has been reversed.

Reversibility turns out to be a desirable property if we want to study molecular evolution. Consider a most-recent common ancestor, with two daughter lineages. If the processes describing sequence evolution are reversible, then we do not need to consider the two lineages separately. The reversibility means that we can treat the daughters as endpoints of one long Markov chain that goes "up" one lineage to the MRCA, and back "down" the other lineage. If the model was not reversible, then this would not be a valid simplification. We would need to model each lineage discretely, in the correct orientation from ancestor to descendant. It turns out that JC, along with all the other models we will discuss, is reversible.

Exercise 12.4. Show that the JC rate matrix satisfies the detailed balance condition. □

Recall that $P_{ij}(d)$ represents the probability of ending in state j when starting in state i, if the distance between the sequences is d. Because the model is reversible, we can use this matrix to find the likelihood of observing a particular pair of characters in a sequence alignment, assuming the sequences are separated by distance d. If the sites in the alignment are independent, then the likelihood of the entire alignment is the product of the individual site likelihoods. That is, if the character pair i, j occurs in the alignment n_{ij} times, then the likelihood of the entire alignment is given by

$$L(d) = \prod_{\forall i,j} P_{ij}(d)^{n_{ij}}.$$

For a variety of reasons, both theoretical and relating to numerical stability, it is more common to work with the log-likelihood

$$l(d) = \log L(d) = \sum_{\forall i,j} n_{ij} \log P_{ij}(d) \tag{12.2}$$

Readers familiar with statistical methods will most likely anticipate the next step: we are now in a position to use a sequence alignment to estimate the evolutionary distance separating the two sequences. We will do this by attempting to find the distance *d* which maximizes the likelihood, given the observed alignment data.

First, we implement a function that counts the occurrences of the various substitutions between two sequences, and another that calculates the log-likelihood, given *d* and the substitution counts.

```
nMatrix <- function(alignment, i, j) {
    alignment <- alignment[c(i, j), ]
    x <- apply(alignment, 2:1, function(y) as.DNAbin(DNA.alphabet) %in%
                        y)
    dimnames(x) <- list(DNA.alphabet, NULL, labels(alignment))
    tcrossprod(x[, , 1], x[, , 2])
}
JC.lnL <- function(d, nm) sum(nm * log(P(d)))
```

Applying the `nMatrix` function to the hemoglobin data, the substitution count matrix for the HBA1 and HBA2 sequences is obtained.

A plot of the JC log-likelihood for the HBA1 and HBA2 alignment is shown in Figure 12.5.

```
N12 <- nMatrix(hemo.aligned, "HBA1", "HBA2")
### substitution pair count for HBA1 & HBA2
N12

##      a   g   c  t
## a  95   2   1  0
## g   2 150   0  3
## c   2   3 207  4
## t   0   2   6 98
```

We can now use numerical optimization to find the maximum likelihood (ML) estimate for the distance separating the two sequences.

```
optimize(JC.lnL, 0:1, nm = N12, maximum = TRUE)

## $maximum
## [1] 0.04478
##
## $objective
## [1] -130.3
```

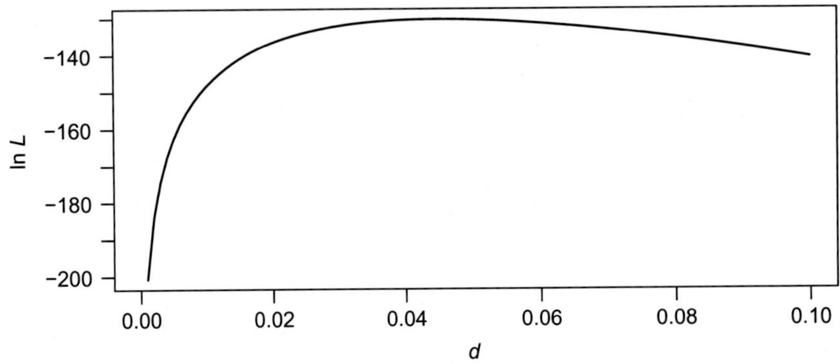

FIGURE 12.5 Log-likelihood function for the JC model and the sequences HBA1 and HBA2.

We estimate that each character site in the alignment has experienced about 0.045 substitutions, total, along both lineages leading from the common ancestral hemoglobin gene to the modern day HBA1 and HBA2.

In fact, it is fairly easy to obtain a closed-form solution for the JC problem. This solution is presented in many other books (e.g., [4]), and we will not discuss it at length here, except to note the final equation for calculating the JC distance between two sequences:

$$d = -\frac{3}{4}\ln\left(1 - \frac{4}{3}p\right). \tag{12.3}$$

Here, p is the proportion of sites in the alignment that have different characters.

The `dist.dna` function from the `ape` package computes the JC distance using the closed-form solution and agrees with our numerical solution.

```
dist.dna(hemo.aligned[c("HBA1", "HBA2"), ], model = "JC")
```

```
##          HBA1
## HBA2 0.04479
```

Exercise 12.5. The matrix exponential of the JC rate matrix is given by the formula

$$P_{ij}(d) = \begin{cases} 0.25 + 0.75\mathrm{e}^{-4d/3} & \text{if } i = j, \\ 0.25 - 0.25\mathrm{e}^{-4d/3} & \text{if } i \neq j. \end{cases}$$

Show that Eq. (12.3) solves the likelihood optimization problem for the JC model. □

12.2.1.3 Kimura 1980

Kimura's two-parameter model (K80) relaxes the JC assumptions by allowing for two different substitution rates. This generalization is based on the fact that nucleic acid residues can be divided into two classes based on their chemical structure and properties, the purines (adenine and guanine) and the pyrimidines (cytosine and thymine/uracil). Empirical results suggest that it is significantly more likely for a duplication error to result as a substitution with a nucleotide from the same class as the original nucleotide ($A \leftrightarrow G$ or $C \leftrightarrow T$, called transitions), than as a substitution with a nucleotide from the other class (all other substitutions, called transversions). The matrix Q has the form (see e.g., [39]).

$$Q = \frac{1}{\kappa + 2}\begin{pmatrix} - & \kappa & 1 & 1 \\ \kappa & - & 1 & 1 \\ 1 & 1 & - & \kappa \\ 1 & 1 & \kappa & - \end{pmatrix}$$

where κ is a parameter for the rate of the transition and the transversion. It is fairly common to suppress the diagonal elements of transition matrices, because they are completely determined by the off-diagonal elements (so that the rows sum to 0) and are often complicated expressions that do not serve to convey any important information about the system.

Like the JC model, the K80 model is reversible and has a stationary distribution that is uniform for all possible nucleotides.

Below we implement the log-likelihood function for the K80 model. The multidimensional optimizer in R is called `optim` and requires that the function to be optimized accept a vector of parameters as the first argument. Note that in fact we implement the negative of the log-likelihood function, since `optim` attempts to find a minimum of the objective function, and it is simpler to invert the log-likelihood than it is to change the behavior of the optimizer.

```
P.K80 <- function(d, k) {
    Q <- matrix(1, nrow = 4, ncol = 4)
    Q[1, 2] <- Q[2, 1] <- Q[3, 4] <- Q[4, 3] <- k
```

```
    Q <- Q/(k + 2)
    diag(Q) <- -1
    expm(d * Q)
}
K80.lnL <- function(par, nm) -sum(nm * log(P.K80(d = par[1], k = par[2])))
optim(par = c(d = 1, k = 1), fn = K80.lnL, nm = N12)$par
```

```
##       d       k
## 0.04487 2.60444
```

If we let p be the proportion of sites showing a transition substitution and q be the proportion showing a transversion, then a closed-form estimate of the K80 distance between two sequences is given by the expression [40],

$$d = -\frac{1}{2}\ln(1 - 2p - q) - \frac{1}{4}\ln(1 - 2q). \tag{12.4}$$

The `dist.dna` function also implements the K80 model and finds the same estimate for the distance separating the sequences.

```
dist.dna(hemo.aligned[c("HBA1", "HBA2"), ], model = "K80")
```

```
##         HBA1
## HBA2 0.04487
```

12.2.1.4 Hasegawa, Kishino, and Yano 1985

The Hasegawa, Kishino, and Yano (HKY) model is a further generalization of the K80 model. It introduces additional parameters (the vector π) that allow the stationary distribution of character frequencies to depart from uniform. This is an important degree of flexibility, because base frequencies are known to vary significantly in nature, both between organisms as well as within a single genome. For example, in the complete 12-million-character genome of common baker's yeast (*Saccharomyces cerevisiae*), the base frequencies vary from 19% each for cytosine and guanine to 31% each for adenine and thymine [41]. Such a significant variation from a uniform distribution cannot reasonably be attributed to chance and needs to be accommodated by the model. The matrix Q for the HKY model is

$$Q = \beta \begin{pmatrix} - & \kappa\pi_g & \pi_c & \pi_t \\ \kappa\pi_a & - & \pi_c & \pi_t \\ \pi_a & \pi_g & - & \kappa\pi_t \\ \pi_a & \pi_g & \kappa\pi_c & - \end{pmatrix}$$

where κ is a parameter for the rate of transitions and transversions and π_i is the base frequencies for $i \in \{A, C, G, T\}$. Although we could try to simultaneously optimize the log-likelihood over d, k, and π, it is more common to estimate the base frequency distribution separately, and optimize only the parameters d and κ. The reason for this is that we often want to compute pairwise distances between all sequences in a multiple sequence alignment, and in this case the equilibrium frequencies must be shared by all sequences. Thus, it makes more sense to estimate the base frequencies once, using all of the available sequence data. This leads to more precise estimates of the base frequencies, and as well enforces a common equilibrium distribution for the entire multiple alignment.

The `base.freq` function from `ape` calculates the observed base frequencies. We use it below to obtain frequencies using the entire hemoglobin alignment and reorder the resulting vector to match our Q matrix indexing.

```
pi <- base.freq(hemo.aligned)[c("a", "g", "c", "t")]
pi

##      a      g      c      t
## 0.2186 0.2630 0.3046 0.2138
```

For models that allow base frequencies to depart from the uniform distribution, the rate-normalizing constant for the rate matrix, β, is more complicated than in previous models. As before, it must be chosen so that the average rate of substitution is 1, but it must take into account the nonuniform base frequency distribution. For any reversible Q matrix, the normalizing constant is given by the following equation [42].

$$\beta = -1 \Big/ \sum_i \pi_i Q_{ii} \tag{12.5}$$

```
P.HKY <- function(d, k, pi) {
    Q <- matrix(pi, 4, 4, byrow = TRUE)
    Q[1, 2] <- Q[1, 2] * k
    Q[2, 1] <- Q[2, 1] * k
    Q[3, 4] <- Q[3, 4] * k
    Q[4, 3] <- Q[4, 3] * k
    diag(Q) <- -rowSums(Q) + pi
    beta <- 1/-sum(diag(Q) * pi)
    colnames(Q) <- rownames(Q) <- DNA.alphabet
    expm(d * beta * Q)
}
HKY.lnL <- function(par, nm) -sum(nm * log(P.HKY(d = par[1], k = par[2],pi = pi)))
optim(par = c(d = 1, k = 1), fn = HKY.lnL, nm = N12)$par

##       d       k
## 0.04551 2.69942
```

This model can be seen of as an extension of the K80 model that takes base frequencies. It is widely used in phylogenetic analysis.

12.2.1.5 Tamura-Nei 1993

The Tamura-Nei 1993 (TN93) model expands slightly on the HKY model by adding a third substitution rate category. The three rate classes in the TN93 model are: $A \leftrightarrow G$ substitutions, $C \leftrightarrow T$ substitutions, and transversion substitutions. Thus, the TN93 rate matrix has the form

$$Q = \beta \begin{pmatrix} - & \kappa_1\pi_g & \pi_c & \pi_t \\ \kappa_1\pi_a & - & \pi_c & \pi_t \\ \pi_a & \pi_g & - & \kappa_2\pi_t \\ \pi_a & \pi_g & \kappa_2\pi_c & - \end{pmatrix}$$

where κ_1 and κ_2 are parameters for two different types of transition and π_i is the base frequency of the state i.

Exercise 12.6. Extend the example code from the HKY model to implement the TN93 model. Use your extension to find the TN93 distance between HBA1 and HBA2. Compare your answer with that from `dist.dna`. □

12.2.1.6 General Time Reversible Model

The general time reversible (GTR) model is the most flexible model of nucleotide substitution that preserves the time reversibility of the Markov process [43]. It allows for all types of character substitution to occur at distinct rates, and for arbitrary equilibrium frequencies. See [43] for the details.

12.2.2 Common Model Extensions

There are a few extensions that are sometimes added to any of the previously mentioned models of substitutions. These extensions are motivated by features commonly observed in empirical sequences that are not well fitted by any of the probabilistic models.

The first model extension allows for certain character sites to be classified as *invariant*. An invariant site is one where all substitutions are forbidden. This is motivated by the assumption that certain positions in a sequence are more important to the sequence function than others, and thus these sites experience strong purifying selection. An example is the regions of a sequence that initiate or terminate protein transcription. If these regions are disturbed by a mutation, there is little chance that the biological function of the sequence will be preserved in any meaningful way. Thus, mutations at these sites are assumed to be almost totally forbidden.

The second important possible model extension, Γ *rate categories*, is intended to account for the fact that different sites in a sequence might evolve at different rates overall. For example, DNA sequences are translated into amino acid sequences in groups of three characters at a time (e.g., the DNA sequence `ATG` translates into the amino acid methionine). These sets of DNA base triplets are called *codons*. There are $4^3 = 64$ possible codons, but only 20 amino acids plus a signal to terminate transcription that need to be encoded. Thus, the encoding is redundant, with each amino acid encoded by an average of three different codons. (However, the redundancy varies from only a single encoding for the cases of methionine or tryptophan, to six encodings each for arginine, leucine, and serine.)

An interesting fact about the encoding is that the codons are not assigned to the amino acids in a random manner. When multiple codons encode a single amino acid, it is quite likely that the redundant encodings share common first and second characters, only varying in the third position. Conversely, changes at the second position are almost certain to result in a change in the translated sequence. For example, consider the codon `CTT`, which encodes the amino acid leucine. The third character can be freely substituted and the new codon will still translate to leucine. However, a substitution at the second position always changes the encoded amino acid.

Thus, substitutions at the second position in a codon should be subject to greater selective pressures than changes at the third position. These differences in selective pressure between the three positions should logically lead to differences in the overall substitution rates as well. In the Γ model extension, a mixture of several scaled-rate processes is used to model these processes.

When calculating transition probabilities, each category is allowed to scale the substitution rate matrix, Q, by a different constant, with the constraint that the combined total rate of substitution across all sites must remain equal to 1. The name of the extension is a reference to the fact that the scaling constants are obtained from quantiles of a mean-1 Gamma distribution. The user typically must specify both the number of categories as well as a constant (possibly named α) that governs the variance of the mean-one Γ distribution used.

The use of these model extensions is typically indicated by the presence of "+I" or "+G" after a model code. For example, HKY + I + G means the HKY model was used both invariant sites and the Γ categories extension. For more information, see, for example, [4].

12.3 RECONSTRUCTING THE TREE

In this section, we consider several methods for phylogenetic tree reconstruction.

12.3.1 Distance-Based Methods

Distance-based methods of tree reconstruction work by first computing the pairwise distances between the sequences in an alignment and then producing a tree using only these summary statistics. This is in contrast to the other methods we discuss, which involve the entire sequence alignment in the tree reconstruction algorithms. Here, we give a brief introduction to a few important methods.

TABLE 12.1 The *p*-Distance Matrix Corresponding to the Hemoglobin Sequence Alignment

	HBA1	HBA2	HBG1	HBG2	HBE1	HBZ
HBA2	0.016					
HBG1	0.427	0.425				
HBG2	0.429	0.427	0.006			
HBE1	0.435	0.437	0.206	0.206		
HBZ	0.337	0.337	0.439	0.437	0.425	
HBB	0.410	0.408	0.271	0.267	0.241	0.461

Note: The matrix is symmetric, so the upper triangular portion (and diagonal, which contains zeros) has been suppressed.

12.3.1.1 Distance Matrices

A distance matrix is a nonnegative, square, symmetric matrix with elements corresponding to estimates of some pairwise distance between the sequences in a set. The simplest definition distance uses the proportion of homologous sites in an alignment with differing characters and is called the *p*-distance, or *Hamming* distance. Although the *p*-distance is simple to calculate and understand, it is more common to use one of the probabilistic definitions of evolutionary distance discussed in the previous section.

Below, we use the `dist.dna` function to obtain distance matrices for the hemoglobin alignment for several different distance choices. Although we could extend our functions from the previous section to perform these calculations, in our opinion there is little to be learned from doing so. Furthermore, the `dist.dna` function is much faster, because it employs closed-form solutions whenever possible, rather than trying to optimize likelihoods. The *p*-distance matrix for the hemoglobin alignment is presented in Table 12.1.

```
p.dist <- dist.dna(hemo.aligned, model = "raw") # p-distance is called ''raw''
jc.dist <- dist.dna(hemo.aligned, model = "JC")
t93.dist <- dist.dna(hemo.aligned, model = "TN93")
```

12.3.1.2 Neighbor Joining

The NJ method of tree reconstruction begins with a completely unresolved (star) tree and attempts at each step to further resolve the tree by adding a node that joins the most closely related nodes in the tree, as determined by a distance matrix. A new distance matrix is computed where the rows and columns associated with the two newly joined taxa are replaced with new entries relating to the new interior node, and the process is repeated until the tree is fully resolved.

The NJ method was developed by Saitou and Nei [44], and has been discussed extensively in other publications. We will not present details of the algorithm here, but rather refer interested readers to [45]. In `ape`, the `nj` function implements the NJ algorithm, while `bionj` implements the related BIONJ algorithm [46] of Gascuel, which claims to offer improved performance when used on highly divergent alignments, as well as the ability to handle distance matrices with missing elements.

```
pnj.tree <- nj(p.dist)
```

This code produces a tree reconstructed by the NJ algorithm.

The NJ algorithm is among the fastest available methods of tree reconstruction. However, it does suffer from some drawbacks; particularly problematic is a lack of statistical consistency in certain situations. (A method is statistically consistent if it is almost certain to converge on the correct tree as the alignment length grows to infinity.) Thus, NJ is most often used to quickly obtain reasonable trees that can be used as starting locations for more computationally intensive tree reconstruction algorithms, such as the ML or Bayesian methods.

Exercise 12.7. Suppose we have a distance matrix as follows:

	1	2	3	4	5	6
1	0	6	8	9	12	11
2	6	0	6	7	10	9
3	8	6	0	3	6	5
4	9	7	3	0	5	4
5	12	10	6	5	0	5
6	11	9	5	4	5	0

Reconstruct a tree using the NJ algorithm. □

Exercise 12.8. In this exercise, you are asked to reconstruct a tree by the NJ algorithm from the alignment you computed in Exercise 12.2.

1. Go to the website http://www.trex.uqam.ca/index.php?action=trex&menuD=1&method=2
2. Save the alignment you computed in Exercise 12.2 in PHYLIP format.
3. Upload your alignment in PHYLIP format.
4. Set substitution model at Kimura 2 parameters. □
5. Use the NJ algorithm to reconstruct a phylogenetic tree.

12.3.1.3 Balanced Minimum Evolution

Balanced minimum evolution [47] is a tree reconstruction method that is roughly analogous to the least-squares method of fitting curves to observed data points. As the name suggests, candidate trees produced by the BME computations are compared using the sum of their branch lengths (a measure of the total amount of evolution required to produce the tree), with smaller trees being considered superior. The principle of Occam's razor is typically cited as a rationale for this method of comparison.

The BME method describes a method of assigning lengths to the branches of an arbitrary tree topology in a way most compatible with a given (fixed) distance matrix, taking into account the fact that the variance of the pairwise distance estimates is smaller for closely related sequences than for highly divergent ones. Finding the optimal tree then involves finding the tree topology on which the total sum of the branch lengths is minimized. Fortunately, there is a simple and fast method of computing the total length of the branches on any given topology (actually computing all branch lengths is not required), known as Pauplin's Formula.

Unfortunately, unless the number of taxa in the tree is very small, a complete census of the possible topologies is utterly infeasible: the number of possible tree topologies grows factorially with the number of taxa, quickly rendering even terms such as "astronomical" nothing more than comical understatements. Not only is a complete search of the space of topologies impossible, but there is no known way to reduce the computational complexity of the BME search to a reasonable level while also guaranteeing that the optimal solution is found [48].

Despite this, several fast algorithms have been developed that provide fairly good solutions to the BME problems, but without the guarantee that the globally optimal topology has been found. The FastME algorithm is implemented in the `ape` function `fastme.bal`.

```
bme.tree <- fastme.bal(p.dist)
```

This procedure will produce a tree reconstructed via the BME.

Exercise 12.9. Suppose we have a distance matrix as follows:

	1	2	3	4	5	6
1	0	6	8	9	12	11
2	6	0	6	7	10	9
3	8	6	0	3	6	5
4	9	7	3	0	5	4
5	12	10	6	5	0	5
6	11	9	5	4	5	0

Reconstruct a tree using the BME algorithm. □

Exercise 12.10. In this exercise, you are asked to reconstruct a tree using the BME algorithm from the alignment you computed in Exercise 12.2.

1. Save the your alignment you computed in Exercise 12.2 in PHYLIP format, named "lolC_align.phy."
2. Open R.
3. Type in R as follows:

```
## Load directory
setwd("YOUR DIRECTORY")

## Load library
library(ape)

## Read in DNA sequences from Phylip
phy.data<-read.dna('lolC_align.phy', format='interleaved')

## Reconstruct NJ tree with raw pair-wise distances
dist.mat<-dist.dna(phy.data,model='K80')
fastme.tree<-fastme.bal(dist.mat)
write.tree(fastme.tree, file="outtree")
```

where "YOUR DIRECTORY" is the directory where the PHYLIP formatted file is located.
4. Using the FASTME algorithm to reconstruct a phylogenetic tree. □

12.3.2 Maximum Parsimony

The maximum parsimony (MP) method, like BME, is a method that attempts to select a topology by minimizing the amount of evolution required to explain the observed alignment. As such, it shares the drawbacks associated with the need to search the entire space of possible tree topologies. (Namely, there is no known method of easily obtaining a definitive solution.) However, unlike BME, it is not a distance-based method, but rather uses the entire sequence alignment in the calculation.

The principle underlying the MP tree is simple: For any given topology we assign sequences to the internal nodes of the tree in a way that minimizes the total number of base substitutions that are required to occur on the entire tree. This total number of base substitutions is used as the criterion by which a tree topology is selected. It should be noted that the MP method has the additional advantage of producing an estimate of the ancestral sequences as a byproduct of the computation, something that the distance-based methods do not do.

The `ape` package does not implement any function that attempts to solve the MP problem, but the `phangorn` package does provide implementations of several MP algorithms, and the documentation suggests that the pratchet function is "the preferred way to search for the best tree." However, the phangorn functions (including pratchet) require that the alignment first be converted into phyDat objects.

```
hemo.phyDat <- phyDat(hemo.aligned)
mp.tree <- pratchet(hemo.phyDat)
```

The commands above produce an MP tree for the sequence hemo.aligned.

12.3.3 Methods Based on Probability Models

The tree reconstruction methods discussed up to this point are not explicitly based on any probabilistic models of sequence evolution, although they may do so implicitly through the pairwise distance matrix. Conversely, the methods in this section are explicitly based on probability models, and the techniques used to obtain phylogenies are similarly grounded in statistical theory.

12.3.3.1 Maximum Likelihood

The ML method of tree estimation shares much of the theoretical machinery introduced in Section 12.2.1, probabilistic models of molecular evolution. However, instead of attempting to model a single Markov process connecting a pair of homologous characters, the ML methods posit a collection of Markov processes, one for each branch on a tree. Given a tree topology and an alignment, the ML method attempts to assign branch lengths to the tree in such a way that the likelihood of the collection of Markov processes is maximized.

The calculation of the likelihood over an entire tree is significantly more complicated than in the case of a single pair of sequences, because the effects on the likelihood of branch lengths and internal character states are all highly interdependent. Fortunately, Joseph Felsenstein developed an efficient *pruning algorithm* that greatly simplifies the calculation of the likelihoods (see [49]). The pml function from phangorn implements this (log)likelihood calculation for a variety of substitution models. The following code evaluates the likelihood of the NJ tree we obtained in Section 12.3.1 using the JC substitution model (the default).

```
pml.obj <- pml(pnj.tree, hemo.phyDat)
logLik(pml.obj)

## 'log Lik.' -3625 (df=11)
```

Like the BME and MP algorithms, an ML tree is produced by searching for the combination of topology and branch lengths that maximizes the likelihood value. Although no fast algorithm is known that guarantees location of the global maximum value, several heuristic search methods are commonly used. The optim.pml function implements several of these methods. In the following example, we use the NJ tree (stored as part of the pml.obj object) as the starting location for the search for a more likely tree.

```
JC <- optim.pml(pml.obj, optNni = TRUE, model = "JC")
```

Of course, pml and optim.pml work with more complicated models than JC. We fit several other substitution models below.

```
F81 <- optim.pml(JC, optNni = TRUE, model = "F81")
HKY <- optim.pml(F81, optNni = TRUE, model = "HKY")
GTR <- optim.pml(HKY, optNni = TRUE, model = "GTR")
```

The Gamma and invariant site model extensions are also implemented.

```
HKY.G <- update(HKY, k = 4) #Add 4 gamma categories
HKY.G <- optim.pml(HKY.G, optNni = TRUE, optGamma = TRUE, model = "HKY")
HKY.I <- update(HKY, inv = 0.1) #Add invariant sites
HKY.I <- optim.pml(HKY.I, optNni = TRUE, optInv = TRUE, model = "HKY")
HKY.IG <- update(HKY, k = 4, inv = 0.1) #Add both
HKY.IG <- optim.pml(HKY.IG, optNni = TRUE, optGamma = TRUE, optInv=TRUE, model = "HKY")
```

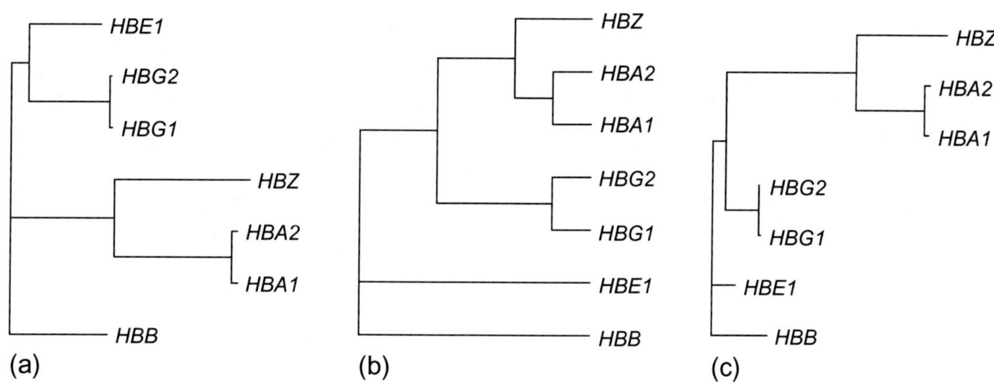

FIGURE 12.6 Trees reconstructed from the hemoglobin gene data using three different methods.

```
plot(pnj.tree, no.margin = TRUE)
plot(mp.tree)
plot(HKY.G)
```

Exercise 12.11. In this exercise, you are asked to reconstruct a tree using the MLE algorithm from the alignment you computed in Exercise 12.2.

1. Save the alignment in PHYLIP format.
2. Go to the website http://www.phylogeny.fr/.
3. Upload your alignment in PHYLIP format.
4. Set HKY model (noted as HKY85 model on the website).
5. Write your e-mail address and reconstruct an MLE tree via PHYML by submitting it. □

As one can see from Figure 12.6, we may get different tree topologies as well as branch lengths if we use different tree reconstruction methods even though we use the same input alignment. If we want to reconstruct a phylogenetic tree from an empirical dataset, it is always good practice to reconstruct trees using different methods because we do not know the true tree.

12.3.3.2 Bayesian Methods

The primary alternative statistical approach to tree reconstruction is the Bayesian method. Like all Bayesian methods, the basic premise is to make inferences based on a *posterior* distribution of the relevant parameters: in this case, the tree topology and branch lengths. The posterior distribution is computed from a model of substitution and a *prior* distribution π on the parameters using Bayes' formula for reversing conditional probabilities. The prior distribution is intended to encode the user's prior beliefs about what form the correct tree might take, but in practice little is known about how to reasonably specify prior beliefs about trees. As such, a few basic default distributions, such as uniform and exponential distributions, are almost always used.

Although direct computation of the posterior distribution is usually impossible due to the presence of a thoroughly intractable integral over the entire space of tree topologies, there is a fairly good method of obtaining a sample from the posterior known as the Metropolis-Hasting algorithm. The Metropolis-Hasting algorithm describes a method of implementing a discrete time Markov chain that generates (correlated) samples from arbitrary density functions, *even if the normalizing constant for the density function is unknown* [42]. This algorithm forms the backbone of the class of methods known as MCMC methods.

The MCMC methods generate a sample of trees from the posterior distribution, and then use this sample as the basis of inferences about the true tree [50]. For example, a common Bayesian method of inferring the topology of a tree is to select the topology that occurs most commonly in the posterior sample.

The programs BEAST [51] and Mr Bayes [50] are the most commonly used implementations of Bayesian tree reconstruction at the present time [52].

12.4 MODEL SELECTION

In previous sections, we presented a number of possible probability models of sequence evolution. We also saw that the choice of probability model can change not only the branch lengths but also the topology of the reconstructed trees. This naturally leads to the question: which substitution model should one use?

The question of model selection is a problem that has a rich history in statistical literature. Although one can always improve the fit to observed data by adding degrees of freedom to the model, such flexibility comes at a cost: a decrease in the precision of the model parameter estimates, and an increase in the amount of computational effort required to obtain them. Although a model that is too simple will typically result in biased inferences, a model that is too complex also often fails to be useful. In such "overfitted" models, the behavior of the parameter estimates is dominated by the statistical noise present in the data, making it very difficult to observe any systematic patterns that may be present. In addition, models with many parameters are very difficult for a human to meaningfully interpret, and usually perform quite poorly when used for predictive purposes, greatly limiting their utility.

Two popular methods of model selection, both of which have long histories in statistics, are the LRT (also known as an χ^2 test, due to the asymptotic distribution of the test statistic) and the Akaike Information Criterion (AIC). A third popular method, the Bayesian Information Criterion (BIC), is sufficiently similar to AIC, both in calculation and use, that we will omit any further discussion of it in the interest of brevity [53].

The hierarchical LRT is essentially a forward-selection method. Beginning with the simplest JC model, a sequence of increasingly complicated models is proposed, and a test is performed to determine whether the resulting improved fit of the model, as determined by the log-likelihood, is sufficient to justify the increase in the number of parameters. One drawback of the LRT method is that the series of models to be tested must nest into each other in some way, with each model in the sequence being a generalization of the previous one.

Below we use the `anova` function to test the sequence of models: JC, F81, HKY, and GTR.

```
anova(JC, F81, HKY, GTR)

## Likelihood Ratio Test Table
##   Log lik. Df Df change Diff log lik. Pr(>|Chi|)
## 1    -3555 11
## 2    -3540 14         3          30.4    1.1e-06
## 3    -3518 15         1          42.9    5.8e-11
## 4    -3518 19         4           0.7       0.95
```

The rows correspond to the four models inputted, and the p-values for the significance tests are found in the final column of the output. These values indicate the probability that the improvement in the likelihood due to the complexity of the given model is attributable only to statistical noise, and not to a true improvement in the fit of the model.

The small p-values in lines 2 and 3 indicate that F81 generates a significantly better fit to the observed data than does the JC model, and likewise HKY provides a better fit than F81. However, the improvement in the fit of the GTR model over HKY is not enough to justify the four additional parameters required by GTR. However, we can still see if either of the substitution model extensions, invariant sites or Gamma categories, might improve the fit. HKY.G represents HKY + G and HKY.IG represents HKY + I + G.

```
anova(HKY, HKY.G, HKY.IG) #Accept HKY.G

## Likelihood Ratio Test Table
##   Log lik. Df Df change Diff log lik. Pr(>|Chi|)
## 1    -3518 15
## 2    -3493 16         1          50.5    1.2e-12
## 3    -3493 17         1           0.0          1
```

The LRT procedure indicates that although the Gamma extension produces a significant improvement in the fit of the model to the hemoglobin alignment, we do not need to include the invariant sites extension in our model of substitution; the final model selected is HKY + G.

The use of the AIC in model selection is simpler, because the requirement that the models nest within each other is not needed. Any model for which we can obtain a likelihood can be tested against any other, and the model with the smallest AIC value is deemed the best.

```
AIC(JC, F81, HKY, GTR, HKY.I, HKY.G, HKY.IG)
```

```
##          df  AIC
## JC       11 7132
## F81      14 7108
## HKY      15 7067
## GTR      19 7074
## HKY.I    16 7025
## HKY.G    16 7018
## HKY.IG   17 7020
```

Like the LRT, the AIC selection method selects the HKY.G model for the hemoglobin alignment. Although it is certainly not guaranteed that AIC and LRT will always agree on the best model to use, it is always nice to have a conclusion that is robust against changes in methodology.

Exercise 12.12. In this exercise, we will use the ape and phangorn libraries in R.

1. Install ape and phangorn libraries in R.
2. Type the following

```
library(ape)
library(phangorn)
data(chloroplast)
(mTAA <- modelTest(chloroplast, model=c("JTT", "WAG", "LG")))
```

What output did you get? □

12.5 STATISTICAL METHODS TO TEST CONGRUENCY BETWEEN TREES

In systematic biology, one of the fundamental questions is how to estimate differences between phylogenetic trees. For example, conflicting phylogenies arise when different phylogenetic reconstruction methods are applied to the same dataset, or even with one reconstruction method applied to multiple different genes. Gene trees may be codivergent by virtue of congruence (identical trees) or insignificantly incongruent Or they may be significantly incongruent [54]. All of these outcomes are fundamentally interesting. Congruence of gene trees (or subtrees) is often considered the most desirable outcome of phylogenetic analysis, because such a result indicates that all sequences in the clade are orthologs (homologs derived from the same ancestral sequence without a history of gene duplication or lateral transfer) and that discrete monophyletic clades can be unambiguously identified, perhaps supporting novel or previously described taxa. In contrast, gene trees that are incongruent are often considered problematic because the precise resolution of speciation events seems to be obscured. Thus, it would also be useful to identify significant incongruencies in gene trees because these represent noncanonical evolutionary processes (e.g., [9, 10, 55, 56]).

The common pattern of gene trees in genome evolution is one of codivergence, the parallel divergence of ecologically associated lineages [57–61]. However, deviations from codivergence can include gene duplications, lateral interspecific gene transfers between species, retention of ancestral nucleotide sequence polymorphisms maintained by balancing selection, and accelerated evolution by neofunctionalization, that is, the gain of novel gene function through sequence divergence by a duplicate copy of a progenitor (ancestral) gene.

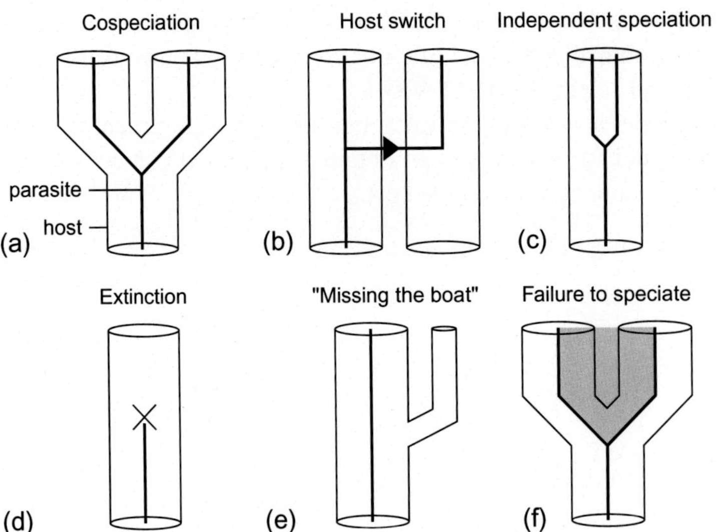

FIGURE 12.7 Evolutionary events that can occur during gene-genome (host-parasite) coevolution [62]. These are equivalent to the following processes associated with gene-genome coevolution: (a) "Tracking": Gene trees track species trees, or gene trees track each other. (b) Lateral gene transfer (LGT). (c) Duplication. (d) Extinction of a gene lineage or a gene lineage is unsampled. (e) Missing the boat: a gene lineage is unrepresented in a new species. (f) Trans-species polymorphism. Multiple gene lineages are carried through multiple speciation events, either because of selection or because of insufficient time (depending on N_e, the effective population size) for fixation.

Thus, as genes within a genome coevolve, there are six commonly recognized types of events that can occur along lineages [62]. These are shown in Figure 12.7. In addition to these six scenarios, there is another phenomenon in gene trees that might cause incongruence between trees, that is, a long-branch attraction (LBA). LBA is an erroneous grouping of two or more long branches as sister groups due to methodological artifacts [63]. It was shown in [64] that the problem of LBA may be severe in the case of trees for four sequences but it might not be a problem for a large tree. However, recently Fares et al. [65] showed positive selection for increased expression and consequent rapid cell growth after whole-genome duplication (WGD). Subsequent rearrangements and gene loss might be the cause of LBA artifacts in phylogenetic trees in the yeast *Saccharomyces cerevisiae*, and this might be the cause of conflicting topologies among NJ trees drawn from different loci. Different selection pressures might affect tree topologies as well [66, 67]. In addition, Kück et al. [68] showed that LBA affects the ML estimation of a phylogenetic tree even with use of the correct model.

Deviations from strict phylogenetic codivergence of genes in genomes generally elicit considerable interest in the scientific community and the public. The notion of lateral (horizontal) gene transfer (LGT) is an excellent example. Claims of LGT between very distantly related organisms (e.g., taxonomic domains or kingdoms) regularly appear in high profile publications, and this possibility also underlies public concerns about genetically modified organisms [69] and the potential emergence of new pathogenic "superbugs" that defy modern cures [70]. Given these important concerns for medicine, agriculture, and the environment, definitive assessment of evolutionary processes such as LGT and others diagrammed in Figure 12.7 is needed.

Much of the evidence for LGT so far has been controversial, especially for eukaryotes [71, 72]. Such claims generally involve first an assessment of whether the sequence groups in a clade that is dominated by homologs from another kingdom. Subsequent statistical tests typically compare alternative tree topologies [73, 74], but the underlying assumption remains untested; that is, it remains unclear whether the tree is actually indicative of LGT rather than some other evolutionary process that would also cause deviation from the species tree. For example, paralogy and gene loss are not tested statistically even though such processes are consistently evident in gene trees touted as evidence of LGT. In prokaryotes, LGT of plasmids and mobile genomic islands (e.g., pathogenicity islands) evident from phylogenetics have been experimentally substantiated [75], but even in these cases a variety of other evolutionary processes undoubtedly operate on these elements. Thus, it would be highly beneficial to move beyond the identification of disparities between phylogenies of genes and genomes or hosts and parasites and elucidate the most likely causes of those disparities by means of explicit statistical tests for different processes underlying gene/genome coevolution and host/parasite coevolution, as proposed herein.

While there has been a well-established understanding of the discordant phylogenetic relationships that can exist among independent gene trees drawn from a common species tree [54, 76–78], phylogenetic studies have only recently begun to shift away from single gene or concatenated gene estimates of phylogeny toward these multi-locus approaches (e.g., [79–83]). These newer approaches focus on the effect of genetic drift in producing patterns of incomplete lineage sorting and gene tree/species tree discordance, largely using coalescent theory (e.g., [84–90]). These theoretical developments have been used to reconstruct species trees from distributions of estimated gene trees (e.g., [10, 55, 91–97]).

In statistics, the relationship between gene and species trees is well understood in terms of coalescent processes [98]. However, coalescent models usually assume that genes cannot be transferred between members of different species. Just as host switching can cause parasite trees to disagree with host trees, LGT can cause gene trees to disagree with species trees. Combinatorially, these mechanisms correspond to *subtree prune and regraft* (SPR) operations [99]. Many techniques have been developed to compare gene trees [9–11, 13–17] and host and parasite trees [19, 22, 100, 101].

Cophylogenetics is the study of concomitantly evolving entities such as groups or sets of genes in genomes, and of syntenic genes or gene blocks. Fundamentally, this area differs from phylogenetics in that the appropriate objects of study in cophylogenetics are tuples of related trees, rather than individual trees.

Definition 12.1. Let $\mathcal{T}_{\mathrm{H}}$ and $\mathcal{T}_{\mathrm{P}}$ be spaces of trees for two sets of taxa H and P. (We use subscript H and subscript P for host and parasite, but these can also represent different genes in a genome; or they may indicate a species tree T_{H} and gene tree T_{P}.) A *cophylogeny* is any pair of trees $(T_{\mathrm{H}}, T_{\mathrm{P}}) \in \mathcal{T}_{\mathrm{H}} \times \mathcal{T}_{\mathrm{P}}$. □

More generally, for tree spaces $\mathcal{T}_{G_1}, \ldots, \mathcal{T}_{G_\ell}$, a cophylogeny is any tuple $(T_{G_1}, \ldots, T_{G_\ell}) \in \mathcal{T}_{G_1} \times \cdots \times \mathcal{T}_{G_\ell}$. We end this chapter with a few open problems related to cophylogenetics.

Space of Cophylogenies

1. Theoretical questions
   (a) Study the geometry and combinatorics of various spaces of cophylogenetic trees, in the sense of [102]. Recently Devadoss et al. [103] reinterpreted the space of phylogenetic trees as a cone over a simplicial complex, constructed by local gluing of the associahedron and the permutohedron. It would be interesting to describe the tree of cophylogenetic trees based on covering by classical polytopes that encapsulate algebraic information, notably the associahedron and the permutohedron.
   (b) Find linear or tropical characterizations, as the extended *Four Point Condition* [104], for spaces of cophylogenetic trees.
2. Methodological questions
   (a) Develop distance-based methods to reconstruct a cophylogeny, that is, extending the balanced minimum evolution (BME) [105] and NJ [44] methods for reconstructing a phylogeny to reconstructing a cophylogeny.
   (b) Show explicit description of BMC polytopes and joint BME polytopes. Eickmeyer et al. [106] developed the geometric view of the BME and NJ methods and defined *BME polytopes* in order to compare the accuracy of the NJ method with the BME method. Similarly, Davidson and Sullivant [107] used the polyhedral subdivisions to compare the accuracy of the NJ, UPGMA (Unweighted Pair-Group Method with Arithmetic Mean) and Least Squares methods. Here we propose to extend this approach to cophylogenies, namely to *joint BME polytopes*. Can we compute and study the face structures of joint BME polytopes for specific spaces of cophylogenetic tree? In particular, one would like to discover whether their vertices and edges can be determined or not.
   (c) Develop a novel method to compute the Bayes estimator with the l_1 loss from the posterior distribution of cophylogeny. Huggins et al. [108] have developed a tree reconstruction method using the Bayes estimator from the posterior distribution of the tree over the tree space. This estimates the "median" of the trees

from the posterior distribution of trees over the tree space. Simulations by [108] showed that the results were better or equal to the ML trees. Can we extend the result of [109], which computes explicitly the posterior mean and median over the tree space, to the posterior distribution of cophylogenies with LGT introduced by [101]?

Cophylogenetic Invariant

1. Theoretical questions: Extend the phylogenetic invariants to cophylogeny. Cavender and Felsenstein [110] and Lake [111] independently introduced *phylogenetic invariants*, that is, polynomial relationships that must hold between the frequencies of various base patterns in idealized data. They were introduced to test whether such polynomials for various trees are "nearly zero" when evaluated on the observed frequencies of patterns in real data sequences, in order to infer the best tree to explain the observed dataset.
2. Methodological questions: Develop a method to reconstruct a cophylogenetic tree using cophylogenetic invariants. Eriksson [112] developed a tree reconstruction method with phylogenetic invariants via a hierarchical clustering method. Rusinko and Hipp [113] developed an invariant-based quartet method and returned more accurate trees than the quartet puzzling with the maximum likelihood estimator (MLE).

REFERENCES

[1] Just W. Computational complexity of multiple sequence alignment with SP-score. J Comput Biol 2001;8:615-23.
[2] Edgar RC. MUSCLE: multiple sequence alignment with high accuracy and high throughput. Nucleic Acids Res 2004;32:1792-7.
[3] Katoh K, Standley DM. MAFFT multiple sequence alignment software version 7: improvements in performance and usability. Mol Biol Evol 2013;30:772-80.
[4] Pachter L, Sturmfels B. Algebraic statistics for computational biology. Cambridge: Cambridge University Press; 2005.
[5] Felsenstein J. Cases in which parsimony or compatibility methods will be positively misleading. Syst Zool 1978;22:240-9.
[6] DeBry RW. The consistency of several phylogeny-inference methods under varying evolutionary rates. Mol Biol Evol 1992;9: 537-51.
[7] Denis F, Gascuel O. On the consistency of the minimum evolution principle of phylogenetic inference. Discret Appl Math 2003;127: 63-77.
[8] Bordewich M, Gascuel O, Huber KT, Moulton V. Consistency of topological moves based on the balanced minimum evolution principle of phylogenetic inference. IEEE/ACM Trans Comput Biol Bioinform 2009;6:110-7.
[9] Liu L, Pearl DK. Species trees from gene trees. Syst Biol 2007;56(3):504-14.
[10] Edwards SV, Liu L, Pearl DK. High-resolution species trees without concatenation. Proc Natl Acad Sci 2007;104:5936-41.
[11] Ane C, Larget B, Baum DA, Smith SD, Rokas A. Bayesian estimation of concordance among gene trees. Mol Biol Evol 2007;24: 412-26.
[12] Swofford DL. PAUP*. Phylogenetic analysis using parsimony (* and other methods). Sunderland, MA: Sinauer Associates, Inc.; 1998.
[13] Ge S, Sang T, Lu B, Hong D. Phylogeny of rice genomes with emphasis on origins of allotetraploid species. PNAS 1999;96: 14400-5.
[14] Voigt K, Cicelnik E, O'Donnel K. Phylogeny and PCR identification of clinically important zygomycetes based on nuclear ribosomal-DNA sequence data. J Clin Microbiol 1999;37:3957-64.
[15] Kishino H, Hasegawa M. Evaluation of the maximum likelihood estimate of the evolutionary tree topologies from DNA sequence data. J Mol Evol 1989;29:170-9.
[16] Vilaa M, Vidal-Romani JR, Björklund M. The importance of time scale and multiple refugia: incipient speciation and admixture of lineages in the butterfly *Erebia triaria* (Nymphalidae). Mol Phylogenet Evol 2005;36:249-60.
[17] Lee MSY, Hugall AF. Partitioned likelihood support and the evaluation of data set conflict. Syst Biol 2003;52:15-22.
[18] Henning W. Phylogenetic systematics. Urbana, IL: Univ. of Illinois Press; 1966.
[19] Dowling APG, Veller MGP, Hoberg EP, Brooks DR. A priori and a posteriori methods in comparative evolutionary studies of host-parasite associations. Cladistics 2003;19:240-53.
[20] Huelsenbeck JP, Rannala B, Yang Z. Statistical tests of host-parasite cospeciation. Evolution 1997;51:410-9.
[21] Huelsenbeck JP, Larget B, Swofford DL. A compound Poisson process for relaxing the molecular clock. Genetics 2000;154: 1879-92.

[22] Hafner MS, Nadler SA. Cospeciation in host parasite assemblages: comparative analysis of rates of evolution and timing of cospeciation events. Syst Zool 1990;39:192-204.

[23] Legendre P, Desdevises Y, Bazin E. A statistical test for host-parasite coevolution. Syst Biol 2002;51:217-34.

[24] Schardl CL, Craven KD, Lindstrom A, Stromberg A, Yoshida R. A novel test for host-symbiont codivergence indicates ancient origin of fungal endophytes in grasses. Syst Biol 2008;57(3):483-98.

[25] Brooks DR. Parsimony analysis in historical biogeography and coevolution: methodological and theoretical update. Syst Zool 1990; 39:14-30.

[26] Brooks DR, McLennan DA. Phylogeny, ecology and behavior: a research program in comparative biology. Chicago, IL: Univ. of Chicago Press; 1991.

[27] Brooks DR, McLennan DA. Parascript: parasites and the language of evolution. Washington, D.C.: Smithsonian Institution Press; 1993.

[28] Brooks DR, Van Veller MGP, McLennan DA. How to do BPA, really. J Biogeogr 2001;28:343-58.

[29] Brooks DR, McLennan DA. The nature of diversity: an evolutionary voyage of discovery. Chicago, IL: Univ. of Chicago Press; 2002.

[30] Dowling APG. Testing the accuracy of TreeMap and Brooks parsimony analyses of coevolutionary patterns using artificial associations. Cladistics 2002;18:416-35.

[31] Page RDM. Component 2.0: tree comparison software for Microsoft Windows. Program and user's manual; 1993.

[32] Page RDM. Treemap 1.0. Program and user's manual; 1995.

[33] Spiering MJ, Moon CD, Wilkinson HH, Schardl CL. Gene clusters for insecticidal loline alkaloids in the grass-endophytic fungus Neotyphodium uncinatum. Genetics 2005;169:1403-14.

[34] Sadava D, Heller HC, Orians GH, Purves WK, Hillis DM. Life: the science of biology. 8th ed. Sunderland, MA: Sinauer Associates, Inc.; 2008.

[35] Elias I. Settling the intractability of multiple alignment. J Comput Biol 2006;13:1323-39.

[36] Durbin R, Eddy S, Korgh A, Mitchison G. Biological sequence analysis: probabilistic models of proteins and nucleic acids. Cambridge: Cambridge University Press; 1998.

[37] Stroock DW. An introduction to Markov processes, vol. 230. New York: Springer; 2005.

[38] Jukes TH, Cantor C. Evolution of protein molecules. In: Munro HN, editor. Mammalian protein metabolism. New York: Academic Press; 1969. p. 21-32.

[39] Zhang Z, Gerstein M. Patterns of nucleotide substitution, insertion and deletion in the human genome inferred from pseudogenes. Nucleic Acids Res 2003;31:5338-48.

[40] Zhang J, Nei M. Evolutionary distance: estimation. In: Encyclopedia of the Human Genome; 2003. p. 1-4.

[41] Benson DA, Karsch-Mizrachi I, Lipman DJ, Ostell J, Wheeler DL. GenBank. Nucleic Acids Res 2008;36:D25-30.

[42] Lawler GF. Introduction to stochastic processes. 2nd ed. New York, NY: Chapman and Hall; 2006.

[43] Tavaré S. Some probabilistic and statistical problems in the analysis of DNA sequences. Lect Math Life Sci 1986;17:57-86.

[44] Saitou N, Nei M. The neighbor joining method: a new method for reconstructing phylogenetic trees. Mol Biol Evol 1987;4:406-25.

[45] Haws D, Hodge T, Yoshida R. Phylogenetic tree reconstruction: geometric approaches. In: Robeva R, Hodge T, editors. Mathematical concepts and methods in modern biology: using modern discrete models. New York: Academic Press; 2013.

[46] Gascuel O. BIONJ: an improved version of the NJ algorithm based on a simple model of sequence data. Mol Biol Evol 1997;14: 685-95.

[47] Desper R, Gascuel O. Fast and accurate phylogeny reconstruction algorithms based on the minimum-evolution principle. J Comput Biol 2002;9:687-705.

[48] Day WHE. Computational complexity of inferring phylogenies from dissimilarity matrices. Bull Math Biol 1987;49:461-7.

[49] Felsenstein J. Inferring phylogenies. Sunderland, MA: Sinauer Associates, Inc.; 2003.

[50] Huelsenbeck JP, Ronquist F. Mr Bayes: Bayesian inference in phylogenetic trees. Bioinformatics 2001;17:754-5.

[51] Bouckaert R, Heled J, Kühnert D, Vaughan T, Wu C-H, Xie D, et al. BEAST 2: a software platform for Bayesian evolutionary analysis. PLoS Comput Biol 2014;10:e1003537. doi:10.1371/journal.pcbi.1003537.

[52] Drummond AJ, Rambaut A. BEAST: Bayesian evolutionary analysis by sampling trees. BMC Evol Biol 2007;7. doi:10.1186/1471-2148-7-214.

[53] Burnham KP, Anderson DR. Multimodel inference: understanding AIC and BIC in model selection. Sociol Methods Res 2004;33: 261-304.

[54] Maddison WP. Gene trees in species trees. Syst Biol 1997;46:523-36.

[55] Maddison WP, Knowles LL. Inferring phylogeny despite incomplete lineage sorting. Syst Biol 2006;55:21-30.

[56] Liu L, Pearl D, Brumfield R, Edwards S. Estimating species trees using multiple-allele DNA sequence data. Evolution 2008;62: 2080-91.

[57] Page RDM. Maps between trees and cladistic analysis of historical associations among genes, organisms, and areas. Syst Biol 1994;43:58-77.

[58] Page RDM, Charleston MA. Trees within trees: phylogeny and historical associations. TREE 1998;13:356-9.

[59] Stolzer M, Lai H, Xu M, Sathaye D, Vernot B, Durand D. Inferring duplications, losses, transfers and incomplete lineage sorting with nonbinary species trees. Bioinformatics 2012;28:i409-15.

[60] Cuthill JH, Charleston M. Phylogenetic codivergence supports coevolution of mimetic heliconius butterflies. PLoS ONE 2012;7:e36464. doi:10.1371/journal.pone.0036464.

[61] Knowles LL, Klimov PB. Estimating phylogenetic relationships despite discordant gene trees across loci: the species tree of a diverse species group of feather mites (Acari: Proctophyllodidae). Parasitology 2011;138:1750-9.

[62] Page R. Tangled trees. Chicago: University of Chicago Press; 2003.

[63] Bergsten J. A review of long-branch attraction. Cladistics 2005;21:163-93.

[64] Page RDM, Holmes EC. Molecular evolution: a phylogenetic approach. Malden, MA: Blackwell Publishing Ltd; 1998.

[65] Fares MA, Byrne KP, Wolfe KH. Rate asymmetry after genome duplication causes substantial long-branch attraction artifacts in the phylogeny of *Saccharomyces* species. Mol Biol Evol 2006;23:245-53.

[66] Rokas A, Carroll SB. Frequent and widespread parallel evolution of protein sequences. Mol Biol Evol 2008;25:1943-53.

[67] Castoe TA, Koning APJ, Kim HM, Gua W, Noonanb BP, Naylorc G, et al. Evidence for an ancient adaptive episode of convergent molecular evolution. Proc Natl Acad Sci USA 2009;106:8986-91.

[68] Kück P, Mayer C, Wágele J-W, Misof B. Long branch effects distort maximum likelihood phylogenies in simulations despite selection of the correct model. PLoS ONE 2012. doi:10.1371/journal.pone.0036593.

[69] Keese P. Risks from GMOs due to horizontal gene transfer. Environ Biosafety Res 2008;7:123-49.

[70] Nordmann P, Naas T, Fortineau N, Poirel L. Superbugs in the coming new decade; multidrug resistance and prospects for treatment of *Staphylococcus aureus*, *Enterococcus* spp. and *Pseudomonas aeruginosa* in 2010. Curr Opin Microbiol 2007;10: 436-40.

[71] Salzberg SL, White O, Peterson J, Eisen JA. Microbial genes in the human genome: lateral transfer or gene loss? Science 2001;292: 1903-6.

[72] Stanhope MJ, Lupas A, Italia MJ, Koretke KK, Volker C, Brown JR. Phylogenetic analyses do not support horizontal gene transfers from bacteria to vertebrates. Nature 2001;411:940-4.

[73] Richards TA, Soanes DM, Jones MDM, Vasieva O, Leonard G, Paszkiewicz K, et al. Horizontal gene transfer facilitated the evolution of plant parasitic mechanisms in the oomycetes. Proc Natl Acad Sci USA 2011;108:15258-63.

[74] Kishore S, Stiller J, Deitsch K. Horizontal gene transfer of epigenetic machinery and evolution of parasitism in the malaria parasite *Plasmodium falciparum* and other apicomplexans. BMC Evol Biol 2013;13. doi:10.1186/1471-2148-13-37.

[75] Lindsay J, Holden MG. Understanding the rise of the superbug: investigation of the evolution and genomic variation of *Staphylococcus aureus*. Funct Integr Genomics 2006;6:186-201.

[76] Pamilo P, Nei M. Relationships between gene trees and species trees. Mol Biol Evol 1988;5:568-83.

[77] Takahata N. Gene genealogy in 3 related populations: consistency probability between gene and population trees. Genetics 1989;122:957-66.

[78] Bollback JP, Huelsenbeck JP. Parallel genetic evolution within and between bacteriophage species of varying degrees of divergence. Genetics 2009;181:225-34.

[79] Carling M, Brumfield R. Integrating phylogenetic and population genetic analyses of multiple loci to test species divergence hypotheses in Passerina buntings. Genetics 2008;178:363-77.

[80] Yu Y, Warnow T, Nakhleh L. Algorithms for MDC-based multi-locus phylogeny inference: beyond rooted binary gene trees on single alleles. J Comput Biol 2011;18:1543-59.

[81] Betancur R, Li C, Munroe TA, Ballesteros JA, Ortí G. Addressing gene tree discordance and non-stationarity to resolve a multi-locus phylogeny of the flatfishes (Teleostei: Pleuronectiformes). Syst Biol 2013. doi:10.1093/sysbio/syt039.

[82] Heled J, Drummond AJ. Bayesian inference of species trees from multilocus data. Mol Biol Evol 2011;27:570-80.

[83] Thompson KL, Kubatko L. Using ancestral information to detect and localize quantitative trait loci in genome-wide association studies. BMC Bioinformatics 2013;14:200.

[84] Rosenberg N. The probability of topological concordance of gene trees and species trees. Theor Popul Biol 2002;61:225-47.

[85] Rosenberg NA. The shapes of neutral gene genealogies in two species: probabilities of monophyly, paraphyly, and polyphyly in a coalescent model. Evolution 2003;57:1465-77.

[86] Degnan JH, Salter LA. Gene tree distributions under the coalescent process. Evolution 2005;59:24-37.

[87] Liua L, Yub L, Kubatkoc L, Pearlc DK, Edwards SV. Coalescent methods for estimating phylogenetic trees. Mol Phylogenet Evol 2009;53:320-8.

[88] Knowles LL. Statistical phylogeography. Annu Rev Ecol Evol Syst 2009;40:593-612.

[89] Yu Y, Than C, Degnan JH, Nakhieh L. Coalescent histories on phylogenetic networks and detection of hybridization despite incomplete lineage sorting. Syst Biol 2011;60:138-49.
[90] Tian Y, Kubatko L. Gene tree rooting methods give distributions that mimic the coalescent process. Mol Phylogenet Evol 2014;70: 63-69.
[91] Carstens BC, Knowles LL. Estimating species phylogeny from gene-tree probabilities despite incomplete lineage sorting: an example from Melanoplus grasshoppers. Syst Biol 2007;56:400-11.
[92] Mossel E, Roch S. Incomplete lineage sorting: consistent phylogeny estimation from multiple loci; 2007. arXiv q-bio.PE.
[93] RoyChoudhury A, Felsenstein J, Thompson EA. A two-stage pruning algorithm for likelihood computation for a population tree. Genetics 2008;180:1095-105.
[94] Knowles LL. Estimating species trees: methods of phylogenetic analysis when there is incongruence across genes. Syst Biol 2009;58:463-7.
[95] Yang Z, Rannala B. Bayesian species delimitation using multilocus sequence data. PNAS 2009;107:9264-9.
[96] Leaché AD, Rannala B. The accuracy of species tree estimation under simulation: a comparison of methods. Syst Biol 2011;60: 126-37.
[97] Hovmoller R, Knowles LL, Kubatko LS. Effects of missing data on species tree estimation under the coalescent. Mol Phylogenet Evol 2013;69:1057-62.
[98] Hein J, Schierup MH, Wiuf C. Gene genealogies, variation and evolution: a primer in coalescent theory. Oxford: Oxford University Press; 2005.
[99] Semple C, Steel M. Phylogenetics, vol. 24 of Oxford lecture series in mathematics and its applications. Oxford: Oxford University Press; 2003.
[100] Schardl CL, Craven KD, Speakman S, Lindstrom A, Stromberg A, Yoshida R. A novel test for host-symbiont codivergence indicates ancient origin of fungal endophytes in grasses. Syst Biol 2008;57:483-98.
[101] Huelsenbeck JP, Rannala B, Larget B. A Bayesian framework for the analysis of cospeciation. Evolution 2000;54:352-64.
[102] Billera LJ, Holmes SP, Vogtmann K. Geometry of the space of phylogenetic trees. Adv Appl Math 2001;27:733-67.
[103] Devadoss SL, Huang D, Spadacene D. Polyhedral Covers of Tree Space. SIAM J. Discrete Math. 2014;28(3):1508-1514.
[104] Buneman P. The recovery of trees from measures of similarity. In: Hodson FR, Kendall DG, Tautu P, editors. Mathematics of the archaeological and historical sciences. Edinburgh: Edinburgh University Press; 1971. p. 387-95.
[105] Desper R, Gascuel O. Fast and accurate phylogeny reconstruction algorithms based on the minimum-evolution principle. J Comput Biol 2002;19:687-705.
[106] Eickmeyer K, Huggins P, Pachter L, Yoshida R. On the optimality of the neighbor-joining algorithm. Algorithms Mol Biol 2008;3:PMC2430562.
[107] Davidson R, Sullivant S. Distance-based phylogenetic methods around a polytomy. IEEE ACM Trans Comput Biol Bioinformatics (TCBB) 2014;11(2):325-335.
[108] Huggins P, Li W, Haws D, Friedrich T, Liu J, Yoshida R. Bayes estimators for phylogenetic reconstruction. Syst Biol 2010;60: 528-40.
[109] Bacák M. Computing medians and means in Hadamard spaces. SIAM J. Optim 2014;24(3):1542-1566
[110] Cavender J, Felsenstein J. Invariants of phylogenies in a simple case with discrete states. J Classif 1987;4:57-71.
[111] Lake JA. A rate-independent technique for analysis of nucleic acid sequences: evolutionary parsimony. Mol Biol Evol 1987;4: 167-91.
[112] Eriksson N. Tree construction with singular value decomposition. In: Pachter L, Sturmfels B, editors. Algebraic statistics for computational biology. Cambridge: Cambridge University Press; 2005. p. 347-58.
[113] Rusinko JP, Hipp B. Invariant based quartet puzzling. Algorithms Mol Biol 2012;7. doi:10.1186/1748-7188-7-35.

Chapter 13

RNA Secondary Structures: Combinatorial Models and Folding Algorithms

Qijun He[1], Matthew Macauley[1] and Robin Davies[2]

[1]*Department of Mathematical Sciences, Clemson University, Clemson, SC, USA,* [2]*Department of Biology, Sweet Briar College, Sweet Briar, VA, USA*

13.1 INTRODUCTION

Deoxyribonucleic acid (DNA) and ribonucleic acid (RNA) are examples of *nucleic acids*—large organic molecules that are polymers of smaller molecules called *nucleotides*. Nucleotides consist of a five-carbon sugar to which are linked a nitrogenous base and a phosphate. The nucleotides are linked together through phosphodiester bonds that link the sugars to phosphates on adjacent nucleotides, resulting in a sugar-phosphate backbone. DNA is better known than RNA to the general public, as it encodes the essential genetic information for all cells, is standard material in most high school biology classes, and has even made its way into common parlance, such as "it is in my DNA." In cells, as well as in the double-stranded DNA viruses, DNA consists of two chains of nucleotides that pair to each other via hydrogen bonds, forming a double-helix structure. In contrast, RNA consists of a single strand of nucleotides that can fold and bond to itself. The specific shape into which RNAs fold plays a major role in their function, which makes RNA folding of prime interest to scientists.

Initially, RNA was regarded as a simple messenger—the conveyor of genetic information from its repository in DNA to the ribosomes. The information encoded in the RNA was then used to direct the construction of proteins, which were then thought to be the only actively functional molecules of the cell. Over the last several decades, however, researchers have discovered an increasing number of important roles for RNA. RNAs have been found to have catalytic activities, to participate in processing of messenger RNAs, to help maintain the telomers (ends) of eukaryotic chromosomes, and to influence gene expression in multiple ways. Clearly, RNA is more than a simple messenger. For more information on the growing recognition of the role of RNA, see the account by Darnell [1].

Not surprisingly, as is true of proteins, the three-dimensional structure of RNA is critically important to its function. Structure determines function, so an understanding of RNA's three-dimensional structure will allow a greater understanding of RNA function. This should lead to the discovery of additional RNA-encoding genes and to the development of RNA-based therapeutic agents.

One can casually think of a nucleic acid chain as a length of ribbon, to which squares of a hook and loop fastener, such as Velcro®, are attached. Furthermore, suppose this ribbon has some thickness and rigidity—like that of a stiff belt one would wear, rather than a decorative ribbon used to wrap a gift. In a DNA model, two separate strands of ribbon are attached together, through the linking of hooks on one ribbon and loops in the corresponding position on the other ribbon. In an RNA model, the position of hooks and loops would allow the ribbon to fold and attach to itself. In this metaphor, the ribbon represents the sugar-phosphate backbone, the squares of Velcro® represent nucleotides, and the attachment of the hooks and loops represents the formation

Algebraic and Discrete Mathematical Methods for Modern Biology. http://dx.doi.org/10.1016/B978-0-12-801213-0.00013-7
Copyright © 2015 Elsevier Inc. All rights reserved.

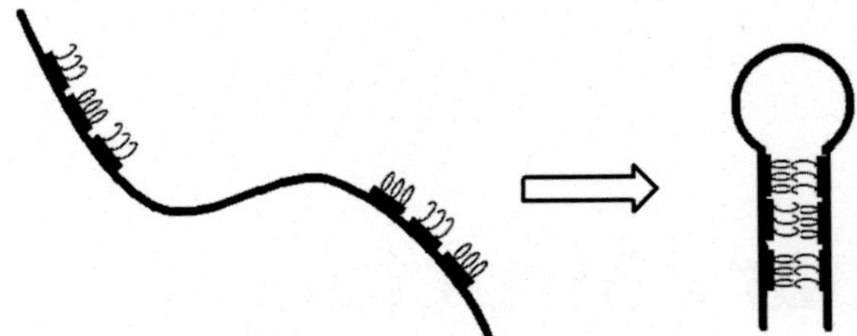

FIGURE 13.1 A cartoon of the ribbon and Velcro® analogy.

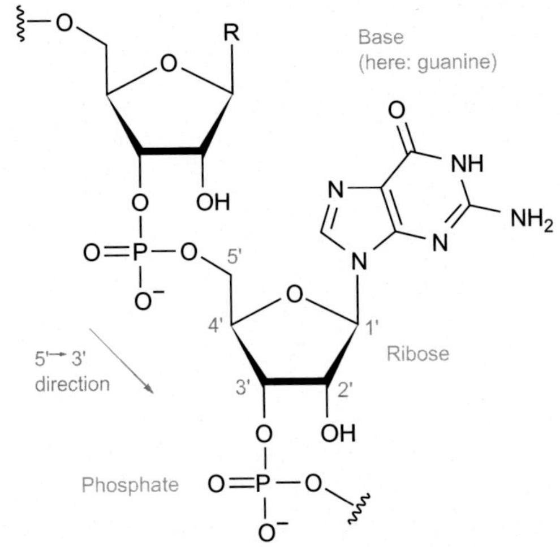

FIGURE 13.2 Two nucleotides attached together via a phosphodiester bond, forming the sugar-phosphate backbone. Figure created by Narayanese at the English language Wikipedia and provided under the Creative Commons Attribution-Share Alike 3.0 Unported license (http://creativecommons.org/licenses/by-sa/3.0/legalcode).

of hydrogen bonds. This analogy is useful because it helps model certain physical constraints of nucleic acids. A cartoon depicting it is shown in Figure 13.1.

Each nucleic acid strand is a polymer consisting of a long sequence of nucleotides linked together to form the sugar-phosphate backbone. Figure 13.2 shows the nature of the sugar-phosphate backbone in an RNA strand. The nucleotides of DNA differ from the nucleotides of RNA in that DNA contains deoxyribose sugar and RNA contains ribose. Although the phosphates are identical, the nucleotides also differ in their nitrogenous bases. There are only five common nitrogenous bases, and they come in two types: *adenine* (A) and *guanine* (G) are purines; *cytosine* (C), *thymine* (T), and *uracil* (U) are pyrimidines. DNA molecules only use A, C, G, T, whereas RNA molecules use A, C, G, U. Each nucleic acid strand has a "front" end, called the 5′-end, and a "back" end, called the 3′-end. The two strands in a DNA molecule run in opposite directions from each other, and each base on one strand bonds with the corresponding base on the other strand. Two base pairs arise in DNA: adenine pairs with thymine via two hydrogen bonds to form AT, and cytosine pairs with guanine via three hydrogen bonds to form CG. Knowing the nucleotide sequence on one strand completely determines the sequence on the complementary strand. This redundancy has a biological advantage in that it is more robust to changes such as mutation.

In this chapter, we will focus our attention on RNA and how it can fold and bond to itself. As nucleotides are determined by their nitrogenous base, the part of the molecule at which hydrogen bonding occurs, they will be called *bases* for short. Two bases that bond are called a *base pair*. Not all bonds between bases are possible; the most common are the *Watson-Crick* base pairs AU (two hydrogen bonds) and CG (three hydrogen bonds). Also possible, but less common and weaker, is the *wobble pair* GU (two hydrogen bonds). These three base pairs are shown in Figure 13.3. Other bonds are either chemically impossible (e.g., GT, AC), or thermodynamically

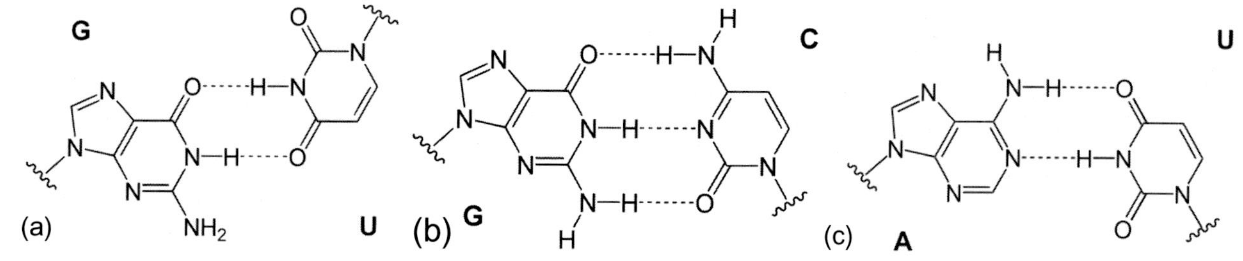

FIGURE 13.3 The three base pairs found in RNA molecules.

unstable (purine-purine, pyrimidine-pyrimidine) and thus very rare. Returning to the ribbon analogy, an unstable bond can be thought of as the ribbon being attached to itself by a piece of very worn Velcro®. It takes only a slight jostle to break the bond, and so while it may exist temporarily, it is highly unlikely for such a bond to persist, and so it is rarely observed.

The raw sequence of an RNA strand is called the *primary structure* and can be thought of as a string over the alphabet {A, C, G, U}. The bonding of the nucleotides, irrespective of spacial embedding, is the *secondary structure*. Finally, the actual spacial embedding, such as how the strand is twisted or knotted in 3D-space, is the *tertiary structure*. For example, the fact that tRNAs form four base-paired stems and three loops is a property of their secondary structure, but the fact that the tRNA folds over into its final functional shape is a property of the tertiary structure. The primary, secondary, and tertiary structures of a tRNA are shown in Figure 13.4. The focus in this chapter is on the secondary structure of RNA. We are interested in the different ways an RNA strand can fold, and how can we model this process to predict the most likely structure.

RNA folding is a sequential process. The unbonded strand does not magically snap into place in a single step. Rather, as a strand begins to fold, base pairs form via hydrogen bonds—some of these bonds are formed and then break apart, whereas some are more stable and persist. Eventually, a strand reaches a stable folded

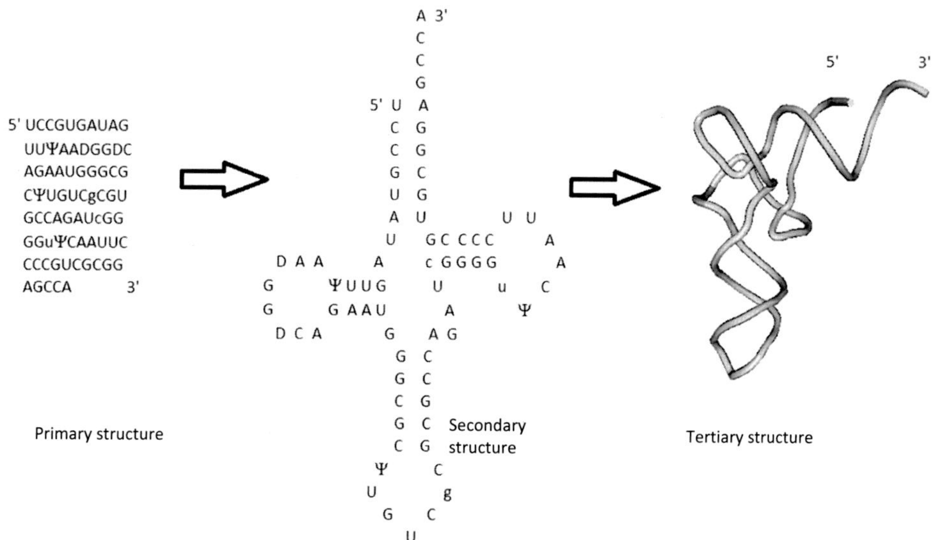

FIGURE 13.4 Primary, secondary, and tertiary structures of the yeast aspartic acid transfer RNA. Nonstandard symbols are as follows: Ψ is pseudouracil, D is dihydrouracil, c is 5-methyl cytosine, g is 1-methyl guanine, and u is 5-methyl uridine. Primary sequence data obtained from the Nucleotide Data Base ndbserver.rutgers.edu [2, 3]. Tertiary structure was prepared using Cn3D (www.ncbi.nlm.nih.gov) with data from [4].

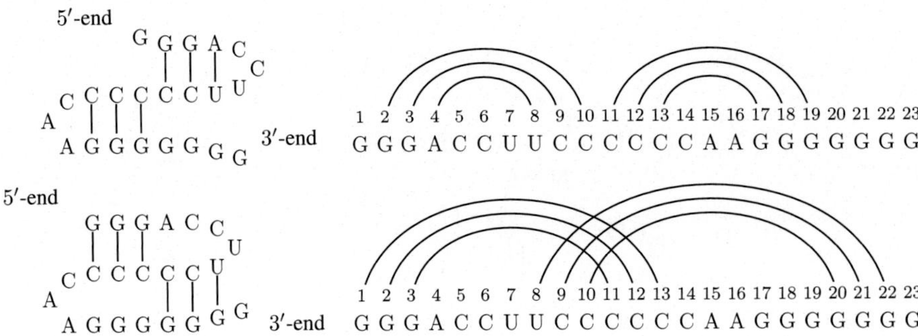

FIGURE 13.5 Two different folds of an RNA strand, and the corresponding arc diagrams. The first structure has a noncrossing arc diagram. The second structure is a pseudoknot because its arc diagram has crossings—it is a 3-noncrossing structure.

configuration. Scientists are interested in understanding what folded configurations can and will arise given a particular sequence, and how to algorithmically model the folding process.

The stability of a folded structure can be quantified by its *free energy*, which represents the amount of energy needed to maintain its structural integrity. A structure with negative free energy is stable because one needs to add energy to break it apart. A (theoretical) structure with positive free energy would be unstable because it would take energy to keep it together. In general, the lower the free energy, the more stable the structure. Any such calculation is merely an approximation, but roughly speaking, base pairs make a negative contribution because they keep the structure together, as the Velcro® stabilizes our toy model. On the other hand, loops make positive contributions because bending the RNA strand in a loop introduces a destabilizing restoring force. In our toy model, it takes physical energy to bend the rigid ribbon, and bending it too sharply could negate the effects of the Velcro® bonds.

There are many models for calculating the free energy of an RNA structure. As frequently happens, there is a tradeoff between complexity and usability. Some models take into account differences such as the number of hydrogen bonds in a base pair (either two or three). Others assign the free energy score of a loop based on its size, because smaller loops have a larger restoring force. Other models distinguish between loops of different types, such as hairpin, interior, or multiloops. The free energy of a structure is also determined by the ambient temperature, though this detail is largely unimportant to us in this chapter. It gets even more difficult when there is knotting in an RNA structure, because the thermodynamics of pseudoknots are poorly understood.

In this chapter, we will begin in Section 13.2 by introducing a combinatorial framework to describe RNA folding. The most useful approach is to use arc diagrams, which are shown in Figure 13.5. In Section 13.3, we will introduce several folding algorithms that are based on free energy minimization and use a recursive technique called dynamic programming (DP). In Section 13.4, we will present a model for RNA folding that uses a stochastic context-free grammar (SCFG), an object out of formal language theory from computational linguistics. These are the two main but very different approaches to RNA folding, and there are many variants of both. In Section 13.5, we will consider the problem of RNA folding into pseudoknot structures—arc diagrams that have crossings, such as the second example in Figure 13.5. The methods that incorporate pseudoknots are considerably more complicated, so in this chapter, we only provide an introduction by showing how the combinatorial models generalize and summarizing the approaches to the folding algorithms.

13.2 COMBINATORIAL MODELS OF NONCROSSING RNA STRUCTURES

13.2.1 Partial Matchings and Physical Constraints

The primary structure of an RNA strand can be encoded as a length-n sequence over the alphabet {A, C, G, U}. As an RNA strand folds, its nucleotides form hydrogen bonds with each other, yielding base pairs: the Watson-Crick pairs AU and CG, and the less common and weaker wobble pair GU. This can be represented by an *arc diagram*.

Definition 13.1. Given a vertex set $V = \{1, 2, \ldots, n\}$, an *arc* is a pair (i,j) with $i < j$. The *length* of an arc is $j - i$, and an arc of length k is called a *k-arc*. A *partial matching* is a collection A of arcs with the requirement that each vertex is contained in at most one arc. This can be represented visually by a graph $G = (V, A)$ called an *arc diagram*. □

In an arc diagram, it is conventional to arrange the vertices $V = \{1, \ldots, n\}$ sequentially in a horizontal line, and then draw the arcs in the upper half plane. One may draw a straight line through the vertices, representing the sugar-phosphate backbone, though we usually omit this practice. Figure 13.5 shows examples of two different folds of an RNA strand, and the resulting arc diagrams. We will return to this example to illustrate some of the key concepts.

An arc diagram represents a possible way an RNA strand can fold without regard to physical constraints or spacial embedding. However, realistically we need to impose certain constraints, in addition to the only allowed base pairs being AU, CG, and GU. Below, we describe three common constraints: minimum arc length, minimum stack size, and maximum crossing number.

One physical constraint that arises is *minimum arc length*. Due to the rigidity of the sugar-phosphate backbone, it is physically impossible for adjacent nucleotides to bond. This means that an arc diagram of a folded RNA structure will never have a 1-arc $(i, i + 1)$. Other small arcs can form but are rare and are easily broken. For example, 2-arcs may form, but they require more energy and thus are easier to break apart and never persist. Returning to the ribbon analogy, and remembering that our ribbon is rigid like a stiff belt, it takes energy to bend it enough so that two sites a distance-2 apart can stick together. Such a bond may last for a little while, but in the end, the configuration is unstable and will likely break apart. Thus, it is common to only consider arc diagrams with a prescribed minimum length λ. While 3-arcs have been observed to occur, they are not as common as larger arcs. Some scientists use the parameter $\lambda = 3$ in models, whereas others use $\lambda = 4$. In the two approaches on which we will focus, the energy minimization via DP assumes $\lambda = 4$, whereas the particular SCFG approach that we will study assumes $\lambda = 3$. Though it would be nice to use the same parameter for both, we are simply sticking with what exists in the literature.

Another natural physical constraint that arises is *minimum stack size*. A *stack* (also frequently called a *helix*, or *stem*) is a sequence of "parallel" nested arcs:

$$(i,j), (i + 1, j - 1), \ldots, (i + (\sigma - 1), j - (\sigma - 1)),$$

and the maximal such σ is its *size*. Stacks look like rainbows in the arc diagrams—each diagram in Figure 13.5 has two stacks of size 3. A stack of size 1 is simply an isolated base pair, and these are seldom observed in actual folded RNA strands. It does not take much energy to break apart an isolated base pair, and so when one forms, it is usually only temporary. Returning to our ribbon analogy, imagine a folded ribbon attached by a single pair of hook and loop squares, versus another ribbon attached by four parallel paired squares of hooks and loops. Naturally, the second one would be more stable and less likely to break apart. An arc diagram is *σ-canonical* if the minimum stack size is at least σ. One common value for this parameter is $\sigma = 2$, which simply means there are no isolated base pairs. However, not all models use this parameter—sometimes there are other ways that models give strong preference to stacks over isolated base pairs.

Finally, the last natural restriction imposed upon arc diagrams is the crossing number. Two arcs (i_1, j_1) and (i_2, j_2) with $i_1 < i_2$ are *crossing* if $i_1 < i_2 < j_1 < j_2$. An arc diagram is *noncrossing* if it has no crossing arcs. More generally, an arc diagram is k-noncrossing if there is no subset of k arcs that are all pairwise crossing. Note that 2-noncrossing simply means noncrossing. An RNA strand whose arc diagram has crossings is called a *pseudoknot*. Most RNA pseudoknots that are observed are 3-noncrossing, though higher crossing numbers do exist. The top structure in Figure 13.5 has a noncrossing arc diagram, but the bottom one is a 3-noncrossing pseudoknot.

There are several different combinatorial objects that have been used to represent noncrossing secondary structures. Arc diagrams are the most common, but others have their own advantages. We will briefly introduce a few. A *Motzkin path* of length n is a path in the plane from the origin $(0, 0)$ in $\mathbb{N}^2$ to $(n, 0)$, consisting of three types of steps: $\nearrow$, $\searrow$, and $\longrightarrow$, where each has width 1. It is easy to construct a Motzkin path from an arc

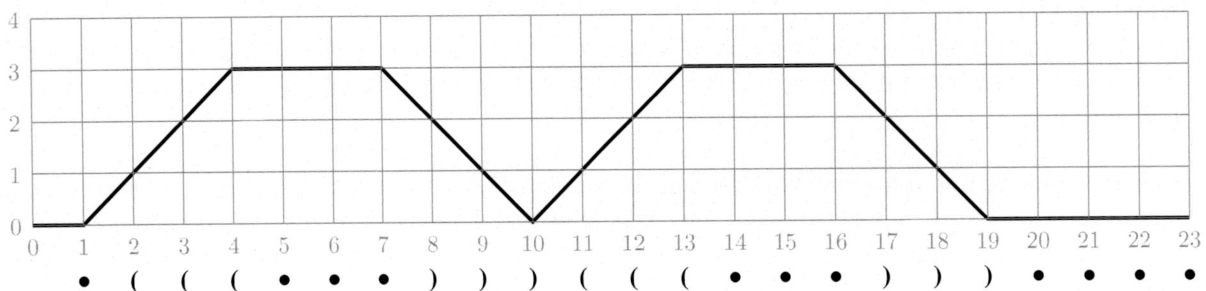

FIGURE 13.6 The Motzkin path and point-bracket representation of the noncrossing arc diagram from Figure 13.5.

diagram: Start the path at the origin and read off the vertices in the diagram from left to right. For each vertex that is the left endpoint of an arc, extend the path by $(1, 1)$ (i.e., travel $\nearrow$). For each right endpoint, extend the path by $(1, -1)$, and for each isolated vertex, extend the path by $(1, 0)$. Figure 13.6 shows the Motzkin path for the noncrossing arc diagram in Figure 13.5. Another equivalent representation is the *point-bracket* notation, where we write each up-step as "(", each down-step as ")", and each horizontal-step as "•". This is also shown in Figure 13.6; we leave the derivation as an exercise for the reader.

Finding equivalent representations of combinatorial objects is a commonly used technique by mathematicians. A feature that is difficult to handle in one representation might become very natural in another. Sometimes, a new representation leads to deep new theorems, and other times it simply gives an alternative perspective. Writing a secondary structure as a Motzkin path falls more in this latter category—though it is never necessary, it is a useful way to visualize several combinatorial features. For example, consider the restriction of having minimum arc length $\geq \lambda$, for some fixed constant $\lambda \geq 2$. In the language of Motzkin paths, this means that the length of each plateau is at least $\lambda - 1$. This is arguably easier than checking every individual arc of a partial matching.

Motzkin paths are also well suited to helping us understand the inductiveness of noncrossing secondary structures, which aids in their enumeration. Specifically, how many noncrossing length-n arc diagrams are there with the constraint that all arcs have length at least λ? Let $T^{\lambda}(n)$ denote the number. Though we do not have a simple closed formula for this, we can derive the next best thing: a recursive relation. Waterman showed that $T^{\lambda}(n)$ satisfies the following recursion [5]:

$$T^{\lambda}(n) = T^{\lambda}(n-1) + \sum_{k=0}^{n-(\lambda+1)} T^{\lambda}(n-2-k)T^{\lambda}(k), \tag{13.1}$$

where $T^{\lambda}(k) = 1$ for $0 \leq k \leq \lambda$. The best way to understand this recursion is to employ a Motzkin path interpretation. An n-step Motzkin path starts either with a horizontal-step $\longrightarrow$ or with an up-step $\nearrow$. There are clearly $T^{\lambda}(n-1)$ Motzkin paths that begin with $\longrightarrow$, which is why this term appears in Eq. (13.1). Next, we claim that the last term in Eq. (13.1) counts the number of Motzkin paths that begin with a $\nearrow$ step.

To see this (follow this explanation using Figure 13.7), first notice that because each Motzkin path starts and ends on the x-axis, concatenating two Motzkin paths yields another Motzkin path without decreasing the minimal plateau length. Thus we can try to decompose a Motzkin path beginning with $\nearrow$ into two shorter Motzkin paths (though one may be empty). Let us look at where the path returns to the x-axis for the first time. Call this vertex $n-k$; it is k steps (possibly $k = 0$) from the end. To the left of that vertex is a Motzkin path of length $n-k$ that begins with a $\nearrow$ step and ends with a $\searrow$ step; note that it must contain a plateau of length at least $\lambda - 1$, and thus $k \leq n - (\lambda + 1)$. Removing the first and last steps from this Motzkin path yields an arbitrary Motzkin path with $n-k-2$ steps. For this particular choice of k, there are $T^{\lambda}(n-2-k)$ possible Motzkin paths to the left, and $T^{\lambda}(k)$ possible Motzkin paths to the right. Summing over all k gives the last term in Eq. (13.1).

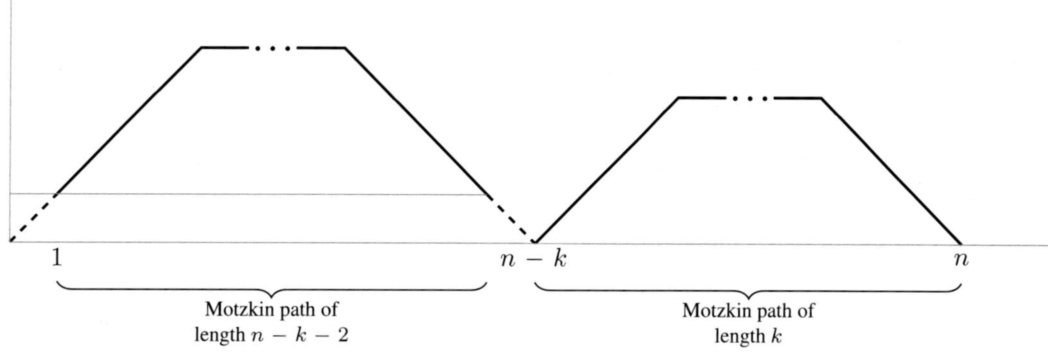

FIGURE 13.7 A picture illustrating the recursion in Eq. (13.1).

Exercise 13.1. Consider the RNA structures shown below.

```
5'-end                                      5'-end
              G                                        G G A  C
       G A A G   G  U U  U                             | | |    C
                          U                 C C C C C C U  U  \
       | |         | |                    A   | | |          \  G
       U U U        A A   U                 A G G G G G G  G  G
             A   A       U
               A                                              3'-end
3'-end
```

For each, draw an RNA diagram and find its minimum arc length, minimum stack size, and crossing number. □

Exercise 13.2. Draw the Motzkin path and point-bracket notation for the first RNA structure in the previous exercise. Attempt to do this for the second RNA structure. What goes wrong, and why? □

Exercise 13.3. Draw an RNA strand that exhibits a pseudoknot structure, as a 5-noncrossing (but not 4-noncrossing) arc diagram. Draw the diagram and an actual RNA strand that it could represent. □

13.2.2 Loop Decomposition

Unfortunately, the term "secondary structure" can refer to one of two different concepts of RNA structures. As previously mentioned, the secondary structure of a particular RNA strand is the bonding of the nucleotides irrespective of spacial embedding. This definition potentially allows crossings of arcs that lead to pseudoknots. In contrast, many scientists use the term "secondary structure" to refer only to folded RNA structures that have noncrossing diagrams. This discrepancy arose for several reasons: First, only noncrossing diagrams were initially studied, in which case these two concepts coincided. Only since people began considering pseudoknots did this ambiguity arise. Second, pseudoknots are considerably more complicated, as their loop structure cannot be recursively generated and their thermodynamics are poorly understood. As a result, pseudoknotting is often considered part of the molecule's tertiary structure. In light of this, we will reluctantly stick to the consensus in the literature and say that an "RNA secondary structure" is a folded RNA strand whose arc diagram is noncrossing.

An RNA secondary structure can be described by its loops, which come in different types. Because many energy-minimization folding algorithms are based on the loop structure and then carried out by a computer, it is imperative to establish a solid mathematical framework for defining loops of a secondary structure by their types and sizes. We follow the "loop decomposition" convention of Sankoff [6] from the 1980s. Given a base pair (i,j) with $i < v < j$, say that v is *accessible* from (i,j) if there is no base pair (i',j') such that

$i < i' < v < j' < j$. Loosely speaking, v is accessible from (i,j) if it can "look up" and see that arc, without any other nested arcs between. A base pair (v, w) is accessible from (i,j) if both v and w are accessible. The *k-loop* closed by (i,j) is the set of $(k-1)$ base pairs and the isolated bases that are accessible from (i,j). We do not include either i or j in the k-loop closed by (i,j). Finally, the *size* of a loop is the number of isolated bases in it.

The unique 0-loop $\mathbf{L}_0$ consists of the base pairs and isolated bases accessible from no base pair. Note that in the noncrossing arc diagram, these are the base pairs that are "on top," and the isolated bases that are not underneath any arcs. Because every base not in $\mathbf{L}_0$, whether it is paired or isolated, is accessible from exactly one base pair (i,j), every base is contained in a unique k-loop, denoted $\mathbf{L}_{i,j}$. This means that given a secondary structure S of a sequence $\mathbf{b}$, there is a well-defined decomposition of the bases of $\mathbf{b}$ into loops:

$$\mathbf{b} = \mathbf{L}_0 \bigcup \left(\bigcup_{(i,j)} \mathbf{L}_{i,j} \right),$$

where the inner union is taken over the disjoint k-loops, for $k \geq 1$. While the term k-loop is natural for mathematicians, biochemists had already developed their own names for these. A 1-loop is called a *hairpin loop*; see Figures 13.13 and 13.14. A k-loop for $k \geq 3$ is called a *multiloop*. Finally, there are several types of 2-loops. Suppose (i', j') is the (unique) base pair accessible from (i,j). The 2-loop is a

2(a) *Stacked pair* if $i' - i = j - j' = 1$;
2(b) *Bulge loop* if exactly one of $i' - i$ and $j - j'$ is > 1;
2(c) *Interior loop* if both $i' - i$ and $j - j'$ are > 1.

Note that the 2-loops of size 0 are precisely the stacked pairs. Figure 13.8 shows an example of a bulge loop (at left) and an interior loop (at right). In both of these, each capital letter is assumed to represent a base from $\{A, C, G, U\}$

In the secondary structure on the left of Figure 13.8, the 2-loop consisting of the vertices {C, D, E, F, I} is a bulge loop of size 3, the number of isolated bases. The sets {B, J} and {G, H} are both 2-loops that are stacked pairs. We do not have enough information to know what type of loop the vertices A and K belong to.

In the secondary structure on the right of Figure 13.8, the 2-loop {C, D, E, F, G, J, K, L, M} is an interior loop of size 7. The sets {B, N} and {H, I} are both 2-loops that are stacked pairs. We do not know what type of loop the vertices A and O belong to.

Exercise 13.4. Write out the loop decomposition of the noncrossing RNA structure from Exercise 13.1 of Section 13.2.1. □

Exercise 13.5. Draw a secondary structure that has a 3-loop, and then draw its arc diagram and write out its loop decomposition. □

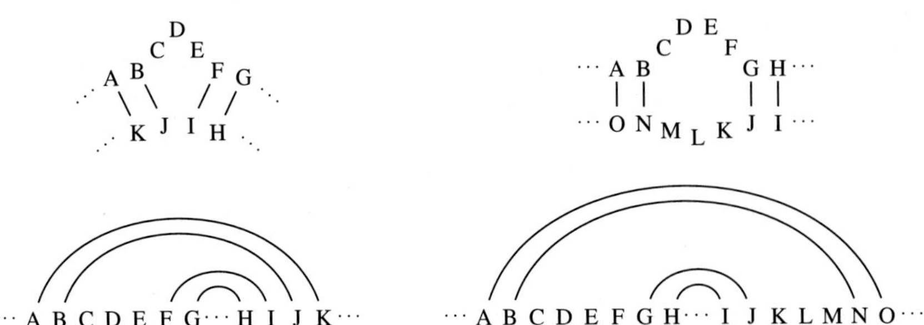

FIGURE 13.8 Two 2-loops: a bulge loop (left) and an interior loop (right). Each secondary structure also contains two 2-loops that are stacked pairs.

13.3 ENERGY-BASED FOLDING ALGORITHMS FOR SECONDARY STRUCTURE PREDICTION

One central problem in nucleic acid research is the *secondary structure prediction* problem: determining how an RNA molecule folds given only its raw sequence of bases. This is analogous to the protein folding problem—the determination of how a protein folds given only its amino acid sequence. While in theory, any noncrossing partial matching of an RNA strand in which only legal base pairs (AU, CG, GU) arise is possible, most of these are very unlikely because they are either thermodynamically unstable or physically impossible. Additionally, RNA strands with the same raw sequences have been observed to fold into multiple different structures, so the best that a deterministic (=nonrandom) algorithm could do is identify the most likely structure. In theory, the best solution to this prediction problem would be a stochastic algorithm such that the distribution of predicted structures is statistically indistinguishable from the distribution of structures that are observed experimentally. Not surprisingly, no such algorithm is known to exist. Additionally, the best deterministic and stochastic algorithms to date perform equally well, so both are worthy of study. The focus of this current section is on deterministic algorithms that minimize free energy via dynamic programming. In the next section, we will look at stochastic approaches using formal grammars and language theory.

It is well known that the most commonly observed secondary structures (recall: the arc diagram must be noncrossing) are very close to the most "thermodynamically stable" foldings, which are defined by having the minimum free energy (mfe) over all possible folds. The free energy of a secondary structure, measured in kilocalories (kcal) per mole, can be thought of as the amount of energy needed to break up the structure. A positive free energy means that it takes energy to maintain its structural integrity, and a negative free energy means that it takes energy to break it apart. The thermodynamics of secondary structures are well understood; they mostly depend on the number and type of base pairs, as well as the number, type, and size of loops. Base pairs are stabilizing and thus contribute negative energy, whereas loops are destabilizing and thus contribute positive energy. In contrast, the thermodynamics of pseudoknots are not well understood by biophysicists. Further complicating matters is the fact that k-noncrossing diagrams cannot be recursively generated in the way that noncrossing diagrams can (recall the Motzkin path recursion from Eq. (13.1) and Figure 13.7). Because of this, pseudoknotting is often considered part of the molecule's tertiary structure, which is one reason why the nucleic acid community uses the term "secondary structure" synonymously with having a noncrossing arc diagram.

In this section, we will see several different deterministic methods for secondary structure prediction. First, we begin with a simple dynamic program that maximizes the bond strength, or equivalently, minimizes the free energy when one ignores the loop structure. The second approach is a complicated but more accurate thermodynamic model that accounts for both bond strengths and the loop structure in its free energy calculation. The accuracy and difficulty of this algorithm are mainly due to the numerous cases of loop type and size, as well as the many specific energy values that need to be experimentally determined beforehand. As a result, the calculations are prohibitively difficult to do by hand, but there are free software programs that carry out these methods. The algorithm that we describe here is implemented by the free software program UNAFold (Unified Nucleic Acid Folding and hybridization package), developed by Michael Zuker and colleagues at Rensselaer Polytechnic Institute. However, we only describe a rough outline and leave the interested reader to check out the details on the DINAMelt (DI-Nucleic Acid hybridization and melting prediction) Web Server, which hosts UNAFold [7].

13.3.1 Maximizing Bond Strengths via Dynamic Programming

Dynamic programming is a method for solving complex problems by solving simpler subproblems and combining their solutions to obtain the overall solution. In the 1970s, Nussinov proposed a simple DP algorithm for secondary structure prediction by seeking to maximize the number of base pairs [8]. However, the three different types of base pairs have different relative strengths, so a more accurate model can be achieved by maximizing a weighted sum of the base pairs. This is equivalent to minimizing the free energy if one ignores the positive energy contributions due to the loops.

The CG pairing uses three hydrogen bonds, whereas AU and UG each use two. However, the wobble pair UG is relatively unstable, having roughly half the strength of an AU bond. To account for this, we define an energy function $e(i,j)$ that represents the strength of the potential bond between b_i and b_j in $\mathbf{b} = b_1, b_2, \ldots, b_n$. Because we are not considering the loop structure, we will assume all energies are positive numbers for convenience. Thus, given an RNA sequence $\mathbf{b} = b_1, b_2, \ldots, b_n$, define

$$e(i,j) = \begin{cases} 3 & \{b_i, b_j\} = \{\mathrm{C}, \mathrm{G}\} \text{ and } i \leq j-4 \\ 2 & \{b_i, b_j\} = \{\mathrm{A}, \mathrm{U}\} \text{ and } i \leq j-4 \\ 1 & \{b_i, b_j\} = \{\mathrm{G}, \mathrm{U}\} \text{ and } i \leq j-4 \\ 0 & \text{otherwise.} \end{cases} \tag{13.2}$$

We assume that the energy of each base pair is independent of all other base pairs and the loop structure. As a consequence, the total energy $E_S(\mathbf{b})$ of a secondary structure S with sequence $\mathbf{b}$ is the sum of the individual energies of the base pairs. The goal is now to *maximize* this energy score.

The DP process is broken up into two steps:

1. Use the optimal energy score of subsequences of $\mathbf{b}$ to determine the optimal (maximum) energy score of $\mathbf{b}$.
2. "Traceback" to reconstruct the actual secondary structure that realizes this maximum.

The major limitation of this model is that it does not take into account the destabilizing loop structures that counteract the stabilizing effect of the bonds. This is handled by the algorithm described in the next section, though the simplicity and ability to do small examples by hand immediately goes out the window.

As an example, suppose we have a fixed RNA sequence $\mathbf{b} = b_1, b_2, \ldots, b_n$ with minimum loop size 4. This means that each base pair (i,j) must satisfy $i \leq j-4$. Our goal is to find the secondary structure that maximizes the sum of the individual energies of the base pairs. Although this may seem like a daunting task that involves checking all possible secondary structures, the idea of DP is that it can easily be determined from knowing the optimal solution on its subsequences.

In particular, we will consider folds of a subsequence $\mathbf{b}_{i,j} = b_i, b_{i+1}, \ldots, b_{j-1}, b_j$ of $\mathbf{b}$. That is, we "chop off" all vertices outside this range and only allow base pairs (i', j') that have both endpoints satisfying $i \leq i' \leq j' \leq j$. Let $E(i,j)$ be the maximum energy score of this subsequence. In other words, $E(i,j)$ is the maximum sum of the individual energies of the base pairs, taken over all the secondary structures that are legal folds of $\mathbf{b}_{i,j}$. Note that our end goal is to find $E(1,n)$.

We compute $E(i,j)$ recursively. If $i > j-4$, then we return $E(i,j) = 0$. Otherwise, there are four ways to recurse on the sequence $\mathbf{b}_{i,j}$, described below and pictured in Figure 13.9.

1. *(i,j) forms a base pair*: Recurse on the sequence $\mathbf{b}_{i+1,j-1}$.
2. *i is unpaired but (k,j) is a base pair*: Recurse on the sequence $\mathbf{b}_{i+1,j}$.
3. *(i,k) is a base pair but j is unpaired*: Recurse on the sequence $\mathbf{b}_{i,j-1}$.
4. *(i,k_i) and (k_j,j) are paired for some $k_i < k_j$*: Recurse on the two subsequences $\mathbf{b}_{i,k}$ and $\mathbf{b}_{k+1,j}$, for some $k_i \leq k < k_j$.

In this last case, we have to check all possible values of $k = i, \ldots, j$, although technically we could get away with only considering $i+4 \leq k < j-4$ (why?). Although simultaneously recursing on all possible k at each step may seem computationally difficult, this is precisely what the DP paradigm is built to handle.

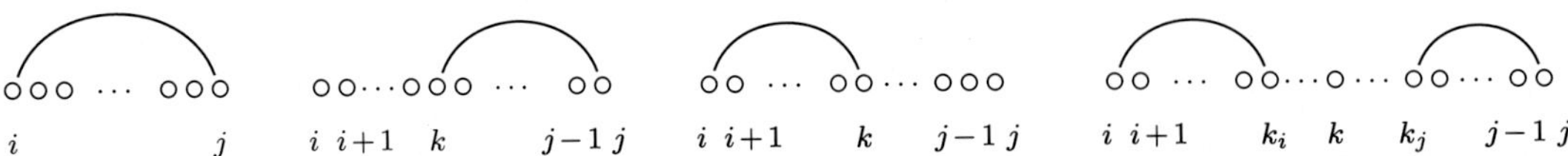

FIGURE 13.9 The four cases that need to be considered when the bases b_i and b_j can bond with each other.

$j \longrightarrow$ (columns), i ↓ (rows)

	G	G	G	A	C	C	U	U	C	C
G	0	0	0	0	3					
G		0	0	0	0	3				
G			0	0	0	0	1			
A				0	0	0	0	2		
C					0	0	0	0	0	
C						0	0	0	0	0
U							0	0	0	0
U								0	0	0
C									0	0
C										0

	G	G	G	A	C	C	U	U	C	C
G	0	0	0	0	3	3	4	4	6	(8)
G		0	0	0	0	3	3	3	5	8
G			0	0	0	0	1	2	5	5
A				0	0	0	0	2	2	2
C					0	0	0	0	0	0
C						0	0	0	0	0
U							0	0	0	0
U								0	0	0
C									0	0
C										0

FIGURE 13.10 Recording the optimal scores in a table during a dynamic programming routine.

Taking the maximum energy score over each of the four ways to recurse on the subsequence $\mathbf{b}_{i,j}$ yields a recurrence for the maximum score $E(i,j)$:

$$E(i,j) = \max \begin{cases} E(i+1,j-1) + e(i,j) \\ E(i+1,j) \\ E(i,j-1) \\ \max\limits_{i<k<j} [E(i,k) + E(k+1,j)]. \end{cases}$$

The optimal energy score of the original strand $\mathbf{b} = \mathbf{b}_{1,n}$ is $E(1,n)$, and this can be determined algorithmically by arranging the $E(i,j)$ values in a table, where the (i,j) entry is $E(i,j)$. See Figure 13.10 for an example of this. Because $i < j$, we can ignore all entries below the main diagonal. Additionally, $E(i,i) = 0$ means that all diagonal entries are 0. Next, notice that all squares directly above the main diagonal are 0 because $E(i,i+1) = 0$. Likewise, we can fill in the next two "diagonal layers" with 0s, because $E(i,i+2) = E(i,i+3) = 0$. The next diagonal layer above these two represents pairs of bases that are a distance four apart, and hence can pair if the bases are compatible. We fill in each (i,j) entry with $e(i,j)$, representing the strength of that bond. A value of 0 means that b_i and b_j do not form a legal base pair. The table on the left in Figure 13.10 shows an example of the partially filled table up to this point, for the sequence $\mathbf{b}$ = GGGACCUUCC.

Once all of the values in the first nonzero diagonal layer are entered, we fill up the remainder of the table by diagonal layers, one by one, until we reach the upper-right corner, where the $E(1,n)$ entry is located. Naturally, we can ignore everything below the main diagonal. The rule for $E(i,j)$, which is the entry that belongs in the box (i,j), makes it simple to calculate from the previously filled-in boxes. One needs only to determine which is bigger: (1) $e(i,j)$ plus the value directly to the bottom left of the box, (2) the value directly below the box, (3) the value directly to the left of the box, or (4) any of the sums of the "complementary entries" from the ith row and jth column, as shown in Figure 13.11.

Once the table of energy scores is filled out, the $E(1,n)$ entry in the upper-right corner tells us exactly the maximum bond strength of a folded secondary structure. This is only half of the algorithm—next, we need to determine which secondary structure realizes this maximum. This is called the *traceback* step. It is best described by an example, so we continue our example using the sequence $\mathbf{b}$ = GGGACCUUCC from Figure 13.10.

We begin the traceback step by circling the $E(1,n)$ entry, which in Figure 13.10 is $E(1,10) = 8$. Next, we need to determine from which entry in the table that value of 8 came. In other words, which of the four recursion cases got us there? In our example, because $E(1,10) = 8$ is not equal to $E(1,9) = 6$, we can discard

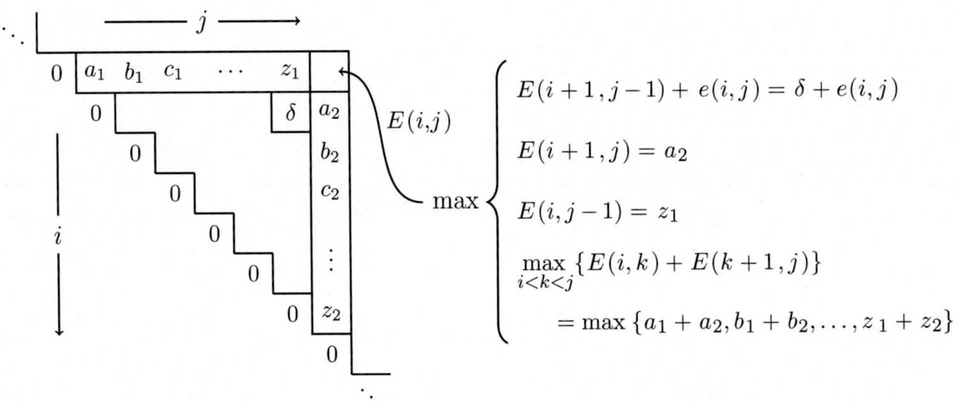

FIGURE 13.11 A visual of how the energy score $E(i,j)$ is calculated.

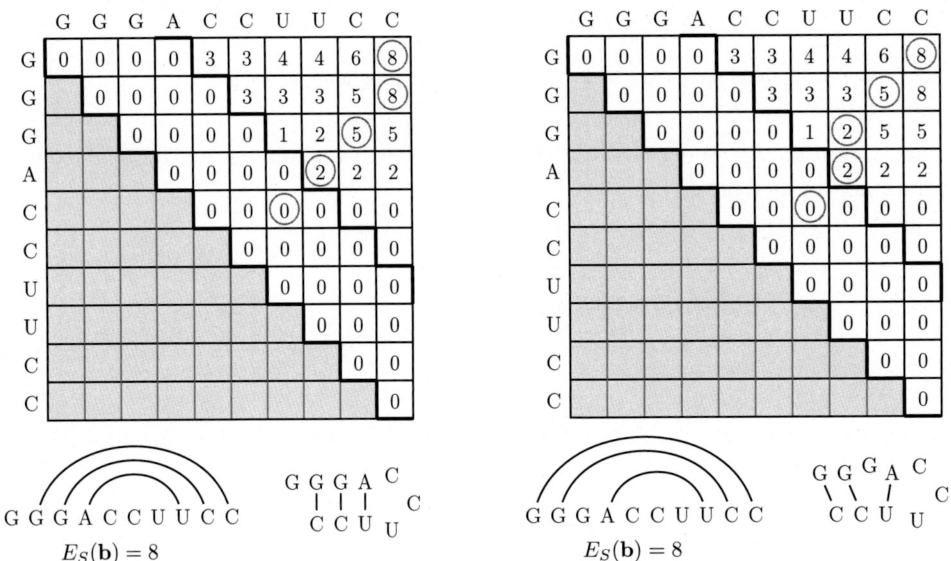

FIGURE 13.12 Two different tracebacks from the same sequence and the associated arc diagrams and secondary structures.

the second case ($E(i,j) = E(i+1,j)$), which means that b_{10} is not an unpaired base. However, $E(1,10)$ could have arisen from $E(2,10) = 8$; this means that b_1 is unpaired. Alternatively, $E(1,10) = 8$ could have arisen from $E(2,9) + e(\mathrm{G},\mathrm{C}) = 5 + 3$; this happens if the bases $b_1 = \mathrm{G}$ and $b_{10} = \mathrm{C}$ bond. The last case to check is that $E(1,10)$ is not equal to any of the sums $E(1,2) + E(3,10)$, $E(1,3) + E(4,10)$, ..., $E(1,8) + E(9,10)$. In this example, all of these sums are less than 8.

In summary, there are two possible values from which $E(1,10)$ could have arisen; we pick *either one of these* and circle it, and then repeat the process on that new value. Different choices will lead to different secondary structures that yield the same maximum value. When we circle an entry from an entire diagonal row of zeros in this traceback process, we may stop. Figure 13.12 shows two possible tracebacks for the DP algorithm to fold $\mathbf{b}$ = GGGACCUUCC. Note that we can recover the structure straight from the circled entries: each "jump" in the circled values signifies a base pair whose strength is equal to the difference in values.

Notice that while both structures in Figure 13.12 have the same total bond strength of $E_S(\mathbf{b}) = 8$, their loop structure is different. The first structure has a size-3 (the number of isolated bases) hairpin loop and a stack of size-3. From a thermodynamic viewpoint, this should have a lower free energy (be more stable) than the second structure, which has a destabilizing bulge loop. Two more foldings of $\mathbf{b}$ = GGGACCUUCC are shown in Figure 13.13. It is reasonable to conjecture that because they have a single hairpin loop and a size-3 stack and

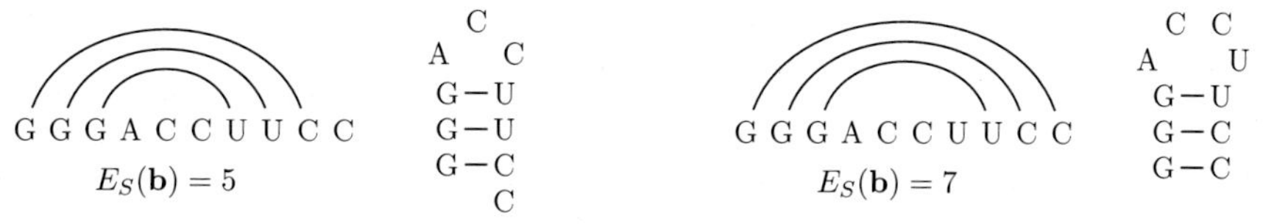

FIGURE 13.13 Two more foldings of the same RNA sequence of length 10. The sum $E_S(\mathbf{b})$ of the bond strengths as defined in Eq. (13.2), is given.

no destabilizing bulge loop, they might be more stable and hence more common than the structure shown on the right in Figure 13.12, despite having a lower bond strength sum. Though these are just toy examples, they illustrate the shortcomings of a DP folding algorithm that does not incorporate loop structure.

Exercise 13.6. Say that a secondary structure S of the RNA strand $\mathbf{b} =$ GGGACCUUCC is *saturated* if no more arcs can be added without introducing crossings. Find all possible foldings S of $\mathbf{b} =$ GGGACCUUCC that are saturated and have at least two base pairs, and compute $E_S(\mathbf{b})$ of each. □

Exercise 13.7. There is one more valid traceback for the table in Figure 13.12. Find it, and draw the corresponding RNA structure and arc diagram. □

Exercise 13.8. The optimal fold of the sequence $\mathbf{b} =$ GAAACAAAAU is a secondary structure with two nonnested arcs. Use a dynamic program to fill out the table, and then traceback. Find the arc diagram and RNA structure. How is the feature of two nonnested arcs reflected in the DP table? □

Exercise 13.9. The example from the text of folding $\mathbf{b} =$ GGGACCUUCC uses the requirement that the minimum arc length is $\lambda = 4$. Repeat this example, except allow arc lengths of size $\lambda = 3$. That is, fill out the table to find $E(1, 10)$ and then traceback to find all secondary structures that achieve this maximum. □

13.3.2 Minimum Free Energy Folding

In this section, we describe a secondary structure prediction algorithm that attempts to minimize the free energy of a folded structure. In this setting, the bond strengths are counted as a negative (stabilizing) energy score, and the loops contribute positive free energy because they are destabilizing. Calculating the free energy of loops is complicated because one has to consider not only the type of the k-loop (hairpin, bulge, interior, etc.) but also its size. To make matters worse, the actual bonds (e.g., whether they are GC, AU, or GU) that close off a loop can make a small difference in the energy score, and there are many other special cases to consider. All in all, this leads to a model with hundreds of parameters, many of which are experimentally determined. As a result, accurate free energy models are quite complicated; even the simplest cases require a computer. The mfold web server is one of the oldest web servers in computational molecular biology, dating back to 1995. It has since been replaced by the UNAFold web server (Unified Nucleic Acid Folding and hybridization package [7]). While the underlying DP algorithms are similar to what we have seen, there are many more details and cases in computing energy scores. These numbers are preprocessed using complex biophysical models, and then stored in text files that the UNAFold software calls while it runs. For example, there is a text file containing the destabilizing energies of 1-loops and 2-loops based on their sizes. A snippet of this is shown in Table 13.1. There are other files storing the (negative) energies of stacked base pairs, which, for the utmost accuracy, depend on the base pair under which they are stacked. Other text files contain energy values of specific interior loops, or contain energy "bonuses" or "penalties" to account for unusual cases, such as asymmetric interior loops, extremely large loops, or special hairpin loops, to name a few. We will not reveal all of these intricate details; the interested reader is encouraged to visit the UNAFold web server at http://mfold.rna.albany.edu, and/or consult the documentation and supporting articles [7, 9, 10]. Instead, we will summarize some of the basics to give the reader a flavor of

TABLE 13.1 Destabilizing Energies of Various Types of 1- and 2-Loops by Size Are Stored by UNAFold in "Loop Files"

Size	1	2	3	4	5	6	7	8	...	30
Internal	∞	∞	∞	1.7	1.8	2.0	2.2	2.3	...	3.7
Bulge	3.8	2.8	3.2	3.6	4.0	4.4	4.6	4.7	...	6.1
Hairpin	∞	∞	5.6	5.5	5.6	5.3	5.8	5.4	...	7.7

the model. Further details on all of these can be found in the book chapter *Algorithms and thermodynamics for RNA secondary structure prediction: a practical guide* by Zuker et al. [10].

An important assumption of the loop-based free energy model are that the free energies of the loops are independent. In other words, because an RNA strand can be decomposed into a disjoint union of loops, its free energy is a sum of the individual free energies of these loops. Notice that because stacked base pairs are simply 2-loops of size 0, such a model can simultaneously take into account both the stabilizing bonds (negative contributions) and the destabilizing loops (positive contributions).

Given a loop $\mathbf{L}$, let $l_s(\mathbf{L})$ and $l_d(\mathbf{L})$ denote the number of isolated bases and base pairs in $\mathbf{L}$, respectively. Recall that $l_s(\mathbf{L})$ is defined to be the size of $\mathbf{L}$, and that $\mathbf{L}_0$ is the unique 0-loop of $\mathbf{L}$, consisting of the bases and base pairs that are not "under" any other arcs. We assume that the energy contributions are additive, and thus

$$E_S(\mathbf{b}) = \sum_{\text{all loops } \mathbf{L}} e(\mathbf{L}) = e(\mathbf{L}_0) + \left(\sum_{(i,j)} e(L_{i,j}) \right), \tag{13.3}$$

where the lowercase e's stand for energy functions of specific loops. Next, we need a model for the free energy of a k-loop given its size and, if $k = 2$, its type (stacked pair, bulge, or interior). These values have been proposed by biochemists, and their derivation is irrelevant to the scope of this chapter. There are many details and special cases. We will only highlight a few main features, and briefly summarize the energy calculations of 1-loops (hairpins), 2-loops, and multiloops, respectively.

For example, in the theory of polymers, the free energy of a hairpin loop $\mathbf{L}$ is approximately

$$e(\mathbf{L}) = 1.75 \cdot R \cdot T \cdot \ln(l_s), \tag{13.4}$$

where T is temperature, and $R \approx 1.987\ \text{cal}^{-1}\ \text{mol}^{-1}\ \text{K}^{-1}$ is a universal gas constant. However, this is not entirely accurate for small hairpin loops (of size $l_s = 3$ or 4), or for very large loops (of size $l_s > 30$). Small tight loops take extra free energy to close off, but other than that, the free energy generally increases with loop size due to the energy cost of not having any of the bases in the loop bond to each other. Additionally, there are certain cases where the free energy of the hairpin depends on the specific composition of the base pair (i,j) that closes it off. Two examples of this are *GGG-loops*, defined by having $b_{i-2} = b_{i-1} = b_i = G$ and $b_j = U$; and the *poly-C loops*, defined by having all isolated bases $b_k = C$. These are shown in Figure 13.14. UNAFold stores the precomputed energy of hairpin loops in several text files. Most of these use Eq. (13.4), but there are also some minor penalties and bonuses to account for in special cases, such as being a GGG- or poly-C hairpin loop, or having size larger than 30.

The 2-loops of size 0 are precisely the stacked base pairs, and these loops make a *negative* free energy contribution. The free energy of a stacked base pair $(b_{i+1}, b_{j-1}) = (X, Y)$ depends not only on the six possible pairs, but also on the pairing (b_i, b_j) that is stacked above it. Figure 13.15 shows an example of the energy contribution of (X, Y) stacked below a $(b_i, b_j) = (\text{C}, \text{G})$ base pair, and the table of energy values stored by UNAFold. One can assume that any disallowed pairing has infinite free energy. Also, slight tweaking is done to account for whether the stacked base pair is in the middle or at the end of a stack.

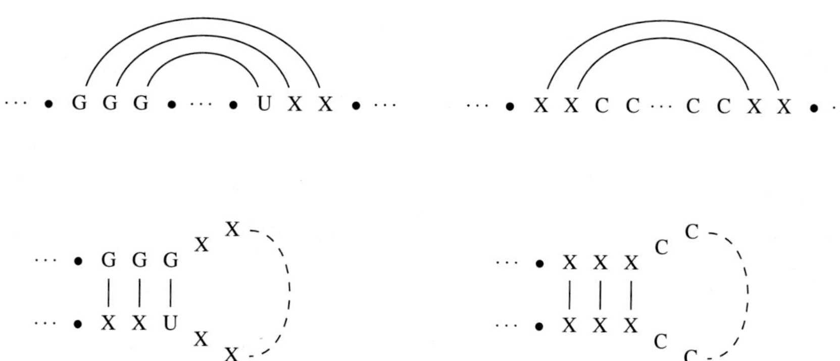

FIGURE 13.14 A GGG-loop (left) and a poly-C hairpin loop (right). The arc diagrams are shown at top and the secondary structure shown below.

X \ Y	A	C	G	U
A	∞	∞	∞	-2.1
C	∞	∞	-3.3	∞
G	∞	-2.4	∞	-1.4
U	-2.1	∞	-2.1	∞

FIGURE 13.15 The base pair $(b_{i+1}, b_{j-1}) = (X, Y)$ is a stacked base pair because (i,j) is also paired. UNAFold stores the energy of (X, Y) in a table in a text file. The value not only depends on what bond (X, Y) is, but also on the fact that it is stacked beneath a CG-bond.

The bulges and interior loops are the two types of destabilizing 2-loops. Except for very large $(n > 30)$ sizes, the energy values of a bulge loop are taken from a "loop file," as shown in Table 13.1. The energy values of interior loops are also taken from this file, but are given bonuses or penalties due to having mismatching base pairs that close off the loop, or having an asymmetric loop.

Finally, UNAFold needs to know the energy of multiloops. Though the thermodynamics of multiloops are poorly understood, biochemists have proposed the following model:

$$e(\mathbf{L}) = a + b \cdot l_s(\mathbf{L}) + c \cdot l_d(\mathbf{L}) + e(\mathbf{L}_{\text{stack}}),$$

where $e(\mathbf{L}_{\text{stack}})$ is an error term to account for stacking interactions, and a, b, and c are computationally determined constants.

13.4 STOCHASTIC FOLDING ALGORITHMS VIA LANGUAGE THEORY

13.4.1 Languages and Grammars

In formal language theory, a *language* is defined to be a set of finite strings constructed from an alphabet Σ of *terminal symbols*. A *grammar* is a collection of production rules that dictate how to change temporary *nonterminal symbols* into strings of symbols. It is conventional to use lowercase letters for terminal symbols, uppercase letters for nonterminals, and Greek letters for strings. One begins with a distinguished (nonterminal) *start symbol* S, and nonterminal symbols are repeatedly turned into strings until no nonterminals remain. The language L generated by such a grammar is the set of all strings over Σ that can be generated in a finite number of steps from the start symbol S. A *derivation* of a string $\alpha \in L$ is a sequence of rules that takes S to α.

Before we formally define a grammar, we will illustrate the concept by an example. Consider the alphabet of terminal symbols $\Sigma = \{a, b\}$ and nonterminal symbols $N = \{S, A\}$ with production rules

$$\begin{aligned} S &\longrightarrow aAa \\ A &\longrightarrow bbA \\ A &\longrightarrow bb. \end{aligned} \tag{13.5}$$

The first rule ensures that the final string starts and begins with the terminal symbol a. The next two rules say that every time a (nonterminal) A appears, it can be replaced with the string bbA or the string bb. This can be written in a more condensed form by $A \longrightarrow bbA|bb$. The following sequence of rules is a derivation of the string $abbbbbba$:

$$S \Longrightarrow aAa \Longrightarrow abbAa \Longrightarrow abbbbAa \Longrightarrow abbbbbba. \tag{13.6}$$

It is customary to use a double arrow to denote individual steps in a derivation. It is easy to see that the language generated by the grammar in Eq. (13.5) is precisely the set $L = \{ab^{2n}a \mid n \geq 0\}$. For another example, consider the grammar with $\Sigma = \{a, b\}$, $N = \{S\}$ and $S \longrightarrow aSb|ab$. This generates the language of strings $L = \{a^n b^n \mid n \geq 1\}$.

Now that we have some insight, the rigorous definition of a grammar should be better motivated. For more details, readers can consult one of the many textbooks on formal language theory, for example, [11]. A *formal grammar* G is a 4-tuple (Σ, N, S, P), where Σ is a finite alphabet of terminal symbols, N is a finite set of nonterminals containing a distinguished start symbol S, and P is a collection of production rules that dictate how to replace the nonterminal symbols with strings of symbols. The *Chomsky hierarchy* classifies grammars by the restrictions placed on their rules. For example, a grammar is *regular*, or "type 3," if all rules have the form $A \rightarrow a$ or $A \rightarrow aB$. A language is regular if it can be generated by a regular grammar. These are precisely the languages that can be generated by *finite state automata*, which are common structures in the field of computer science. More generally, if the left-hand side of each production rule consists of a single nonterminal symbol and nothing else (no restrictions on the right-hand side), the grammar is *context-free*, or "type 2." The resulting languages are those that can be generated by *push-down automata*. These are the types of grammars that are primarily used for RNA folding algorithms. An example of a noncontext-free language is $\{0^n 1^n 2^n \mid n \geq 1\}$. There are also more general "type 1" and "type 0" grammars, the latter consisting of the class with no restrictions on the rules. The resulting languages are precisely those that can be generated by a Turing machine [11].

In a context-free grammar (CFG), every derivation of a string can be represented using a *parse tree*. This is a tree whose leaves are all terminal symbols, whose interior nodes are all nonterminals, and that is rooted at the start symbol S. The children of an internal node are the productions of that nonterminal node, in left-to-right order. Each level describes an application of a rule in the derivation. This is best seen by an example; consider the derivation of $\alpha = abbbbbba$ in Eq. (13.6). The parse tree for α is shown in Figure 13.16. Notice how one can determine the string α from its parse tree—start at the root S, and the "walk around" the tree in a counterclockwise direction. The order in which the leaves are reached is precisely the string α. The grammar defined in Eq. (13.5) has the special property that every string in the language L has a unique derivation, and hence a unique parse tree. We will not prove this, but the reader should take a moment to verify that indeed, there is only one way to generate the string $ab^{2n}a$. This grammar is said to be *unambiguous*. We will wait to formally define what it means for a grammar to be ambiguous and unambiguous, as there are some intricacies that arise in more complicated examples.

Exercise 13.10. Find a derivation for the string $\alpha = aaaabbbb$ from the grammar $S \rightarrow aSb|ab$ and draw the parse tree. Is this grammar unambiguous? □

Exercise 13.11. Construct a regular grammar that generates the language $\{b^n a \mid n \geq 0\}$. □

Exercise 13.12. Construct a regular grammar that generates the language $\{ab^n a \mid n \geq 0\}$. Try to construct a regular grammar that generates the language $\{a^n b^n \mid n \geq 0\}$. What goes wrong? □

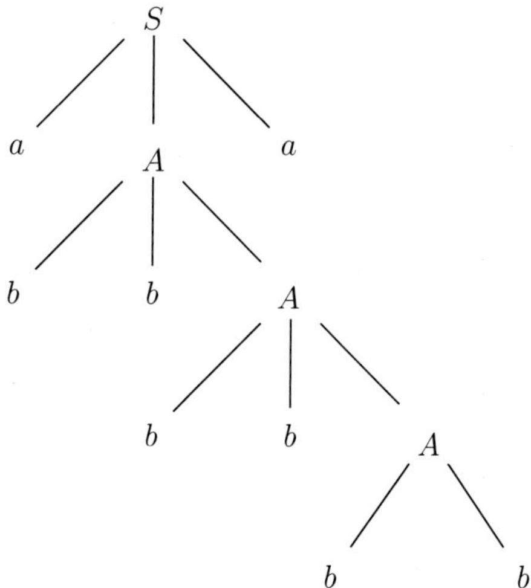

FIGURE 13.16 Parse tree for the derivation of the string $\alpha = abbbbbba$ from the context-free grammar $S \to aAa, A \to bbA|bb$. Notice that α can be read off from the tree by starting at S and "walking around" the tree in a counterclockwise order.

13.4.2 The Knudsen-Hein Grammar for RNA Secondary Structures

In his famous 1859 book *On the origin of species* [12], Charles Darwin wrote that "the formation of different languages and of distinct species, and the proofs that both have been developed through a gradual process, are curiously parallel." Little did Darwin know that decades later, scientists would discover a macromolecule called DNA that encoded genetic instructions for life in a mysterious language over the alphabet $\Sigma = \{a, c, g, t\}$. Though this would eventually lead to the fields of molecular biology and linguistics becoming intertwined, major developments were needed in both fields before this could happen.

Noam Chomsky is widely considered to be the father of modern linguistics. In the 1950s, he helped popularize the universal grammar theory, which asserts that there are structural grammatical similarities and rules underlying all languages, and these are instinctual to human nature. Formal grammars are simply mathematical abstractions of this idea, and Chomsky's work led to a more rigorous mathematical treatment of formal languages, revolutionizing the field of linguistics. Also in the 1950s, the structure of DNA, the newly discovered fundamental building block of life, was finally understood. The use of formal language theory to study molecular biology began in the 1980s [13, 14]. David Searls' wonderful article titled "The Language of Genes," published in *Nature* in 2002, gives a nice overview of how the fields of linguistics and molecular biology became intertwined [15].

The earliest work of linguistics in biology involved using regular grammars to model biological sequences. Assigning probabilities to the production rules yields *hidden Markov models* (HMMs), and these have been widely used in sequence analysis. For example, Chapter 9 of Robeva and Hodge [16], titled *Identifying CpG islands: sliding windows and hidden Markov models*, describes how this framework can be used to identity regions in a genome with a disproportionally high percentage of adjacent C and G bases. These areas, called CpG islands, are found near the beginning of many genes in humans and other mammals and can be used to help identify potential genes in DNA sequence data.

One restriction of regular grammars is that the location of terminal symbols in the final string is uncorrelated. However, the location of bases in RNA strands are not uncorrelated, so a larger class of grammars needs to be used to account for this. This requirement leads to the use of CFGs to construct biological sequences. Assigning probabilities to the production rules defines SCFGs. Naturally, for each nonterminal symbol A, the sum $\sum_\alpha P(A \to \alpha)$ of the probabilities of all rules with A on the left-hand side must be 1.

A number of SCFGs have been proposed for the RNA secondary structure prediction problem, ranging from the simple to the very detailed. Though one would expect a trade-off between simplicity and accuracy, a very simple grammar proposed in 1999 by Knudsen and Hein [17] has been shown to perform just as well as or better than other more complicated grammars [18]. The *Knudsen-Hein grammar* has been implemented into the secondary structure prediction software Pfold [19], and their prediction algorithms have comparable performance to the energy minimization algorithms described in the previous section, though both have individual strengths and weaknesses.

The Knudsen-Hein grammar consists of the nonterminal symbols $\{S, L, F\}$ and terminal symbols $\{d, d', s\}$. The s denotes an isolated base, and (d, d') denotes a base pair. *In this model, we do not distinguish between different chemical bases*; we will see how to account for this in Section 13.4.3. The production rules of the Knudsen-Hein grammar are

$$\begin{array}{llll} S \longrightarrow LS & \text{with probability } p_1, & \text{or } L & \text{with probability } q_1 \\ L \longrightarrow dFd' & \text{with probability } p_2, & \text{or } s & \text{with probability } q_2 \\ F \longrightarrow dFd' & \text{with probability } p_3, & \text{or } LS & \text{with probability } q_3. \end{array}$$

The start symbol S produces loops and the nonterminal symbol F makes stacks. The rule $F \to LS$ ensures that each hairpin loop has a size of at least 3. If one wanted hairpin loops to have a size of at least 4, as was considered in the previous section, this would have to be changed to $F \to LLS$.

Because the p_is and q_is are probabilities, they must satisfy $p_i + q_i = 1$. Specific values of these probability parameters were estimated in [17] by analyzing the statistical properties of actual secondary structures such as the distribution of lengths of loops, stems, loop types, and so on. This data was used as a "training set" for a more sophisticated algorithm designed to estimate such unknown parameters in stochastic grammars [20]. We will return to this problem in the next section.

For an explicit example of the Knudsen-Hein grammar in action, consider the sequence $\mathbf{b} =$ GGACUGC, which can fold into seven secondary structures, if one allows loop sizes of minimum length 3. One of these is the trivial structure S_0 with no base pairs, and the other six are shown in Figure 13.17. Below each structure S_i is the probability $P(S_i)$ of its derivation using the Knudsen-Hein grammar. Because each step is assumed to be independent, $P(S_i)$ is just the product of the probabilities of each step. Figure 13.18 shows a step-by-step derivation of the structure S_1, if one ignores the actual bases. The nonterminal symbol highlighted in boldface is the one being changed in the next step. Arcs are drawn to show the base pairs. This is not necessary, but we include them because it helps illustrate the process.

Unlike the other CFGs we have seen, the Knudsen-Hein grammar has rules with multiple nonterminals on the right-hand side. This means that at times, one will have a string containing multiple nonterminals, and there will be a choice of which nonterminal to convert into a string first. The question arises of whether order matters, and the answer depends on what one means by "matters." For example, if a string λ contains nonterminals A and B, the context-free property implies that the replacements $A \to \alpha$ and $B \to \beta$ done in either order yield the same string. On the other hand, changing the order by definition changes the derivation, and also changes the parse tree. Therefore, it is desirable to establish a convention for the order in which nonterminals will be replaced. A canonical way to do this is to replace remaining nonterminals in left-to-right order. The resulting derivation

GGACUGC GGACUGC GGACUGC GGACUGC GGACUGC GGACUGC

$P(S_1) = p_2^2 q_1^3 q_2^3 q_3^2$ $P(S_2) = p_1^3 p_2 q_1^2 q_2^5 q_3$ $P(S_3) = p_1^3 p_2 q_1^2 q_2^5 q_3$ $P(S_4) = p_1^3 p_2 q_1^2 q_2^5 q_3$ $P(S_5) = p_1^3 p_2 q_1^2 q_2^5 q_3$ $P(S_6) = p_1^3 p_2 q_1^2 q_2^5 q_3$

FIGURE 13.17 The six legal folds of the sequence $\mathbf{b} =$ GGACUGC that have at least one base pair and the corresponding probability of being generated by the Knudsen-Hein grammar. The "unfolded" secondary structure has probability $P(S_0) = p_1^6 q_1 q_2^7$.

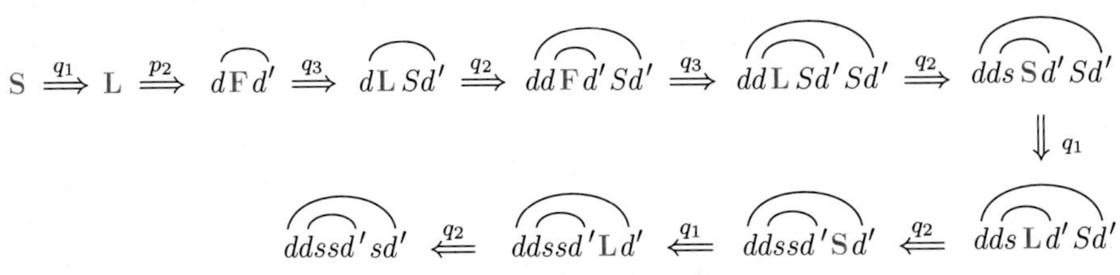

$$S \xrightarrow{q_1} L \xrightarrow{p_2} d\mathbf{F}d' \xrightarrow{q_3} d\mathbf{L}Sd' \xrightarrow{q_2} dd\mathbf{F}d'Sd' \xrightarrow{q_3} dd\mathbf{L}Sd'Sd' \xrightarrow{q_2} dds\mathbf{S}d'Sd'$$

$$\Downarrow q_1$$

$$ddssd'sd' \xleftarrow{q_2} ddssd'\mathbf{L}d' \xleftarrow{q_1} ddssd'\mathbf{S}d' \xleftarrow{q_2} dds\mathbf{L}d'Sd'$$

FIGURE 13.18 Example derivation using the Knudsen-Hein grammar.

$S \Rightarrow SS \Rightarrow aS \Rightarrow aSS \Rightarrow aaS \Rightarrow aaa \qquad S \Rightarrow SS \Rightarrow SSS \Rightarrow aSS \Rightarrow aaS \Rightarrow aaa$

FIGURE 13.19 Two left derivations, and the corresponding parse trees, for the string $\alpha = aaa$ derived from the ambiguous grammar $S \to SS \mid a$.

is called the *leftmost derivation*, and the corresponding parse tree is called the *left parse tree*. A grammar is *unambiguous* if every string in L has only one leftmost derivation, or equivalently, a unique left parse tree. For a simple example of an ambiguous grammar, consider the grammar defined by the rules $S \to SS \mid a$. Two derivations of the string $\alpha = aaa$, and the resulting parse trees, are shown in Figure 13.19.

For a general CFG grammar, there is no algorithm that can determine whether or not it is ambiguous—it is a computationally *undecidable* problem. However, it is known that the Knudsen-Hein grammar is unambiguous. This allows us to simplify certain derivations. For example, if a rule is applied repetitively, we can "collapse" these into a single step. This is illustrated by the following example of the generation of the structure S_2 from Figure 13.17.

$$\mathrm{S} \overset{q_1}{\Longrightarrow} \mathrm{L} \overset{p_2}{\Longrightarrow} \overset{\frown}{d\mathrm{F}d'} \overset{q_3}{\Longrightarrow} \overset{\frown}{dL\mathrm{S}d'} \overset{p_1^3}{\Longrightarrow} \overset{\frown}{dLLLL\mathrm{S}d'} \overset{q_1}{\Longrightarrow} \overset{\frown}{d\mathrm{LLLLL}\,d'} \overset{q_2^5}{\Longrightarrow} \overset{\frown}{dsssssd'}$$

At this point, the Knudsen-Hein grammar generates RNA secondary structures only—the actual bases are not specified. From here, one could assign bases randomly, via some distribution. One natural way to do this would be to look at the actual distribution of bases and base pairs from real data, and assign these to the s and (d, d') pairs accordingly. In the next section, we will revisit the goal of using the SCFG framework for the secondary structure prediction problem.

Exercise 13.13. Use the Knudsen-Hein grammar to construct the next three secondary structures, S_3, S_4, and S_5, from Figure 13.17. □

Exercise 13.14. Use the Knudsen-Hein grammar to construct a derivation of the hairpin loop $ssddsssd'd'ss$, and compute its probability. □

Exercise 13.15. Modify the rules to make the minimum loop size $j - i \geq 4$, and repeat the previous exercise. □

Exercise 13.16. Consider the following stochastic CFG for RNA folding:

$$S \longrightarrow sS(p_1) \mid dSd'S(p_2) \mid \varepsilon(p_3),$$

where ε is the empty string. This grammar was proposed by Ivo Hofacker but unpublished, and it was one of the eight grammars compared to the Knudsen-Hein grammar in [18]. What types of structures does this grammar produce? What are its weaknesses? □

Exercise 13.17. The following grammar proposed in [18] is ambiguous:

$$S \longrightarrow dSd' \mid dS \mid Sd \mid SS \mid \varepsilon.$$

Find a secondary structure that has multiple left parse trees. □

Exercise 13.18. Consider the following "mystery grammar" from [21]:

$$\begin{aligned} S &\longrightarrow aAu \mid cAg \mid gAc \mid uAa \\ A &\longrightarrow aBu \mid cBg \mid gBc \mid uBa \\ B &\longrightarrow aCu \mid cCg \mid gCc \mid uCa \\ C &\longrightarrow gaaa \mid gcaa. \end{aligned}$$

What is the language L derived from this grammar? Describe it in terms of RNA secondary structures. □

Exercise 13.19. Draw the left parse tree of the string $ddssd'sd'$ in the Knudsen-Hein grammar. The left derivation of this string is shown in Figure 13.18. □

Exercise 13.20. Compute the right derivation of the string $ddssd'sd'$ and draw its right parse tree. □

13.4.3 Secondary Structure Prediction Using SCFGs

Thus far, we have seen how CFGs and SCFGs can create different secondary structures, but that is far from being able to predict the most common structure given a primary sequence **b**. One simple solution is to pick the most likely derivation. However, this assumes that the probability parameters are known, and even then, just finding the most likely derivation could be a computationally challenging problem. The former problem can be done by the Cocke-Younger-Kasami (CYK) algorithm [22] or the inside-outside algorithm [23], and then the most likely derivation can be found by DP. Both of these problems have hidden Markov model (HMM) analogs, which are described in Chapter 9 of Robeva and Hodge [16]. The HMM version of the CYK algorithm is called the Viterbi algorithm, and the inside-outside algorithm is analogous to the *forward-backward* algorithm used for the HMM posterior decoding problem.

If one knows all possible secondary structures $\{S_i : i \in I\}$ that a fixed **b** can fold into, and its associated probability distribution, then the most likely derivation is the one with the maximal probability. In formal mathematical terms, this is *the probability of getting S_i, conditional on the sequence being* **b**, and is denoted $P(S_i \mid \mathbf{b})$. In our toy example from the previous section, this is

$$P(S_i \mid \mathbf{b} = \text{GGACUGC}) = \frac{P(S_i)}{P(S_0)+P(S_1)+P(S_2)+P(S_3)+P(S_4)+P(S_5)+P(S_6)}.$$

The solution to the prediction problem is simply the structure S_i such that $P(S_i \mid \mathbf{b})$, or equivalently $P(S_i)$, is maximized. Though this maximum could be achieved for multiple structures, especially in the small toy examples considered in this chapter, it is unlikely that this would happen for real RNA strands.

One remaining problem is how to find this structure because $P(S_i)$ highly depends on the actual values of the probabilities p_i and q_i, which, *a priori*, are not known. The set of all feasible probabilities can be thought of as a vector $\mathbf{p} = (p_1, q_1, p_2, q_2, p_3, q_3)$ in the unit 6-cube, $[0, 1]^6$. Every such vector that lies on the subset (a three-dimensional affine space) is defined by the relation $p_i + q_i = 1$. The values of $P(S_i)$ for some sample probability vectors are shown in Table 13.2. Notice how sensitive the "most likely structure" is to the choice of probabilities, highlighting the importance of being able to accurately determine them. The last line shows the best known values of the probability parameters, as estimated in [17]. Estimating these probability parameters is a problem from the rich field of *algebraic statistics*; the interested reader is encourage to check out *Algebraic Statistics for Computational Biology*, by Pachter and Sturmfels [31].

To summarize, we are now in a "chicken or egg" dilemma—we need the values of the probability parameters p_i and q_i to compute the conditional probabilities, and thereby predict the most likely secondary structure. However, it is difficult to determine the probability parameters without knowing the conditional probabilities of the secondary structures. The solution to this problem is to estimate the probability parameters using actual data. Specifically, given a fixed RNA sequence **b** where the distribution of secondary structures is known, there are "training algorithms" that can determine what values of the p_is and q_is make the conditional probabilities $P(S_i \mid \mathbf{b})$ closest to the observed distribution. One such method is called the *inside-outside algorithm* [23], which

TABLE 13.2 Probabilities of Each Secondary Structure Conditioned on the Fixed Sequence b = GGACUGC

(p_1, q_1)	(p_2, q_2)	(p_3, q_3)	$P(S_0 \mid b)$	$P(S_1 \mid b)$	$P(S_2 \mid b)$
(0.45, 0.55)	(0.5, 0.5)	(0.5, 0.5)	0.01479	0.53883	0.08927
(0.5, 0.5)	(0.5, 0.5)	(0.5, 0.5)	0.02703	0.43243	0.10811
(0.25, 0.75)	(0.75, 0.25)	(0.25, 0.75)	5.30×10^{-6}	0.98856	0.00229
(0.75, 0.25)	(0.25, 0.75)	(0.75, 0.25)	0.74988	0.00325	0.04937
(0.75, 0.25)	(0.75, 0.25)	(0.75, 0.25)	0.07663	0.24219	0.13623
(0.869, 0.131)	(0.788, 0.212)	(0.895, 0.105)	0.33641	0.04543	0.12363

Note: Because $P(S_2 \mid \mathbf{b}) = P(S_3 \mid \mathbf{b}) = P(S_4 \mid \mathbf{b}) = P(S_5 \mid \mathbf{b}) = P(S_6 \mid \mathbf{b})$, only one of these is listed.

is analogous to the *forward-backward* algorithm used for posterior decoding of HMMs, described in Chapter 9 of Robeva and Hodge [16]. The method that is then used for predicting the structure is called the CYK algorithm [22], which employs DP. Specifically, DP is used to find the probability of the most likely structure, and then the traceback step determines which structure achieves that maximum. Neither of these will be described here, as those techniques are quite complicated and specialized, and this chapter is intended to be an introduction to RNA folding and an overview of the main problems and methods.

13.4.4 Summary

Modeling RNA sequences using SCFGs has a number of advantages to the energy minimization approach. For one, the algorithm produces a number of structures and specifies the probability distribution. A second advantage is that it involves less prior knowledge. Some minimal energy DP algorithms involve hundreds of parameters, most of which are experimentally determined.

Additionally, researchers have used combinatorial analysis techniques to determine the average number of structural motifs (such as base pairs, stacks, loops) as a function of the sequence length n. Poznanović and Heitsch recently showed that, independent of the choice of probability parameters, the expected number of each type of loop obeys a normal distribution, that is, a standard bell curve [24]. They have also discovered other relationships between the frequencies of these motifs. For example, the expected number of stacks is four times as large as the expected number of multiloops.

One limitation of the SCFG framework is that it does not handle folding RNA sequences into pseudoknots, because pseudoknots cannot be recursively generated as secondary structures can. Though pseudoknots could be modeled by the larger class of "type 1" (context-sensitive) grammars in the Chomsky hierarchy, computational problems within this class quickly become NP (nondeterministic polynomial time)-complete. Eddy and Rivas have proposed a grammar for folding RNA sequences that includes pseudoknots under restrictions of the complexity of the pseudoknot [25]. In a separate paper, they develop a DP algorithm for folding RNA sequences into 3-noncrossing pseudoknots [26].

13.5 PSEUDOKNOTS

A folded RNA structure is a pseudoknot if its arc diagram has crossings. These are not actual topological knots, but their folded structures are tangled and thus resemble knots–hence their name. Figure 13.20 shows an pseudoknot from RNase P class B found in *Bacillus subtilis*. Pseudoknots were not even discovered until 1982 [27], when they were observed in the RNA of the turnip yellow mosaic virus. In [27], Rietveld et al. used chemical and enzymatic methods to deduce the existence of the base pairing responsible for the pseudoknot structure. Recognizing the existence of base pairs of any type—in classical stems or in pseudoknots—can be

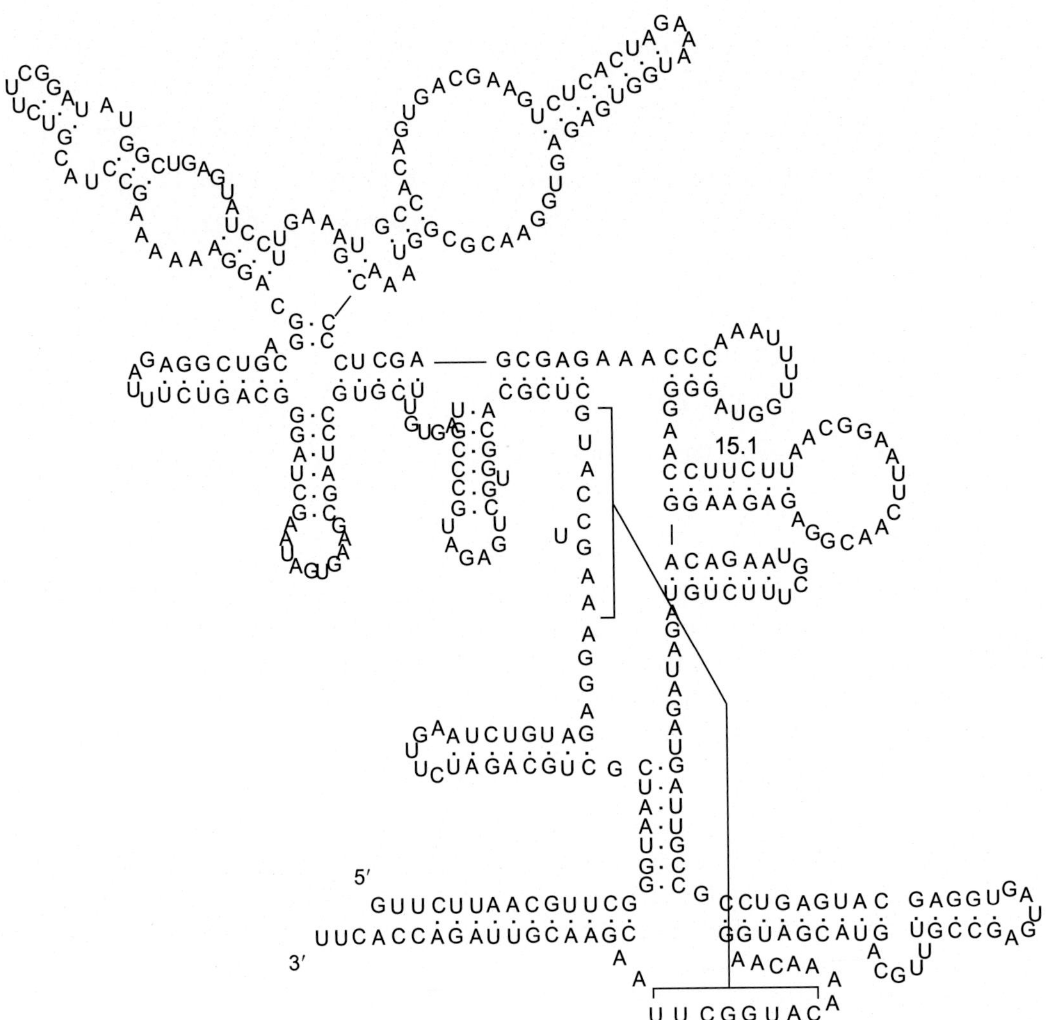

FIGURE 13.20 An RNA pseudoknot structure from RNase P class B found in *Bacillus subtilis*. The secondary structure is drawn in the plane without crossings, and the pseudoknotting is shown by the brackets: the subsequence CAUGGCUU on the multiloop at the bottom (this appears "reversed" because it is read from the 5′-to-3′ direction) bonds with the subsequence AAG[U]CCAUG on another multiloop. The [U] denotes an unpaired base in a size-1 bulge loop. Image from http://en.wikipedia.org/wiki/RNase_P.

facilitated by observing covariation of nucleotide pairs in related RNAs. For example, if one compares two homologous RNAs from different species, if it is observed that a change in a nucleotide from A to C has a strong correlation to a different nucleotide changing from U to G, then that provides evidence of a base pair in these two sites. Because the overall shape of the RNA is the determinant of its function, the identity of the pairing nucleotides is much less important than the fact that a base pair exists. Although we will not pursue it further in this chapter, covariation has been used to identify not only pseudoknots but other conserved RNA structures [28].

Recall from Section 13.2 that an arc diagram is k-noncrossing if there is no subset of k arcs that are mutually crossing. Natural RNA pseudoknots tend to have low crossing numbers [29], and most that have been observed are 3-noncrossing. These are precisely those arc diagrams that can be drawn as noncrossing if one allows arcs to be drawn upside-down (i.e., in the lower half plane).

The structure of pseudoknots can be simplified by reducing them to simpler arc diagrams called "cores" and "shapes." The *core* of an arc diagram is formed by collapsing every stack into a single arc, and then relabeling the vertices accordingly. This core has the same crossing number and minimum arc length as the original structure,

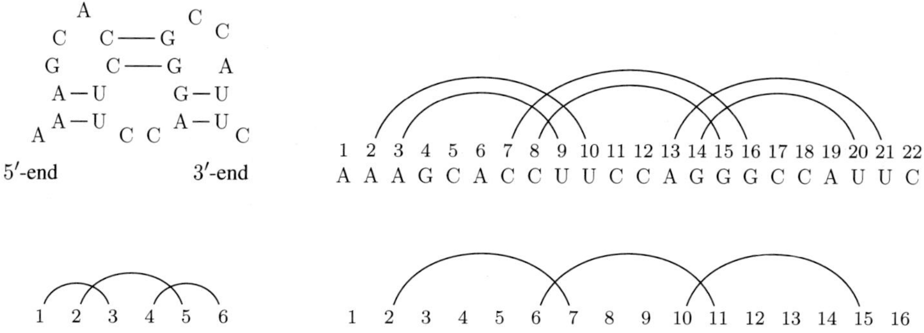

FIGURE 13.21 A pseudoknot arises from two "kissing hairpins." Its arc diagram is shown to the right. Directly below it is its *core*, which is formed by collapsing every stack to a single arc. To the left of the core is its *shape*, which is formed by removing isolated vertices, and then repeating the collapsing process if necessary.

and excluding stacked pairs, it preserves the loop decomposition. An example of this is shown for the "kissing hairpin" pseudoknot shown in Figure 13.21. The core is the structure directly below the arc diagram.

Every arc diagram also has a well-defined *shape*, which is a complete matching that reflects the pseudoknotting structure of the original RNA structure. By definition, a shape is a complete matching with maximum stack size 1. This can be constructed from the core by first removing all isolated vertices. If this process introduces stacks of size $\sigma \geq 2$, then they are iteratively collapsed until one ends up with an arc diagram where every stack has size 1. In the arc diagram in Figure 13.21, removing the isolated vertices from the core yields the shape, which is the diagram in the bottom left. A more complicated example is shown in Figure 13.22, where removing isolated vertices introduces a stack of size 2, and so the shape is found by collapsing this stack into a single arc.

Creating the core and shape of an arc diagram not only helps us better visualize and classify the pseudoknotting structure, but it is necessary for certain algorithms. For example, as with secondary structures, every pseudoknot structure has a unique loop decomposition. Each vertex can be assigned in a well-defined manner to a unique 0-loop, a 1-loop, 2-loop, multiloop, or pseudoknot. To motivate the nontriviality of this, consider the arc diagram shown in Figure 13.22. It should be clear that a loop decomposition should contain three stacked pairs, an interior loop, and a pseudoknot. However, it is not clear exactly which vertices should belong to the pseudoknot. Due to space limitations, we will not describe the loop decomposition algorithm here, but refer the interested reader to Chapter 6.2 in [30].

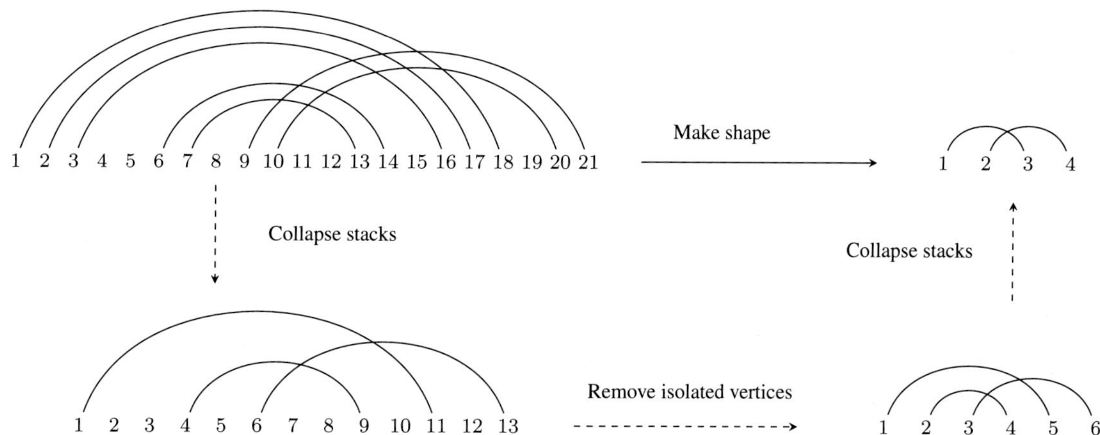

FIGURE 13.22 Given the arc diagram in the upper left, its *core* is formed by collapsing all stacks into single arcs. To find its *shape*, all isolated vertices are removed. However, this introduces new stacks of size greater than 1, so this process must be repeated. The shape is the diagram in the upper right.

As with secondary structures, one main motivation for establishing a well-defined loop decomposition is for using it in other mathematical endeavors, such as folding algorithms. The same is true for pseudoknots. Shapes and cores provide a foundation for combinatorial analysis techniques to enumerate the number of RNA pseudoknots and study their asymptotics. The loop decomposition also plays a role in pseudoknot folding algorithms. Though these are much more limited than those for secondary structures, they do exist, sometimes given extra restrictions. For example, Rivas and Eddy developed an algorithm for folding RNA into pseudoknots whose arc diagrams are 3-noncrossing [25]. Their algorithm is essentially a souped-up version of the DP algorithm from Section 13.3, except there are more cases to handle. Though pseudoknots with arbitrary crossings cannot be generated recursively, 3-noncrossing diagrams can be drawn without crossings by using arcs in the lower half plane, and this provides a framework for all possible recursions that need to be considered. Details can be found in their original paper [25], or in Chapter 6 of Reidy's book on RNA pseudoknots [30]. Another new approach to studying pseudoknots is to view them topologically as so-called "fat graphs." This is the topic of the next and final chapter of this book.

Exercise 13.21. Draw the core and the shape of the pseudoknot shown in Figure 13.5. □

Exercise 13.22. Draw the core and the shape of the pseudoknot from Exercise 13.1. □

Exercise 13.23. Find all shapes on exactly six nodes. That is, find all noncrossing matchings on six nodes with maximum stack length equal to 1. □

ACKNOWLEDGMENTS

The authors thank Svetlana Poznanović for her feedback and advice on the content of this chapter. They also thank Catherine Gurri and Brian Loftus for their help with making some of the figures. The first and second authors were supported by National Science Foundation Grant DMS-#1211691.

REFERENCES

[1] Darnell JE. RNA: life's indispensable molecule. New York: Cold Spring Harbor Press; 2011.
[2] Berman HM, Olson WK, Beveridge DL, Westbrook J, Gelbin A, Demeny T, et al. The nucleic acid database. A comprehensive relational database of three-dimensional structures of nucleic acids. Biophys J 1992;63:751-9.
[3] Narayanan B, Westbrook J, Ghosh S, Petrov AI, Sweeney B, Zirbel CL, et al. The Nucleic Acid Database: new features and capabilities. Nucleic Acids Res 2013;42:D114-22.
[4] Comarmond MB, Giege R, Thierry JC, Moras D, Fischer J. Three-dimensional structure of yeast T-RNA-ASP. I. Structure determination. Acta Crystallogr Sect B 1986;42:272-80.
[5] Waterman MS. Secondary structure of single-stranded nucleic acids. Adv Math Suppl Stud 1978;1:167-212.
[6] Sankoff D, Kruskal JB, Mainville S, Cedergren RJ. Fast algorithms to determine RNA secondary structures containing multiple loops. In: Time warps, string edits, and macromolecules: the theory and practice of sequence comparison. Reading, MA: Addison-Wesley; 1983. p. 93-120.
[7] Markham NR, Zuker M. UNAFold: software for nucleic acid folding and hybridization. In: Bioinformatics, vol. II. Structure, function and applications, vol. 453. Totowa: Springer; 2008. p. 3-31.
[8] Nussinov R, Jacobson AB. Fast algorithm for predicting the secondary structure of single-stranded RNA. Proc Natl Acad Sci USA 1980;77:6309-13.
[9] Markham NR, Zuker M. DINAMelt web server for nucleic acid melting prediction. Nucleic Acids Res 2005;33:W577-81.
[10] Zuker M, Mathews DH, Turner DH. Algorithms and thermodynamics for RNA secondary structure prediction: a practical guide. In: RNA biochemistry and biotechnology. Dordrecht: Springer; 1999. p. 11-43.
[11] Linz P. An introduction to formal languages and automata. Sudbury, MA: Jones & Bartlett Publishers; 2011.
[12] Darwin C. On the origin of species by means of natural selection, or the preservation of favoured races in the struggle for life. London: John Murray; 1859.
[13] Brendel V, Busse HG. Genome structure described by formal languages. Nucleic Acids Res 1984;12:2561-8.

[14] Head T. Formal language theory and DNA: an analysis of the generative capacity of specificrecombinant behaviors. Bull Math Biol 1987;49:737-59.
[15] Searls DB. The language of genes. Nature 2002;420:211-7.
[16] Robeva R, Hodge T. Mathematical concepts and methods in modern biology: using modern discrete models. London: Academic Press; 2013.
[17] Knudsen B, Hein J. RNA secondary structure prediction using stochastic context-free grammars and evolutionary history. Bioinformatics 1999;15:446-54.
[18] Dowell RD, Eddy SR. Evaluation of several lightweight stochastic context-free grammars for RNA secondary structure prediction. BMC Bioinformatics 2004;5:71.
[19] Knudsen B, Hein J. Pfold: RNA secondary structure prediction using stochastic context-free grammars. Nucleic Acids Res 2003;31: 3423-8.
[20] Baker JK. Trainable grammars for speech recognition. J Acoust Soc Am 1979;65:132.
[21] Durbin R, Eddy SR, Krogh A, Mitchison G. Biological sequence analysis: probabilistic models of proteins and nucleic acids. Cambridge: Cambridge University Press; 1998.
[22] Eddy SR. A memory-efficient dynamic programming algorithm for optimal alignment of a sequence to an RNA secondary structure. BMC Bioinformatics 2002;3:18.
[23] Lari K, Young SJ. The estimation of stochastic context-free grammars using the inside-outside algorithm. Comput Speech Lang 1990;4:35-56.
[24] Heitsch C, Poznanović S. Asymptotic distribution of motifs in a stochastic context-free grammar model of RNA folding. J Math Biol 2014;69(6-7):1743-72.
[25] Rivas E, Eddy SR. A dynamic programming algorithm for RNA structure prediction including pseudoknots. J Mol Biol 1999;285: 2053-68.
[26] Rivas E, Eddy SR. The language of RNA: a formal grammar that includes pseudoknots. Bioinformatics 2000;16:326-33.
[27] Rietveld K, Van Poelgeest R, Pleij CW, Van Boom JH, Bosch L. The tRNA-like structure at the 3′ terminus of turnip yellow mosaic virus RNA. Differences and similarities with canonical tRNA. Nucleic Acids Res 1982;10:1929-46.
[28] Parsch J, Braverman JM, Stephan W. Comparative sequence analysis and patterns of covariation in RNA secondary structures. Genetics 2000;154:909-21.
[29] Haslinger C, Stadler PF. RNA structures with pseudo-knots. Bull Math Biol 1999;61:437-67.
[30] Reidys C. Combinatorial computational biology of RNA: pseudoknots and neutral networks. New York: Springer; 2010.
[31] Pachter L, Sturmfels B. Algebraic statistics for computational biology. Vol. 13. Cambridge University Press, 2005.

Chapter 14

RNA Secondary Structures: An Approach Through Pseudoknots and Fatgraphs

Christian M. Reidys

Department of Mathematics and Computer Science, University of Southern Denmark, Odense M, Denmark

14.1 INTRODUCTION

Almost three decades ago, Michael Waterman pioneered the combinatorics and prediction of the ribonucleic acid (RNA) secondary structures. An RNA molecule is described by its primary sequence, a linear string composed by the four nucleotides A, G, U, and C. RNA is structurally less constrained than its chemical relative DNA, and folds into tertiary structures. RNA plays a central role within living cells, facilitating a whole variety of biochemical tasks, all of which are closely connected to its tertiary structure.

RNA acts as a messenger, linking DNA and proteins, and catalyzes reactions just as proteins do. The global conformation of RNA molecules is to a large extent determined by topological constraints encoded at the level of secondary structure, that is, by the mutual arrangements of the base-paired helices [1]. Each nucleotide can interact (base pair) with at most one other nucleotide by means of specific hydrogen bonds. Only the Watson-Crick pairs GC and AU, as well as the wobble GU, are admissible. These base pairs determine the secondary structure. Note that we have neglected here base triples and other types of more complex interactions. Secondary structures can thus be represented as graphs, where nucleotides are represented by vertices, the backbone of the molecule, and the hydrogen bonds are represented by edges; see Figure 14.1a. More conveniently, we use the convention to represent the backbone of the polymer by a horizontally drawn chain. As before, this chain consists of vertices and arcs, respectively, representing the nucleotides and covalent bonds. However, the edges representing the base pairs now are depicted as arcs in the upper half plane; see Figure 14.1b. We call this representation the *diagram* of the molecule.

Although the vast majority of RNAs have simple, that is, pseudoknot-free conformations, `PseudoBase` [2] lists more than 250 records of pseudoknots determined by a variety of experimental and computational techniques, including crystallography, nuclear magnetic resonance (NMR), mutational experiments, and comparative sequence analysis. In many cases, they are crucial for molecular function. Examples include the catalytic cores of several ribozymes [3], programmed frameshifting [4], and telomerase activity [5], reviewed in [6, 7].

Pseudoknots have long been known as important structural elements in RNA [8]. These cross-serial interactions between RNA nucleotides are functionally important in tRNAs, RNaseP [9], telomerase RNA [6], and ribosomal RNAs [10]. Pseudoknots in plant virus RNAs mimic tRNA structures, and *in vitro* selection experiments have produced pseudoknotted RNA families that bind to the HIV-1 reverse transcriptase [11].

In [12], a folding algorithm, `gfold`, for one such class of RNA structures has been presented. Such an algorithm has an RNA sequence of fixed length as an input and outputs the ensemble of energy-weighted potential structures. This ensemble is oftentimes referred to as the partition function. The class of structures realized by `gfold` consists of structures composed by certain building blocks, called shapes, of fixed topological genus. In

Algebraic and Discrete Mathematical Methods for Modern Biology. http://dx.doi.org/10.1016/B978-0-12-801213-0.00014-9
Copyright © 2015 Elsevier Inc. All rights reserved.

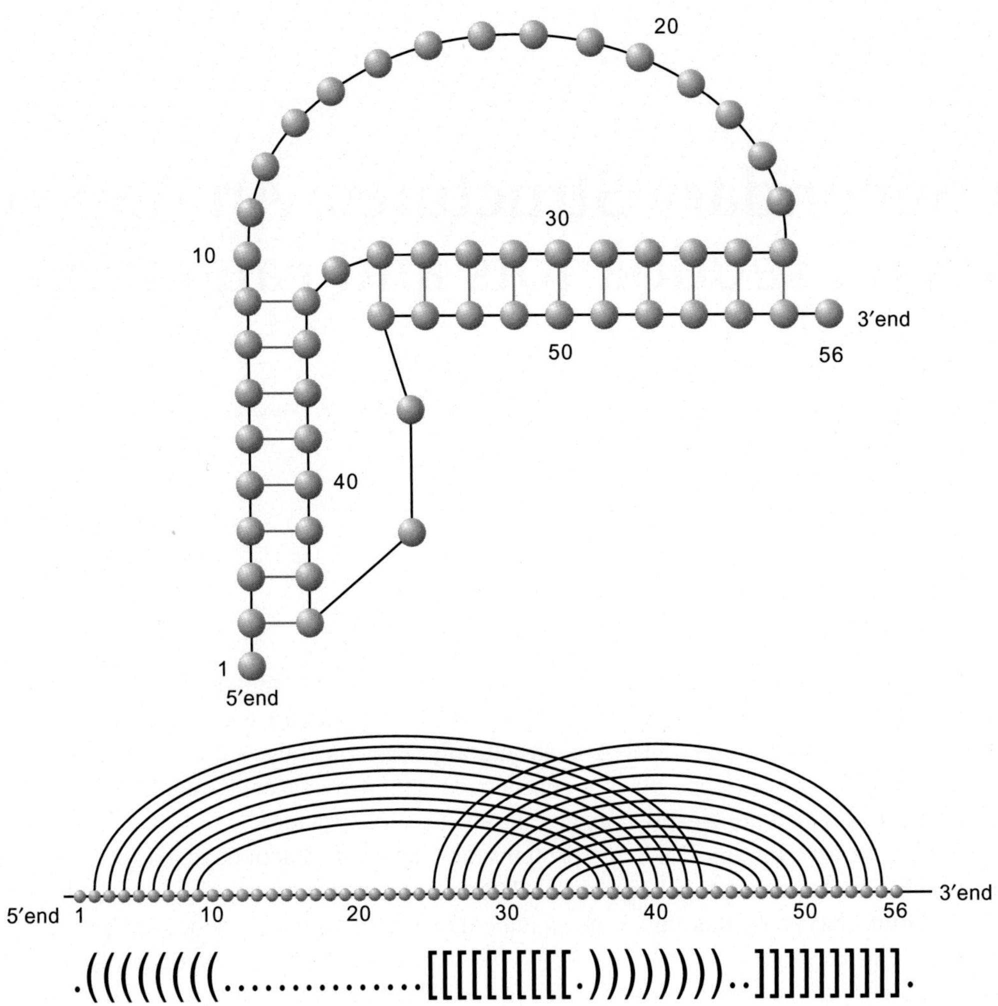

FIGURE 14.1 RNA structure as planar graph (hydrogen bonds (resp. backbone) represented by dark gray (resp. black) edges) and diagram.

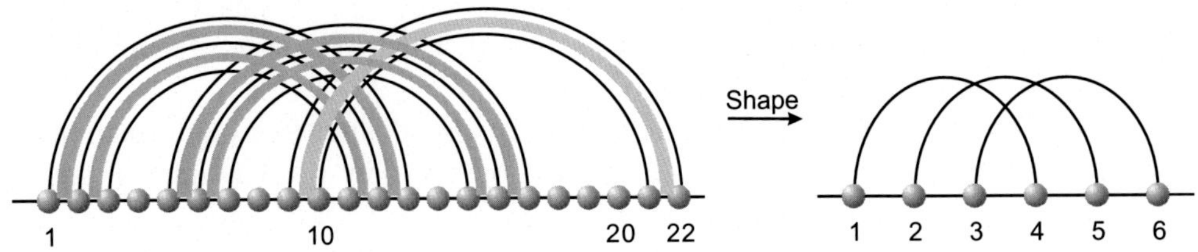

FIGURE 14.2 From a diagram (a) to its shape (b).

this sense, it naturally generalizes the idea behind secondary structure folding algorithms by replacing simple arcs with shapes.

The topological filtration of RNA structures was first studied by Penner and Waterman in [13] and later as an application of the matrix model in [14] and [15]. In [12, 16], a representation theoretic ansatz is employed that traces back to Zagier [17]. [16] connects RNA shapes of fixed topological genus with Riemann's moduli space.

By construction, `gfold` depends crucially on RNA shapes. Shapes represent the "key" topological information that lies within the original structure; see Figure 14.2b. RNA shapes are the central determinant of the multiple-context-free language of topological RNA structures.

FIGURE 14.3 The four shapes employed in the topological folding algorithm gfold.

TABLE 14.1 The Number of Shapes of Fixed Genus g

$g = 1$	2	3	4	5
4	3696	15,214,144	148,120,104,704	2,638,025,019,442,176

Reidys et al. [12] identify a particular topological fact, crucial for folding. That is, for fixed topological genus, there exist only finitely many shapes. This immediately implies that, despite the fact that there are infinitely many RNA structures of fixed topological genus, the generating function can be reduced to a generating polynomial. We shall refer to this polynomial as the shape polynomial. While the situation is fairly easy for genus 1 [12], see Figure 14.3, for higher genera it is not trivial to obtain the shapes.

Interestingly, more than 95% of all known RNA-pseudoknot structures are built very "regularly." They are derived from the aforementioned four shapes by means of concatenation and nesting. This observation has led to the notion of γ-structures [18], obtained as concatenation and nesting of shapes of genus less than γ. Thus, despite the fact that the overall genus of γ-structures is arbitrary, they are composed by finitely many blocks of at most genus γ-complexity.

This fits well with what we know about RNA secondary structures: these are built by concatenation and nesting of simple arcs. Topological RNA structures generalize this in a natural way, utilizing novel building blocks, more complex than simple arcs, that is, RNA shapes described in the following. The problem is thus reduced to finding and analyzing shapes, whose numbers increase rapidly with increasing genus; see Table 14.1.

14.2 FATGRAPHS AND SHAPES

We shall identify a structure with a diagram, that is, a labeled graph over the vertex set $[n] = \{1, 2, \ldots, n\}$ represented by drawing the vertices $1, 2, \ldots, n$ on a horizontal line in the natural order and the arcs (i,j), where $i < j$, in the upper half plane. The backbone of a diagram is the sequence of consecutive integers $(1, \ldots, n)$ together with the edges $\{\{i, i+1\} \mid 1 \leq i \leq n-1\}$. The arcs of a diagram, (i,j), where $i < j$, are drawn in the upper half plane. We shall distinguish backbone edges $\{i, i+1\}$ from arcs $(i, i+1)$, which we refer to as a 1-arc. Two arcs (i,j), (r,s), where $i < r$ are crossing if $i < r < j < s$ holds. Parallel arcs of the form $\{(i,j), (i+1, j-1), \ldots, (i+\ell-1, j-\ell+1)\}$ are called a stack, and ℓ is called the length of this stack. Furthermore, the particular arc, $(1, n)$, is called the rainbow. See Figure 14.4.

In order to understand the topological properties of RNA molecules, we need to pass from the picture of RNA as diagrams or contact graphs to that of topological surfaces. Only the associated surface carries the important invariants leading to a meaningful filtration of RNA structures. The main idea is to "thicken" the edges into (untwisted) bands or ribbons and to expand each vertex to a disk, as shown in Figure 14.5. This inflation of edges leads to a fatgraph $\mathbb{D}$ [19, 20].

A fatgraph, sometimes also called "ribbon graph" or "map," is a graph equipped with a cyclic ordering of the incident half-edges at each vertex. Thus, $\mathbb{D}$ refines its underlying graph D insofar as it encodes the ordering of the ribbons incident on its disks. In the following we will deal with orientable ribbon graphs.[1] Each ribbon has two boundaries. The first one in counterclockwise order is labeled by an arrowhead; see Figure 14.5 and 14.8. A $\mathbb{D}$-cycle, also referred to as a boundary component, is constructed by following these directed boundaries from

1. Ribbons may also be allowed to twist, giving rise to possibly nonorientable surfaces [21].

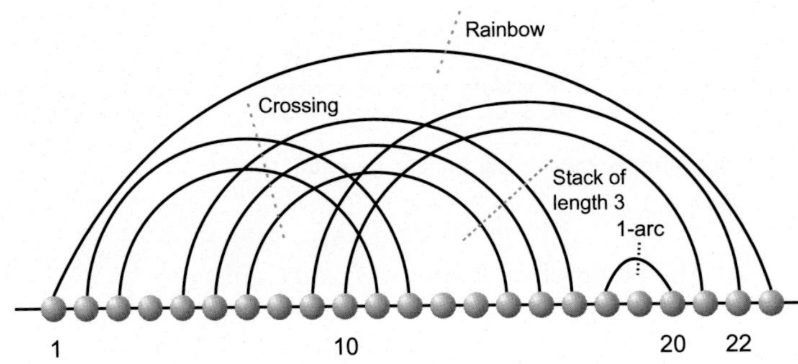

FIGURE 14.4 A diagram with rainbow, 1-arc, and crossing stacks marked.

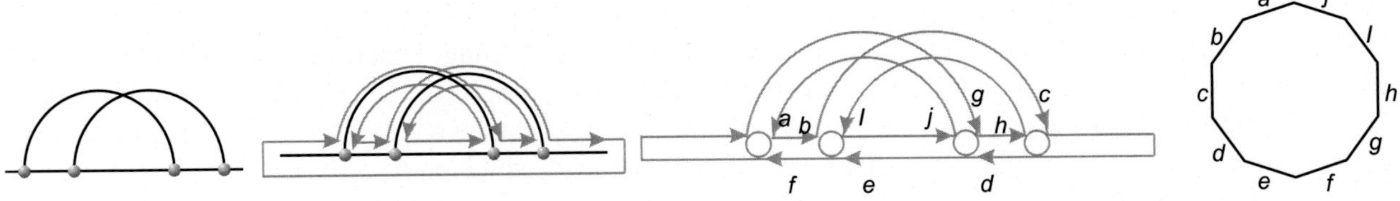

FIGURE 14.5 Inflation of edges and vertices to ribbons and disks. Here, we have four vertices, five edges, and one boundary component $(\vec{a}, \vec{b}, \vec{c}, \vec{d}, \vec{e}, \vec{f}, \vec{g}, \vec{h}, \vec{i}, \vec{j})$. The corresponding surface has Euler characteristic $\chi = v - e + r = 0$ and genus $g = 1$; see Equations (14.1) and (14.2).

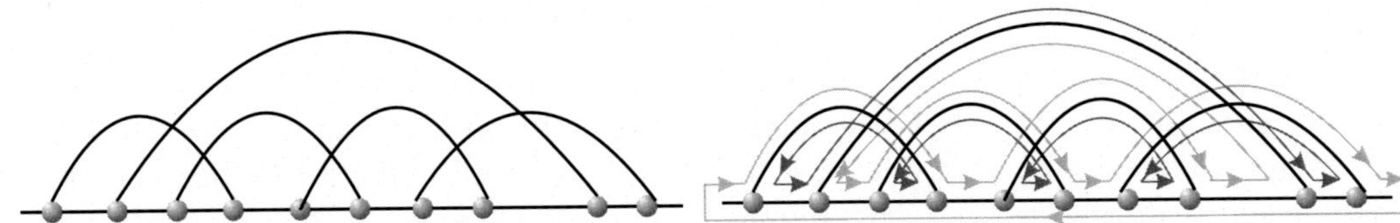

FIGURE 14.6 Computing the number of boundary components. The diagram contains $5 + 9$ edges and 10 vertices. We follow the alternating paths described in the text and observe that there are exactly two boundary components gray and dark gray. According to Equation (14.1), the genus of the diagram is given by $1 - \frac{1}{2}(10 - 14 + 2) = 2$.

disk to disk, thereby alternating between base-pair ribbons and backbone, with the exception of the segment of the boundary component that travels along the bottom of the backbone using only backbone bonds, as shown in Figures 14.5 and 14.6. Topological invariants such as the number of boundary components of the fatgraph $\mathbb{D}$ can thus be computed directly from the underlying diagram D. The fact that we fatten into ribbons without twists implies that each of their sides is traversed in complementary directions. This is implied by the classification theorem of surfaces, which can only be a sphere, a connected sum of tori, or a connected sum of projective planes. Such surfaces exhibit a combinatorial blueprint, which is in fact a fatgraph, in which edge pairs are in case of the sphere complementary and adjacent; in case of the torus complementary, intertwined pairs $aba^{-1}b^{-1}$; or in case of the projective plane twisted pairs, aa. Only the sphere and the connected sum of tori are orientable, whence the appearance of complementary edge pairs in our fatgraphs does not come as a surprise.

The fatgraph $\mathbb{D}$ gives rise to a unique surface $X_{\mathbb{D}}$ with boundary. This means that each point in the ribbon graph has a neighborhood homeomorphic to an open disk or half-disk if it is a point on the boundary. Each $\mathbb{D}$-cycle corresponds to a boundary component of $X_{\mathbb{D}}$, whose Euler characteristic and genus are given by

$$\chi(X_{\mathbb{D}}) = v - e + r \tag{14.1}$$

$$g(X_{\mathbb{D}}) = 1 - \frac{1}{2}\chi(X_{\mathbb{D}}), \tag{14.2}$$

where v, e, r denotes the number of disks, ribbons, and boundary components in $\mathbb{D}$ [21]. The graph D can readily be obtained by continuously contracting the ribbons and disks of $\mathbb{D}$ into the edges and vertices. It requires some work to confirm that Euler's formula provides topological genus. The key point here is that it is independent of

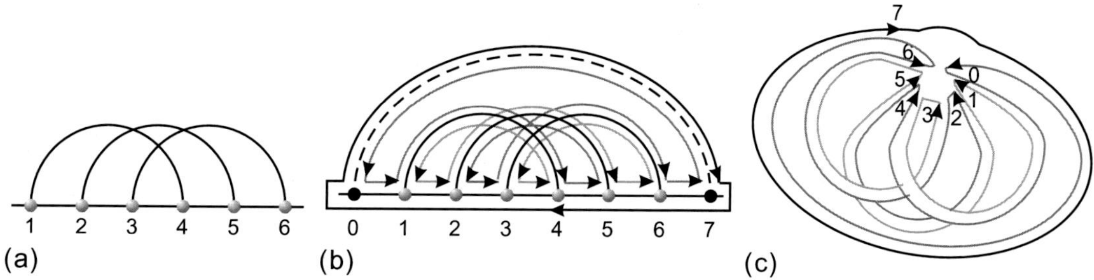

FIGURE 14.7 Reduction to fatgraphs with a single vertex. Contracting the backbone of a diagram into a single vertex decreases the length of the boundary components and preserves the genus. The contracted fatgraph is equivalent to the labeled directed cycle. The backbone of the polymer can be recovered by reinflating the disk into the backbone. The polygon (right-hand side) represents the standard 2D-model of a surface as discussed in [21].

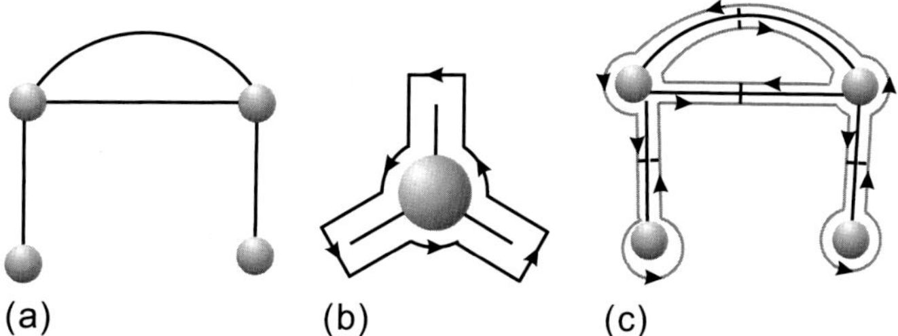

FIGURE 14.8 From graphs to fatgraphs: (a) A graph with four vertices and four edges. (b) Inflation of a vertex. (c) A fatgraph derived from (a) induces a topological surface.

the choice of the fatgraph. This boils down to proving the equivalence of simplicial and singular homology [22], and the concept of cellular homology [22] allows use to express the Euler characteristic as the alternating sum of ranks of relative homology groups, implying that the above combinatorial invariants are in fact invariants of $X_{\mathbb{D}}$.

We next make use of an additional feature of RNA structures, namely, that the backbone forms a unique oriented chain determined by the covalent bonds. Thus, the backbone can be collapsed to a single disk because the surface is orientable: in absence of twisted ribbons, there is no particular information in the backbone itself. Indeed, the procedure can be undone by reinflating the disk and rebuilding the backbone. The contraction of the n vertices to a single one and the removal of the $(n-1)$ covalent bonds therefore preserve the Euler characteristic and genus; see Figure 14.7.

Using the collapsed fatgraph,[2] we see that the relation between the genus of the surface and the number of boundary components is determined by the number of arcs in the upper half plane namely,

$$2 - 2g - r = 1 - n, \tag{14.3}$$

where n is the number of base pairs and r the number of boundary components. The latter can be computed easily and therefore controls the genus of the molecules. Equation (14.3) follows from Equations (14.2) and (14.1), which together yield $2 - 2g - r = v - e$, and the observation that the contracted graph has $e = n$ arcs and a single $(v = 1)$ vertex.

Exercise 14.1. Compute the genus and number of boundary components of the diagrams in Figure 14.9. □

A fatgraph is thus a graph enriched by a cyclic ordering of the incident half-edges at each vertex and consists of the following data: a set of half-edges, H; cycles of half-edges as vertices; and pairs of half-edges as edges.

2. In order to relate this to the standard 2D-models of surfaces derived from triangulations: from the collapsed fatgraph, we can derive the *polygonal model of the surface* $X_{\mathbb{D}}$, that is, a $2n$-gon in which edges are identified in pairs; see Figure 14.7.

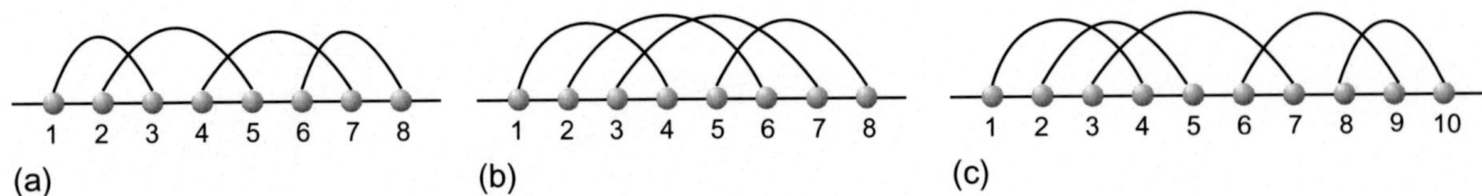

FIGURE 14.9 Exercise 14.1.

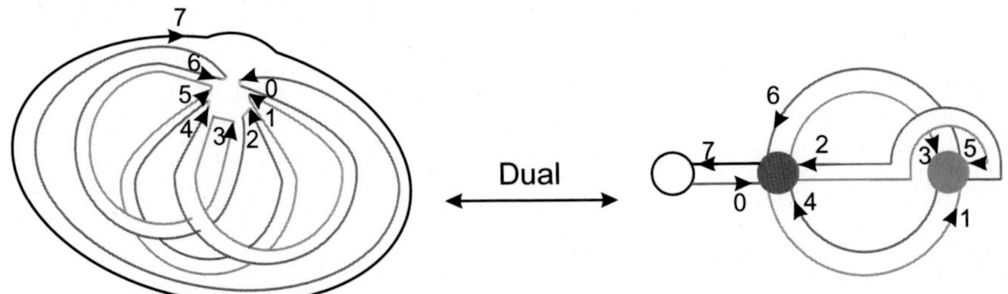

FIGURE 14.10 The Poincaré dual: we map a fatgraph with one vertex and three boundary components into a fatgraph with three vertices and one boundary component.

The idea of half-edges stems from the observation that ribbons have two sides and are traversed in complementary directions. It is then a matter of convention to denote the terminal half of these sides as half-edges.

Definition 14.1. A fatgraph is a triple (H, σ, α), where σ is the vertex permutation and α a fixed-point free involution. □

Furthermore, $\mathbb{D}$-cycles or boundary components, constructed by following these directed boundaries from disk to disk can be expressed algebraically by forming the permutation $\gamma = \alpha \circ \sigma$. In the following, we consider only diagrams with a rainbow. As we shall see, the rainbow arc provides a canonical first boundary component, which travels on top of the rainbow arc and the bottom of the backbone of the diagram.

Definition 14.2. A unicellular map $\mathfrak{m}$ of size n is a fatgraph $\mathfrak{m}(n) = (H, \alpha, \sigma)$ in which the permutation $\alpha \circ \sigma$ is a cycle of length $2n$. □

While unicellular maps are simply particular fatgraphs, they naturally arise in the context of diagrams by two observations. First, as we have seen, in the diagram one may collapse the backbone into a single vertex. Second, the mapping

$$\pi : (H, \sigma, \alpha) \mapsto (H, \alpha \circ \sigma, \alpha)$$

is a bijection between fatgraphs having one vertex and unicellular maps; see Figure 14.10. This mapping is called the *Poincaré dual* and interchanges by construction boundary components by vertices, preserving topological genus. This follows immediately from comparing the Euler characteristic of the fatgraph and its Poincaré dual. In the following, we use π to denote the Poincaré dual.

Exercise 14.2. Compute the Poincaré dual of the fatgraphs in Figure 14.11. □

Given a unicellular map, the permutation σ and γ induces two linear orders of half-edges, see Figure 14.12.

$$r <_\gamma \gamma(r) <_\gamma \cdots <_\gamma \gamma^{2n-1}(r), \quad r <_\sigma \sigma(r) <_\sigma \cdots <_\sigma \sigma^k(r).$$

Let a_1 and a_2 be two distinct half-edges in $\mathfrak{m}$. Then $a_1 <_\gamma a_2$ expresses the fact that a_1 appears before a_2 in the boundary component $\gamma = \alpha \circ \sigma$. Suppose two half-edges a_1 and a_2 belong to the same vertex v. Note that v is effectively a cycle that we assume to originate with the first half-edge along which one enters v traveling γ. Then $a_1 <_\sigma a_2$ expresses the fact that a_1 appears (counterclockwise) before a_2. The Poincaré dual maps the rainbow into a distinguished vertex of degree 1 and provides thereby a natural origin for the cycle γ. We call this vertex

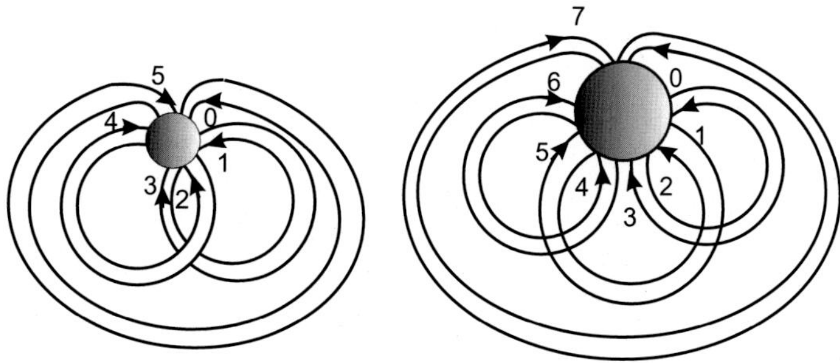

FIGURE 14.11 Exercise 14.2.

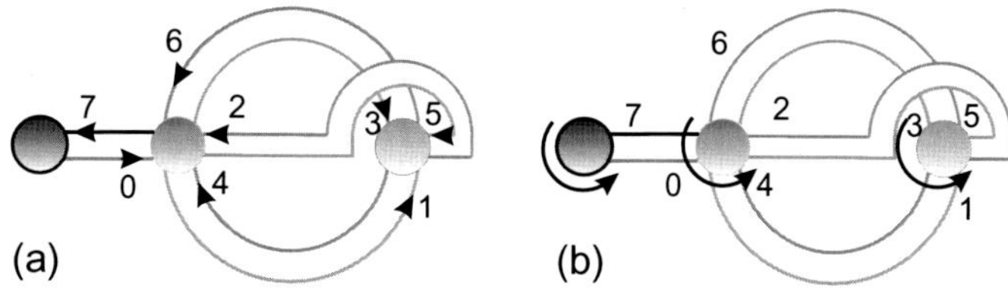

FIGURE 14.12 σ and γ order.

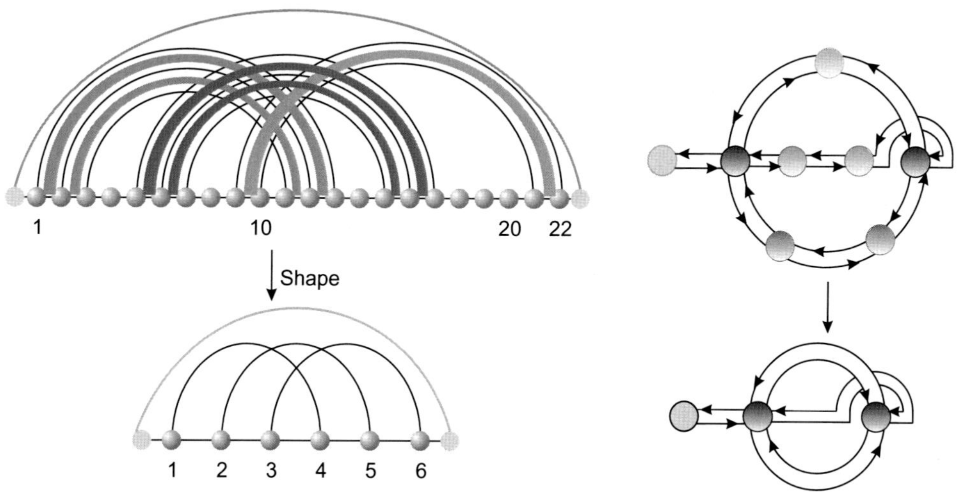

FIGURE 14.13 From a diagram to a shape by removing all 1-arc and parallel arcs. The dash arc is a rainbow arc, where a preshape is nested inside.

the *plant*; see Figure 14.10. Given a unicellular map, we call a half-edge the minimum half-edge of a vertex v if it is the first half-edge via which γ visits v.

Shapes are particular fatgraphs. As mentioned in the introduction, they constitute the building blocks of RNA-pseudoknot structures. Accordingly, let us introduce the idea behind a shape.

An arc is called a 1-arc if it is the form $(i, i+1)$. Two arcs are called parallel if they are of the form (i,j) and $(i+1, j-1)$. A diagram is called a *preshape* if it contains neither 1-arcs nor parallel arcs; see Figure 14.13. A preshape without a rainbow is called pure. Clearly, there is a projection from preshapes to pure preshapes obtained by removing the latter. A shape is then obtained from a pure preshape by adding a rainbow.

Proposition 14.1. *A fatgraph S_g is a shape of genus g having n arcs if and only if its associated unicellular map, $\mathfrak{s}_g$, has only vertices of degree ≥ 3.*

Proof: Suppose v is a vertex in $\mathfrak{s}_g$. The boundary component in S_g associated to v travels $\mathrm{d}(v)$ arcs. The Poincaré dual maps a boundary component to a vertex. In case of $\mathrm{d}(v) = 1$, the boundary component travels

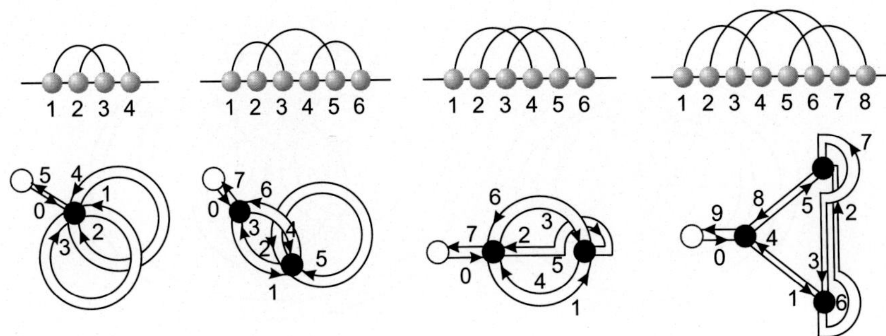
FIGURE 14.14 The four shapes of genus 1.

only one arc and is thus a 1-arc. A boundary component consisting of two arcs is obtained by either parallel arcs or subsequent arcs, where the endpoint of the second arc travels via the backbone to the start point of the first.

Therefore, if $\mathfrak{s}_g$ contains only vertices of degree ≥ 3, the diagram contains no 1-arcs and no parallel arcs, that is, it is a shape. If $\mathfrak{s}_g$ is a shape it has a rainbow arc, which excludes the case of subsequent arcs, where the endpoint of the second arc travels via the backbone to the start point of the first. Thus, $\mathfrak{s}_g$ contains only vertices of degree ≥ 3 and the proposition follows; see Figure 14.14. □

14.3 GENUS RECURSION

In view of the Euler characteristic, reducting topological genus requires us to either remove two edges from the fatgraph while maintaining its unicellularity, or to increase the number of vertices by two. In this section, we present a construction [23] facilitating the latter that is bijective, that is, we present an algorithm allowing us to decrease genus of unicellular maps until we reach a planar tree, together with a collection of additional data, which takes the form of colored vertices of this tree. In fact the algorithm is derived from two mutually inverse processes: a slicing-map Ξ and a gluing-map Λ, which, when restricted to the proper classes, are inverse to each other; see Figure 14.15.

The slicing process splits a vertex into $(2g+1)$ vertices and thereby reduces the genus of the map by g. Gluing effectively inverts slicing, namely, gluing any $(2g+1)$ vertices in a unicellular map increases the genus of the map by g. Slicing and gluing preserve unicellularity.

Definition 14.3. A half-edge h is an *up-step* if $h <_\gamma \sigma(h)$, and a *down-step* if $\sigma(h) \leq_\gamma h$. h is called a *trisection* if h is a down-step and $\sigma(h)$ is not the minimum half-edge of its respective vertex. □

Exercise 14.3. Find all trisections in the fatgraphs displayed in Figure 14.16. □

The number of trisections in a unicellular map is an invariant of a unicellular map with fixed genus g. Moreover, the number is given by the following lemma:

Lemma 1. *Let $\mathfrak{m}_g$ be a unicellular map of genus g. Then $\mathfrak{m}_g$ has exactly $2g$ trisections [23].*

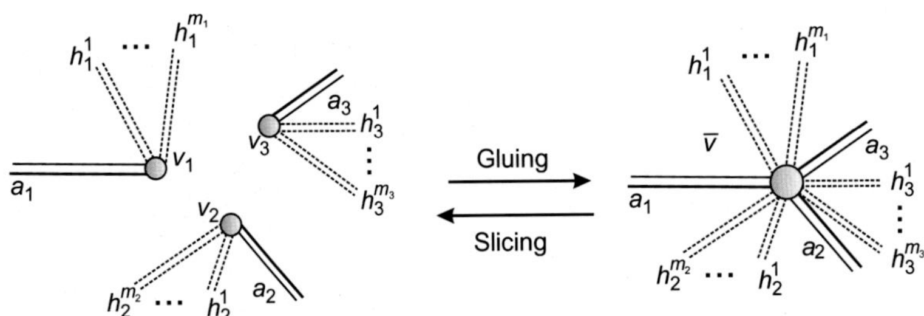

FIGURE 14.15 Gluing and slicing in unicellular maps.

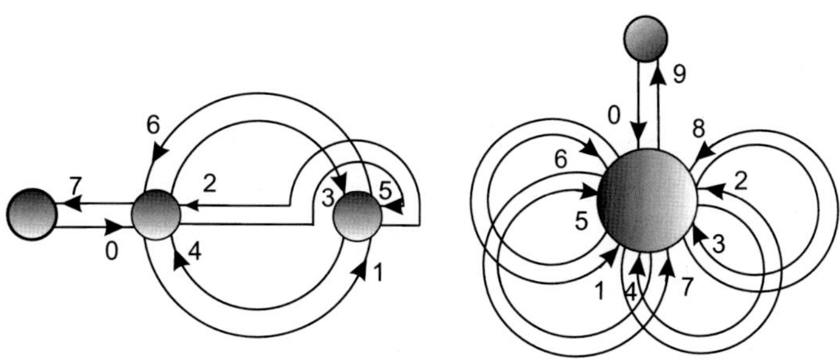

FIGURE 14.16 Exercise 14.3.

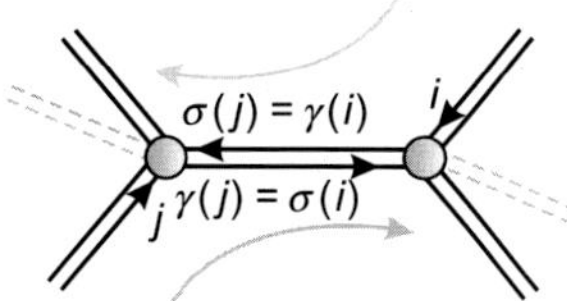

FIGURE 14.17 Pairing of up- and down-steps.

Proof: To prove the lemma, one observes that up-steps and down-steps come in pairs; see Figure 14.17. To count trisections, all we have to do is subtract the trivial down-steps, which equals the number of vertices in the unicellular map, and use the Euler characteristic. □

Slicing reduces the number of trisections in a unicellular map of genus g. First, we pick up a trisection τ and assume it is contained in a vertex v. Let a_1 denote the minimum half-edge in $\overline{v}$, and a_3 denote the half-edge located anticlockwise from τ. We consider for the half-edge between a_1 and a_3, a_2, that it is the minimum half-edge satisfying $a_2 >_\gamma a_3$. We can always find such a half-edge a_2 because τ is a trisection and $\tau >_\gamma a_3$, by definition.

Let

$$\overline{v} = (a_1, h_2^1, \ldots, h_2^{m_2}, a_2, h_3^1, \ldots, h_3^{m_3}, a_3, h_1^1, \ldots, h_1^{m_1}),$$

and

$$\overline{\gamma} = (\ell_1^1, \ldots, \ell_{k_1}^1, a_1, \ell_1^3, \ldots, \ell_{k_3}^3, a_3, \ell_1^2, \ldots, \ell_{k_2}^2, a_2, \ell_1^4, \ldots, \ell_{k_4}^4).$$

We slice $\overline{v}$ into three vertices v_i, $i = 1, 2, 3$, where $v_i = (a_i, h_i^1, \ldots, h_i^{m_i})$. The new boundary is given by

$$\gamma = (\ell_1^1, \ldots, \ell_{k_1}^1, a_1, \ell_1^2, \ldots, \ell_{k_2}^2, a_2, \ell_1^3, \ldots, \ell_{k_3}^3, a_3, \ell_1^4, \ldots, \ell_{k_4}^4).$$

By construction, a_1 and a_2 are the minimum half-edges in v_1 and v_2, respectively. However, a_3 is not necessarily minimal in v_3. If a_3 is the minimum, we have a_1, a_2, and a_3 as the minimum half-edges in v_1, v_2, and v_3, respectively. Otherwise, τ remains a trisection in v_3.

Consequently, we have two mappings, depending on whether or not a_3 is minimal:

$$\rho_1 \colon (\overline{\mathfrak{m}}, \tau) \to (\mathfrak{m}, v_1, v_2, v_3), \quad \rho_2 \colon (\overline{\mathfrak{m}}, \tau) \to (\mathfrak{m}, v_1, v_2, \tau),$$

where $\mathfrak{m}$, $\overline{\mathfrak{m}}$ are unicellular maps of genus g and $g+1$, respectively.

In the first case, τ is no longer a trisection after slicing and is called a *type I*. In the second case, τ remains a trisection of *type II*.

Proposition 14.2. *The mappings ρ_1 and ρ_2 are bijections [23].*

Gluing can be described as follows: given a unicellular map of $\mathfrak{m}_{g-k}$, together with a sequence of vertices $V = \{v_1, \ldots, v_{2k+1}\}$, where $v_i <_\gamma v_{i+1}$, $\forall 1 \leq i < 2k+1$, then

I. We glue the last three vertices v_{2k-1}, v_{2k}, and v_{2k+1} via ρ_1^{-1}, thereby obtaining the unicellular map $\mathfrak{m}_{g-k+1}$ together with a type I trisection τ^I.

II. We apply $\rho_2^{-1}(\mathfrak{m}_{g-k+i}, v_{2k-2i-1}, v_{2k-2i}, \tau^{\mathrm{I}})$ $k-1$ times for $i = 1$ to $i = k-1$. This produces the unicellular map $\mathfrak{m}_g(n)$, together with a trisection τ^{II}. The process defines a mapping

$$\Lambda(\mathfrak{m}_{g-k}, v_1, \ldots, v_{2k+1}) = (\mathfrak{m}_g, \tau).$$

The order of the vertices in V is induced by the boundary component, γ. Thus, V can be considered as a set of vertices in $\mathfrak{m}_{g-k}$, ordered by $< \gamma$. Λ merges vertices from right to left by first applying Φ once, then applying Ψ until all vertices are glued together.

Λ is reversed as follows: given a unicellular map $\mathfrak{m}_g$ of genus g and $i = 0$,

1. If τ is type II trisection in $\mathfrak{m}_{g-i}$, then let $(\mathfrak{m}_{g-i-1}, v_{2i+1}, v_{2i+2}, \tau) = \rho_2(\mathfrak{m}_{g-i}, \tau)$. We increase i to $i+1$ and repeat Step 1.
2. If τ has type I, let $(\mathfrak{m}_{g-i}, v_{2i+1}, v_{2i+2}, v_{2i+3}) = \rho_1^{-1}(\mathfrak{m}_{g-i-1}, \tau)$.

Then we return

$$\Xi(\mathfrak{m}_g, \tau) = (\mathfrak{m}_{g-i}, V_\tau).$$

By construction, Λ and Ξ are inverse to each other.

The bijections Λ and Ξ immediately induce a connection between unicellular maps having higher genus with those of lower genus. See Figure 14.18.

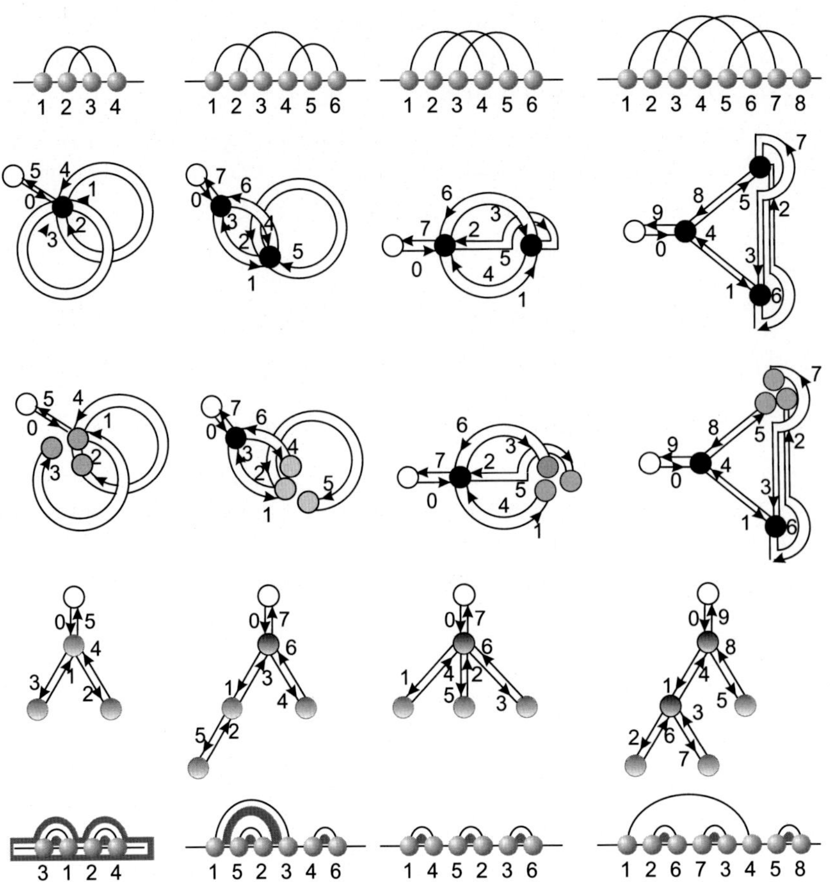

FIGURE 14.18 The four shapes of genus 1 and corresponding secondary structure.

Theorem 14.1. *Let U_g^t denote the set of tuples $(\mathfrak{m}_g, v_1, \ldots, v_t)$, where $v_1, \ldots, v_t$ is a sequence of vertices in $\mathfrak{m}_g$. Furthermore, let D_g denote the set of tuples $(\mathfrak{m}_g, \tau)$, where τ is a trisection of $\mathfrak{m}_g$ [23]. Then*

$$\Lambda\colon \dot{\bigcup}_{k=0}^{g-1} U_k^{2g-2k+1} \to D_g, \quad \Xi\colon D_g \to \dot{\bigcup}_{k=0}^{g-1} U_k^{2g-2k+1},$$

are bijections and $\Lambda \circ \Xi = id$ and $\Xi \circ \Lambda = id$.

The theorem has the following enumerative corollary: let $\varepsilon_g(n)$ denote the number of unicellular maps of genus g having n edges. Then

Corollary 14.1.

$$2g \cdot \varepsilon_g(n) = \binom{n+1-2(g-1)}{3}\varepsilon_{g-1}(n) + \cdots + \binom{n+1}{2g+1}\varepsilon_0(n). \tag{14.4}$$

Here, the $2g$-factor on the left-hand side counts the number of trisections in $\mathfrak{m}_g$, and the binomial coefficients on the right-hand side count the number of distinct selections of subsets of $(2k+1)$ vertices from a unicellular map $\mathfrak{m}_{g-k}$ [23].

Iterating Ξ, we obtain

$$\varepsilon_g(n) = \sum_{0=g_0<g_1<\cdots<g_r=g} \prod_{i=1}^{r} \frac{1}{2g_i}\binom{n+1-2g_{i-1}}{2(g_i-g_{i-1})+1} \cdot \varepsilon_0(n), \tag{14.5}$$

where $\varepsilon_0(n)$ is the number of planar trees having n edges, that is, the Catalan number $\frac{1}{n+1}\binom{2n}{n}$.

14.4 SHAPES OF FIXED TOPOLOGICAL GENUS

In this section, we study shapes of fixed topological genus. Because there are only finitely many shapes for fixed genus g [12], their generating function is a polynomial. We give an explicit formula for the coefficients of the shape polynomial.

We showed in Section 14.2 that a shape corresponds to a unicellular map in which each vertex has degree ≥ 3, $\mathfrak{s}_{g,n}$. Applying Ξ iteratively to $\mathfrak{s}_{g,n}$, we derive a tree. By construction, any unlabeled vertex in this tree originally comes from $\mathfrak{s}_{g,n}$, and thus retains its degree.

Let $\mathbb{S}_{g,n}^{(k)}$ denote the set of unicellular maps $\mathfrak{m}_g^{(k)}$ having k-labeled vertices, in which any unlabeled vertex has degree ≥ 3. In particular, the set of unicellular maps corresponding to shapes of genus g having n edges is $\mathbb{S}_{g,n}^{(0)}$. In the following, let $\mathfrak{s}_{g,n}^{(k)}$ denote an element in $\mathbb{S}_{g,n}^{(k)}$.

Because neither Λ nor Ξ alter unlabeled vertices we have induced bijections

$$\Lambda\colon \left(\mathfrak{s}_{g-t,n}^{(k+2t+1)}, V_{2t+1}\right) \to \left(\mathfrak{s}_{g,n}^{(k)}, \tau\right)\ \Xi\colon \left(\mathfrak{s}_{g,n}^{(k)}, \tau\right) \to \left(\mathfrak{s}_{g-t,n}^{(k+2t+1)}, V_{2t+1}\right).$$

Indeed, Ξ slices a vertex together with a trisection into a sequence of labeled vertices, and thus does not change the degree of unlabeled vertices in the map. Furthermore, Λ glues three or more labeled vertices into one vertex, which has accordingly minimum degree 3.

Let $\eta_g(n,k)$ denote the cardinality of $\mathbb{S}_{g,n}^{(k)}$, $k \geq 0$. In view of the above and Equation (14.5), we have

$$2g \cdot \eta_g(n,k) = \sum_{i=1}^{g}\binom{k+2i+1}{2i+1}\eta_{g-i}(n,k+2i+1) + \sum_{i=1}^{g}\binom{k+2i}{2i+1}\eta_{g-i}(n,k+2i), \tag{14.6}$$

where $\binom{n}{m} = 0$, for $n < m$.

Due to the compatibility of slicing and gluing with the vertex degree of unlabeled vertices, we can conclude

Proposition 14.3. *The number of shapes of genus g is given by*

$$\eta_g(n,0)=\sum_{t=0}^{g-1} a_t^{(g)}\eta_0(n,2g+t+1), \tag{14.7}$$

where the $a_t^{(g)}$ are given by

$$a_t^{(g)}=\sum_{\substack{0=g_0<g_1<\cdots<g_r=g\\ 0=t_0=t_1\leq t_2\leq\cdots\leq t_r=r-t}}\prod_{i=1}^{r}\frac{1}{2g_i}\binom{2g+t-(2g_{i-1}+(i-1))+t_i}{2(g_i-g_{i-1})+1}. \tag{14.8}$$

Proof: Iterating the recursion in Equation (14.6), we reduce the genus. In view of

$$\eta_g(n,0)=\frac{1}{2g}\sum_{g_1=0}^{g-1}\eta_{g_1}(n,2(g-g_1)+1),$$

we then substitute $\eta_{g_1}(n,2(g-g_1)+1)$ and obtain

$$\begin{aligned}\eta_g(n,0)=&\frac{1}{2g}\sum_{g_1=0}^{g-1}\frac{1}{2g_1}\sum_{g_1>g_2}\left(\binom{2(g-g_2)+2}{2(g_1-g_2)+1}\eta_{g_2}(n,2(g-g_2)+2)\right.\\&\left.+\binom{2(g-g_2)+1}{2(g_1-g_2)+1}\eta_{g_2}(n,2(g-g_2)+1)\right).\end{aligned}$$

Continuing this substitution, we arrive at $\eta_0(n,w)$ for some integer w. Because in each substitution the number of labeled vertices increases by either $2i$ or $2i+1$, we derive

$$\eta_g(n,0)=\sum_{t=0}^{g-1} a_t^{(g)}\eta_0(n,2g+t+1),$$

where the coefficients, $a_t^{(g)}$, are given by

$$a_t^{(g)}=\sum_{\substack{0=g_0<g_1<\cdots<g_r=g\\ 0=t_0=t_1\leq t_2\leq\cdots\leq t_r=r-t}}\prod_{i=1}^{r}\frac{1}{2g_i}\binom{2g+t-(2g_{i-1}+(i-1))+t_i}{2(g_i-g_{i-1})+1}. \tag{14.9}$$

□

At this point, we observe that it is possible to analyze the terms $\eta_0(n,2g+t+1)$ further. The idea is to "remove" all unlabeled vertices from any partially labeled tree, thereby reducing the recursion to fully labeled trees. The latter are then enumerated by Catalan numbers.

This removal is facilitated by observing that we can restrict the bijection of Rémy to $\mathbb{S}_{0,n}^{(k)}$-trees.

To this end, let us first recall Rémy's bijection for planar trees [24].

Theorem 14.2. *Let $\varepsilon_0(n)$ denote the number of planar trees with n edges. Then we have the recursion*

$$(n+1)\varepsilon_0(n)=2(2n-1)\varepsilon_0(n-1).$$

The bijection of Theorem 14.2 associates a planar tree having n edges and a labeled vertex to a planar tree with $(n-1)$ edges with a labeled sector. It is constructed as follows: observing that in a planar tree having n edges, there are $n+1$ vertices and $2n-1$ sectors, Rémy's bijection, illustrated in Figure 14.19, entails two ways of inserting a vertex into a labeled sector. This vertex insertion generates from a planar tree with $n-1$ edges and a labeled sector a planar tree with n edges and a labeled vertex. The process can be reversed, that is, a planar tree with n edges and a labeled vertex can be re-tracked to a planar tree with $n-1$ edges and a labeled sector. Depending on the labeled vertex being a leaf or not, one derives a planar tree having two types of labeled sectors.

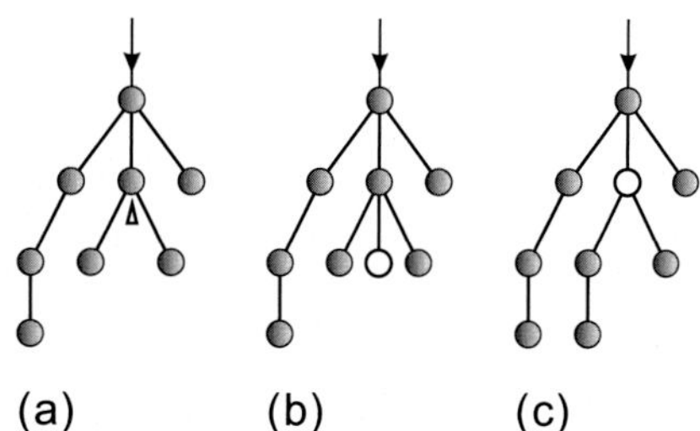

FIGURE 14.19 Rémy's bijection: two ways of obtaining a planar tree with n edges and a labeled vertex from a planar tree with $n-1$ edges with a labeled sector. We pass from (a) to (b) by inserting a labeled vertex as a leaf to the labeled sector and from (a) to (c) by replacing the vertex containing the sector by the labeled vertex, and carrying the subtree on the left of the sector as its leftmost subtree. This case applies if the labeled vertex is not a leaf.

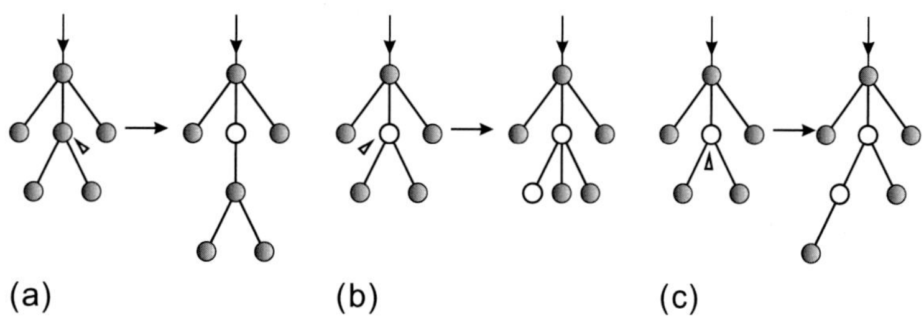

FIGURE 14.20 How to construct nonshape-sectors via Rémy's procedure. The following insertions produce an unlabeled vertex of degree 2: (a) the sector is in a labeled vertex and $\{v_{\ell+1},\ldots,v_m\}=\varnothing$, (b) ($\ell=0$), and (c) ($\ell=1$) if the sector is in an unlabeled vertex.

We shall prove that Rémy's procedure contracts unlabeled vertices of an $\mathbb{S}^{(k)}_{0,n+1}$-tree into a particular type of sector in the resulting $\mathbb{S}^{(k)}_{0,n}$-tree. These sectors are referred to as *shape-sector* and are defined as follows: suppose we are given an $\mathbb{S}^{(k)}_{0,n}$-tree. A shape-sector is a sector for which Rémy's procedure, inserting a nonleaf unlabeled vertex, generates an $\mathbb{S}^{(k)}_{0,n+1}$-tree.

Lemma 2. *An $\mathbb{S}^{(k)}_{0,n}$-tree contains exactly $(2k-n-2)$ shape-sectors.*

Proof: In order to construct an $\mathbb{S}^{(k)}_{0,n+1}$ from an $\mathbb{S}^{(k)}_{0,n}$-tree by Rémy's procedure, we need to ensure that the newly inserted, unlabeled vertex has at least degree 3 and that it does not reduce the degree of the other unlabeled vertex. We shall consider only the insertion not producing a leaf, because it is a vertex of degree 1. Given a sector τ in a vertex v, assume the children of v are indexed counterclockwise $v_1,\ldots,v_m$, where $m\geq 2$. The sector τ partitions the v-children into two blocks:

$$\{v_1,\ldots,v_\ell\} \quad \text{and} \quad \{v_{\ell+1},\ldots,v_m\}.$$

We make two observations: (a) The newly inserted vertex has at least degree 3 if $\{v_{\ell+1},\ldots,v_m\}\neq\varnothing$; see Figure 14.20. (b) The vertex that is pushed "down" by the newly inserted vertex retains degree ≥ 3, if and only if $\ell\geq 2$, that is, if τ is to the right of the second v-child in counterclockwise order.

The latter applies by construction only to the case where the pushed-down vertex is unlabeled.

Therefore, a sector in a vertex v is a shape-sector if and only if

- τ has the property $\{v_{\ell+1},\ldots,v_m\}\neq\varnothing$.
- If v is unlabeled, τ satisfies furthermore $|\{v_1,\ldots,v_\ell\}|\geq 2$.

There are in total $2n+1$ sectors in a tree. The first criterion rules out $n+1$ sectors, because each vertex has one such sector. The second criterion rules out two sectors from unlabeled vertices, and there are $n+1-k$ of them. Accordingly, the number of shape-sectors is given by

$$2n+1-(n+1)-2(n+1-k)=2k-n-2.$$

Any of the $2k-n-2$ shape-preserved sectors produces an $\mathbb{S}^{(k)}_{0,n+1}$-tree by Rémy's procedure when inserting a nonleaf vertex and the lemma follows. □

TABLE 14.2 The Coefficients $\kappa_t^{(g)}$

	g				
t	1	2	3	4	5
0	1	21	1485	225,225	59,520,825
1		105	18,018	4,660,227	1,804,142,340
2			50,050	29,099,070	18,472,089,636
3				56,581,525	78,082,504,500
4					117,123,756,750

Exercise 14.4. Show the following corollary. □

Corollary 14.2. *Let τ denote a shape-preserved sector and $\mathfrak{m} \in \mathbb{S}_{0,n}^{(k)}$. Then*

$$\rho\colon (\mathfrak{m}, \tau) \to (\mathfrak{m}', v)$$

is a bijection, where v is an unlabeled vertex in $\mathfrak{m}'$ and $\mathfrak{m}' \in \mathbb{S}_{0,n+1}^{(k)}$. In particular, we have

$$(2k - n - 2)\eta_0(n, k) = (n + 1 - k)\eta_0(n + 1, k)$$

$$\eta_0(n, k) = \binom{2k - (k-1) - 2}{n + 1 - k}\eta_0(k - 1, k) = \binom{k-1}{n+1-k}\mathrm{Cat}(k - 1).$$

In particular, the number of shape-sectors decreases by 1, upon insertion of one unlabeled vertex, and there are at most $2k - (k - 1) - 2 = k - 1$ insertions into a fully labeled tree having n edges. This provides another proof that for fixed topological genus, there are only finitely many shapes.

We next compute the shape polynomial $S_g(z) = \sum_n s_g(n)z^g$, where s_g is the number of shapes having n arcs. Note that $s_g(n) = \eta_g(n, 0)$.

Theorem 14.3. *The shape-generating function is given by*

$$S_g(z) = \sum_{t=0}^{g-1} \kappa_t^{(g)} z^{2g+t}(1 + z)^{2g+t}, \tag{14.10}$$

where $\kappa_t^{(g)} = a_t^{(g)}\mathrm{Cat}(2g + t)$.

We present the coefficients $\kappa_t^{(g)}$ for genera $g \leq 5$ in Table 14.2.

Proof: Because $s_g(n) = \eta_g(n, 0)$, we have

$$S_g(z) = \sum_n \eta_g(n, 0)z^n = \sum_n \left(\sum_{t=0}^{g-1} a_t^{(g)}\eta_0(n, 2g + t + 1)\right) z^n.$$

By Corollary 14.2, we can express the terms $\eta_0(n, 2g + t + 1)$,

$$S_g(z) = \sum_n \left(\sum_{t=0}^{g-1} a_t^{(g)}\eta_0(n, 2g + t + 1)\right) z^n$$

$$= \sum_n \sum_{t=0}^{g-1} a_t^{(g)} \binom{2g+t}{n - (2g + t)}\mathrm{Cat}(2g + t)z^n$$

$$= \sum_{t=0}^{g-1} \kappa_t^{(g)} \left(z^{2g+t} \sum_{n=2g+t}^{2(2g+t)} \binom{2g+t}{n-(2g+t)} z^{n-(2g+t)} \right)$$

$$= \sum_{t=0}^{g-1} \kappa_t^{(g)} z^{2g+t} (1+z)^{2g+t},$$

where $\kappa_t^{(g)} = a_t^{(g)} \cdot \mathrm{Cat}(2g+t)$. □

Exercise 14.5. Show

$$s_g(n) = \sum_{t=1}^{g} \kappa_t^{(g)} \binom{2g+t-1}{n-(2g+t-1)}, \tag{14.11}$$

where $\binom{n}{k} = 0$ if $k < 0$ or $k > n$. □

Exercise 14.6. Show that the number $\kappa_t^{(g)}$ is a positive integer. □

ACKNOWLEDGMENTS

We acknowledge the financial support of the Future and Emerging Technologies (FET) program within the Seventh Framework Program (FP7) for Research of the European Commission, under the FET-Proactive grant agreement TOPDRIM, number FP7-ICT-318121. We wish to thank Benjamin Fu for his help preparing this chapter.

REFERENCES

[1] Bailor MH, Sun X, Al-Hashimi HM. Topology links RNA secondary structure with global conformation, dynamics, and adaptation. Science 2010;327:202-6.
[2] Taufer M, Licon A, Araiza R, Mireles D, van Batenburg FHD, Gultyaev A, et al. PseudoBase++: an extension of PseudoBase for easy searching, formatting and visualization of pseudoknots. Nucleic Acids Res 2009;37:D127-35.
[3] Doudna JA, Cech TR. The chemical repertoire of natural ribozymes. Nature 2002;418:222-8.
[4] Namy O, Moran SJ, Stuart DI, Gilbert RJC, Brierley I. A mechanical explanation of RNA pseudoknot function in programmed ribosomal frameshifting. Nature 2006;441:244-7.
[5] Theimer CA, Blois CA, Feigon J. Structure of the human telomerase RNA pseudoknot reveals conserved tertiary interactions essential for function. Mol Cell 2005;17:671-82.
[6] Staple DW, Butcher SE. Pseudoknots: RNA structures with diverse functions. PLoS Biol 2005;3:e213.
[7] Giedroc DP, Cornish PV. Frameshifting RNA pseudoknots: structure and mechanism. Virus Res 2009;139:193-208.
[8] Westhof E, Jaeger L. RNA pseudoknots. Curr Opin Struct Biol 1992;2:327-33.
[9] Loria A, Pan T. Domain structure of the ribozyme from eubacterial ribonuclease. RNA 1996;2:551-63.
[10] Konings D, Gutell R. A comparison of thermodynamic foldings with comparatively derived structures of 16s and 16s-like rRNAs. RNA 1995;1:559-74.
[11] Tuerk C, MacDougal S, Gold L. RNA pseudoknots that inhibit human immunodeficiency virus type 1 reverse transcriptase. Proc Natl Acad Sci USA 1992;89(15):6988-92.
[12] Reidys CM, Huang F, Andersen JE, Penner RC, Stadler PF, Nebel ME. Topology and prediction of RNA pseudoknots. Bioinformatics 2011;27:1076-85.
[13] Penner RC, Waterman MS. Spaces of RNA secondary structures. Adv Math 1993;101:31-49.
[14] Orland H, Zee A. RNA folding and large *n* matrix theory. Nucl Phys B 2002;620:456-76.
[15] Bon M, Vernizzi G, Orland H, Zee A. Topological classification of RNA structures. J Mol Biol 2008;379:900-11.
[16] Andersen JE, Penner RC, Reidys CM, Waterman MS. Topological classification and enumeration of RNA structures by genus. J Math Biol 2013;67(5):1261-78.
[17] Zagier D. On the distribution of the number of cycles of elements in symmetric groups. Nieuw Arch Wisk IV 1995;13:489-95.

[18] Han HSW, Li TJX, Reidys CM. Combinatorics of γ-structures. J Comput Biol 2014;21(8):591–608. doi:10.1089/cmb.2013.0128. Epub 2014 Apr 1.

[19] Loebl M, Moffatt I. The chromatic polynomial of fatgraphs and its categorification. Adv Math 2008;217:1558-87.

[20] Penner RC, Knudsen M, Wiuf C, Andersen JE. Fatgraph models of proteins. Comm Pure Appl Math 2010;63:1249-97.

[21] Massey WS. Algebraic topology: an introduction. New York: Springer-Verlag; 1967.

[22] Hatcher A. Algebraic topology. New York, NY: Cambridge University Press; 2002.

[23] Chapuy G. A new combinatorial identity for unicellular maps, via a direct bijective approach. Adv Appl Math 2011;47(4):874-93.

[24] Rémy JL. Un procédé itératif de dénombrement dárbres binaires et son application á leurgénération aléatoire. RAIRO Inform Théor 1985;19(2):179-95.

Index

Note: Page numbers followed by *f* indicate figures and *t* indicate tables.

D

E

F

N

O

P